Solution Manual

Intermediate Algebra

Naoki Sato
Mathew Crawford
Richard Rusczyk

Art of Problem Solving

Books • Online Classes • Videos • Interactive Resources

www.artofproblemsolving.com

Published by: AoPS Incorporated
                  10865 Rancho Bernardo Rd Ste 100
                  San Diego, CA 92127
                  books@artofproblemsolving.com

ISBN #: 978-1-934124-05-5

Visit the Art of Problem Solving website at http://www.artofproblemsolving.com

Cover image designed by Vanessa Rusczyk using KaleidoTile software.

Printed in the United States of America.

Fifth Printing 2018.

Foreword

This book contains the full solution to every Exercise, Review Problem, and Challenge Problem in the text *Intermediate Algebra*.

In most problems, the final answer is contained in a box, $\boxed{\text{like this}}$. However, we strongly recommend against just looking up the final answer and moving on to the next problem. Instead, even if you got the right answer, read the solution in this book. It might show you a different way of solving the problem that you might not have thought of.

If you don't understand a solution, or you think you have a better way of solving the problem, or (gasp!) find an error in one of our solutions, we invite you to come to our message board at

www.artofproblemsolving.com

and discuss it. Our message board is free to use and includes thousands of the world's most eager mathematical problem-solvers.

Contents

19 Functional Equations

20 Some Advanced Strategies

CHAPTER 1

Basic Techniques for Solving Equations

Exercises for Section 1.1

1.1.1

(a) Simplifying both sides, we get $-4x+13 = -6x+3$. Adding $6x$ to both sides gives $2x+13 = 3$, then subtracting 13 from both sides gives $2x = -10$. Dividing both sides by 2, we find $x = \boxed{-5}$.

(b) Multiplying both sides by 2, we get $5(y + 2) = 2(7y - 4)$. Expanding both sides, we get $5y + 10 = 14y - 8$. Then, subtracting $5y$ from both sides, we get $10 = 9y - 8$. Adding 8 to both sides, we get $18 = 9y$. Finally, dividing both sides by 9, we find $y = \boxed{2}$.

1.1.2

(a) Adding $2t$ to both sides, we get $8 \geq 11 + 3t$, and subtracting 11 from both sides, we get $-3 \geq 3t$. Dividing both sides by 3, we find $t \leq -1$. In interval notation, we have $\boxed{t \in (-\infty, -1]}$.

(b) Adding $2x$ to all three parts of the inequality chain, we get $7 \leq 8x - 1 < 23$. Then adding 1 to all three parts, we get $8 \leq 8x < 24$. Dividing all three parts by 8, we find $1 \leq x < 3$. In interval notation, we have $\boxed{x \in [1, 3)}$.

1.1.3 Subtracting $4x^2 + 2x$ from both sides, we get $x^2 - 1 = 7$. Adding 1 to both sides, we get $x^2 = 8$. Therefore, $x = \pm \sqrt{8} = \boxed{\pm 2\sqrt{2}}$.

1.1.4 The first step is to express b in terms of a. Multiplying both sides by $3 - b$, we get

$$2 - b = 5a(3 - b) = 15a - 5ab.$$

Then to isolate b, we add $5ab - 2$ to both sides, to get $5ab - b = 15a - 2$. We can factor b from $5ab - b$ to give $b(5a - 1) = 15a - 2$. Finally, dividing both sides by $5a - 1$, we find $b = (15a - 2)/(5a - 1)$.

We see that there is a value of b for every value of a, except when the denominator $5a - 1$ is zero. The denominator $5a - 1$ is zero when $5a = 1$, or $a = 1/5$.

Therefore, the only value of a for which there no value of b that satisfies the given equation is $a = \boxed{1/5}$.

1.1.5 We know that if a and b are the legs of a right triangle and c is the hypotenuse, then $a^2 + b^2 = c^2$. In this problem, we are not given which length is the hypotenuse, so we must check every case.

First, note that the hypotenuse is always the longest side of a right triangle. Since $t + 1$ is greater than $t - 1$, the length $t - 1$ cannot be the hypotenuse. Therefore, either $t + 1$ or 4 is the hypotenuse.

If $t + 1$ is the hypotenuse, then $(t - 1)^2 + 4^2 = (t + 1)^2$. Expanding both sides, we get $t^2 - 2t + 1 + 16 = t^2 + 2t + 1$, which simplifies as $16 = 4t$. Dividing both sides by 4, we find $t = 4$. We check this answer. If $t = 4$, then $t - 1 = 3$ and $t + 1 = 5$, and $3^2 + 4^2 = 25 = 5^2$.

If 4 is the hypotenuse, then $(t-1)^2 + (t+1)^2 = 4^2$. Expanding both sides, we get $t^2 - 2t + 1 + t^2 + 2t + 1 = 16$. Simplifying both sides, we get $2t^2 + 2 = 16$. Subtracting 2 from both sides, we get $2t^2 = 14$, and finally dividing both sides by 2, we get $t^2 = 7$. Therefore, $t = \pm\sqrt{7}$. But since $t - 1$ must be positive, t cannot be $-\sqrt{7}$, so $t = \sqrt{7}$. We check this answer: If $t = \sqrt{7}$, then

$$(t-1)^2 + (t+1)^2 = t^2 - 2t + 1 + t^2 + 2t + 1 = 2t^2 + 2 = 2 \cdot 7 + 2 = 16 = 4^2.$$

Therefore, the values of t such that $t-1$, $t+1$, and 4 are the lengths of the sides of a right triangle are $\boxed{4 \text{ and } \sqrt{7}}$.

1.1.6 To get rid of the fractions, we can multiply both sides by the product of the denominators, namely $(1 - \sqrt{a-2})(1 + \sqrt{a-2})$. This gives us

$$\left(\frac{3}{1-\sqrt{a-2}} + \frac{3}{1+\sqrt{a-2}}\right)(1-\sqrt{a-2})(1+\sqrt{a-2}) = 6(1-\sqrt{a-2})(1+\sqrt{a-2}).$$

The left side becomes

$$\left(\frac{3}{1-\sqrt{a-2}} + \frac{3}{1+\sqrt{a-2}}\right)(1-\sqrt{a-2})(1+\sqrt{a-2})$$
$$= \frac{3}{1-\sqrt{a-2}} \cdot (1-\sqrt{a-2})(1+\sqrt{a-2}) + \frac{3}{1+\sqrt{a-2}} \cdot (1-\sqrt{a-2})(1+\sqrt{a-2})$$
$$= 3(1+\sqrt{a-2}) + 3(1-\sqrt{a-2})$$
$$= 3 + 3\sqrt{a-2} + 3 - 3\sqrt{a-2}$$
$$= 6,$$

and the right becomes

$$6(1-\sqrt{a-2})(1+\sqrt{a-2}) = 6(1+\sqrt{a-2}-\sqrt{a-2}-\sqrt{a-2}\sqrt{a-2}) = 6[1-(a-2)] = 6(3-a) = 18-6a,$$

so we get $6 = 18 - 6a$. Adding $6a$ to both sides, we get $6a + 6 = 18$, and subtracting 6 from both sides, we get $6a = 12$. Finally, dividing both sides by 6, we find $a = 2$. We check this answer. If $a = 2$, then

$$\frac{3}{1-\sqrt{a-2}} + \frac{3}{1+\sqrt{a-2}} = \frac{3}{1} + \frac{3}{1} = 6.$$

Therefore, $a = \boxed{2}$.

Exercises for Section 1.2

1.2.1 From the second equation, we have $y = x + 5$. Substituting this expression for y into the first equation, we get $-x + 2(x+5) = 8$, so $x = -2$. Then $y = -2 + 5 = 3$. Therefore, $(x, y) = \boxed{(-2, 3)}$.

1.2.2 If $3ax - 8b = 4x + 6a$ is true for all values of x, then it must be true in particular when $x = 0$, so $-8b = 6a$. It must also be true when $x = 1$, so $3a - 8b = 4 + 6a$. But $-8b = 6a$, so by substitution, we have $3a + 6a = 4 + 6a$. Solving for a, we find $a = 4/3$. Then $-8b = 6a = 8$, so $b = \boxed{-1}$.

We could also solve this problem by noting that if $3ax - 8b = 4x + 6a$ is true for all x, then the coefficients of x on both sides must be the same, and the constant terms on both sides must be the same. This gives us $3a = 4$ and $-8b = 6a$. The first equation gives $a = 4/3$ and the second gives $b = 6a/(-8) = -1$.

1.2.3 From the first equation, we have $y = 6 - x$. Substituting this expression for y into the second equation, we get $x^2 + x(6 - x) + 2(6 - x)^2 = 32$. This simplifies as $2x^2 - 18x + 40 = 0$. Dividing by 2 and factoring gives $(x - 4)(x - 5) = 0$. So, $x = 4$ or $x = 5$. Since $y = 6 - x$, we have $y = 2$ when $x = 4$ and $y = 1$ when $x = 5$. Therefore, the solutions are $(x, y) = \boxed{(4, 2) \text{ and } (5, 1)}$.

1.2.4 The first and third equations have similar terms, though the terms are squared in the third equation. From the first equation, we have $x + 2y = z + 5$. Substituting this into the third equation makes the left side $(x + 2y)^2 - z^2 = (z + 5)^2 - z^2 = z^2 + 10z + 25 - z^2 = 10z + 25$, so we have $10z + 25 = 15$. Solving this equation gives $z = -1$.

Substituting this value into the first two equations, we obtain the equations $x + 2y + 1 = 5$ and $3x + 2y - 1 = 11$. From the first equation, we have $x = 4 - 2y$. Substituting this expression for x into the second equation, we get $3(4 - 2y) + 2y - 1 = 11$, so $y = 0$. Therefore, $x = 4 - 2y = 4$, and the solution is $(x, y, z) = \boxed{(4, 0, -1)}$.

Exercises for Section 1.3

1.3.1 Multiplying the first equation by 2, we get $6a - 4b = -16$. Adding this to $5a + 4b = 5$, we get $11a = -11$, so $a = -1$. Substituting this value into the first equation, we get $-3 - 2b = -8$, so $b = 5/2$. Therefore, $(a, b) = \boxed{(-1, 5/2)}$.

1.3.2 We are given $\frac{2}{x} = \frac{y}{3}$ and $\frac{2}{x} = \frac{x}{y}$. Multiplying these two equations eliminates y and gives $\frac{4}{x^2} = \frac{x}{3}$. Cross-multiplying this equation gives $x^3 = \boxed{12}$.

1.3.3 To eliminate y, we can square the second equation to get $x^8 y^2 = 2^{64}$. Dividing this by the first equation, $x^3 y^2 = 2^{34}$, we get $x^5 = 2^{64}/2^{34} = 2^{30}$, so $x = 2^6$. Then from the second equation, $y = 2^{32}/x^4 = 2^{32}/2^{24} = 2^8$. Therefore, $(x, y) = (2^6, 2^8) = \boxed{(64, 256)}$.

1.3.4 Adding the second and third equations eliminates the xy term, and we get $z^2 + 2z = -1$. This simplifies to $(z + 1)^2 = 0$, so $z = -1$. Substituting this value of z into the first equation, we get $x - 2 = 1$, so $x = 3$. Substituting these values of x and z into the third equation, we get $3y + 1 = -14$, so $y = -5$. Therefore, $(x, y, z) = \boxed{(3, -5, -1)}$.

1.3.5 We can rewrite the first equation as $r(r - 2s) = 27$, and the second equation as $s(r - 2s) = 9$. We know that $r - 2s$, r, and s are nonzero (otherwise, $r^2 - 2rs$ or $rs - 2s^2$ will equal 0), so we can divide the first equation by the second to get $r/s = 27/9$, which means $r = 3s$.

Substituting this expression for r into the second equation, we get $3s^2 - 2s^2 = 9$, so $s = 3$ or $s = -3$. These give us $r = 9$ and $r = -9$, respectively. We check that both solutions work, so the solutions are $(r, s) = \boxed{(9, 3) \text{ and } (-9, -3)}$.

Exercises for Section 1.4

1.4.1

(a) Note that the third equation does not contain the variable b. Furthermore, we can eliminate the variable b from the first and second equations by adding them, to get $7a + 4c = 7$. But the third equation says that $7a + 4c = 8$, which gives us two different values of $7a + 4c$. Therefore, the system is inconsistent and has $\boxed{\text{no solutions}}$.

(b) This system is nearly the same system as the one in part (a). The difference is that the right side of the first equation is 3 instead of 2. Now, when we add the first and second equations, we get $7a + 4c = 8$, which is the same as the third equation (a redundant equation). Solving for a, we find $a = (8 - 4c)/7$. Substituting this expression for a into the first equation, we get

$$2 \cdot \frac{8 - 4c}{7} + b + c = 3.$$

Solving for b, we find $b = (c + 5)/7$. Hence, all solutions to the system are of the form

$$(a, b, c) = \boxed{\left(\frac{8 - 4c}{7}, \frac{c + 5}{7}, c \right)},$$

where c is any real number. Note that this is not the only form in which we can write the solution. For example, we can solve $a = (8 - 4c)/7$ for c in terms of a to get $c = (8 - 7a)/4$. Substituting this into our solution above, we see that we can write all solutions in the form $(a, b, c) = \left(a, \frac{4-a}{4}, \frac{8-7a}{4} \right)$. There are also infinitely many other ways we can write the solution. If you have something different, try substituting your solution back into the given equations to see if it works.

(c) Adding the second and third equations, we get $2x = 8$, so $x = 4$. Then the first equation becomes $4 + y + z = 13$, so $y + z = 9$, and the second equation becomes $4 + y - z = -1$, so $y - z = -5$. Adding these equations, we get $2y = 4$, so $y = 2$. The first equation is now $4 + 2 + z = 13$, so $z = 7$. Therefore, $(x, y, z) = \boxed{(4, 2, 7)}$.

1.4.2 We seek constants k_1 and k_2 such that

$$k_1(a + b + c) + k_2(a + 2b + 3c) = 2a + 5b + 8c.$$

We want the coefficients of a, b, and c to match, which gives us the system of equations

$$k_1 + k_2 = 2,$$
$$k_1 + 2k_2 = 5,$$
$$k_1 + 3k_2 = 8.$$

Subtracting the first equation from the second, we get $k_2 = 3$. Then the first equation becomes $k_1 + 3 = 2$, so $k_1 = -1$. Note that $k_1 + 3k_2 = -1 + 3 \cdot 3 = 8$, so the third equation is also satisfied. Therefore, the constants we seek are $\boxed{k_1 = -1 \text{ and } k_2 = 3}$.

1.4.3 Since we are given the values $a^3 b^2 c$ and $a^2 b^3 c^5$, we can try to express $a^3 b^7 c^{14}$ as a product of powers of $a^3 b^2 c$ and $a^2 b^3 c^5$. In other words, we seek constants x and y such that

$$a^3 b^7 c^{14} = (a^3 b^2 c)^x (a^2 b^3 c^5)^y = a^{3x+2y} b^{2x+3y} c^{x+5y}.$$

We want the exponents of a, b, and c to match, which gives us the system of equations

$$3x + 2y = 3,$$
$$2x + 3y = 7,$$
$$x + 5y = 14.$$

From the third equation, we have $x = 14 - 5y$. Substituting this expression for x into the second equation, we get $2(14 - 5y) + 3y = 7$, so $y = 3$. Then $x = 14 - 5 \cdot 3 = -1$. Since $3x + 2y = 3 \cdot (-1) + 2 \cdot 3 = 3$, the first equation is also satisfied. Therefore, the solution to this system is $(x, y) = (-1, 3)$.

Hence, $a^3 b^7 c^{14} = (a^3 b^2 c)^{-1} (a^2 b^3 c^5)^3 = \dfrac{240^3}{108} = \boxed{128{,}000}$.

1.4.4

(a) We can solve the system by eliminating the variables one by one. We begin by eliminating d. Adding the second equation and third equation, we get $a - 2b - 3c = 7$. Subtracting the third equation from the fourth equation, we get $a + b - 2c = 12$. We now have the system of equations

$$a + 3b - 2c = 18,$$
$$a - 2b - 3c = 7,$$
$$a + b - 2c = 12.$$

Now, subtracting the third equation from the first equation, we get $2b = 6$, so $b = 3$. Substituting this value for b, the above equations become

$$a - 2c = 9,$$
$$a - 3c = 13.$$

Subtracting the second equation from the first equation, we get $c = -4$. Then from the first equation, $a = 2c + 9 = 1$. Finally, the equation $3b - 7c + d = 35$ in the problem becomes $9 + 28 + d = 35$, so $d = -2$. Therefore, the solution is $(a, b, c, d) = \boxed{(1, 3, -4, -2)}$.

(b) Substituting the third equation, $x_2 + x_3 = 0$, into the first equation gives $x_1 + x_4 = 1$, so $x_1 = 1 - x_4$. Then from the second equation, we have $x_2 - x_3 = 2 - x_1 = 2 - (1 - x_4) = x_4 + 1$. Adding this to the third equation, we get $(x_2 + x_3) + (x_2 - x_3) = x_4 + 1$, so $x_2 = (x_4 + 1)/2$. Again from the third equation, we have $x_3 = -x_2 = -(x_4 + 1)/2$. Therefore, all solutions to the system are of the form

$$(x_1, x_2, x_3, x_4) = \boxed{\left(1 - x_4, \frac{x_4 + 1}{2}, -\frac{x_4 + 1}{2}, x_4\right)},$$

where x_4 is any real number. As with part (b) of Exercise 1.4.1, there are infinitely many ways we can describe these solutions. If you have a different one, substitute it back into the three equations to confirm that yours is correct.

1.4.5 Let E_1, E_2, and E_3 denote the first, second, and third equations, respectively. Then from the given information, we have $E_3 = aE_1 + bE_2$. Solving for E_1, we get

$$E_1 = \frac{E_3 - bE_2}{a} = -\frac{b}{a}E_2 + \frac{1}{a}E_3.$$

Thus, the first equation is c times the second equation plus d times the third equation, where $\boxed{c = -b/a}$ and $\boxed{d = 1/a}$. (These values are well-defined since we know that a is nonzero.)

Review Problems

1.19

(a) Multiplying the first equation by 2, we get $8x + 10y = 86$. Multiplying the second equation by 5, we get $45x - 10y = 285$. Adding these, we get $53x = 371$, so $x = 7$. Then the first equation becomes $28 + 5y = 43$, so $y = 3$. Therefore, $(x, y) = \boxed{(7, 3)}$.

(b) Dividing the second equation by 3, we get $x - y = -4$, so $y = x + 4$. Substituting this expression for y into the first equation, we get $2x + 4(x + 4) = 18$, so $x = 1/3$. Then $y = 1/3 + 4 = 13/3$. Therefore, $(x, y) = \boxed{(1/3, 13/3)}$.

(c) Dividing the second equation by 4, we get $3x - y = 1$. This is the same as the first equation, so the second equation is redundant. Solving for x, we find $x = (y + 1)/3$. Therefore, all solutions to the system are of the form $(x, y) = \boxed{((y+1)/3, y)}$, where y is any real number. We can also solve $x = (y + 1)/3$ for y to get $y = 3x - 1$, then use this to write our solutions as $(x, y) = (x, 3x - 1)$, where x is any real number. There are also infinitely many other ways to write the solutions.

1.20 Let x be Jeff's daughter's current age. Since Jeff is 4 times older than his daughter, Jeff's age is $4x$. Five years ago, Jeff was $4x - 5$ and his daughter was $x - 5$, so $4x - 5 = 9(x - 5) = 9x - 45$. Solving for x, we find $x = \boxed{8}$.

1.21

(a) We can eliminate both y and z by adding the first and third equations, which gives $12x = 60$, so $x = 5$. Then the first equation becomes $40 + y - z = 46$, so $y - z = 6$, and the second equation becomes $15 + 4y - 2z = 27$, so $4y - 2z = 12$. Dividing both sides by 2, we get $2y - z = 6$. Subtracting the equation $y - z = 6$, we get $(2y - z) - (y - z) = 6 - 6 = 0$, so $y = 0$. Finally, substituting this value of y into the equation $y - z = 6$, we get $-z = 6$, so $z = -6$. Therefore, $(x, y, z) = \boxed{(5, 0, -6)}$.

(b) We can eliminate the variable a by using the third equation. Multiplying the third equation by 2, we get $2a + 8b - 22c = -30$. Subtracting the first equation, $2a - 3b + 5c = 17$, we get

$$(2a + 8b - 22c) - (2a - 3b + 5c) = -30 - 17 = -47,$$

so $11b - 27c = -47$.

Multiplying the third equation by 3, we get $3a + 12b - 33c = -45$. Subtracting the second equation, $3a + b - 6c = -4$, we get $(3a + 12b - 33c) - (3a + b - 6c) = -45 - (-4) = -41$, so $11b - 27c = -41$. We have found two different values of $11b - 27c$. Therefore, the system is inconsistent and has $\boxed{\text{no solutions}}$.

Alternatively, we could have observed that the adding the first equation and third equation gives us $3a + b - 6c = 2$, but by the second equation, we have $3a + b - 6c = -4$, which also tells us that there are no solutions.

(c) Multiplying the first equation by 3, we get $3x + 3y - 3z = 33$. Subtracting this from the third equation, $5x + 3y - 3z = 43$, we get
$$(5x + 3y - 3z) - (3x + 3y - 3z) = 43 - 33 = 10,$$

so $x = 5$.

Then the first equation becomes $5 + y - z = 11$, so $y - z = 6$. The second equation becomes $15 - 2y + 2z = 3$, so $-2y + 2z = -12$. Dividing both sides by -2, we get $y - z = 6$. Finally, the third equation becomes $25 + 3y - 3z = 43$, so $3y - 3z = 18$. Dividing both sides by 3, we get $y - z = 6$.

Thus, every equation in the system reduces to $y - z = 6$, or $y = z + 6$. Therefore, all solutions to the system are of the form $(x, y, z) = \boxed{(5, z + 6, z)}$, where z is any real number.

1.22 There isn't any obvious way to isolate any of the variables with elimination. (As an extra challenge, see if you can find a way to use elimination to isolate a variable!) Instead, we substitute for the variables in sequence, as follows. From the first equation, $b = 2 - a$. Substituting this expression for b into the second equation, we get $2 - a + c = 13$, so $c = a + 11$. Substituting this expression for c into the third equation, we get $a + 11 + 3d = 37$, so $3d = 26 - a$.

Substituting this expression for $3d$ into the fourth equation, we get $26 - a + 4e = -23$, so $4e = a - 49$, which means $8e = 2a - 98$. Finally, substituting this expression for $8e$ into the fifth equation, we get $2a - 98 + 9a = -43$. Solving for a, we find $a = 5$.

Then $b = 2 - a = -3$, $c = a + 11 = 16$, $d = (26 - a)/3 = 7$, and $e = (a - 49)/4 = -11$. Therefore, $(a, b, c, d, e) = \boxed{(5, -3, 16, 7, -11)}$.

1.23

(a) Multiplying both sides by 18, we get $2 \cdot 2(x-1) \le x + 2 \cdot 2$, so $4x - 4 \le x + 4$. Simplifying, we get $3x \le 8$, so $x \le 8/3$. In interval notation, we have $x \in \boxed{(-\infty, 8/3]}$.

(b) We tackle the two ends of the inequality chain separately. Adding $-x - 2$ to both sides of $x - 1 \le 3x + 2$ gives $-3 \le 2x$, so $-\frac{3}{2} \le x$. Adding $-2 - 2x$ to both sides of $3x + 2 \le 2x + 6$ gives $x \le 4$. Combining the results of both ends of the inequality chain gives $-\frac{3}{2} \le x \le 4$, so $x \in \boxed{\left[-\frac{3}{2}, 4\right]}$.

(c) Simplifying, we get $4x < 20$, so $x < 5$. In interval notation, we have $x \in \boxed{(-\infty, 5)}$.

1.24 Letting $a = \sqrt{x}$ and $b = \sqrt{y}$, we obtain the system of equations

$$a + b = 7,$$
$$3a - 4b = -14.$$

Multiplying the first equation by 4 and adding the second equation, we get $4(a + b) + (3a - 4b) = 4 \cdot 7 - 14$, so $7a = 14$, which means $a = 2$. Then the first equation becomes $2 + b = 7$, so $b = 5$. Then $x = a^2 = 4$ and $y = b^2 = 25$, so $(x, y) = \boxed{(4, 25)}$.

Note that we didn't have to perform a substitution. We could have added 4 times the first original equation to the second original equation to give $7\sqrt{x} = 14$, from which we find $\sqrt{x} = 2$, so $x = 4$. From here, we can use either of the original equations to find $y = 25$.

1.25 We can eliminate both x_1 and x_4 by subtracting the first equation from the second equation, which gives $x_2 + 4x_3 = 1$. We can also eliminate both x_1 and x_4 by subtracting three times the first equation from the third equation, which gives $(3x_1 + 5x_2 + 5x_3 - 3x_4) - 3(x_1 + x_2 - x_3 - x_4) = 6 - 3$, so $2x_2 + 8x_3 = 3$ But from $x_2 + 4x_3 = 1$, we have $2x_2 + 8x_3 = 2(x_2 + 4x_3) = 2$. Therefore, the system is inconsistent and has $\boxed{\text{no solutions}}$.

1.26 Cross-multiplying the first equation, we get $3r = 2(s - r) = 2s - 2r$, so $2s = 5r$, or $s = 5r/2$. Substituting this expression for s into $\frac{1}{6} = \frac{1}{r} - \frac{1}{s}$, we get

$$\frac{1}{6} = \frac{1}{r} - \frac{1}{s} = \frac{1}{r} - \frac{2}{5r} = \frac{5}{5r} - \frac{2}{5r} = \frac{3}{5r}.$$

From $1/6 = 3/(5r)$, we find so $r = 18/5$. Then $s = (5/2)(18/5) = 9$. Therefore, $(r, s) = \boxed{(18/5, 9)}$.

1.27 Since we are given the values of $2a - b + 5c$ and $2a + 3b + c$, we can try to express $a + b + c$ as a linear combination of these expressions. In other words, we seek constants x and y such that

$$a + b + c = x(2a - b + 5c) + y(2a + 3b + c) = (2x + 2y)a + (-x + 3y)b + (5x + y)c.$$

We want the coefficients of a, b, and c to match, which gives us the system of equations

$$2x + 2y = 1,$$
$$-x + 3y = 1,$$
$$5x + y = 1.$$

From the third equation, we have $y = 1 - 5x$. Substituting this expression for y into the second equation, we get $-x + 3(1 - 5x) = 1$, so $x = 1/8$. Then $y = 1 - 5/8 = 3/8$. Note that $2x + 2y = 2/8 + 6/8 = 1$, so the first equation is also satisfied. Therefore, the solution to this system is $(x, y) = (1/8, 3/8)$.

Hence, $a + b + c = \frac{1}{8}(2a - b + 5c) + \frac{3}{8}(2a + 3b + c) = \frac{1}{8} \cdot 13 + \frac{3}{8} \cdot 75 = \boxed{\frac{119}{4}}$.

1.28   Squaring the second equation, we get $a^2/9^b = 18^2 = 324$. Dividing the first equation by this equation, we get

$$\frac{a^2 9^b}{a^2/9^b} = \frac{4}{324} = \frac{1}{81},$$

so $81^b = \frac{1}{81}$, which means $b = -1$. Then the second equation becomes $3a = 18$, so $a = 6$. Therefore, $(a, b) = \boxed{(6, -1)}$.

1.29   Adding the two equations, we get $x^2 + 2xy + y^2 = (x + y)^2 = 28 - 12 = 16$, so $x + y = 4$ or $x + y = -4$.

If $x + y = 4$, then $y = 4 - x$. Substituting this expression for y into the first equation, we get $x^2 + x(4 - x) = 28$, which simplifies to $4x = 28$, so $x = 7$. Then $y = 4 - 7 = -3$.

If $x + y = -4$, then $y = -4 - x$. Substituting this expression for y into the first equation, we get $x^2 + x(-4 - x) = 28$, which simplifies to $-4x = 28$, so $x = -7$. Then $y = -4 + 7 = 3$.

Therefore, the solutions are $(x, y) = \boxed{(7, -3) \text{ and } (-7, 3)}$.

We can also tackle this problem by factoring the left sides of both equations, which gives $x(x + y) = 28$ and $y(x + y) = -12$. We clearly can't have $x + y$, x, or y equal to 0, so we can safely divide the first equation by the second to give $x/y = -7/3$, from which we have $x = -7y/3$. Substituting this into the second equation gives $y^2 - 7y^2/3 = -12$, from which we find $y^2 = 9$. This means $y = \pm 3$, and using $x = -7y/3$, we find the same solutions as before.

1.30   From the first equation, we have $a = 2b - 4$. Substituting this expression for a into the second equation, we get $(2b - 4)^2 - 2b^2 = -14$, which simplifies to $2b^2 - 16b + 30 = 0$. Factoring gives $2(b - 3)(b - 5) = 0$, so $b = 3$ or $b = 5$. Since $a = 2b - 4$, we have $a = 2$ if $b = 3$ and $a = 6$ if $b = 5$. Therefore, the solutions are $(a, b) = \boxed{(2, 3) \text{ and } (6, 5)}$.

Challenge Problems

1.31   Let n be the number of games that the tennis player has won at the start of the weekend. Since her win ratio is exactly 0.5, she has played exactly $2n$ games. During the weekend, she won three games and lost one game, so her win ratio is now

$$\frac{n + 3}{2n + 4} > 0.503.$$

Multiplying both sides by $2n + 4$, we get $n + 3 > 1.006n + 2.012$, so $0.006n < 0.988$. Therefore,

$$n < \frac{0.988}{0.006} = \frac{988}{6} = 164 + \frac{2}{3}.$$

Since n is an integer, the largest possible value of n is $\boxed{164}$.

1.32   It is tempting to multiply both sides of $1/x < 3$ by x to get $1 < 3x$. However, this step is valid if and only if x is positive. If x is negative, then we must change the direction of the inequality, to get $1 > 3x$. To deal with this possibility, we divide into the cases where x is positive and x is negative. (Note that x cannot be zero.)

If x is positive, then we get $1 < 3x$, so $x > 1/3$. The second inequality, $1/x > -4$, is always satisfied, since $1/x$ is positive. So, if $x > 0$, we must have $x > 1/3$.

If x is negative, then the first inequality, $1/x < 3$, is always satisfied, since $1/x$ is negative. Multiplying the second inequality by x (and changing the direction of the inequality because x is negative), we get $1 < -4x$. Dividing both sides by -4 (and again changing the direction of the inequality), we get $-1/4 > x$.

Therefore, the set of all values x such that $1/x < 3$ and $1/x > -4$ is all values of x such that either $\boxed{x > 1/3 \text{ or } x < -1/4}$.

1.33 We are given that $ax^2 + bx + c = dx^2 + ex + f$ for all values of x. Substituting the values of $x = -1$, $x = 0$, and $x = 1$, we obtain the system of equations

$$a - b + c = d - e + f,$$
$$c = f,$$
$$a + b + c = d + e + f.$$

Adding the first and third equations, we get $2a + 2c = 2d + 2f$, so $a + c = d + f$. From the second equation, we have $c = f$, so $a = d$. Then from the third equation, we have $b = e$, as desired.

1.34 Let a and b be the legs and c be the hypotenuse of the triangle, so $a^2 + b^2 = c^2$. The perimeter and area are both 30, so $a + b + c = ab/2 = 30$. Since one of our equations does not involve c, we choose to substitute for c. From $a + b + c = 30$, we get $c = 30 - (a + b)$. Substituting this expression for c into the equation $a^2 + b^2 = c^2$, we get

$$a^2 + b^2 = [30 - (a + b)]^2 = 900 - 60(a + b) + a^2 + 2ab + b^2,$$

which simplifies to $60(a + b) = 900 + 2ab$. Since $ab/2 = 30$, we have $2ab = 120$, so $60(a + b) = 900 + 2ab = 1020$. Therefore, we have $a + b = 1020/60 = 17$ and $c = 30 - (a + b) = 30 - 17 = 13$.

Now we must solve for a and b. Since $a + b = 17$, we have $b = 17 - a$. Then $ab = 60$ gives us $a(17 - a) = 60$, which simplifies to $a^2 - 17a + 60 = 0$. Factoring gives $(a - 5)(a - 12) = 0$. This means that $a = 5$ or $a = 12$. Since $b = 17 - a$, we have $b = 12$ when $a = 5$ and $b = 5$ when $a = 12$. Therefore, the sides of the triangle are $\boxed{5, 12, \text{ and } 13}$.

1.35 Expanding the given equation, we get

$$x^2 - 2ax + a^2 + 4x^2 - 4bx + b^2 = x^2 - 6x + 9 + 4x^2$$
$$\Rightarrow \quad 5x^2 - (2a + 4b)x + a^2 + b^2 = 5x^2 - 6x + 9$$
$$\Rightarrow \quad -(2a + 4b)x + a^2 + b^2 = -6x + 9.$$

This equation holds for all x if and only if $-(2a + 4b) = -6$ and $a^2 + b^2 = 9$. Dividing the first equation by -2, we get $a + 2b = 3$, so $a = 3 - 2b$. Substituting this expression for a into the second equation, we get $(3 - 2b)^2 + b^2 = 9$, which simplifies to $5b^2 - 12b = 0$. Factoring gives $b(5b - 12) = 0$, so $b = 0$ or $b = 12/5$. Since $a = 3 - 2b$, the corresponding values of a are $a = 3$ and $a = -9/5$. Therefore, the solutions are $(a, b) = \boxed{(3, 0) \text{ and } (-9/5, 12/5)}$.

(It follows that $(-9/5, 12/5)$ is the reflection of the point $(3, 0)$ in the line $y = 2x$. Why?)

1.36 Since we are given the values of x^2y/z and y^4z/x, we can try to express $x^8/(yz)^5$ as a product of powers of x^2y/z and y^4z/x. In other words, we seek constants a and b such that

$$\frac{x^8}{y^5z^5} = \left(\frac{x^2y}{z}\right)^a \left(\frac{y^4z}{x}\right)^b = \frac{x^{2a}y^a}{z^a} \cdot \frac{y^{4b}z^b}{x^b} = x^{2a-b}y^{a+4b}z^{-a+b}.$$

We want the exponents of x, y, and z to match, which gives us the system of equations

$$2a - b = 8,$$
$$a + 4b = -5,$$
$$-a + b = -5.$$

Adding the second and third equations, we get $5b = -10$, so $b = -2$. Then the third equation becomes $-a - 2 = -5$, so $a = 3$. Note that $2a - b = 2 \cdot 3 - (-2) = 8$, so the first equation is also satisfied. Therefore, the solution to this system is $(a, b) = (3, -2)$.

Hence, $\dfrac{x^8}{y^5z^5} = \left(\dfrac{x^2y}{z}\right)^3 \left(\dfrac{y^4z}{x}\right)^{-2} = \dfrac{24^3}{30^2} = \boxed{\dfrac{384}{25}}$.

1.37 For $1 \le n \le 10$, let x_n denote the number picked by the person who announced the average n. We seek the value of x_6. We can find the numbers picked by other people in the circle in terms of x_6, by going around the circle as follows.

The average of x_6 and x_8 is 7, so $x_8 = 2 \cdot 7 - x_6 = 14 - x_6$.

The average of x_8 and x_{10} is 9, so $x_{10} = 2 \cdot 9 - x_8 = 18 - x_8 = 18 - (14 - x_6) = x_6 + 4$.

The average of x_{10} and x_2 is 1, so $x_2 = 2 \cdot 1 - x_{10} = 2 - x_{10} = 2 - (x_6 + 4) = -x_6 - 2$.

The average of x_2 and x_4 is 3, so $x_4 = 2 \cdot 3 - x_2 = 6 - x_2 = 6 - (-x_6 - 2) = x_6 + 8$.

Finally, the average of x_4 and x_6 is 5. But $x_4 = x_6 + 8$, so $x_6 + x_6 + 8 = 2 \cdot 5 = 10$, which gives $x_6 = \boxed{1}$.

1.38 From the second equation, we have $a = 4/d$. From the fourth equation, we have $b = 6/d$. Then from the first equation, we have $c = \frac{6}{a} = \frac{6}{4/d} = \frac{3}{2}d$. Note that $bc = (6/d)(3d/2) = 9$, so the third equation is also satisfied. Therefore, all solutions to the system are of the form

$$(a, b, c, d) = \boxed{\left(\frac{4}{d}, \frac{6}{d}, \frac{3}{2}d, d \right)},$$

where d is any nonzero real number. (There are infinitely many other ways we can express the solutions. If you found a different solution, substitute it back into the equations to check it. However, if your solution is just a single set of numbers, as opposed to an expression with variables, then your solution is not complete, even if it satisfies the given equations.)

1.39 We can easily isolate either x or y in the first equation. So, we take a look at the second equation and note that substituting for y will take less effort, since there is only one instance of y in that equation.

We solve the first equation for y in terms of x to find $y = \dfrac{2 - 9x}{3}$. We use this equation to replace y in the second equation in our system:

$$9x^2 + 3 \left(\frac{2 - 9x}{3} \right)^2 - 7x = 0.$$

After some careful manipulation and simplification of this equation, we have $108x^2 - 57x + 4 = 0$. Factoring this quadratic gives us $(9x - 4)(12x - 1) = 0$, so we have $x = 4/9$ or $x = 1/12$. These give us the solutions $(x, y) = \boxed{(4/9, -2/3) \text{ and } (1/12, 5/12)}$.

1.40 We can expand the first equation as $\frac{2}{x} - \frac{3}{y} = -\frac{7}{12}$. Similarly, we can expand the second equation as $\frac{3}{x} + \frac{5}{y} = \frac{25}{4}$.

Then if we let $a = 1/x$ and $b = 1/y$, we obtain the system of equations

$$2a - 3b = -\frac{7}{12},$$
$$3a + 5b = \frac{25}{4}.$$

Multiplying the first equation by 5 and the second equation by 3, we get

$$10a - 15b = -\frac{35}{12},$$
$$9a + 15b = \frac{75}{4}.$$

Adding these, we get $19a = -\frac{35}{12} + \frac{75}{4} = \frac{95}{6}$, so $a = \frac{5}{6}$. Then the first equation becomes $\frac{5}{3} - 3b = -\frac{7}{12}$, so $b = \frac{3}{4}$. Therefore, $(x, y) = (1/a, 1/b) = \boxed{(6/5, 4/3)}$.

1.41 We multiply the first equation by d and the second by b to set up eliminating the y terms:

$$adx + bdy = de,$$
$$bcx + bdy = bf.$$

Subtracting the second equation from the first gives

$$(ad - bc)x = de - bf.$$

We recognize the coefficient of x and the right side of the equation as having the form of the determinants described in the problem, so we have

$$\begin{vmatrix} a & b \\ c & d \end{vmatrix} x = \begin{vmatrix} e & b \\ f & d \end{vmatrix} \quad \Rightarrow \quad x = \frac{\begin{vmatrix} e & b \\ f & d \end{vmatrix}}{\begin{vmatrix} a & b \\ c & d \end{vmatrix}}.$$

Similarly, if we multiply our first original equation by c and the second by a, we are set up to eliminate the x terms:

$$acx + bcy = ce,$$
$$acx + ady = af.$$

Subtracting the first equation from the second gives

$$(ad - bc)y = af - ce \quad \Rightarrow \quad \begin{vmatrix} a & b \\ c & d \end{vmatrix} y = \begin{vmatrix} a & e \\ c & f \end{vmatrix} \quad \Rightarrow \quad y = \frac{\begin{vmatrix} a & e \\ c & f \end{vmatrix}}{\begin{vmatrix} a & b \\ c & d \end{vmatrix}}.$$

In both our solution for x and our solution for y, we run into trouble if $ad - bc = 0$, since then our equation after elimination has no variable. We've seen this before in solving systems of equations. If combining the equations leads to an equation with no variables, then either the system has infinitely many solutions (if the no-variable equation is true), or the system has no solutions (if the no-variable equation is false). So, we look at the two equations we found with elimination:

$$(ad - bc)x = de - bf,$$
$$(ad - bc)y = af - ce.$$

If $ad - bc = 0$, then the left sides of both are 0. Therefore, if we have $de - bf = af - ce = 0$, then we have infinitely many solutions to this system. Rearranging $de - bf = 0$ gives $b/d = e/f$, and rearranging $af - ce = 0$ gives $a/c = e/f$. Similarly, rearranging $ad - bc = 0$ gives $a/c = b/d$. So, if $a/c = b/d = e/f$, then we have infinitely many solutions. (In this case, one equation is just a constant multiple of the other.)

However, if $ad - bc = 0$ and either $de - bf$ or $af - ce$ is not zero, then there are no solutions to the system.

1.42 We can take advantage of the symmetry in the equations by adding them. This gives us $(k+2)(x+y+z) = 3k$. If $k = -2$, then this equation becomes $0 = -6$, so the system is inconsistent and has no solutions. Otherwise, $k \neq -2$, and we have $x + y + z = \frac{3k}{k+2}$. Subtracting this from the first equation, we get

$$(kx + y + z) - (x + y + z) = k - \frac{3k}{k+2}$$
$$\Rightarrow \quad (k-1)x = \frac{k^2 + 2k - 3k}{k+2} = \frac{k^2 - k}{k+2} = \frac{k(k-1)}{k+2}.$$

If $k = 1$, then the original system of equations becomes

$$x + y + z = 1,$$
$$x + y + z = 1,$$
$$x + y + z = 1.$$

This system clearly has an infinite number of solutions. Otherwise, $k \neq 1$, and we have $x = \frac{k}{k+2}$. By the same calculations, we find $x = y = z = \frac{k}{k+2}$.

Hence, to summarize, the system

(a) has no solutions when $k = \boxed{-2}$,

(b) has an infinite number of solutions when $k = \boxed{1}$, and

(c) has a unique solution, namely $(x, y, z) = (\frac{k}{k+2}, \frac{k}{k+2}, \frac{k}{k+2})$, $\boxed{\text{for all } k \text{ such that } k \neq -2 \text{ and } k \neq 1}$.

1.43 Solving for y in the first equation, we get $y = \frac{c-bx}{a}$. Solving for z in the second equation, we get $z = \frac{b-cx}{a}$.

Substituting these expressions for y and z into the third equation, we get

$$
\begin{aligned}
& b \cdot \frac{b-cx}{a} + c \cdot \frac{c-bx}{a} = a \\
\Rightarrow\quad & b(b-cx) + c(c-bx) = a^2 \\
\Rightarrow\quad & b^2 - bcx + c^2 - bcx = a^2 \\
\Rightarrow\quad & 2bcx = b^2 + c^2 - a^2 \\
\Rightarrow\quad & x = \frac{b^2+c^2-a^2}{2bc}.
\end{aligned}
$$

Then we have $y = \dfrac{c-bx}{a} = \dfrac{c - b \cdot \frac{b^2+c^2-a^2}{2bc}}{a} = \dfrac{c - \frac{b^2+c^2-a^2}{2c}}{a} = \dfrac{2c^2-b^2-c^2+a^2}{2ac} = \dfrac{a^2+c^2-b^2}{2ac}$, and $z = \dfrac{b-cx}{a} = \dfrac{b - c \cdot \frac{b^2+c^2-a^2}{2bc}}{a} = \dfrac{b - \frac{b^2+c^2-a^2}{2b}}{a} = \dfrac{2b^2-b^2-c^2+a^2}{2ab} = \dfrac{a^2+b^2-c^2}{2ab}$.

Therefore, the solution is $(x, y, z) = \boxed{\left(\dfrac{b^2+c^2-a^2}{2bc}, \dfrac{a^2+c^2-b^2}{2ac}, \dfrac{a^2+b^2-c^2}{2ab}\right)}$.

If you have studied trigonometry, then you might recognize these expressions as being the cosines of a triangle. So why do these expressions appear here?

1.44 Our goal is to find a positive integer n such that for some integer k, $0 \leq k \leq n - 2$, we have

$$
\binom{n}{k} : \binom{n}{k+1} : \binom{n}{k+2} = 3 : 4 : 5.
$$

We can obtain linear equations in n and k by taking the ratios of the binomial coefficients as follows:

$$
\frac{3}{4} = \frac{\binom{n}{k}}{\binom{n}{k+1}} = \frac{\frac{n!}{k!(n-k)!}}{\frac{n!}{(k+1)!(n-k-1)!}} = \frac{(k+1)!(n-k-1)!}{k!(n-k)!} = \frac{k+1}{n-k},
$$

and

$$
\frac{4}{5} = \frac{\binom{n}{k+1}}{\binom{n}{k+2}} = \frac{\frac{n!}{(k+1)!(n-k-1)!}}{\frac{n!}{(k+2)!(n-k-2)!}} = \frac{(k+2)!(n-k-2)!}{(k+1)!(n-k-1)!} = \frac{k+2}{n-k-1}.
$$

From $\frac{3}{4} = \frac{k+1}{n-k}$, we have $4k + 4 = 3n - 3k$, and from $\frac{4}{5} = \frac{k+2}{n-k-1}$, we have $5k + 10 = 4n - 4k - 4$. Thus, we obtain the system of equations $7k = 3n - 4$, $9k = 4n - 14$.

Multiplying the first equation by 9, we get $63k = 27n - 36$. Multiplying the second equation by 7, we get $63k = 28n - 98$. Therefore, $27n - 36 = 28n - 98$, so $n = \boxed{62}$. (The corresponding value of k is 26.)

CHAPTER 2

Functions Review

Exercises for Section 2.1

2.1.1

(a) $f(2) = 2^2 - 2 - 6 = \boxed{-4}$.

(b) Since $f(x) = x^2 - x - 6 = (x - 3)(x + 2)$, $f(x) = 0$ when $\boxed{x = 3 \text{ and } x = -2}$.

(c) $f(x - 1) = (x - 1)^2 - (x - 1) - 6 = x^2 - 2x + 1 - x + 1 - 6 = \boxed{x^2 - 3x - 4}$.

(d) If $f(x) = 6$, then $x^2 - x - 6 = 6$, so $x^2 - x - 12 = 0$. Factoring gives $(x - 4)(x + 3) = 0$, which means $x = 4$ or $x = -3$. Hence, $f(4) = f(-3) = 6$, so $\boxed{\text{yes}}$, we can have $f(x) = 6$.

(e) Completing the square, we get $f(x) = x^2 - x - 6 = \left(x - \dfrac{1}{2}\right)^2 - \dfrac{25}{4}$. Hence, the range of f is $\boxed{[-25/4, +\infty)}$. (If you don't know completing the square, don't worry, it is covered in Chapter 4 in the text.)

2.1.2

(a) We can evaluate the absolute value of $1 - x$ for any real number x, so the domain of f is $\boxed{\mathbb{R}}$. The absolute value of a real number is always nonnegative, and $1 - x$ can equal any nonnegative real number, so the range of f is $\boxed{[0, +\infty)}$.

(b) We can only take the square root of a nonnegative real number, so $2 - t$ must be nonnegative. Hence, the domain of f is $\boxed{(-\infty, 2]}$. Also, the square root of a nonnegative real number is always nonnegative, so the range of f is $\boxed{[0, +\infty)}$.

(c) We can evaluate $1 - x^2$ for any real number x, so the domain of h is $\boxed{\mathbb{R}}$. Since x^2 is always nonnegative, $1 - x^2$ is always less than or equal to 1, so the range of h is $\boxed{(-\infty, 1]}$.

(d) First, we note that we cannot have $u = 0$, since this makes the denominator of $1/u$ equal to 0. We also cannot have the denominator of the whole function, $1 + 1/u$, equal to 0. If $1 + 1/u = 0$, then $u = -1$, so the domain of g is $\boxed{\text{all real numbers except 0 and } -1}$.

   To find the range of g, let $v = g(u) = \frac{1}{1+1/u}$. Taking the reciprocal of both sides and rearranging gives

$$1 + \frac{1}{u} = \frac{1}{v} \quad \Rightarrow \quad \frac{1}{u} = \frac{1}{v} - 1 = \frac{1 - v}{v} \quad \Rightarrow \quad u = \frac{v}{1 - v}.$$

Hence, for every value of v, there is a corresponding value of u, except when $v = 1$. Furthermore, as determined above, u cannot be equal to 0 or -1. There is no real number v that corresponds to $u = -1$, but the real number $v = 0$ corresponds to $u = 0$, so v cannot be equal to 0 either. Therefore, the range of g is $\boxed{\text{all real numbers except 0 and 1}}$.

2.1.3 The function $(f \cdot g)(x) = \sqrt{4-x} \cdot \sqrt{2x-6}$ is defined only where both the functions $f(x) = \sqrt{4-x}$ and $g(x) = \sqrt{2x-6}$ are defined. The function $\sqrt{4-x}$ is defined only for $x \le 4$, and the function $\sqrt{2x-6}$ is defined only for $x \ge 3$. Hence, the domain of the function $f \cdot g$ is $\boxed{[3,4]}$.

As with $f \cdot g$, the function

$$\left(\frac{f}{g}\right)(x) = \frac{\sqrt{4-x}}{\sqrt{2x-6}}$$

is defined only where both the functions $f(x) = \sqrt{4-x}$ and $g(x) = \sqrt{2x-6}$ are defined. However, we must also exclude points where the denominator is 0. In this case, the denominator is 0 when $x = 3$. Hence, the domain of the function f/g is $\boxed{(3,4]}$.

The domains of $f \cdot g$ and f/g are different because we must exclude those values where $g(x) = 0$ from the domain of f/g, but we include these values in the domain of $f \cdot g$ if they are also in the domain of f.

2.1.4 The function

$$f(x) = \sqrt{\frac{2x-5}{x-8}}$$

is defined only where $(2x-5)/(x-8) \ge 0$. For this inequality to hold, either both $2x-5$ and $x-8$ must be positive, both of them must be negative, or $2x-5 = 0$.

If both $2x-5$ and $x-8$ are positive, then $x > 5/2$ and $x > 8$, so in this case, we get $x > 8$. If both $2x-5$ and $x-8$ are negative, then $x < 5/2$ and $x < 8$, so in this case, we get $x < 5/2$. Finally, if $2x-5 = 0$, then $x = 5/2$. Therefore, the set of x such that $(2x-5)/(x-8) \ge 0$ is $(-\infty, 5/2] \cup (8, +\infty)$, which is the domain of f.

On the other hand, the function

$$g(x) = \frac{\sqrt{2x-5}}{\sqrt{x-8}}$$

is defined only where both the functions $\sqrt{2x-5}$ and $\sqrt{x-8}$ are defined, and the denominator is nonzero. The function $\sqrt{2x-5}$ is defined for $x > 5/2$, and the function $\sqrt{x-8}$ is defined for $x > 8$. The denominator is zero when $x = 8$, so the domain of g is $(8, +\infty)$.

Thus, the functions $\boxed{f \text{ and } g \text{ have different domains}}$, because those values of x for which $2x-5$ and $x-8$ are negative are in the domain of f, but not in the domain of g.

2.1.5 $f(x) \cdot f(-x) = \dfrac{x+1}{x-1} \cdot \dfrac{-x+1}{-x-1} = \dfrac{x+1}{x-1} \cdot \dfrac{x-1}{x+1} = \boxed{1}$.

2.1.6 We break up the interval $[-3, 4]$ into the intervals $[-3, -2]$ and $[-2, 4]$. If $-3 \le x \le -2$, then $-1 \le x+2 \le 0$, so $0 \le (x+2)^2 \le 1$. If $-2 \le x \le 4$, then $0 \le x+2 \le 6$, so $0 \le (x+2)^2 \le 36$. Hence, every value in the range of g is in the interval $[0, 36]$.

Intuitively, it seems clear that every value in this interval is in the range. To prove it we let $y = (x+2)^2$ and solve for x. Taking the square root of both sides gives $x + 2 = \pm\sqrt{y}$, so $x = \pm\sqrt{y} - 2$. Consider the solution $x = \sqrt{y} - 2$. For any y such that $0 \le y \le 36$, we have $0 \le \sqrt{y} \le 6$, so $-2 \le \sqrt{y} - 2 \le 4$. Therefore, the number $\sqrt{y} - 2$ is in the domain of g if $0 \le y \le 36$. So, if we let $x = \sqrt{y} - 2$ for any y such that $0 \le y \le 36$, we have $f(x) = y$. Therefore, every value in $[0, 36]$ is in the range, and our range is $\boxed{[0, 36]}$.

2.1.7

(a) We have $T(2, 3, -5) = 3 \cdot 2^3 - (-5) = 3 \cdot 8 + 5 = \boxed{29}$.

(b) Since $T(x, 2, 6) = 3x^2 - 6$, we have $3x^2 - 6 = 21$. Isolating x^2 gives $x^2 = 9$, from which we find $x = \boxed{\pm 3}$.

Exercises for Section 2.2

2.2.1

(a)   If $3x - 7 = 0$, then $x = 7/3$, so the x-intercept is $\boxed{(7/3, 0)}$. Since $f(0) = -7$, the y-intercept is $\boxed{(0, -7)}$.

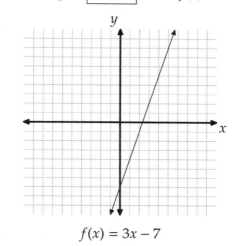

$$f(x) = 3x - 7$$

(b)   If $2 - |x| = 0$, then $|x| = 2$, so $x = \pm 2$. Hence, the x-intercepts are $\boxed{(2, 0) \text{ and } (-2, 0)}$. Since $f(0) = 2$, the y-intercept is $\boxed{(0, 2)}$.

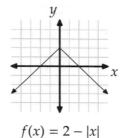

$$f(x) = 2 - |x|$$

(c)   Note that $f(x) = x^2 - 5x + 6$ factors as $(x - 2)(x - 3)$. If $(x - 2)(x - 3) = 0$, then $x = 2$ or $x = 3$, so the x-intercepts are $\boxed{(2, 0) \text{ and } (3, 0)}$. Since $f(0) = 6$, the y-intercept is $\boxed{(0, 6)}$.

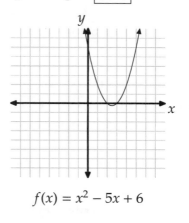

$$f(x) = x^2 - 5x + 6$$

2.2.2

(a)   Every vertical line intersects the graph at most once, so the graph represents a function.

(b) Every vertical line intersects the graph at most once, so the graph represents a function. Note that every vertical line does not need to intersect the graph, such as the line $x = 0$.

(c) There are many vertical lines that intersect the graph more than once, such as the line $x = 1$, so the graph does not represent a function.

2.2.3

(a) All we know about the function f is that $f(9) = 2$. Hence, if we set $x - 3 = 9$, then we have $x = 12$, and $y = f(12 - 3) + 5 = f(9) + 5 = 7$. Therefore, a point on the graph of $y = f(x - 3) + 5$ is $\boxed{(12, 7)}$. We can also find this point by noting that the graph of $y = f(x - 3) + 5$ results from shifting the graph of $y = f(x)$ to the right by 3 and up by 5. So, because $(9, 2)$ is on the graph of $y = f(x)$, the point $(9 + 3, 2 + 5) = (12, 7)$ is on the graph of $y = f(x - 3) + 5$.

(b) If we set $x/4 = 9$, then $x = 36$, and $y = 2f(9) = 4$. Therefore, a point on the graph of $y = 2f(x/4)$ is $\boxed{(36, 4)}$. We can also see this by noting that the graph of $y = 2f(x/4)$ results from scaling the graph of $y = f(x)$ vertically by a factor of 2 and horizontally by a factor of 4. So, because $(9, 2)$ is on the graph of $y = f(x)$, the point $(9 \cdot 4, 2 \cdot 2) = (36, 4)$ on the graph of $y = 2f(x/4)$.

(c) If we set $3x - 1 = 9$, then $x = 10/3$, and $y = 2f(9) + 7 = 11$. Therefore, a point on the graph of $y = 2f(3x - 1) + 7$ is $\boxed{(10/3, 11)}$.

2.2.4 The graph of $y = f(2x)$ is the same as the graph of $y = f(x)$ compressed by a factor of 2 horizontally, and the graph of $y = f(x - 6)$ is the same as the graph of $y = f(x)$ shifted 6 units to the right.

Knowing these results, we may think that the graph of $y = f(2x - 6)$ is the same as the graph of $y = f(x)$ compressed by a factor of 2 horizontally to give the graph of $y = f(2x)$, and then shifted 6 units to the right. However, this is not true. The first step is correct; compressing the graph of $y = f(x)$ by a factor of two horizontally gives the graph of $y = f(2x)$. To see how we produce the graph of $y = f(2x - 6)$, we let $h(x) = f(2x)$. Then, the graph of $y = h(x)$ is the same as the graph of $y = f(2x)$. To get the graph of $y = f(2x - 6)$, we note that $h(x - 3) = f(2(x - 3)) = f(2x - 6)$. The graph of $y = h(x - 3)$, which is the same as the graph of $y = f(2x - 6)$, is the result of shifting the graph of $y = h(x)$, which is the same as the graph of $y = f(2x)$, 3 units to the right.

Therefore, the answer is (c). The graph of $y = f(2x - 6)$ is the same as the graph of $y = f(2x)$ shifted 3 units to the right.

Exercises for Section 2.3

2.3.1 If $g(x) = 3x + 7$, then $g(g(x)) = g(3x + 7) = 3(3x + 7) + 7 = \boxed{9x + 28}$. If $g(x) = -3x - 14$, then $g(g(x)) = g(-3x - 14) = -3(-3x - 14) - 14 = \boxed{9x + 28}$. Hence, both solutions work.

2.3.2

(a) First, we note that the domain of g is $(-\infty, 1]$. Since the domain of f is all reals, all outputs of g are in the domain of f. For $x \le 1$, we therefore have $f(g(x)) = f(\sqrt{1 - x}) = \boxed{1 - x - 2\sqrt{1 - x}, \text{ where } x \le 1}$.

(b) The domain of $f(g(x))$ is the set of values of x such that both $g(x)$ is defined, and $g(x)$ lies in the domain of $f(x)$.

From $g(x) = \sqrt{1 - x}$, we see that $g(x)$ is defined only for $x \le 1$. The domain of $f(x)$ consists of all real numbers, so the domain of $f(g(x))$ is $\boxed{(-\infty, 1]}$.

(c) First, $f(-2) = (-2)^2 - 2 \cdot (-2) = 8$. Since $g(x)$ is only defined for $x \le 1$, we see that $g(f(-2))$ is $\boxed{\text{not defined}}$.

2.3.3 We show that $f^3(x)$ is the same as $f^2(f(x))$ by expanding both expressions. By definition, $f^3(x) = f(f(f(x)))$. Since $f^2(x) = f(f(x))$, we may substitute to get $f^2(f(x)) = f(f(f(x)))$, so $f^2(f(x)) = f^3(x)$. Similarly, $f(f^2(x)) =$

$f(f(f(x)))$, so $f^3(x)$ and $f(f^2(x))$ are the same, as well.

2.3.4 To get a feel for the problem, we compute $f^n(x)$ for the first few values of n:

$$f^1(x) = f(x) = ax,$$
$$f^2(x) = f(f(x)) = f(ax) = a^2x,$$
$$f^3(x) = f(f^2(x)) = f(a^2x) = a^3x,$$
$$f^4(x) = f(f^3(x)) = f(a^3x) = a^4x,$$

and so on.

Applying the function f to x multiplies it by a. Therefore, applying the function f to x exactly n times multiplies it by a exactly n times; in other words, it multiplies it by a^n. Hence, $f^n(x) = \boxed{a^n x}$.

2.3.5 All the statement "$(3,0)$ is the only x-intercept of the graph of g" tells us is that $g(3) = 0$. So, we have $h(3) = f(g(3)) = f(0)$. This tells us that $(3, f(0))$ is on the graph of $y = h(x)$, but this doesn't tell us anything about the intercepts of h. We cannot conclude anything about the intercepts of h from the given information.

Exercises for Section 2.4

2.4.1 Let $y = f^{-1}(3)$. So, we have $f(y) = f(f^{-1}(3)) = 3$. This means we have

$$\frac{y-1}{y-2} = 3 \quad \Rightarrow \quad y - 1 = 3(y-2) = 3y - 6 \quad \Rightarrow \quad y = \boxed{\frac{5}{2}}.$$

2.4.2

(a) Let $x = f(y) = 2y - 7$. Then $y = (x+7)/2$, so $f^{-1}(x) = \boxed{(x+7)/2}$.

(b) Let $x = f(y) = \frac{1}{2y+3}$. Then $2y + 3 = \frac{1}{x}$, so $y = (\frac{1}{x} - 3)/2 = (1 - 3x)/(2x)$. Therefore, $f^{-1}(x) = \boxed{\dfrac{1-3x}{2x}}$.

(c) Let $x = f(y) = \sqrt{2-y}$. Then $x^2 = 2 - y$, so $y = 2 - x^2$. Therefore, $f^{-1}(x) = 2 - x^2$. Note that since the range of $f(x)$ is the set of all nonnegative real numbers, the domain of f^{-1} is also all nonnegative real numbers. So, we have $\boxed{f^{-1}(x) = 2 - x^2, \text{ where the domain of } f^{-1} \text{ is } [0, +\infty)}$.

(d) Since $f(4) = |5 - 4| = 1$ and $f(6) = |5 - 6| = 1$, there are two different values of x that give the same value of $f(x)$. Therefore, the function $f(x)$ $\boxed{\text{does not have an inverse}}$.

(e) Since $f(2) = \sqrt{4-2} + \sqrt{2-2} = \sqrt{2}$ and $f(4) = \sqrt{4-4} + \sqrt{4-2} = \sqrt{2}$, there are two different values of x that give the same value of $f(x)$. Therefore, the function $f(x)$ $\boxed{\text{does not have an inverse}}$.

(f) Completing the square gives $f(x) = (x-3)^2 - 6$. We might at first think that this function does not have an inverse, since $f(1) = f(5)$. However, $f(x)$ is only defined for $x \geq 4$. By considering the graph of $f(x)$, or by noting that the function $f(x)$ strictly increases as x increases for $x \geq 4$, we see that no two values of x in the domain of f give the same output from f. In particular, $f(4) = -5$, so the range of f is $[-5, +\infty)$. So, f does have an inverse. But what is it?

   To find the inverse of f, we solve $x = f(y) = (y-3)^2 - 6$ for y in terms of x. Adding 6 to both sides gives $(y-3)^2 = x + 6$. Since x is in the range of f, we have $x \geq -5$, so $x + 6 \geq 1$. Therefore, we take the positive square root to find $y = \sqrt{x+6} + 3$. Hence, $\boxed{f^{-1}(x) = \sqrt{x+6} + 3, \text{ where the domain of } f^{-1} \text{ is } [-5, +\infty)}$. Note that $f^{-1}(x)$ is defined only for $x \geq -5$, because this is the range of f.

2.4.3 To find $f^{-1}(x)$, let $x = f(y) = ay + b$, so $y = (x - b)/a$. Then the equation $f(x) = f^{-1}(x)$ becomes

$$ax + b = \frac{x - b}{a} = \frac{x}{a} - \frac{b}{a}.$$

This holds for all x if and only if $a = 1/a$ and $b = -b/a$. From the first equation, we have $a^2 = 1$, so $a = 1$ or $a = -1$.

If $a = 1$, then the second equation becomes $b = -b$, so $b = 0$. If $a = -1$, then the second equation becomes $b = b$, which is true for any real number b. Therefore, the ordered pairs (a, b) that satisfy the problem are $\boxed{(1, 0) \text{ and } (-1, b) \text{ for any real number } b}$.

2.4.4 First, we find $g^{-1}(x)$. Let $x = g(y) = (ay + b)/(cy + d)$. Then

$$x(cy + d) = ay + b \quad \Rightarrow \quad cxy + dx = ay + b \quad \Rightarrow \quad cxy - ay = b - dx \quad \Rightarrow \quad y(cx - a) = b - dx$$

$$\Rightarrow \quad y = \frac{b - dx}{cx - a}.$$

Therefore, we have $g^{-1}(x) = \dfrac{b - dx}{cx - a}$. We see that $g^{-1}(x)$ is defined for all x except when $cx - a = 0$, or $x = \boxed{a/c}$.

Review Problems

2.23

(a) Since $A(x) = 4x^2 + 1$ is defined for all real numbers, the domain of A is $\boxed{\mathbb{R}}$. Since $4x^2$ can equal any nonnegative real number, but no negative number, the range of $A(x) = 4x^2 + 1$ is $\boxed{[1, +\infty)}$.

(b) The function $o(x) = 3 + \sqrt{16 - (x - 3)^2}$ is defined only when $(x - 3)^2 \le 16$, so $x - 3$ must be between -4 and 4. In other words, $-4 \le x - 3 \le 4$, so $-1 \le x \le 7$. Therefore, the domain of o is $\boxed{[-1, 7]}$.

As x varies from -1 to 3, $\sqrt{16 - (x - 3)^2}$ varies from 0 to 4, so $o(x) = 3 + \sqrt{16 - (x - 3)^2}$ varies from 3 to 7. As x varies from 3 to 7, $\sqrt{16 - (x - 3)^2}$ varies from 4 to 0, so $o(x) = 3 + \sqrt{16 - (x - 3)^2}$ varies from 7 to 3. Therefore, the range of o is $\boxed{[3, 7]}$.

(c) The function $P(x) = 1/(3 + \sqrt{x + 1})$ is defined only if $x + 1 \ge 0$, or $x \ge -1$. Therefore, the domain of P is $\boxed{[-1, +\infty)}$.

Since $\sqrt{x + 1}$ can take on any nonnegative value, we see that $P(x) = 1/(3 + \sqrt{x + 1})$ varies from 1/3 to 0 (without ever reaching 0). Therefore, the range of P is $\boxed{(0, 1/3]}$.

(d) The function $S(x) = (12x - 9)/(6 - 9x) = (4x - 3)/(2 - 3x)$ is defined for all x except when the denominator $2 - 3x$ is zero. We have $2 - 3x = 0$ when $x = 2/3$, so the domain of S is $\boxed{\text{all reals except } 2/3}$.

To find the range of S, we let $y = S(x) = (4x - 3)/(2 - 3x)$. Then we have

$$y(2 - 3x) = 4x - 3 \quad \Rightarrow \quad 2y - 3xy = 4x - 3 \quad \Rightarrow \quad 3xy + 4x = 2y + 3 \quad \Rightarrow \quad x = \frac{2y + 3}{3y + 4}.$$

Hence, for every value of y except $y = -4/3$, there is a corresponding value of x such that $S(x) = y$. Therefore, the range of S is $\boxed{\text{all reals except } -4/3}$.

2.24

(a) The function $\frac{1}{\sqrt{2x-5}}$ is defined only when $2x - 5 > 0$, or $x > 5/2$, and the function $\sqrt{9 - 3x}$ is defined only when $9 - 3x \ge 0$, or $x \le 3$. Combining these, the domain of f is $\boxed{(5/2, 3]}$.

(b)   The function $|\sqrt{x} - 2|$ is defined only when $x \geq 0$ and the function $|\sqrt{x-2}|$ is defined only when $x - 2 \geq 0$, or $x \geq 2$. Therefore, the domain of f is $\boxed{[2, +\infty)}$.

(c)   The function $\sqrt{|x| - 2}$ is defined only when $|x| \geq 2$. This is equivalent to $x \leq -2$ or $x \geq 2$. The function $\sqrt{|x - 2|}$ is defined for all x, since $|x - 2|$ is always nonnegative. Therefore, the domain of g is $\boxed{(-\infty, -2] \cup [2, +\infty)}$.

2.25 $f(g(x)) = f\left(\dfrac{2x}{x+4}\right) = \dfrac{4 \cdot \frac{2x}{x+4}}{\frac{2x}{x+4} + 2} = \dfrac{8x}{2x + 2(x+4)} = \dfrac{8x}{4x + 8} = \boxed{\dfrac{2x}{x+2}}$, where $x \neq -4$, because -4 is not in the domain of g.

2.26   The equation $f(1) = g(1) + 2$ gives us $a + b + c = a - b + c + 2$, so $b = 1$. Then the equation $f(2) = 2$ gives $4a + 2b + c = 2$, so $4a + c = 2 - 2b = 0$. Therefore, $g(2) = 4a - 2b + c = (4a + c) - 2b = 0 - 2 = \boxed{-2}$.

2.27   For part (a), we shift the graph 3 units to the right. For part (b), we scale the graph vertically by a factor of 2, then shift the graph 1 unit upwards. For part (c), we reflect the graph in the y-axis.

For part (d), we reflect the graph in the y-axis, then shift the graph 2 units to the right. To see why we shift to the right rather than the left, let $h(x) = f(-x)$. Then, the graph of $y = f(2 - x)$ is the same as the graph of $y = h(x - 2)$. The graph of $y = h(x - 2)$ is the result of shifting the graph of $y = h(x)$ to the right 2 units. So, the graph of $y = f(2 - x)$ is the result of shifting the graph of $y = f(-x)$ to the right 2 units.

For part (e), we compress the graph horizontally by a factor of 1/2, then shift the graph 1/2 a unit to the right. (As with part (d), we can see that this shift is 1/2 a unit by letting $h(x) = f(2x)$; therefore, $h(x - \frac{1}{2}) = f(2x - 1)$, so we must shift the graph of $h(x)$, which is the same as the graph of $y = f(2x)$, to the right 1/2 unit to get the graph of $y = f(2x - 1)$.) We then compress the graph of $y = f(2x - 1)$ vertically by a factor of 1/2, then shift the graph 3 units upwards.

The results for all five parts are shown below.

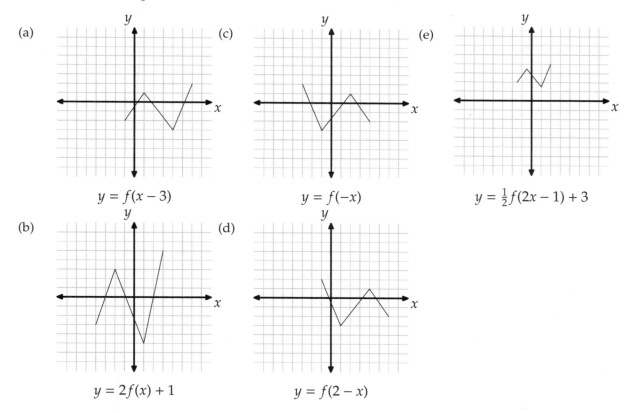

(a) $y = f(x - 3)$

(c) $y = f(-x)$

(e) $y = \frac{1}{2}f(2x - 1) + 3$

(b) $y = 2f(x) + 1$

(d) $y = f(2 - x)$

2.28 The functions f and g need not be the same. For example, let $f(x) = |x|$ and $g(x) = x$. Then $f(f(x)) = |f(x)| = \||x|\| = |x|$, and $g(f(x)) = g(|x|) = |x|$. So, we have $f(f(x)) = g(f(x))$, but $f(x) \neq g(x)$.

2.29

(a) Every vertical line intersects the graph at most once, so the graph can represent a function. We also see that every horizontal line intersects the graph at most once, so the function has an inverse, the graph of which is shown at right. (It is simply the graph in the problem reflected over the line $y = x$.)

(b) There are vertical lines that intersect the graph more than once, so the graph cannot represent a function.

(c) Every vertical line intersects the graph at most once, so the graph can represent a function. However, there are horizontal lines that intersect the graph more than once, so the function does not have an inverse.

2.30 If $f(x) = x^2$, then $f(f(x)) = f(x^2)$, so f does not have to be a constant.

2.31 We have $f(a, b, c) = \frac{a-c}{b-c} = 1$, so $a - c = b - c$, which means $a = b$. Then $f(a, c, b) = \frac{a-b}{c-b} = \boxed{0}$.

2.32 If $f(x) = -x^2 + bx + c$ and $g(x) = dx + e$, then

$$f(g(x)) = f(dx + e) = -(dx + e)^2 + b(dx + e) + c = -d^2x^2 - 2dex - e^2 + bdx + be + c = -d^2x^2 + (bd - 2de)x - (e^2 - be - c).$$

For $f(g(x))$ to be equal to x^2 for all x, the coefficient of x^2 must be 1. However, there is no real number d for which $-d^2 = 1$, so it is $\boxed{\text{not possible}}$ for $f(g(x))$ to be x^2.

2.33 From $f(1) = 0$ and $f(3) = 2$, we obtain the equations $a + b + c = 0$ and $9a + 3b + c = 2$. Then

$$g(2) = (5a + 2b + c)2 = 10a + 4b + 2c = (9a + 3b + c) + (a + b + c) = \boxed{2}.$$

2.34 To find the inverse of $f(x) = ax + b$, set $x = f(y) = ay + b$. Then $y = \frac{x-b}{a} = \frac{x}{a} - \frac{b}{a}$, so we have $f^{-1}(x) = \frac{1}{a}x - \frac{b}{a}$. This is a linear function. We conclude that the inverse of a linear function is always a linear function.

2.35 Consider a point $(x, |f(x)|)$ on the graph of $y = |f(x)|$. If $f(x) \geq 0$, then $|f(x)| = f(x)$. On the other hand, if $f(x) < 0$, then $|f(x)| = -f(x)$, and the point $(x, |f(x)|) = (x, -f(x))$ is the reflection of the point $(x, f(x))$ over the x-axis.

Thus, the graph of $y = |f(x)|$ can be constructed from the graph of $y = f(x)$ by reflecting the portion of the graph of $y = f(x)$ that lies below the x-axis over the x-axis. An example is shown below. The dashed lines in the diagram on the right indicates portions of the graph of $y = f(x)$ that are reflected over the x-axis to produce the graph of $y = |f(x)|$.

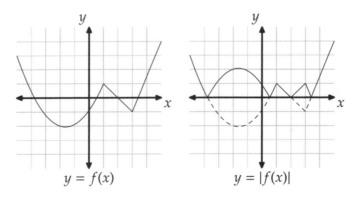

$$y = f(x) \qquad\qquad y = |f(x)|$$

2.36 If $y = f(x) = x/(1 - x)$, then

$$y(1 - x) = x \quad\Rightarrow\quad y - xy = x \quad\Rightarrow\quad xy + x = y \quad\Rightarrow\quad x(y + 1) = y \quad\Rightarrow\quad x = \frac{y}{y + 1}.$$

We compare this to each of the given possibilities:

$$f\left(\frac{1}{y}\right) = \frac{1/y}{1 - 1/y} = \frac{1}{y - 1},$$
$$-f(y) = -\frac{y}{1 - y},$$
$$-f(-y) = -\frac{-y}{1 + y} = \frac{y}{y + 1},$$
$$f(-y) = \frac{-y}{1 + y} = -\frac{y}{y + 1},$$
$$f(y) = \frac{y}{1 - y}.$$

Therefore, we have $x = \dfrac{y}{y + 1} = \boxed{-f(-y)}$.

2.37 We have $f(f(x)) = f\left(\dfrac{cx}{2x + 3}\right) = \dfrac{c \cdot cx/(2x + 3)}{2 \cdot cx/(2x + 3) + 3} = \dfrac{c^2 x}{2cx + 3(2x + 3)} = \dfrac{c^2 x}{(2c + 6)x + 9}.$

If $f(f(x)) = x$ for all x except $-3/2$, then

$$\frac{c^2 x}{(2c + 6)x + 9} = x \quad \Rightarrow \quad c^2 x = x[(2c + 6)x + 9] \quad \Rightarrow \quad c^2 x = (2c + 6)x^2 + 9x.$$

This equation holds for all x except $-3/2$ if and only if $2c + 6 = 0$ and $c^2 = 9$. The only value of c that satisfies both equations is $c = \boxed{-3}$.

2.38 Let $g(x) = f(2x)$. As discussed in the text, we find the inverse of $g(x)$ by solving the equation $x = g(y)$ for y in terms of x. From $x = g(y)$, we have $x = f(2y)$. Because f is invertible, $x = f(2y)$ tells us that $2y = f^{-1}(x)$, so $y = \frac{1}{2}f^{-1}(x)$. Thus, the inverse of $f(2x)$ is $\boxed{\frac{1}{2}f^{-1}(x)}$, not $f^{-1}(2x)$.

2.39 For the functions f and g to be inverses, we also require that $g(f(x)) = x$ for all x in the domain of f. But $g(f(x)) = \sqrt{x^2} = |x|$, which is not equal to x when x is negative, so f and g are not inverse functions.

Since the function $f(x) = x^2$ fails the horizontal line test (for example, $f(-1) = f(1) = 1$), no function g can be an inverse of f.

Challenge Problems

2.40 If $f(x) = f^{-1}(x)$, then $f(f(x)) = f(f^{-1}(x)) = x$ for all x in the domain of f. We find an expression for $f(f(x))$ as follows:

$$f(f(x)) = f\left(\frac{2x + a}{bx - 2}\right) = \frac{2 \cdot (2x + a)/(bx - 2) + a}{b \cdot (2x + a)/(bx - 2) - 2} = \frac{2(2x + a) + a(bx - 2)}{b(2x + a) - 2(bx - 2)}$$
$$= \frac{4x + 2a + abx - 2a}{2bx + ab - 2bx + 4} = \frac{(ab + 4)x}{ab + 4}.$$

This simplifies to x as long as $ab + 4 \neq 0$. If $ab + 4 = 0$, then $a = -4/b$, and

$$f(x) = \frac{2x + a}{bx - 2} = \frac{2x - 4/b}{bx - 2} = \frac{2bx - 4}{b(bx - 2)} = \frac{2(bx - 2)}{b(bx - 2)} = \frac{2}{b},$$

which is a constant, so f does not have an inverse if $ab = -4$.

Hence, $f^{-1}(x)$ exists and $f^{-1}(x) = f(x)$ $\boxed{\text{for all } a \text{ and } b \text{ such that } ab \neq -4}$.

2.41

(a) The function $f(x + 1)$ is defined only when $-1 < x + 1 < 1$, or $-2 < x < 0$. Therefore, the domain of $f(x + 1)$ is $\boxed{(-2, 0)}$.

(b) The function $f(1/x)$ is defined only when $-1 < 1/x < 1$. If x is positive, then multiplying all three parts of the inequality chain by x gives $-x < 1 < x$. Since x is positive, $-x$ is negative, so the inequality $-x < 1$ is always satisfied. Hence, the solution in this case is $x > 1$.

If x is negative, then multiplying all three parts of the inequality chain by x gives $-x > 1 > x$. Since x is negative, the inequality $1 > x$ is always satisfied. Multiplying the inequality $-x > 1$ by -1, we get $x < -1$. Hence, the solution in this case is $x < -1$.

Therefore, the domain of $f(1/x)$ is $\boxed{(-\infty, -1) \cup (1, +\infty)}$.

(c) The function $f(\sqrt{x})$ is defined only when both $x \geq 0$ and $-1 < \sqrt{x} < 1$. Since $\sqrt{x}$ is always nonnegative, the inequality $-1 < \sqrt{x}$ is always satisfied. For the same reason, we can square both sides of the inequality $\sqrt{x} < 1$ to get $x < 1$. Therefore, the domain of the function $f(\sqrt{x})$ is $\boxed{[0, 1)}$.

(d) The function $f(\frac{x+1}{x-1})$ is defined only when $-1 < \frac{x+1}{x-1} < 1$. We take the cases where $x - 1 > 0$ and $x - 1 < 0$ separately.

If $x - 1 > 0$, then multiplying all parts of the inequality $-1 < \frac{x+1}{x-1} < 1$ by $x - 1$, we get

$$-x + 1 < x + 1 < x - 1.$$

Subtracting $x + 1$ from all three parts gives $-2x < 0 < -2$. Therefore, there are no solutions in this case.

If $x - 1 < 0$, then multiplying all parts of the inequality $-1 < \frac{x+1}{x-1} < 1$ by $x - 1$ (and changing the directions of the inequalities because $x - 1$ is negative), we get

$$-x + 1 > x + 1 > x - 1.$$

Subtracting $x + 1$ from all three parts gives $-2x > 0 > -2$, so $x < 0$. Therefore, the solution is $x < 0$ in this case.

Hence, the domain of $f(\frac{x+1}{x-1})$ is $\boxed{(-\infty, 0)}$.

2.42 The function $f(x) = \sqrt{2 - x - x^2}$ is defined only when $2 - x - x^2 \geq 0$. Factoring gives $(2 + x)(1 - x) \geq 0$. Either both factors must be nonnegative, or both factors must be nonpositive.

If both factors are nonnegative, then $2 + x \geq 0$, so $x \geq -2$, and $1 - x \geq 0$, so $x \leq 1$. Hence, the solution in this case is $-2 \leq x \leq 1$. If both factors are nonpositive, then $2 + x \leq 0$, so $x \leq -2$, and $1 - x \leq 0$, so $x \geq 1$. There is no value of x such that $x \leq -2$ and $x \geq 1$ simultaneously, so there is no solution in this case. Therefore, the domain of $f(x)$ is $\boxed{[-2, 1]}$.

To find the range of $f(x)$, we complete the square inside the radical:

$$f(x) = \sqrt{2 - x - x^2} = \sqrt{\frac{9}{4} - \left(x + \frac{1}{2}\right)^2}.$$

As x varies from -2 to $-1/2$, the expression $9/4 - (x + 1/2)^2$ varies from 0 to $9/4$, so $f(x)$ varies from 0 to $3/2$. Then as x varies from $-1/2$ to 1, the expression $9/4 - (x + 1/2)^2$ varies from $9/4$ to 0, so $f(x)$ varies from $3/2$ to 0. Therefore, the range of $f(x)$ is $\boxed{[0, 3/2]}$.

2.43 Let $t = x^2 + 1$, so $x^2 = t - 1$ and $x^4 = (t - 1)^2$. Hence,

$$f(t) = f(x^2 + 1) = x^4 + 5x^2 + 3 = (t - 1)^2 + 5(t - 1) + 3 = t^2 - 2t + 1 + 5t - 5 + 3 = t^2 + 3t - 1.$$

Substituting $t = x^2 - 1$, we get $f(x^2 - 1) = (x^2 - 1)^2 + 3(x^2 - 1) - 1 = x^4 - 2x^2 + 1 + 3x^2 - 3 - 1 = \boxed{x^4 + x^2 - 3}$.

2.44   Consider a point $(x, f(|x|))$ on the graph of $y = f(|x|)$. If $x \geq 0$, then $|x| = x$, so $(x, f(|x|))$ is the same as $(x, f(x))$. On the other hand, if $x < 0$, then $|x| = -x$, and the point $(x, f(|x|)) = (x, f(-x))$ is the reflection of the point $(-x, f(-x))$ over the y-axis.

Thus, the graph of $y = f(|x|)$ consists of the portion of the graph of $y = f(x)$ that is on or to the right of the y-axis, together with the curve formed when reflecting this portion of the graph of $y = f(x)$ over the y-axis. An example is shown below.

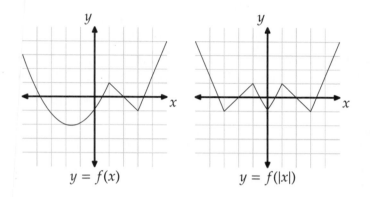

$$y = f(x) \qquad\qquad y = f(|x|)$$

2.45   Let x be a root of $f(x) = 0$. Applying f to both sides, we get $f(f(x)) = f(0) = 2003$. But $f(f(x)) = x$ for all values of x, so $x = 2003$. Therefore, the only root of the equation $f(x) = 0$ is $x = \boxed{2003}$.

2.46   Since $g(g^{-1}(3)) = 3$, we know that if $x = g^{-1}(3)$, then $g(x) = 3$. Hence, we replace $g(x)$ with 3 in the given equation, and we have

$$3 = \frac{3x + 1}{2x + 3} \quad\Rightarrow\quad 3(2x + 3) = 3x + 1 \quad\Rightarrow\quad x = \boxed{-\frac{8}{3}}.$$

2.47   To find the inverse of $f(x)$, set $x = f(y) = (ay + b)/(cy + d)$. Then

$$x(cy + d) = ay + b \quad\Rightarrow\quad cxy + dx = ay + b \quad\Rightarrow\quad cxy - ay = b - dx \quad\Rightarrow\quad y = \frac{b - dx}{cx - a}.$$

Thus, the inverse of $f(x)$, if it exists, must be $f^{-1}(x) = \dfrac{b - dx}{cx - a}$.

To check if this works, we substitute:

$$f(f^{-1}(x)) = f\left(\frac{b - dx}{cx - a}\right) = \frac{a \cdot (b - dx)/(cx - a) + b}{c \cdot (b - dx)/(cx - a) + d} = \frac{a(b - dx) + b(cx - a)}{c(b - dx) + d(cx - a)} = \frac{(bc - ad)x}{bc - ad}.$$

This simplifies to x as long as $ad - bc \neq 0$. Hence, the function f has an inverse if and only if $\boxed{ad - bc \neq 0}$.

Note that the function

$$f(x) = \frac{ax + b}{cx + d}$$

is well-defined as long as c and d are not simultaneously 0, and the function

$$f^{-1}(x) = \frac{b - dx}{cx - a}.$$

is well-defined as long as a and c are not simultaneously 0. However, if $c = d = 0$, then $ad - bc = 0$, and if $a = c = 0$, then $ad - bc = 0$. Hence, the condition $ad - bc \neq 0$ also ensures that both of these functions are well-defined.

2.48   Let $(x, 0)$ be an x-intercept of the graph of f, so $f(x) = 0$. Then $g(x) = f(x)f(|x|) = 0$, so $(x, 0)$ is also an x-intercept of the graph of g. Hence, all x-intercepts of the graph of f must also be x-intercepts of the graph of g.

On the other hand, an x-intercept of the graph of g is not necessarily an x-intercept of the graph of f. For example, take $f(x) = x - 1$, so $g(x) = f(x)f(|x|) = (x - 1)(|x| - 1)$. Note that $g(-1) = 0$, but $f(-1) = -2 \neq 0$. Thus, the point $(-1, 0)$ is an x-intercept of the graph of g, but not an x-intercept of the graph of f.

2.49   To construct the graph of $y = \frac{3}{2}f(2x - 2) - 1$ from the graph of $y = f(x)$, we describe the construction one step at a time, working from the inside of the function out.

First, we look at the expression $f(2x - 2) = f(2(x - 1))$. As described in the solution to Problem 2.2.4, the graph of $y = f(2(x - 1))$ is the result of scaling the graph of $y = f(x)$ horizontally by a factor of $1/2$ and shifting the ensuing graph to the right 1 unit. This produces the middle graph below.

Then, the graph of $y = \frac{3}{2}f(2x - 2)$ is the result of scaling the graph of $f(2x - 2)$ vertically by a factor of $3/2$, and the graph of $y = \frac{3}{2}f(2x - 2) - 1$ results from shifting the graph of $y = \frac{3}{2}f(2x - 2)$ downward 1 unit. This produces the final graph at right below.

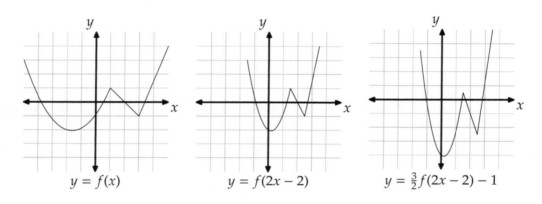

$$y = f(x) \qquad\qquad y = f(2x - 2) \qquad\qquad y = \tfrac{3}{2}f(2x - 2) - 1$$

2.50   To find the unique number that is not in the range of f, let $y = f(x) = (ax + b)/(cx + d)$. Then

$$y(cx + d) = ax + b \quad \Rightarrow \quad cxy + dy = ax + b \quad \Rightarrow \quad cxy - ax = b - dy \quad \Rightarrow \quad x = \frac{b - dy}{cy - a}.$$

Hence, for every value of y, there is a corresponding value of x such that $f(x) = y$, except when the denominator $cy - a$ is equal to 0. Therefore, the unique number that is not in the range of f is a/c.

Now we compute $f(f(x))$:

$$f(f(x)) = f\left(\frac{ax + b}{cx + d}\right) = \frac{a \cdot (ax + b)/(cx + d) + b}{c \cdot (ax + b)/(cx + d) + d} = \frac{a(ax + b) + b(cx + d)}{c(ax + b) + d(cx + d)}$$
$$= \frac{(a^2 + bc)x + (ab + bd)}{(ac + cd)x + (bc + d^2)}.$$

If $f(f(x)) = x$ for all x, then $\dfrac{(a^2 + bc)x + (ab + bd)}{(ac + cd)x + (bc + d^2)} = x$, so $(a^2 + bc)x + (ab + bd) = (ac + cd)x^2 + (bc + d^2)x$ for all x.

Then the coefficient of x^2 must be 0, so $ac + cd = 0$, which means $c(a + d) = 0$. Since c is nonzero, we have $a + d = 0$, so $d = -a$. Furthermore, if $d = -a$, then

$$f(f(x)) = \frac{(a^2 + bc)x + (ab + bd)}{(ac + cd)x + (bc + d^2)} = \frac{(a^2 + bc)x + (ab - ab)}{(ac - ac)x + (a^2 + bc)} = \frac{(a^2 + bc)x}{a^2 + bc} = x,$$

as desired. Therefore, $f(x) = \dfrac{ax+b}{cx+d} = \dfrac{ax+b}{cx-a}$.

Since $f(19) = 19$, we have

$$\frac{19a+b}{19c-a} = 19 \quad \Rightarrow \quad 19^2c - 19a = 19a + b \quad \Rightarrow \quad 19^2c = 2 \cdot 19a + b.$$

Also, $f(97) = 97$, so we have

$$\frac{97a+b}{97c-a} = 97 \quad \Rightarrow \quad 97^2c - 97a = 97a + b \quad \Rightarrow \quad 97^2c = 2 \cdot 97a + b.$$

Subtracting the first equation from the second, we get

$$(97^2 - 19^2)c = 2(97 - 19)a \quad \Rightarrow \quad (97+19)(97-19)c = 2(97-19)a \quad \Rightarrow \quad \frac{a}{c} = \frac{97+19}{2} = \boxed{58}.$$

Here is a faster way to derive the same equations. Since $f(f(x)) = x$ for all x, we have $f(f(0)) = 0$. Since $f(0) = b/d$, we have

$$f(f(0)) = f(b/d) = \frac{a \cdot b/d + b}{c \cdot b/d + d} = \frac{ab + bd}{bc + d^2},$$

so $f(f(0)) = 0$ gives $(ab + bd)/(bc + d^2) = 0$. Therefore, we have $ab + bd = 0$, so $b(a + d) = 0$. Since b is nonzero, we have $a + d = 0$, so $d = -a$.

Now, we know that $x = 19$ and $x = 97$ satisfy the equation $f(x) = x$. Writing this equation out, we get

$$\frac{ax+b}{cx+d} = x \quad \Rightarrow \quad x(cx+d) = ax + b \quad \Rightarrow \quad cx^2 + (d-a)x - b = 0.$$

The sum of the roots of this quadratic is $-(d-a)/c = 2a/c$. (This fact is covered in Chapter 4.) We know that the roots of this quadratic are 19 and 97, since these are the solutions to $f(x) = x$. Therefore, we have $2a/c = 19 + 97 = 116$, which gives us $a/c = 116/2 = 58$.

CHAPTER 3

Complex Numbers

Exercises for Section 3.1

3.1.1 Isolating z^2 gives $z^2 = -3$, and taking the square root of both sides gives $z = \pm\sqrt{-3} = \pm\sqrt{3}\sqrt{-1} = \boxed{\pm i\sqrt{3}}$.

3.1.2

(a) $a - b = (3 + 4i) - (12 - 5i) = 3 - 12 + 4i + 5i = \boxed{-9 + 9i}$.

(b) We have $ab = (3 + 4i)(12 - 5i) = 3(12 - 5i) + 4i(12 - 5i) = 36 - 15i + 48i - 20i^2 = \boxed{56 + 33i}$.

(c) Factoring first makes this calculation easier:

$$a^2 + 3a + 2 = (a + 1)(a + 2) = (4 + 4i)(5 + 4i) = 20 + 16i + 20i - 16 = \boxed{4 + 36i}.$$

(d)

$$\frac{a}{b} + \frac{\bar{a}}{b} = \frac{3 + 4i}{12 + 5i} + \frac{3 - 4i}{12 - 5i} = \frac{3 + 4i}{12 + 5i} \cdot \frac{12 - 5i}{12 - 5i} + \frac{3 - 4i}{12 - 5i} \cdot \frac{12 + 5i}{12 + 5i}$$

$$= \frac{(3 + 4i)(12 - 5i)}{(12 + 5i)(12 - 5i)} + \frac{(3 - 4i)(12 + 5i)}{(12 - 5i)(12 + 5i)} = \frac{56 + 33i}{169} + \frac{56 - 33i}{169} = \boxed{\frac{112}{169}}.$$

3.1.3

(a) $\overline{2i} + \overline{7 - 2i} = -2i + (7 + 2i) = \boxed{7}$.

(b) $\dfrac{1}{\overline{2 + 3i} + \overline{1 + 2i}} = \dfrac{1}{(2 - 3i) + (1 - 2i)} = \dfrac{1}{3 - 5i} = \dfrac{1}{3 - 5i} \cdot \dfrac{3 + 5i}{3 + 5i} = \dfrac{3 + 5i}{(3 - 5i)(3 + 5i)} = \boxed{\dfrac{3}{34} + \dfrac{5i}{34}}$.

3.1.4

(a) Since $i^2 = -1$, we have $i^4 = (i^2)^2 = (-1)^2 = 1$, and $i^6 = (i^2)^3 = (-1)^3 = -1$. Therefore,

$$f(i) = \frac{i^6 + i^4}{i + 1} = \frac{-1 + 1}{i + 1} = \boxed{0}.$$

(b) Since $(-i)^2 = -1$, we have $(-i)^4 = [(-i)^2]^2 = (-1)^2 = 1$, and $(-i)^6 = [(-i)^2]^3 = (-1)^3 = -1$. Therefore,

$$f(-i) = \frac{(-i)^6 + (-i)^4}{-i + 1} = \frac{-1 + 1}{-i + 1} = \boxed{0}.$$

(c) First, we compute $(i - 1)^2 = i^2 - 2i + 1 = -1 - 2i + 1 = -2i$. Then $(i - 1)^4 = [(i - 1)^2]^2 = (-2i)^2 = 4i^2 = -4$, and $(i - 1)^6 = [(1 - i)^2]^3 = (-2i)^3 = -8i^3 = 8i$. Therefore,

$$f(i - 1) = \frac{(i - 1)^6 + (i - 1)^4}{(i - 1) + 1} = \frac{8i - 4}{i} = \frac{(8i - 4)i}{i^2} = \frac{-8 - 4i}{-1} = \boxed{8 + 4i}.$$

3.1.5 Note that $i^{-1} = 1/i = i/i^2 = i/(-1) = -i$. Therefore, $(i - i^{-1})^{-1} = (i + i)^{-1} = \dfrac{1}{2i} = \dfrac{i}{2i^2} = \boxed{-\dfrac{1}{2}i}$.

3.1.6

(a) Multiplying both sides of the equation by $z - 3$, we get $z + 3i = 2z - 6$. Subtracting z from both sides, and adding 6 to both sides, we find that $z = \boxed{6 + 3i}$.

(b) Multiplying both sides of the equation by z, we get $\frac{1+2i}{3} = (4 + 5i)z$. Then dividing both sides by $4 + 5i$, we get

$$z = \frac{1 + 2i}{3(4 + 5i)} = \frac{(1 + 2i)(4 - 5i)}{3(4 + 5i)(4 - 5i)} = \frac{14 + 3i}{3(4^2 + 5^2)} = \boxed{\frac{14}{123} + \frac{1}{41}i}.$$

(c) Multiplying both sides of the equation by $1 + i$, we get $3(1 + i)z + z = (10 - 4i)(1 + i)$, which simplifies to $(4 + 3i)z = 14 + 6i$. Dividing both sides by $4 + 3i$, we get

$$z = \frac{14 + 6i}{4 + 3i} = \frac{(14 + 6i)(4 - 3i)}{(4 + 3i)(4 - 3i)} = \frac{74 - 18i}{4^2 + 3^2} = \boxed{\frac{74}{25} - \frac{18}{25}i}.$$

3.1.7 We have $i^{-n} = (i^{-1})^n = (\frac{1}{i})^n$. Since $\frac{1}{i} = \frac{1}{i} \cdot \frac{i}{i} = \frac{i}{i^2} = -i$, we therefore have $i^{-n} = (-i)^n$. Just as the powers of i repeat every four terms, the powers of $-i$ also repeat every four terms. So, we only have to check $i^n + (-i)^n$ for $n = 0, 1, 2,$ and 3, to determine all possible values of $i^n + (-i)^n$. We have

$$i^0 + (-i)^0 = 2,$$
$$i + (-i)^1 = i - i = 0,$$
$$i^2 + (-i)^2 = -1 - 1 = -2,$$
$$i^3 + (-i)^3 = -i + i = 0.$$

Hence, S has only $\boxed{3}$ distinct possible values, namely $-2, 0,$ and 2.

Exercises for Section 3.2

3.2.1 In the diagram at right, we see each of the points $4 + 7i$, $-6 - 2i$, and $(3 + i)(-2 + 5i) = -11 + 13i$ plotted in the complex plane, similar to the way the points $(4, 7)$, $(-6, -2)$, and $(-11, 13)$ would be plotted in the Cartesian coordinate plane.

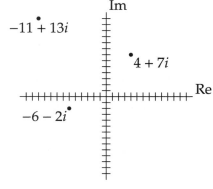

3.2.2

(a) The magnitude of $24 - 7i$ is $\sqrt{24^2 + (-7)^2} = \sqrt{625} = \boxed{25}$.

(b) The magnitude of $2 + 2\sqrt{3}i$ is $\sqrt{2^2 + (2\sqrt{3})^2} = \sqrt{16} = \boxed{4}$.

(c) The magnitude of $(1 + 2i)(2 + i) = 5i$ is $\sqrt{5^2} = \boxed{5}$.

3.2.3 As we see in the diagram at right, the complex numbers w, z, $\overline{w}$, and $\overline{z}$ form an isosceles trapezoid. The lengths of the bases are $|w - \overline{w}| = |10i| = 10$ and $|z - \overline{z}| = |4i| = 4$. The height is $12 - 3 = 9$, so the area of the trapezoid is $9 \cdot \frac{10+4}{2} = \boxed{63}$.

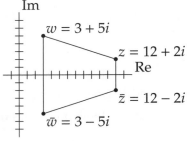

3.2.4 The distance between $4 + 7i$ and $-3 - 17i$ is the magnitude of their difference:
$$|4 + 7i - (-3 - 17i)| = |7 + 24i| = \sqrt{7^2 + 24^2} = \sqrt{625} = \boxed{25}.$$

3.2.5 Let $z_1 = a_1 + b_1 i$ and $z_2 = a_2 + b_2 i$. So, we have
$$\frac{z_1 + z_2}{2} = \frac{a_1 + b_1 i + a_2 + b_2 i}{2} = \frac{a_1 + a_2}{2} + \frac{b_1 + b_2}{2} \cdot i.$$

The midpoint of the segment connecting z_1 and z_2 on the complex plane corresponds to the midpoint of the segment connecting (a_1, b_1) and (a_2, b_2) on the Cartesian plane. This midpoint on the Cartesian plane is $(\frac{a_1 + a_2}{2}, \frac{b_1 + b_2}{2})$, which does indeed correspond to the point $(z_1 + z_2)/2$ on the complex plane as shown by our calculation above.

3.2.6

(a) The magnitude of $\dfrac{1 + 2i}{2 + i} = \dfrac{(1 + 2i)(2 - i)}{(2 + i)(2 - i)} = \dfrac{4 + 3i}{5} = \dfrac{4}{5} + \dfrac{3}{5}i$ is $\sqrt{(4/5)^2 + (3/5)^2} = \sqrt{25/5^2} = \boxed{1}$.

(b) The magnitude of
$$\frac{6 + 11i}{11 + 6i} = \frac{(6 + 11i)(11 - 6i)}{(11 + 6i)(11 - 6i)} = \frac{132 + 85i}{157} = \frac{132}{157} + \frac{85}{157}i$$
is $\sqrt{(132/157)^2 + (85/157)^2} = \sqrt{24649/157^2} = \boxed{1}$.

(c) From parts (a) and (b), the magnitude of $(x + yi)/(y + xi)$ appears to be 1. We can prove this as follows. The magnitude of
$$\frac{x + yi}{y + xi} = \frac{(x + yi)(y - xi)}{(y + xi)(y - xi)} = \frac{2xy + (y^2 - x^2)i}{x^2 + y^2} = \frac{2xy}{x^2 + y^2} + \frac{y^2 - x^2}{x^2 + y^2}i$$

is

$$\sqrt{\left(\frac{2xy}{x^2 + y^2}\right)^2 + \left(\frac{y^2 - x^2}{x^2 + y^2}\right)^2} = \sqrt{\frac{4x^2y^2}{(x^2 + y^2)^2} + \frac{x^4 - 2x^2y^2 + y^4}{(x^2 + y^2)^2}}$$
$$= \sqrt{\frac{x^4 + 2x^2y^2 + y^4}{(x^2 + y^2)^2}} = \sqrt{\frac{(x^2 + y^2)^2}{(x^2 + y^2)^2}} = \boxed{1}.$$

3.2.7 Every square is also a parallelogram, and the diagonals of a parallelogram bisect each other. Plotting the points $1 + 2i$, $-2 + i$, and $-1 - 2i$ in the complex plane, we see that the numbers $1 + 2i$ and $-1 - 2i$ must lie at opposite corners of the square. Hence, the midpoint of the diagonal between these points is $[(1 + 2i) + (-1 - 2i)]/2 = 0$.

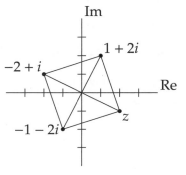

Therefore, the midpoint of z and $-2+i$ is also 0. In other words, $[z+(-2+i)]/2 = 0$, so $z = \boxed{2 - i}$.

There are many other ways we might have approached this problem; we could have used slopes, distances, or even congruent triangles.

Exercises for Section 3.3

3.3.1 Instead of expanding the product, we can use the fact that the magnitude of the product is equal to the product of the magnitudes:

$$|(10 + 24i)(8 - 6i)| = |10 + 24i| \cdot |8 - 6i| = \sqrt{10^2 + 24^2} \cdot \sqrt{8^2 + 6^2} = \sqrt{676} \cdot \sqrt{100} = 26 \cdot 10 = \boxed{260}.$$

3.3.2 Let $z = a + bi$, where a and b are real numbers. Then $|z| = |a + bi| = \sqrt{a^2 + b^2}$, and $|\bar{z}| = |a - bi| = \sqrt{a^2 + (-b)^2} = \sqrt{a^2 + b^2}$. Hence, $|z| = |\bar{z}|$.

3.3.3 Let $z = a + bi$ for some real numbers a and b. Then $\bar{z} = a - bi$, and the given equation becomes

$$3(a + bi) + 4(a - bi) = 12 - 5i \quad \Rightarrow \quad 7a - bi = 12 - 5i.$$

Two complex numbers are equal if and only if their real parts are equal and their imaginary parts are equal, so we get $7a = 12$ and $b = 5$. Therefore, $z = a + bi = \boxed{\dfrac{12}{7} + 5i}$.

We also could have solved for z without writing out the real and imaginary parts as follows. Taking the conjugate both sides of the given equation $3z + 4\bar{z} = 12 - 5i$, we get $3\bar{z} + 4\bar{\bar{z}} = \overline{12 - 5i}$, so $3\bar{z} + 4z = 12 + 5i$. This gives us the system of equations

$$3z + 4\bar{z} = 12 - 5i,$$
$$4z + 3\bar{z} = 12 + 5i.$$

Multiplying the first equation by 3 and the second equation by 4, we get

$$9z + 12\bar{z} = 36 - 15i,$$
$$16z + 12\bar{z} = 48 + 20i.$$

Subtracting the first equation from the second equation, we get $7z = 12 + 35i$, so $z = \frac{12}{7} + 5i$.

3.3.4 Let $z = a + bi$, for some real numbers a and b.

(a) If $z^2 = 2i$, then $2i = (a + bi)^2 = a^2 - b^2 + 2abi$. Equating the real parts and equating the imaginary parts, we get the system of equations $a^2 - b^2 = 0$ and $2ab = 2$. From the second equation, we have $b = 1/a$. Substituting this expression for b into the first equation, we get

$$a^2 - \frac{1}{a^2} = 0 \quad \Rightarrow \quad a^4 = 1.$$

Since a is real, we have $a = \pm 1$. The corresponding values of $b = 1/a$ are ± 1. Therefore, the two square roots of $2i$ are $\boxed{1 + i \text{ and } -1 - i}$.

(b) If $z^2 = -5 + 12i$, then $-5 + 12i = (a + bi)^2 = a^2 - b^2 + 2abi$. Equating the real parts and equating the imaginary parts, we get the system of equations $a^2 - b^2 = -5$ and $2ab = 12$. From the second equation, we have $b = 6/a$. Substituting this expression for b into the first equation, we get

$$a^2 - \frac{36}{a^2} = -5 \quad \Rightarrow \quad a^4 + 5a^2 - 36 = 0 \quad \Rightarrow \quad (a^2 + 9)(a^2 - 4) = 0.$$

Since a is real, we have $a = \pm 2$. The corresponding values of b are $6/a = \pm 3$. Therefore, the two square roots of $-5 + 12i$ are $\boxed{2 + 3i \text{ and } -2 - 3i}$.

CHAPTER 3. COMPLEX NUMBERS

(c) If $z^2 = 24 - 10i$, then $24 - 10i = (a + bi)^2 = a^2 - b^2 + 2abi$. Equating the real parts and equating the imaginary parts, we get the system of equations $a^2 - b^2 = 24$ and $2ab = -10$. From the second equation, we have $b = -5/a$. Substituting this expression for b into the first equation, we get

$$a^2 - \frac{25}{a^2} = 24 \quad \Rightarrow \quad a^4 - 24a^2 - 25 = 0 \quad \Rightarrow \quad (a^2 - 25)(a^2 + 1) = 0.$$

Since a is real, we have $a = \pm 5$. The corresponding values of b are $b = -5/a = \mp 1$. (The $\mp$ symbol indicates that $b = -1$ goes with $a = 5$ and $b = 1$ goes with $a = -5$.) Therefore, the two square roots of $24 - 10i$ are $\boxed{5 - i \text{ and } -5 + i}$.

3.3.5

(a) The magnitude of

$$\frac{1}{z} = \frac{1}{3 + 4i} = \frac{3 - 4i}{(3 + 4i)(3 - 4i)} = \frac{3 - 4i}{25} = \frac{3}{25} - \frac{4}{25}i$$

is $|1/z| = \sqrt{(3/25)^2 + (4/25)^2} = \sqrt{25/25^2} = \boxed{1/5}$.

(b) The magnitude of z is $|z| = \sqrt{3^2 + 4^2} = \sqrt{25} = 5$, so $1/|z| = \boxed{1/5}$.

(c) First, we make the denominator of z/w real:

$$\frac{z}{w} = \frac{3 + 4i}{5 - 12i} = \frac{(3 + 4i)(5 + 12i)}{(5 - 12i)(5 + 12i)} = \frac{(3 + 4i)(5 + 12i)}{169}.$$

We leave the numerator as a product, because it makes the magnitude easier to compute:

$$\left|\frac{z}{w}\right| = \left|\frac{1}{169} \cdot (3 + 4i) \cdot (5 + 12i)\right| = \left|\frac{1}{169}\right| \cdot |3 + 4i| \cdot |5 + 12i|$$
$$= \frac{1}{169} \cdot \sqrt{3^2 + 4^2} \cdot \sqrt{5^2 + 12^2} = \frac{1}{169} \cdot \sqrt{25} \cdot \sqrt{169} = \frac{1}{169} \cdot 5 \cdot 13 = \boxed{\frac{5}{13}}.$$

(d) As computed in part (b), the magnitude of z is $|z| = 5$. The magnitude of w is $|w| = \sqrt{5^2 + (-12)^2} = \sqrt{169} = 13$. Therefore, $|z|/|w| = \boxed{5/13}$.

It appears that $|z/w| = |z|/|w|$. This holds for all complex numbers z and w, where $w \neq 0$. This is because

$$\left|\frac{z}{w}\right| \cdot |w| = \left|\frac{z}{w} \cdot w\right| = |z|.$$

Dividing both sides by $|w|$, we get $\left|\dfrac{z}{w}\right| = \dfrac{|z|}{|w|}$.

3.3.6 Let $z = a + bi$, where a and b are real numbers. Then

$$\frac{z}{\overline{z}} = \frac{a + bi}{a - bi} = \frac{(a + bi)(a + bi)}{(a - bi)(a + bi)} = \frac{(a^2 - b^2) + 2abi}{a^2 + b^2} = \frac{a^2 - b^2}{a^2 + b^2} + \frac{2ab}{a^2 + b^2}i.$$

(a) If $z/\overline{z}$ is real, then the imaginary part $2ab/(a^2 + b^2)$ must be 0. This is zero if and only if $a = 0$ or $b = 0$. Since $z = a + bi$ is real if and only if $b = 0$ and z is imaginary if and only if $a = 0$, we conclude that $z/\overline{z}$ is real if and only if $\boxed{z \text{ is real or imaginary (and not equal to 0)}}$.

(b) If $z/\overline{z}$ is imaginary, then the real part $(a^2 - b^2)/(a^2 + b^2)$ must be 0. This is zero if and only if $a^2 = b^2$, or $a = \pm b$. Therefore, $z/\overline{z}$ is imaginary if and only if z is of the form $a \pm ai = a(1 \pm i)$ for some nonzero real number a.

30

3.3.7 Let k be a real number such that $x = ki$ is an imaginary solution to the given equation. Substituting $x = ki$ into the left side gives

$$(ki)^4 - 3(ki)^3 + 5(ki)^2 - 27(ki) - 36 = k^4 + 3k^3 i - 5k^2 - 27ki - 36 = (k^4 - 5k^2 - 36) + (3k^3 - 27k)i.$$

If this expression equals 0, then the real part and imaginary part equal 0, so we have the system of equations

$$k^4 - 5k^2 - 36 = 0,$$
$$3k^3 - 27k = 0.$$

The first equation factors as $(k^2 - 9)(k^2 + 4) = 0$, so $(k - 3)(k + 3)(k^2 + 4) = 0$. The second equation factors as $3k(k - 3)(k + 3) = 0$. The common roots are $k = 3$ and $k = -3$. Therefore, the imaginary solutions of the given equation are $x = \boxed{\pm 3i}$. (What are the other two roots?)

Exercises for Section 3.4

3.4.1

(a) Dividing the given equation by 2, we get $(z + \bar{z})/2 = 2$, so $\text{Re}(z) = 2$. The graph is a vertical line 2 units to the right of the imaginary axis, shown on the left below.

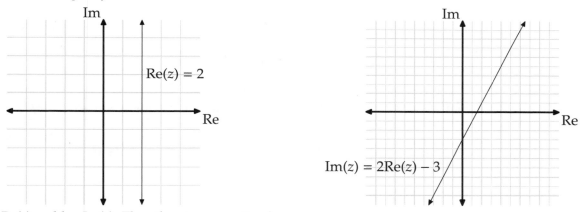

(b) Let $a = \text{Re}(z)$ and $b = \text{Im}(z)$. Then the given equation becomes

$$
\begin{aligned}
& (1 - 2i)(a + bi) + (-1 - 2i)(a - bi) = -6i \\
\Rightarrow\quad & a + 2b - 2ai + bi - a - 2b - 2ai + bi = -6i \\
\Rightarrow\quad & -4ai + 2bi = -6i \\
\Rightarrow\quad & -2a + b = -3 \\
\Rightarrow\quad & b = 2a - 3.
\end{aligned}
$$

Thus, the graph is a line with slope 2 that intersects the imaginary axis at $-3i$, shown on the right above.

(c) Geometrically speaking, the quantity $|z - 2 + i| = |z - (2 - i)|$ is the distance from the complex number z to the point $2 - i$ in the complex plane. Hence, the graph of the equation is the circle centered at $2 - i$ of radius 6, shown at right.

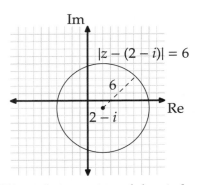

(d) Geometrically speaking, the equation $|z - i| = |z + i|$ states that the complex number z is equidistant from both i and $-i$. The set of points that are equidistant from both i and $-i$ is the perpendicular bisector of the line segment joining them. This perpendicular bisector is the real axis.

3.4.2

(a) The distance between the complex numbers z and $-2 + 5i$ is $|z - (-2 + 5i)|$, so the equation of the circle centered at $-2 + 5i$ with radius $3\sqrt{2}$ is $|z - (-2 + 5i)| = 3\sqrt{2}$, or $\boxed{|z + 2 - 5i| = 3\sqrt{2}}$.

(b) The complex number z lies on the perpendicular bisector of the line segment joining $4 + i$ and $7 - 2i$ if and only if z is equidistant from both points. Hence, the equation is of the perpendicular bisector is given by $|z - (4 + i)| = |z - (7 - 2i)|$, or $\boxed{|z - 4 - i| = |z - 7 + 2i|}$.

(c) Let $a = \text{Re}(z) = (z + \bar{z})/2$ and $b = \text{Im}(z) = (z - \bar{z})/(2i)$. For the complex number $3 + i$, we have $a = 3$ and $b = 1$, so the equation of the line through $3 + i$ with slope 2, in terms of a and b, is given by $b - 1 = 2(a - 3)$. This simplifies to $b = 2a - 5$, which we can write as $\text{Im}(z) = 2\text{Re}(z) - 5$.

Substituting $a = (z + \bar{z})/2$ and $b = (z - \bar{z})/(2i)$, we get

$$\frac{z - \bar{z}}{2i} = 2 \cdot \frac{z + \bar{z}}{2} - 5 \quad \Rightarrow \quad z - \bar{z} = 2i(z + \bar{z}) - 10i \quad \Rightarrow \quad (1 - 2i)z - (1 + 2i)\bar{z} = -10i.$$

Therefore, an equation of the line is $\boxed{(1 - 2i)z - (1 + 2i)\bar{z} = -10i}$.

3.4.3 Let $z = a + bi$. Then $(3 + 4i)z = (3 + 4i)(a + bi) = 3a - 4b + (4a + 3b)i$. Hence, $(3 + 4i)z$ is a real number if and only if $4a + 3b = 0$, or $b = -4a/3$. Thus, S is $\boxed{\text{the line passing through the origin with slope } -4/3}$.

3.4.4 Let $w = a + bi$ and $z = c + di$, where a, b, c, and d are real numbers. Then the slope of the line joining O to w is b/a, and the slope of the line joining O to z is d/c. Hence, these two lines are perpendicular if and only if $(b/a) \cdot (d/c) = -1$, which is the same as $ac + bd = 0$.

The complex number w/z is equal to

$$\frac{a + bi}{c + di} = \frac{(a + bi)(c - di)}{(c + di)(c - di)} = \frac{ac + bd + (bc - ad)i}{c^2 + d^2}.$$

We see that the real part of w/z is $(ac + bd)/(c^2 + d^2)$. Hence, w/z is imaginary for nonzero w and z if and only if $ac + bd = 0$, which is the same condition as above.

3.4.5 Let $u = a + bi$, where a and b are real numbers. Then the reflection of u in the line $\text{Re}(z) = \text{Im}(z)$ is simply $v = b + ai$.

We know that $a = \text{Re}(u) = (u + \bar{u})/2$ and $b = \text{Im}(u) = (u - \bar{u})/(2i)$. Therefore,

$$v = b + ai = \frac{u - \bar{u}}{2i} + \frac{u + \bar{u}}{2} \cdot i = \frac{(u - \bar{u})i}{2i^2} + \frac{iu + i\bar{u}}{2} = -\frac{iu - i\bar{u}}{2} + \frac{iu + i\bar{u}}{2} = \boxed{i\bar{u}}.$$

Review Problems

3.23

(a) $2w - 3z = 2(2 + 3i) - 3(4 - 5i) = 4 + 6i - 12 + 15i = \boxed{-8 + 21i}$.

(b) $\dfrac{1}{w} = \dfrac{1}{2 + 3i} = \dfrac{(2 - 3i)}{(2 + 3i)(2 - 3i)} = \dfrac{2 - 3i}{2^2 + 3^2} = \dfrac{2 - 3i}{13} = \boxed{\dfrac{2}{13} - \dfrac{3}{13}i}$.

(c) We begin by computing the denominator: $\overline{w} + z = (2 - 3i) + (4 - 5i) = 6 - 8i$. Then

$$\frac{2}{\overline{w} + z} = \frac{2}{6 - 8i} = \frac{1}{3 - 4i} = \frac{3 + 4i}{(3 - 4i)(3 + 4i)} = \frac{3 + 4i}{3^2 + 4^2} = \boxed{\frac{3}{25} + \frac{4}{25}i}.$$

(d) First, we can factor the numerator and denominator:

$$\frac{w^3 + 2w^2z + wz^2}{w^2z + wz^2} = \frac{w(w^2 + 2wz + z^2)}{wz(w + z)} = \frac{w(w + z)^2}{wz(w + z)} = \frac{w + z}{z}.$$

Then

$$\frac{w + z}{z} = \frac{(2 + 3i) + (4 - 5i)}{4 - 5i} = \frac{6 - 2i}{4 - 5i} = \frac{(6 - 2i)(4 + 5i)}{(4 - 5i)(4 + 5i)} = \frac{34 + 22i}{41} = \boxed{\frac{34}{41} + \frac{22}{41}i}.$$

(e) $\dfrac{-y + xi}{x + yi} = \dfrac{(-y + xi)(x - yi)}{(x + yi)(x - yi)} = \dfrac{-xy + x^2i + y^2i + xy}{x^2 + y^2} = \dfrac{(x^2 + y^2)i}{x^2 + y^2} = \boxed{i}$. We could have also observed that $-y + xi = i(x + yi)$.

(f) Taking a cue from part (e), we can observe that $1 - i = -i(1 + i)$. Therefore,

$$\frac{(1 - i)^4}{(1 + i)^3} = \frac{(-i)^4(1 + i)^4}{(1 + i)^3} = \boxed{1 + i}.$$

3.24 Grouping the x^2 terms on one side and constants on the other gives $2x^2 = -12$. Dividing by 2 gives $x^2 = -6$. Taking the square root of both sides gives $x = \pm\sqrt{-6} = \pm\sqrt{6}\sqrt{-1} = \boxed{\pm i\sqrt{6}}$.

3.25 We evaluate the expression one step at a time, starting from the bottom:

$$1 - \frac{1}{1 + i} = \frac{1 + i - 1}{1 + i} = \frac{i}{1 + i} = \frac{i(1 - i)}{(1 + i)(1 - i)} = \frac{1 + i}{2},$$

so

$$1 + \cfrac{1}{1 - \cfrac{1}{1 + i}} = 1 + \cfrac{1}{\cfrac{1 + i}{2}} = 1 + \frac{2}{1 + i} = \frac{1 + i + 2}{1 + i} = \frac{3 + i}{1 + i} = \frac{(3 + i)(1 - i)}{(1 + i)(1 - i)} = \frac{4 - 2i}{2} = 2 - i.$$

Then

$$\cfrac{1}{1 + \cfrac{1}{1 - \cfrac{1}{1 + i}}} = \frac{1}{2 - i} = \frac{2 + i}{(2 - i)(2 + i)} = \frac{2 + i}{5} = \boxed{\frac{2}{5} + \frac{1}{5}i}.$$

3.26 Multiplying both sides by $1 + z$, we get $z = (-1 + i)(1 + z) = (-1 + i) + (-1 + i)z$. Then $(2 - i)z = -1 + i$, so

$$z = \frac{-1 + i}{2 - i} = \frac{(-1 + i)(2 + i)}{(2 - i)(2 + i)} = \frac{-3 + i}{5} = \boxed{-\frac{3}{5} + \frac{1}{5}i}.$$

3.27 Let $z = a + bi$, where a and b are real numbers. Then $\bar{z} = a - bi$, which has the same real part as z, but opposite imaginary part. Hence, $\bar{z}$ is the reflection of z in the real axis. Similarly, $-\bar{z} = -a + bi$ has the same imaginary part as z, but opposite real part. Hence, $-\bar{z}$ is the reflection of z in the imaginary axis. Finally, $-z = -a - bi$ has both opposite real part and opposite imaginary part as z, so $-z$ is the reflection of z through the origin. We can then label $-z$, $\bar{z}$, and $-\bar{z}$ as shown at the right.

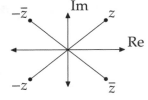

3.28 Let $w = a + bi$ and $z = c + di$, where a, b, c, and d are all real.

(a) In terms of a, b, c, and d, we have

$$w\bar{z} + \bar{w}z = (a + bi)(c - di) + (a - bi)(c + di)$$
$$= ac + bd + (-ad + bc)i + ac + bd + (ad - bc)i = 2ac + 2bd,$$

which is real.

(b) In terms of a, b, c, and d, we have

$$w\bar{z} - \bar{w}z = (a + bi)(c - di) - (a - bi)(c + di)$$
$$= ac + bd + (-ad + bc)i - (ac + bd) - (ad - bc)i = (-2ad + 2bc)i,$$

which is imaginary.

We can also prove these results by observing that the conjugate of $w\bar{z}$ is $\overline{w\bar{z}} = \bar{w} \cdot \bar{\bar{z}} = \bar{w}z$. Hence, $w\bar{z} + \bar{w}z$ is the sum of $w\bar{z}$ and its conjugate, which we know to be real. (In particular, it is twice the real part of $w\bar{z}$.)

Similarly, $w\bar{z} - \bar{w}z$ is the difference between $w\bar{z}$ and its conjugate, which we know to be imaginary. (In particular, it is $2i$ times the imaginary part of $w\bar{z}$.)

3.29 Let $z = a + bi$, for some real numbers a and b. If $z^2 = 5 - 12i$, then $5 - 12i = (a + bi)^2 = a^2 - b^2 + 2abi$. Equating the real parts and equating the imaginary parts, we get the system of equations $a^2 - b^2 = 5$ and $2ab = -12$. From the second equation, we have $b = -6/a$. Substituting this expression for b into the first equation, we get

$$a^2 - \frac{36}{a^2} = 5 \quad \Rightarrow \quad a^4 - 5a^2 - 36 = 0 \quad \Rightarrow \quad (a^2 - 9)(a^2 + 4) = 0.$$

Since a is real, we have $a = \pm 3$. The corresponding values of $b = -6/a$ are ∓ 2. (The $\mp$ symbol indicates that $b = -2$ goes with $a = 3$ and $b = 2$ goes with $a = -3$.) Therefore, the square roots of $5 - 12i$ are $\boxed{3 - 2i \text{ and } -3 + 2i}$.

3.30 If $7 + i$ is 5 units from $10 + ci$, then from the distance formula, we have

$$\sqrt{(7 - 10)^2 + (1 - c)^2} = 5 \quad \Rightarrow \quad 9 + (c - 1)^2 = 25 \quad \Rightarrow \quad (c - 1)^2 = 16 \quad \Rightarrow \quad c - 1 = \pm 4,$$

so $c = \boxed{5 \text{ or } -3}$.

3.31 The graph of $|z - (4 - 5i)| = 2\sqrt{3}$ is a circle centered at $4 - 5i$ with radius $2\sqrt{3}$, which has area $(2\sqrt{3})^2\pi = \boxed{12\pi}$.

3.32 Let $z = a + bi$ be the complex number corresponding to the point F, where a and b are real numbers. From the diagram, we can say two things about z. First, since F lies to the right of the imaginary axis and above the real axis, $a > 0$ and $b > 0$. Second, F lies outside the unit circle, so $|z| > 1$. We must deduce similar facts about $1/z$.

First we express $\frac{1}{z}$ as a complex number:

$$\frac{1}{z} = \frac{1}{a + bi} = \frac{a - bi}{(a + bi)(a - bi)} = \frac{a - bi}{a^2 + b^2} = \frac{a}{a^2 + b^2} - \frac{b}{a^2 + b^2}i.$$

Since a and b are positive, the real part of $1/z$ is positive, and the imaginary part is negative. Therefore, $1/z$ lies to the right of the imaginary axis, and below the real axis. This means $1/z$ must be A or C.

Next, we see that $\left|\dfrac{1}{z}\right| = \dfrac{1}{|z|} < 1$. Hence, $1/z$ lies inside the unit circle. Therefore, the reciprocal of F is $\boxed{C}$.

3.33 Let $z = a + bi$, where a and b are real numbers, so the equation becomes

$$|a + 1 + (b - 1)i| = |a - 2 + bi|$$
$$\Rightarrow \quad \sqrt{(a + 1)^2 + (b - 1)^2} = \sqrt{(a - 2)^2 + b^2}$$
$$\Rightarrow \quad (a + 1)^2 + (b - 1)^2 = (a - 2)^2 + b^2$$
$$\Rightarrow \quad a^2 + 2a + 1 + b^2 - 2b + 1 = a^2 - 4a + 4 + b^2$$
$$\Rightarrow \quad 6a = 2b + 2$$
$$\Rightarrow \quad b = 3a - 1.$$

Hence, all solutions are of the form $z = \boxed{a + (3a - 1)i}$, where a is any real number.

Note that we could also have solved this problem by noting that $|z + 1 - i|$ can be written as $|z - (-1 + i)|$, so this expression represents the distance from z to $-1 + i$ on the complex plane. The expression $|z - 2|$ equals the distance from z to 2, so the solutions to the equation $|z + 1 - i| = |z - 2|$ represent all z equidistant from $-1 + i$ and 2 on the complex plane. Therefore, z must be on the perpendicular bisector of the segment connecting $-1 + i$ and 2 on the complex plane. This segment has slope $-\frac{1}{3}$, and its midpoint is $\frac{1}{2} + \frac{i}{2}$. So, z is on the line through $\frac{1}{2} + \frac{i}{2}$ with slope 3. See if you can finish from here to get the same answer as above.

3.34 We are given that $(a + bi)(c + di) = ac - bd + (ad + bc)i$ is real, so $ad + bc = 0$. Then

$$(b + ai)(d + ci) = bd - ac + (bc + ad)i = bd - ac + (ad + bc)i = bd - ac,$$

so the product of $b + ai$ and $d + ci$ is also real.

3.35

(a) Dividing the given equation by $2i$, we get $(z - \bar{z})/(2i) = -4$, so $\text{Im}(z) = -4$. The graph is a horizontal line 4 units below the real axis, shown to the left below.

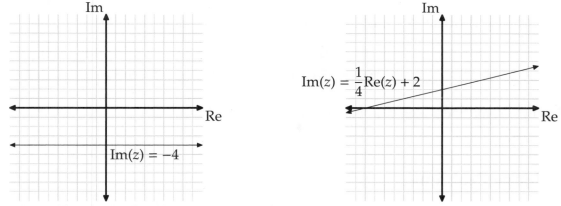

(b) Let $a = \text{Re}(z)$ and $b = \text{Im}(z)$. Then the given equation becomes

$$(4 - i)(a + bi) - (4 + i)(a - bi) = 16i$$
$$\Rightarrow \quad 4a + b - ai + 4bi - 4a - b - ai + 4bi = 16i$$
$$\Rightarrow \quad -2ai + 8bi = 16i$$
$$\Rightarrow \quad -a + 4b = 8$$
$$\Rightarrow \quad b = \dfrac{a}{4} + 2.$$

Thus, the graph is a line with slope 1/4 that intersects the imaginary axis at $2i$, shown to the right above.

(c) The given equation can be rewritten as $|7 + i - 2z| = |-(7 + i - 2z)| = |2z - 7 - i| = 4$. Dividing both sides by 2, we get

$$\left|z - \dfrac{7 + i}{2}\right| = 2.$$

The graph is then a circle centered at $\frac{7}{2} + \frac{1}{2}i$ with radius 2.

3.36 For a complex number z, $|z - 5 + 2i| = |z - (5 - 2i)|$ is the distance between z and $5 - 2i$. Hence, the graph of $|z - (5 - 2i)| \le 4$ is the set of all complex numbers z that are within 4 units of $5 - 2i$. This is the disc (which is both the circle and its interior) centered at $5 - 2i$ with radius 4, shown at right.

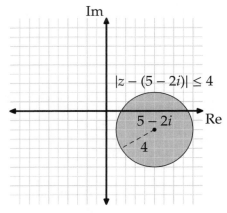

3.37 To deal with such large exponents, we compute the first few powers of $i + 1$ to see if we can find a pattern:

$$(i + 1)^1 = i + 1,$$
$$(i + 1)^2 = i^2 + 2i + 1 = 2i,$$
$$(i + 1)^3 = (i + 1)^2 \cdot (i + 1) = 2i(i + 1) = -2 + 2i,$$
$$(i + 1)^4 = (i + 1)^3 \cdot (i + 1) = (-2 + 2i)(i + 1) = -4.$$

We note that $(1 + i)^4$ is a real number, which makes it easy to work with. Similarly, we find that $(i - 1)^4 = -4$. Therefore,

$$(i + 1)^{3200} - (i - 1)^{3200} = [(i + 1)^4]^{800} - [(i - 1)^4]^{800} = (-4)^{800} - (-4)^{800} = \boxed{0}.$$

3.38 Let $z = a + bi$, for some real numbers a and b. We are given that $|z| = 1$, so $|z| = |a + bi| = \sqrt{a^2 + b^2} = 1$, or $a^2 + b^2 = 1$. Then

$$|z - 1|^2 + |z + 1|^2 = |(a - 1) + bi|^2 + |(a + 1) + bi|^2 = (a - 1)^2 + b^2 + (a + 1)^2 + b^2$$
$$= a^2 - 2a + 1 + b^2 + a^2 + 2a + 1 + b^2 = 2a^2 + 2b^2 + 2$$
$$= 2(a^2 + b^2) + 2 = 2 + 2 = 4.$$

Alternatively, we could have used the fact that $z\bar{z} = |z|^2$ for all complex numbers z:

$$|z - 1|^2 + |z + 1|^2 = (z - 1) \cdot \overline{(z - 1)} + (z + 1) \cdot \overline{(z + 1)} = (z - 1)(\bar{z} - 1) + (z + 1)(\bar{z} + 1)$$
$$= z\bar{z} - z - \bar{z} + 1 + z\bar{z} + z + \bar{z} + 1 = 2z\bar{z} + 2 = 2|z|^2 + 2 = 4.$$

Finally, can you find a geometric interpretation of the result?

Challenge Problems

3.39 The powers of i repeat every four terms: $i^0 = 1$, $i^1 = i$, $i^2 = -1$, $i^3 = -i$, $i^4 = 1$, $i^5 = i$, $i^6 = -1$, $i^7 = -i$, $i^8 = 1$, and so on. Hence, the given sum becomes

$i + 2i^2 + 3i^3 + 4i^4 + \cdots + 64i^{64}$
$$= i - 2 - 3i + 4 + 5i - 6 - 7i + 8 + 9i - 10 - 11i + 12 + \cdots + 61i - 62 - 63i + 64$$
$$= (-2 + 4 - 6 + 8 - 10 + 12 + \cdots - 62 + 64) + (1 - 3 + 5 - 7 + 9 - 11 + \cdots + 61 - 63)i$$
$$= [(-2 + 4) + (-6 + 8) + (-10 + 12) + \cdots + (-62 + 64)] + [(1 - 3) + (5 - 7) + (9 - 11) + \cdots + (61 - 63)]i$$
$$= 2 \cdot 16 - 2 \cdot 16i = \boxed{32 - 32i}.$$

3.40 Let $z = a + bi$, where a and b are real numbers. From the equation $|z - 1| = |z + 3|$, we obtain

$$
\begin{aligned}
|(a - 1) + bi| &= |(a + 3) + bi| \\
\Rightarrow \quad \sqrt{(a - 1)^2 + b^2} &= \sqrt{(a + 3)^2 + b^2} \\
\Rightarrow \quad (a - 1)^2 + b^2 &= (a + 3)^2 + b^2 \\
\Rightarrow \quad a^2 - 2a + 1 + b^2 &= a^2 + 6a + 9 + b^2 \\
\Rightarrow \quad 8a &= -8 \\
\Rightarrow \quad a &= -1.
\end{aligned}
$$

Then from the equations $|z - 1| = |z - i|$ and $a = -1$, we obtain

$$
\begin{aligned}
|-2 + bi| &= |-1 + (b - 1)i| \\
\Rightarrow \quad \sqrt{(-2)^2 + b^2} &= \sqrt{(-1)^2 + (b - 1)^2} \\
\Rightarrow \quad 4 + b^2 &= 1 + (b - 1)^2 \\
\Rightarrow \quad 4 + b^2 &= 1 + b^2 - 2b + 1 \\
\Rightarrow \quad 2b &= -2 \\
\Rightarrow \quad b &= -1.
\end{aligned}
$$

Therefore, $z = a + bi = \boxed{-1 - i}$.

We can interpret the problem geometrically as follows. The equations $|z - 1| = |z + 3| = |z - i|$ state that the complex number z is equidistant from the numbers 1, -3, and i. Hence, z is the circumcenter of the triangle formed by the complex numbers 1, -3, and i in the complex plane. From $|z - 1| = |z + 3|$, we find that z is on the perpendicular bisector of the segment connecting 1 and -3 on the complex plane. This perpendicular bisector is $\operatorname{Re}(z) = -1$, so the real part of z is -1. From $|z - 1| = |z - i|$, we find that z is on the perpendicular bisector of the segment connecting 1 and i on the complex plane. This line is simply $\operatorname{Re}(z) = \operatorname{Im}(z)$. Combining this with $\operatorname{Re}(z) = -1$, we have $z = -1 - i$, as before.

3.41 The first inequality is satisfied when z is *inside* the circle centered at $-4 + 4i$ of radius $4\sqrt{2}$, and the second inequality is satisfied when z is *outside* the circle centered at $4 + 4i$ of radius $4\sqrt{2}$. Hence, the region of interest is the set of points that lie inside the first circle but outside the second circle, as shown below.

Let A be the point $-4 + 4i$, the center of the first circle, and let B be the point $4 + 4i$, the center of the second circle. Then

$$AB = |(4 + 4i) - (-4 + 4i)| = |8| = 8.$$

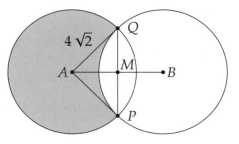

Let M be the midpoint of $\overline{AB}$, so $AM = BM = AB/2 = 4$.

Let the two circles intersect at P and Q. Since $\overline{PQ}$ is a common chord of the two circles, $\overline{PQ}$ is perpendicular to $\overline{AB}$. Furthermore, M is also the midpoint of $\overline{PQ}$.

By the Pythagorean Theorem, $AM^2 + QM^2 = AQ^2$, so $16 + QM^2 = (4\sqrt{2})^2 = 32$. Therefore, $QM^2 = 16$, so $QM = 4$. Then $PQ = 2QM = 8$.

Now, to find the area of the shaded region, we can take the area of the first circle and subtract two times the area of the circular segment cut off by $\overline{PQ}$. To find the area of the circular segment cut off by $\overline{PQ}$, we can find the area of sector PAQ, and subtract the area of triangle PAQ.

Since $AM = QM$ and $\angle AMQ = 90°$, we have $\angle QAM = 45°$. Also, $AM = PM$ and $\angle AMP = 90°$, so $\angle PAM = 45°$. Therefore, $\angle PAQ = 45° + 45° = 90°$. Hence, the area of sector PAQ is one quarter the area of the circle, which is $\frac{1}{4}\pi(4\sqrt{2})^2 = 8\pi$. The area of triangle PAQ is $\frac{1}{2}AM \cdot PQ = \frac{1}{2} \cdot 4 \cdot 8 = 16$. Hence, the area of the circular segment cut off by $\overline{PQ}$ is $8\pi - 16$.

Finally, the area of the shaded region is $\pi(4\sqrt{2})^2 - 2(8\pi - 16) = \boxed{16\pi + 32}$.

3.42 Note that

$$\frac{w}{\overline{z}} + \frac{\overline{w}}{z} = \frac{wz + \overline{wz}}{z\overline{z}}.$$

Since $\overline{wz}$ is the conjugate of wz, the sum $wz + \overline{wz}$ is a real number. Also, $z\overline{z} = |z|^2$ is a real number. Therefore, the given expression is a real number.

Alternatively, we could have observed that $\overline{w}/z$ is the conjugate of $w/\overline{z}$. Therefore, their sum is a real number.

3.43 Let w, x, y, and z be the vertices of a parallelogram in the complex plane, where w and y are opposite vertices, so x and z are also opposite vertices. We know that in every parallelogram, the diagonals bisect each other; in other words, the midpoints of each diagonal coincide. Therefore,

$$\frac{w + y}{2} = \frac{x + z}{2} \quad \Rightarrow \quad w + y = x + z.$$

Thus, the sum of w and y is equal to the sum of x and z.

Conversely, let the sum of two of w, x, y, and z be equal to the sum of the other two. Suppose $w + y = x + z$. Then $(w + y)/2 = (x + z)/2$. This means the midpoint of the line segment between w and y and the midpoint of the line segment between x and z coincide. In other words, in the quadrilateral formed by the complex numbers w, x, y, and z, the diagonals bisect each other. Hence, the quadrilateral formed by these four points is a parallelogram.

Make sure you see why we need to include both parts of this proof.

3.44 We compute the first few terms of the sequence:

$$z_2 = z_1^2 + i = i,$$
$$z_3 = z_2^2 + i = -1 + i,$$
$$z_4 = z_3^2 + i = (-1 + i)^2 + i = 1 - 2i - 1 + i = -i,$$
$$z_5 = z_4^2 + i = (-i)^2 + i = -1 + i,$$
$$z_6 = z_5^2 + i = (-1 + i)^2 + i = 1 - 2i - 1 + i = -i,$$

and so on.

We observe that the sequence repeats every two terms beginning with z_3. We know that this pattern persists, because each term in the sequence depends only on the previous term. We conclude that $z_{111} = z_3 = -1 + i$.

Therefore, the distance between z_{111} and the origin is $|-1 + i| = \sqrt{(-1)^2 + 1^2} = \boxed{\sqrt{2}}$.

3.45 From the equation $|a + bi| = 8$, we get $\sqrt{a^2 + b^2} = 8$, so $a^2 + b^2 = 64$.

We are also given that the image of each point in the complex plane is equidistant from that point and the origin. In other words, for all complex numbers z, we have $|f(z) - z| = |f(z)|$. Substituting $f(z) = (a + bi)z$, we get $|(a - 1 + bi)z| = |(a + bi)z|$, which we can rewrite as $|a - 1 + bi| \cdot |z| = |a + bi| \cdot |z|$.

Since this relation holds for all complex numbers z, we can assume that $|z| \neq 0$. Therefore, dividing both sides by $|z|$, we get

$$|a - 1 + bi| = |a + bi| \quad \Rightarrow \quad \sqrt{(a - 1)^2 + b^2} = \sqrt{a^2 + b^2} \quad \Rightarrow \quad (a - 1)^2 + b^2 = a^2 + b^2 \quad \Rightarrow \quad a = \frac{1}{2}.$$

Hence, $b^2 = 64 - a^2 = 64 - \left(\frac{1}{2}\right)^2 = \boxed{\frac{255}{4}}$.

Can you find a geometric approach to this problem?

3.46

(a) If $z \neq 0$, then $f(f(z)) = f\left(\frac{1}{\bar{z}}\right) = \dfrac{1}{\overline{1/\bar{z}}} = \dfrac{1}{1/z} = z.$

(b) Since $w = f(z) = 1/\bar{z}$, we have $\bar{z} = 1/w$. Taking the conjugate of both sides, we get $\bar{\bar{z}} = 1/\overline{w}$, so $z = 1/\overline{w}$. Substituting these expressions into the equation of the line $(1 + 2i)z - (1 - 2i)\bar{z} = i$, we get

$$\frac{1 + 2i}{\overline{w}} - \frac{1 - 2i}{w} = i \quad \Rightarrow \quad (1 + 2i)w - (1 - 2i)\overline{w} = iw\overline{w}.$$

Let $w = a + bi$, where a and b are real numbers. Then the equation above becomes

$$(1 + 2i)(a + bi) - (1 - 2i)(a - bi) = i(a + bi)(a - bi)$$
$$\Rightarrow \quad a - 2b + (2a + b)i - (a - 2b) + (2a + b)i = (a^2 + b^2)i$$
$$\Rightarrow \quad (4a + 2b)i = (a^2 + b^2)i$$
$$\Rightarrow \quad 4a + 2b = a^2 + b^2$$
$$\Rightarrow \quad a^2 - 4a + b^2 - 2b = 0$$
$$\Rightarrow \quad a^2 - 4a + 4 + b^2 - 2b + 1 = 5$$
$$\Rightarrow \quad (a - 2)^2 + (b - 1)^2 = 5.$$

Hence, it appears that w traces a circle centered at $2 + i$ with radius $\sqrt{5}$. However, we must be careful to note that w cannot be 0, since there is no complex number z for which $0 = 1/\bar{z}$. So, w traces the entire circle centered at $2 + i$ with radius $\sqrt{5}$, except the origin.

3.47 We have seen the expression $a^2 + b^2$ arise when computing the reciprocal of the complex number $z = a + bi$, where a and b are real numbers:

$$\frac{1}{z} = \frac{1}{a + bi} = \frac{a - bi}{(a + bi)(a - bi)} = \frac{a - bi}{a^2 + b^2}.$$

Thus, it may be possible to express the given equations in terms of the complex number $z = a + bi$.

We see that the first equation contains the term a, which is the real part of $a + bi$. The second equation contains the term b, which is the imaginary part of $a + bi$. Furthermore, the first equation contains the term $\frac{a}{a^2+b^2}$, which is the real part of $1/z$, as seen above, and the second equation contains the term $-\frac{b}{a^2+b^2}$, which is the imaginary part of $1/z$.

Hence, by the taking the expression in the first equation as the real part of a complex number, and the expression in the second equation as the imaginary part, we get

$$\left(a + \frac{a + 8b}{a^2 + b^2}\right) + \left(b + \frac{8a - b}{a^2 + b^2}\right)i = a + bi + \frac{a - bi}{a^2 + b^2} + \frac{8(b + ai)}{a^2 + b^2}$$
$$= z + \frac{1}{z} + \frac{8(b + ai)}{a^2 + b^2},$$

where $z = a + bi$. This leaves us with the task of expressing $\frac{8(b + ai)}{a^2 + b^2}$ in terms of z.

Since $i(a - bi) = b + ai$, we have $\frac{8(b + ai)}{a^2 + b^2} = \frac{8i(a - bi)}{a^2 + b^2} = \frac{8i}{z}$. Thus, we have the equation $z + \frac{1}{z} + \frac{8i}{z} = 2.$

Multiplying both sides by z, we get $z^2 + 1 + 8i = 2z$. Rearranging this equation gives $z^2 - 2z + 1 = -8i$, so $(z - 1)^2 = -8i$. Hence, we must find the numbers whose squares are $-8i$.

Let $w = c + di$, where c and d are real numbers and $w^2 = -8i$. Then, we have $w^2 = (c + di)^2 = c^2 - d^2 + 2cdi = -8i$. Equating the real parts and equating the imaginary parts, we get the system of equations $c^2 - d^2 = 0$ and $2cd = -8$. From the second equation, $d = -4/c$. Substituting this expression for d into the first equation, we get

$$c^2 = \frac{16}{c^2} \quad \Rightarrow \quad c^4 = 16 \quad \Rightarrow \quad c = \pm 2.$$

The corresponding values of $d = -4/c$ are ∓ 2. (The $\mp$ symbol indicates that $d = -2$ goes with $c = 2$ and $d = 2$ goes with $c = -2$.) Therefore, the square roots of $-8i$ are $2 - 2i$ and $-2 + 2i$.

Hence, $z - 1 = 2 - 2i$ or $z - 1 = -2 + 2i$, so $z = 3 - 2i$ or $z = -1 + 2i$. The corresponding values of (a, b) are $(3, -2)$ and $(-1, 2)$. We check that both of these solutions work, so the solutions are $(a, b) = \boxed{(3, -2) \text{ and } (-1, 2)}$.

3.48 Let

$$z = \frac{4uv}{(u + v)^2}.$$

To show that z is real, it suffices to show that $z = \bar{z}$ (as described in the text), so we compute the conjugate of z. Since $|u| = 1$, we have $u\bar{u} = |u|^2 = 1$, so $\bar{u} = 1/u$. Similarly, $\bar{v} = 1/v$. Then

$$\bar{z} = \overline{\left(\frac{4uv}{(u + v)^2}\right)} = \frac{\overline{4uv}}{\overline{(u + v)^2}} = \frac{4\bar{u} \cdot \bar{v}}{(\bar{u} + \bar{v})^2} = \frac{4 \cdot \frac{1}{u} \cdot \frac{1}{v}}{(\frac{1}{u} + \frac{1}{v})^2}.$$

Multiplying the numerator and denominator by $u^2 v^2$, this becomes $\bar{z} = \frac{4uv}{(u+v)^2} = z$. Therefore, $z = 4uv/(u + v)^2$ is real.

Now, let $w = \dfrac{u + v}{u - v}$. To show that w is imaginary, it suffices to show that $w = -\bar{w}$. We have

$$\bar{w} = \overline{\left(\frac{u + v}{u - v}\right)} = \frac{\overline{u + v}}{\overline{u - v}} = \frac{\bar{u} + \bar{v}}{\bar{u} - \bar{v}} = \frac{\frac{1}{u} + \frac{1}{v}}{\frac{1}{u} - \frac{1}{v}} = \frac{v + u}{v - u} = -w.$$

Therefore, $w = (u + v)/(u - v)$ is imaginary.

CHAPTER 4

Quadratics

Exercises for Section 4.1

4.1.1

(a) Let $x^2 + 19x + 90 = (x + p)(x + q) = x^2 + (p + q)x + pq$, so $p + q = 19$ and $pq = 90$. Searching for positive factors of 90 that sum to 19, we find 9 and 10, so $x^2 + 19x + 90 = \boxed{(x + 9)(x + 10)}$. Therefore, the roots are $x = \boxed{-9 \text{ and } -10}$.

(b) If the given quadratic factors nicely, then one of the factors has leading coefficient 1, and the other has leading coefficient 2. So, we write

$$2y^2 - 15y + 18 = (y + a)(2y + b) = 2y^2 + (2a + b)y + ab,$$

for some integers a and b. Equating the corresponding coefficients on the far left and the far right, we get $2a + b = -15$ and $ab = 18$. The equation $ab = 18$ tells us that a and b are factors of 18 with the same sign. Since $2a + b$ is negative, both a and b must be negative. Searching among these factors of 18, we find the solution $(a, b) = (-6, -3)$, so $2y^2 - 15y + 18 = \boxed{(y - 6)(2y - 3)}$. Therefore, the roots are $y = \boxed{6 \text{ and } 3/2}$.

(c) Let $6a^2 + 13a - 8 = (Pa + Q)(Ra + S) = PRa^2 + (PS + QR)a + QS$, for some integers P, Q, R, and S. Equating the corresponding coefficients, we get

$$PR = 6,$$
$$PS + QR = 13,$$
$$QS = -8.$$

From the second equation, we see that either PS is odd and QR is even, or vice-versa. This greatly limits the possibilities we must test. Considering the divisors of 6 and 8, we see that the only way for PS to be odd is if S is ± 1 and P is either 1 or 3. With so few possibilities to try, we quickly find $6a^2 + 13a - 8 = \boxed{(2a - 1)(3a + 8)}$. Therefore, the roots are $a = \boxed{1/2 \text{ and } -8/3}$.

(d) If the given quadratic factors nicely, then one of the factors has leading coefficient 1, and the other has leading coefficient 7. So, we write

$$7x^2 + 296x - 215 = (x + a)(7x + b) = 7x^2 + (7a + b)x + ab,$$

for some integers a and b. Equating the corresponding coefficients on the far left and the far right, we get $7a + b = 296$ and $ab = -215$. The equation $ab = -215$ tells us that a and b are factors of 215 with opposite signs. The prime factorization of 215 is $5 \cdot 43$, so it is not difficult to search through all the factors to find the solution $(a, b) = (43, -5)$, so $7x^2 + 296x - 215 = \boxed{(x + 43)(7x - 5)}$. Therefore, the roots are $x = \boxed{-43 \text{ and } 5/7}$.

4.1.2

(a) Let $y = x^2$. Then the given equation becomes $3y^2 + 26 = 19y$, so $3y^2 - 19y + 26 = 0$, which is a quadratic equation in y. If this quadratic factors nicely, then one of the factors has leading coefficient 1, and the other has leading coefficient 3. So, we write

$$3y^2 - 19y + 26 = (y + a)(3y + b) = 3y^2 + (3a + b) + ab,$$

for some integers a and b. Equating the corresponding coefficients on the far left and the far right, we get $3a + b = -19$ and $ab = 26$. The equation $ab = 26$ tells us that a and b are factors of 26 with the same sign. The equation $3a + b = -19$ tells us that this common sign is negative. Searching among the factors of 26, we find the solution $(a, b) = (-2, -13)$, so $3y^2 - 19y + 26 = (y - 2)(3y - 13)$. Therefore, we must have $(x^2 - 2)(3x^2 - 13) = 0$, so the solutions are $x = \boxed{\pm \sqrt{2} \text{ and } \pm \sqrt{13/3}}$. We can rationalize the denominator of $\sqrt{\frac{13}{3}}$ by multiplying by $\frac{\sqrt{3}}{\sqrt{3}}$ to find $\sqrt{\frac{13}{3}} = \frac{\sqrt{39}}{3}$.

(b) Multiplying both sides by x and rearranging, we obtain the quadratic equation $x^2 - 104x + 679 = 0$. Searching among the factors of 679 that sum to 104, we find 7 and 97, so $x^2 - 104x + 679 = (x - 7)(x - 97)$. Therefore, the solutions are $x = \boxed{7 \text{ and } 97}$.

(c) Squaring both sides, we get $x^2 + x - 7 = x - 3$, so $x^2 = 4$. Hence, $x = \pm 2$. However, since one of our steps was squaring the equation, we must check our answers. We find that neither $x = 2$ nor $x = -2$ satisfies the original equation, so there are $\boxed{\text{no solutions}}$.

4.1.3 Rearranging the given equation, we get $a^2 - 4ab + 4b^2 = 0$. Factoring gives $(a - 2b)^2 = 0$, so $a = 2b$. Therefore, the only possible value of a/b is $\boxed{2}$.

4.1.4 Instead of dividing both sides by $2x - 3y$, we can move everything to one side, to get

$$3x(2x - 3y) + 5y(2x - 3y) = 0.$$

Factoring out $(2x - 3y)$ gives $(3x + 5y)(2x - 3y) = 0$. Hence, $x/y = -5/3$ or $x/y = 3/2$, as before.

4.1.5 We can simplify the given expression to obtain a quartic equation, but we have not yet covered how to solve quartic equations. Instead, it may be worth considering why the given terms have been divided into two groups.

The first group immediately factors as $x^2(x^2 - 11x + 24)$. Factoring the quadratic gives $x^2 - 11x + 24 = (x - 3)(x - 8)$.

The second group factors as $4x^2 - 44x + 96 = 4(x^2 - 11x + 24) = 4(x - 3)(x - 8)$. Hence,

$$(x^4 - 11x^3 + 24x^2) - (4x^2 - 44x + 96) = x^2(x - 3)(x - 8) - 4(x - 3)(x - 8)$$
$$= (x - 3)(x - 8)(x^2 - 4)$$
$$= (x - 3)(x - 8)(x - 2)(x + 2).$$

Therefore, the solutions are $x = \boxed{3, 8, 2, \text{ and } -2}$.

Exercises for Section 4.2

4.2.1

(a) The sum of the roots is $-(-5)/1 = \boxed{5}$. The product of the roots is $7/1 = \boxed{7}$.

(b) Rearranging the equation, we get $2x^2 - 7x - 120 = 0$. The sum of the roots is $-(-7)/2 = \boxed{7/2}$. The product of the roots is $-120/2 = \boxed{-60}$.

(c)   Simplifying the equation, we get $3x^2 + 3x + 3 = 0$. Dividing this by 3, we get $x^2 + x + 1 = 0$. The sum of the roots is $-1/1 = \boxed{-1}$. The product of the roots is $1/1 = \boxed{1}$.

4.2.2   Expanding the left side, we get $x^2 - 3x + 2 + x^2 - 7x + 12 + x^2 - 11x + 30 = 0$, so $3x^2 - 21x + 44 = 0$. By Vieta's Formulas, the product of the roots is $\boxed{44/3}$.

4.2.3   Let r be the other root. Vieta's Formulas for the sum and product of the roots give us two different ways to compute the other root of the given quadratic in terms of a, b, and c.

Solution 1: The sum of the roots is $1 + r = -\dfrac{b(c-a)}{a(b-c)}$, so

$$r = -\frac{b(c-a)}{a(b-c)} - 1 = \frac{-b(c-a) - a(b-c)}{a(b-c)} = \frac{-bc + ab - ab + ac}{a(b-c)} = \frac{ac - bc}{a(b-c)} = \boxed{\frac{c(a-b)}{a(b-c)}}.$$

Solution 2: The product of the roots is $1 \cdot r = \dfrac{c(a-b)}{a(b-c)}$, so $r = \boxed{\dfrac{c(a-b)}{a(b-c)}}$.

4.2.4   By Vieta's Formulas, we have $r_1 + r_2 = -7/2$, and $r_1 r_2 = c/2$. Then $\dfrac{1}{r_1} + \dfrac{1}{r_2} = \dfrac{r_1 + r_2}{r_1 r_2} = \dfrac{-7/2}{c/2} = \boxed{-\dfrac{7}{c}}$.

4.2.5   By Vieta's Formulas, we have $m + n = -b/a$ and $mn = c/a$. To find an expression for $m^2 + n^2$, we square the equation $m + n = -b/a$ to get $m^2 + 2mn + n^2 = \frac{b^2}{a^2}$. Then subtracting $2mn$, which equals $2c/a$, we get

$$m^2 + n^2 = \frac{b^2}{a^2} - 2mn = \frac{b^2}{a^2} - \frac{2c}{a} = \frac{b^2 - 2ac}{a^2}.$$

Since $m^2 + n^2 = 1$, we have $\dfrac{b^2 - 2ac}{a^2} = 1$, which means $b^2 - 2ac = a^2$, so $b^2 - a^2 = 2ac$.

4.2.6   We are given that $ax^2 + bx + c = dx^2 + ex + f$ for all values of x. Substituting the values of $x = -1$, $x = 0$, and $x = 1$, we obtain the system of equations

$$a - b + c = d - e + f,$$
$$c = f,$$
$$a + b + c = d + e + f.$$

Adding the first and third equations, we get $2a + 2c = 2d + 2f$, so $a + c = d + f$. From the second equation, we have $c = f$, so $a = d$. Then from the third equation, we have $b = e$, as desired.

Exercises for Section 4.3

4.3.1

(a) $x^2 + 8x = x^2 + 8x + 16 - 16 = \boxed{(x+4)^2 - 16}$.

(b) $y^2 - 6y + 1 = y^2 - 6y + 9 - 9 + 1 = \boxed{(y-3)^2 - 8}$.

(c) $2z^2 + 20z + 3 = 2(z^2 + 10z) + 3 = 2(z^2 + 10z + 25) - 2(25) + 3 = \boxed{2(z+5)^2 - 47}$.

(d) $9x^2 + \dfrac{1}{2}x + \dfrac{4}{3} = 9\left(x^2 + \dfrac{1}{18}x\right) + \dfrac{4}{3} = 9\left(x^2 + \dfrac{2}{36}x + \dfrac{1}{36^2}\right) - \dfrac{9}{36^2} + \dfrac{4}{3} = \boxed{9\left(x + \dfrac{1}{36}\right)^2 + \dfrac{191}{144}}$.

4.3.2

(a) From part (a) of the last problem, we can rewrite the equation as

$$(x+4)^2 - 16 = 12 \quad \Rightarrow \quad (x+4)^2 = 28,$$

so $x+4 = \pm\sqrt{28} = \pm 2\sqrt{7}$. Therefore, the solutions are $x = \boxed{-4 \pm 2\sqrt{7}}$.

(b) From part (b) of the last problem, we can rewrite our equation as

$$(y-3)^2 - 8 = -4 \quad \Rightarrow \quad (y-3)^2 = 4,$$

so $y - 3 = \pm 2$. Therefore, the solutions are $y = \boxed{1 \text{ and } 5}$.

(c) From part (c) of the last problem, we can rewrite our equation as

$$2(z+5)^2 - 47 = -30 \quad \Rightarrow \quad (z+5)^2 = \frac{17}{2} = \frac{34}{4}.$$

so $z+5 = \pm\sqrt{34}/2$. Therefore, the solutions are $z = \boxed{-5 \pm \sqrt{34}/2}$.

(d) From part (d) of the last problem, we can rewrite our equation as

$$9\left(x+\frac{1}{36}\right)^2 + \frac{191}{144} = 2$$
$$\Rightarrow \quad 9\left(x+\frac{1}{36}\right)^2 = \frac{97}{144}$$
$$\Rightarrow \quad \left(x+\frac{1}{36}\right)^2 = \frac{97}{9 \cdot 144}$$
$$\Rightarrow \quad x+\frac{1}{36} = \pm\frac{\sqrt{97}}{36}.$$
$$\Rightarrow \quad x = \boxed{-\frac{1}{36} \pm \frac{\sqrt{97}}{36}}.$$

4.3.3 Let x be the number in question. If x is 8 more than its reciprocal, then $x = 8 + \frac{1}{x}$. Multiplying by x gives $x^2 = 8x + 1$ and rearranging gives $x^2 - 8x - 1 = 0$ By the quadratic formula, $x = \frac{8 \pm \sqrt{8^2+4}}{2} = \frac{8 \pm \sqrt{68}}{2} = \frac{8 \pm 2\sqrt{17}}{2} = 4 \pm \sqrt{17}$. Since x is positive, we have $x = \boxed{4 + \sqrt{17}}$.

4.3.4 We have $g(3) = 9a^2 - 63a + 13 = -59$, which simplifies to $9a^2 - 63a + 72 = 0$. Dividing this by 9, we get $a^2 - 7a + 8 = 0$. By the quadratic formula, we have

$$a = \frac{7 \pm \sqrt{49 - 4 \cdot 8}}{2} = \boxed{\frac{7 \pm \sqrt{17}}{2}}.$$

4.3.5

(a) By the quadratic formula, the roots of $2x^2 + 17x + 21 = 0$ are

$$x = \frac{-17 \pm \sqrt{17^2 - 4\cdot 2 \cdot 21}}{2\cdot 2} = \frac{-17 \pm \sqrt{121}}{4} = \frac{-17 \pm 11}{4}.$$

Therefore, the roots are $x = \boxed{-7 \text{ and } -3/2}$.

(b) By the quadratic formula, we have $y = \frac{16 \pm \sqrt{(-16)^2 - 4\cdot 1 \cdot 51}}{2} = \frac{16 \pm \sqrt{52}}{2} = \frac{16 \pm 2\sqrt{13}}{2} = \boxed{8 \pm \sqrt{13}}$.

(c)   By the quadratic formula, we have $z = \dfrac{-4 \pm \sqrt{4^2 - 4 \cdot 1 \cdot 5}}{2} = \dfrac{-4 \pm \sqrt{-4}}{2} = \dfrac{-4 \pm 2i}{2} = \boxed{-2 \pm i}$.

(d)   To make the given equation easier to work with, we multiply it by $\sqrt{5}$, to get

$$5x^2 - (40 + 10\sqrt{5})x + (100 + 40\sqrt{5}) = 0.$$

This step at least makes the coefficient of x^2 an integer. Now we can divide the equation by 5, to get

$$x^2 - (8 + 2\sqrt{5})x + (20 + 8\sqrt{5}) = 0.$$

By the quadratic formula, we have

$$x = \frac{8 + 2\sqrt{5} \pm \sqrt{(8 + 2\sqrt{5})^2 - 4 \cdot (20 + 8\sqrt{5})}}{2} = \frac{8 + 2\sqrt{5} \pm \sqrt{(84 + 32\sqrt{5}) - (80 + 32\sqrt{5})}}{2}$$

$$= \frac{8 + 2\sqrt{5} \pm \sqrt{4}}{2} = \frac{8 + 2\sqrt{5} \pm 2}{2} = 4 + \sqrt{5} \pm 1.$$

Therefore, the roots are $x = \boxed{3 + \sqrt{5} \text{ and } 5 + \sqrt{5}}$.

4.3.6   We want all values of t such that $2t^2 - 8\sqrt{5}t + 25 = -13$, which simplifies to $2t^2 - 8\sqrt{5}t + 38 = 0$. Dividing by 2, we get $t^2 - 4\sqrt{5}t + 19 = 0$. By the quadratic formula, we have

$$t = \frac{4\sqrt{5} \pm \sqrt{(4\sqrt{5})^2 - 4 \cdot 19}}{2} = \frac{4\sqrt{5} \pm 2}{2} = \boxed{2\sqrt{5} \pm 1}.$$

4.3.7   By the quadratic formula, we have $z = \dfrac{3 - 8i \pm \sqrt{(3 - 8i)^2 + 4(14 + 12i)}}{2} = \dfrac{3 - 8i \pm \sqrt{1}}{2} = \dfrac{3 - 8i \pm 1}{2}$. Therefore, the solutions are $z = \boxed{2 - 4i \text{ and } 1 - 4i}$.

Exercises for Section 4.4

4.4.1   Since the coefficients of the quadratic are real, nonreal roots of the quadratic must be a conjugate pair. Hence, if $6 + i\sqrt{5}$ is one root, then the other root must be $\overline{6 + i\sqrt{5}} = \boxed{6 - i\sqrt{5}}$.

4.4.2   The quadratic $ax^2 - 5x + 9$ has only one distinct root when it has a double root, which occurs only when its discriminant is 0. Hence, $5^2 - 4 \cdot a \cdot 9 = 0$, so $a = \boxed{25/36}$.

4.4.3   Let $ax^2 + bx + c = 0$ be a quadratic with real coefficients. If it has a double root, then its discriminant is 0, so by the quadratic formula, the double root is

$$x = \frac{-b \pm \sqrt{b^2 - 4ac}}{2a} = -\frac{b}{2a}.$$

Hence, the double root must be real.

4.4.4   If the coefficients of a quadratic are real, then nonreal roots must be a conjugate pair. However, the coefficients of the quadratic in part (c) of Problem 4.16 are not real, so this condition does not apply.

4.4.5   The coefficients of the quadratic are real, and we are given that one of the roots is $3 + 2i$, a nonreal number. Therefore, the other root must be $\overline{3 + 2i} = 3 - 2i$. Then by Vieta's Formulas, the product of the roots is $s/2$, so $s/2 = (3 + 2i)(3 - 2i) = 13$. Therefore, $s = \boxed{26}$.

4.4.6 Expanding the right side and organizing all the terms on one side gives $K^2 x^2 - (3K^2 + 1)x + 2K^2 = 0$. This equation is a quadratic in x. (We treat K as a constant.) The roots of this quadratic are real when the discriminant is nonnegative, so we must have $(-(3K^2 + 1))^2 - 4(K^2)(2K^2) \geq 0$. Simplifying the left side gives us $K^4 + 6K^2 + 1 \geq 0$. Because K is real, both K^4 and K^2 are nonnegative. Therefore, $K^4 + 6K^2 + 1$ is nonnegative for all real values of K. So, there are real roots to the equation for all real values of K.

Exercises for Section 4.5

4.5.1 Bringing all the terms to one side gives $t^2 - 13t + 40 \leq 0$. Factoring the quadratic gives $(t - 5)(t - 8) \leq 0$. If $t < 5$, both $t - 5$ and $t - 8$ are negative, so their product is positive. If $t > 8$, both $t - 5$ and $t - 8$ are positive, so their product is positive. But if $5 \leq t \leq 8$, then $t - 5 \geq 0$ and $t - 8 \leq 0$, so $(t - 5)(t - 8) \leq 0$. Therefore, the solution is $\boxed{t \in [5, 8]}$.

4.5.2 Multiplying the inequality by -1 (to make the coefficient of x^2 equal to 1), we get $x^2 - 7x + 13 < 0$. The discriminant of the quadratic is $7^2 - 4 \cdot 13 = -3$, so it has no real roots.

We then complete the square, which gives us

$$x^2 - 7x + 13 = \left(x - \frac{7}{2}\right)^2 - \left(\frac{7}{2}\right)^2 + 13 = \left(x - \frac{7}{2}\right)^2 + \frac{3}{4}.$$

This expression is always at least $3/4$, so in particular, it is never negative. Therefore, there are $\boxed{\text{no solutions}}$.

4.5.3 The real number k is in the range of $f(x)$ if and only if $\frac{7x^2 - 4x + 4}{x^2 + 1} = k$ for some real number x. Multiplying both sides by $x^2 + 1$ and rearranging, we obtain the equation $(k - 7)x^2 + 4x + k - 4 = 0$.

This quadratic must have a real root in x, so its discriminant must be nonnegative. In other words,

$$4^2 - 4(k - 7)(k - 4) \geq 0 \quad \Rightarrow \quad 4 - (k - 7)(k - 4) \geq 0 \quad \Rightarrow \quad k^2 - 11k + 24 \leq 0 \quad \Rightarrow \quad (k - 3)(k - 8) \leq 0.$$

If $k < 3$, then $(k - 3)(k - 8) > 0$. If $3 \leq k \leq 8$, then $(k - 3)(k - 8) \leq 0$. If $k > 8$, then $(k - 3)(k - 8) > 0$. Hence, the inequality is satisfied for $3 \leq k \leq 8$. We conclude that the range of $f(x)$ is $\boxed{[3, 8]}$.

4.5.4 For all three parts let $f(x) = ax^2 + bx + c$.

(a) Completing the square gives $f(x) = a(x - h)^2 + k$ for some constants h and k. The roots of $f(x)$ are the solutions to the equation $a(x - h)^2 + k = 0$. Rearranging this equation gives $(x - h)^2 = -k/a$. We are given that $a > 0$ and that the roots of $f(x)$ are nonreal. So, we must have $k > 0$, as well. Since $k > 0$ and $a(x - h)^2 \geq 0$ for all x, we have $f(x) = a(x - h)^2 + k > 0$ for all x.

(b) Completing the square again gives $f(x) = a(x - h)^2 + k$ for some constants h and k. Rearranging this equation gives $(x - h)^2 = -k/a$. We are given that $a < 0$ and that the roots of $f(x)$ are nonreal. So, we must have $k < 0$, as well. Since $k < 0$ and $a(x - h)^2 \leq 0$ for all x (remember, a is negative), we have $f(x) = a(x - h)^2 + k < 0$ for all x.

Alternatively, we could have used part (a). If $f(x)$ is a quadratic with nonreal roots, and the coefficient of x^2 is positive, then $-f(x)$ is also quadratic with nonreal roots, and the coefficient of x^2 is negative. Therefore, by part (a), $-f(x) > 0$ for all x, so $f(x) < 0$ for all x.

(c) If $f(x)$ has a double root, then $f(x) = a(x - r)^2$ for some real constants a and r. Since $(x - r)^2$ is nonnegative for all x, the sign of $f(x)$ matches the sign of a for all x except $x = r$, which makes $f(x) = 0$. So, if $a > 0$, we have $f(x) \geq 0$ for all x, and if $a < 0$, we have $f(x) \leq 0$ for all x.

4.5.5 As a quadratic in y, the equation has real roots only when the discriminant is nonnegative. So, we must have

$$(4x)^2 - 4 \cdot 4 \cdot (x + 6) \geq 0 \quad \Rightarrow \quad 16x^2 - (16x + 96) \geq 0 \quad \Rightarrow \quad x^2 - x - 6 \geq 0 \quad \Rightarrow \quad (x + 2)(x - 3) \geq 0.$$

If $x \le -2$, then $(x + 2)(x - 3) \ge 0$. If $-2 < x < 3$, then $(x + 2)(x - 3) < 0$. If $x \ge 3$, then $(x + 2)(x - 3) \ge 0$. Hence, y is real for $\boxed{x \le -2 \text{ and } x \ge 3}$.

4.5.6 We first factor an x out the expression on the left to find $x(x^2 - 6x + 9) > 0$. We then factor the quadratic to find $x(x - 3)^2 > 0$. The expression $(x - 3)^2$ is positive for all x except $x = 3$, and x is positive for $x > 0$, so the product $x(x - 3)^2$ is positive if and only if $\boxed{x \in (0,3) \cup (3, +\infty)}$.

Review Problems

4.25

(a) Factoring the quadratic gives $(x + 3)(x + 9) = 0$. Therefore, the roots are $x = \boxed{-9 \text{ and } -3}$.

(b) By the quadratic formula, we have $x = \dfrac{6 \pm \sqrt{6^2 - 4}}{2} = \dfrac{6 \pm \sqrt{32}}{2} = \dfrac{6 \pm 4\sqrt{2}}{2} = \boxed{3 \pm 2\sqrt{2}}$.

(c) Factoring the quadratic gives $(2y - 1)(y - 2) = 0$. Therefore, the roots are $y = \boxed{1/2 \text{ and } 2}$.

(d) By the quadratic formula, the roots of the quadratic $3x^2 + 15x - 7 = 0$ are

$$x = \frac{-15 \pm \sqrt{15^2 - 4 \cdot 3 \cdot (-7)}}{2 \cdot 3} = \frac{-15 \pm \sqrt{225 + 84}}{6} = \boxed{\frac{-15 \pm \sqrt{309}}{6}}.$$

4.26 We have $f(2) = 4a^2 + 5a + 3$, so $4a^2 + 5a + 3 = 2$. Rearranging and factoring gives $(4a + 1)(a + 1) = 0$. Therefore, all possible values of a are $\boxed{-1 \text{ and } -1/4}$.

4.27 We are given that $(x - 1)\left(x + \frac{1}{2}\right) = 1$. Expanding and rearranging then gives $2x^2 - x - 3 = 0$. Factoring the quadratic gives $(x + 1)(2x - 3) = 0$. Therefore, the roots of the quadratic are $x = -1$ and $3/2$. Since x is positive, $x = \boxed{3/2}$.

4.28 Let r and s be the roots of the quadratic $x^2 + 2hx - 3 = 0$. Then by Vieta's Formulas, we have $r + s = -2h$ and $rs = -3$. Then
$$r^2 + s^2 = (r + s)^2 - 2rs = (-2h)^2 - 2 \cdot (-3) = 4h^2 + 6,$$
so we have $4h^2 + 6 = 10$, which gives us $h^2 = 1$. Hence, $|h| = \boxed{1}$.

4.29

(a) Squaring both sides of the equation, we get $6a^2 + 5a + 21 = 196$, which becomes $6a^2 + 5a - 175 = 0$. By the quadratic formula, we have

$$a = \frac{-5 \pm \sqrt{5^2 - 4 \cdot 6 \cdot (-175)}}{2 \cdot 6} = \frac{-5 \pm \sqrt{4225}}{12} = \frac{-5 \pm 65}{12}.$$

Therefore, we find $a = \boxed{5 \text{ and } -35/6}$. A quick check reveals that neither solution is extraneous. Note that we could also have factored the quadratic as $(6a + 35)(a - 5) = 0$ to find our solutions.

(b) Squaring both sides of the equation, we get $4x^2 + 20x + 25 = 4x^2 + 20x + 25$, which is true for all x. The expression $\sqrt{4x^2 + 20x + 25}$ is nonnegative for all x, so we must have $2x + 5 \ge 0$ in order for the two sides to be the same. (If $2x + 5 < 0$, then we have $\sqrt{4x^2 + 20x + 25} = -(2x + 5)$.) Solving the inequality $2x + 5 \ge 0$ gives us $\boxed{x \ge -5/2}$.

4.30 In order to get an expression for $f(3z)$, we must choose an expression for x to make $f(x/3)$ become $f(3z)$. Solving $x/3 = 3z$ gives $x = 9z$. Making this substitution in $f(x/3) = x^2 + x + 1$ gives $f(3z) = (9z)^2 + 9z + 1 = 81z^2 + 9z + 1$.

So, the equation $f(3z) = 7$ gives us $81z^2 + 9z + 1 = 7$, which simplifies to $81z^2 + 9z - 6 = 0$. Factoring gives $3(9z - 2)(3z + 1) = 0$. Therefore, the solutions are $z = 2/9$ and $z = -1/3$, and their sum is $2/9 - 1/3 = \boxed{-1/9}$. Alternatively, we can use Vieta to note that the sum of the roots of $81z^2 + 9z - 6$ is $-9/81 = -1/9$.

4.31 The function $f(x) = \frac{x+7}{1+\frac{1}{x}+\frac{1}{x^2-6}}$ is defined as long as $x \neq 0$, $x^2 - 6 \neq 0$, and $\frac{1}{x} + \frac{1}{x^2-6} \neq 0$. The second condition is equivalent to $x \neq \sqrt{6}$ and $x \neq -\sqrt{6}$.

To analyze the third condition, we solve the equation $\frac{1}{x} + \frac{1}{x^2-6} = 0$. Writing the left side with a common denominator gives us $\frac{x^2+x-6}{x(x^2-6)} = 0$, and factoring the numerator gives $\frac{(x-2)(x+3)}{x(x^2-6)} = 0$. Therefore, we have $\frac{1}{x} + \frac{1}{x^2-6} = 0$ when $x = 2$ or $x = -3$, so we must exclude these values from the domain of f.

Combining our restrictions, we see that the domain of f is $\boxed{\text{all real numbers except } 0, \sqrt{6}, -\sqrt{6}, 2, \text{ and } -3}$.

4.32 By Vieta's Formulas, the product of the roots of $x^2 - px + q = 0$ is

$$q = \left(a + \frac{1}{b}\right)\left(b + \frac{1}{a}\right) = ab + 2 + \frac{1}{ab}.$$

Again by Vieta's Formulas, the product of the roots of $x^2 - mx + 2 = 0$ is $ab = 2$, so $q = ab + 2 + \frac{1}{ab} = 2 + 2 + \frac{1}{2} = \boxed{\frac{9}{2}}$.

4.33 Let r be one root and let $3r$ be the other. Then by Vieta's Formulas, we have $a = r + 3r = 4r$, and $2a + 3 = r \cdot 3r = 3r^2$. From the first equation, we have $r = a/4$. Substituting this expression for r into the second equation, we get

$$2a + 3 = 3\left(\frac{a}{4}\right)^2 \quad \Rightarrow \quad 3a^2 - 32a - 48 = 0 \quad \Rightarrow \quad (a - 12)(3a + 4) = 0.$$

Therefore, the solutions are $a = \boxed{12 \text{ and } -4/3}$.

4.34 Rearranging gives the inequality $6x^2 + 5x - 4 < 0$. Factoring the quadratic gives $(3x + 4)(2x - 1) < 0$.

If $x \leq -4/3$, then $(3x+4)(2x-1) \geq 0$. If $-4/3 < x < 1/2$, then $(3x+4)(2x-1) < 0$. If $x \geq 1/2$, then $(3x+4)(2x-1) \geq 0$. Therefore, the solution is $\boxed{-4/3 < x < 1/2}$.

4.35 The quadratic $x^2 + kx + 27 = 0$ has two distinct real solutions if and only if its discriminant is positive. In other words, $k^2 - 4 \cdot 27 > 0$, so $k^2 > 108$. Therefore, the values of k for which the quadratic has two distinct real solutions are $\boxed{k < -6\sqrt{3} \text{ and } k > 6\sqrt{3}}$.

4.36 Let $r = x/y$, so $x = ry$. Substituting this expression for x into the given equation, we get $(ry)^2 = (ry) \cdot y + 12y^2$, which becomes $r^2y^2 = ry^2 + 12y^2$. Since y is nonzero, we can divide both sides by y^2 to get $r^2 = r + 12$, so $r^2 - r - 12 = 0$, which gives $(r - 4)(r + 3) = 0$. Therefore, the possible values of r are $\boxed{4 \text{ and } -3}$.

4.37 From the second equation, we have $y = 2 - x$. Substituting this expression for y into the first equation, we get

$$2 - x = \frac{8}{x^2 + 4} \quad \Rightarrow \quad (2 - x)(x^2 + 4) = 8 \quad \Rightarrow \quad 2x^2 + 8 - x^3 - 4x = 8 \quad \Rightarrow \quad x(x^2 - 2x + 4) = 0.$$

Hence, $x = 0$ or $x^2 - 2x + 4 = 0$. By the quadratic formula, the roots of the quadratic are

$$x = \frac{2 \pm \sqrt{2^2 - 4 \cdot 4}}{2} = \frac{2 \pm \sqrt{-12}}{2} = \frac{2 \pm 2i\sqrt{3}}{2} = 1 \pm i\sqrt{3}.$$

Therefore, the possible values of x are $\boxed{0, 1 - i\sqrt{3}, \text{ and } 1 + i\sqrt{3}}$.

4.38 If the Bolts increase their ticket prices by $\$k$, then they sell $20000 - 100k$ tickets. Therefore, their total revenue, in dollars, is

$$(k + 50)(20000 - 100k) = 100(k + 50)(200 - k).$$

Therefore, we seek all k such that

$$100(k + 50)(200 - k) > 1000000$$
$$\Rightarrow \quad (k + 50)(200 - k) > 10000$$
$$\Rightarrow \quad 200k - k^2 + 10000 - 50k > 10000$$
$$\Rightarrow \quad k^2 - 150k < 0$$
$$\Rightarrow \quad k(k - 150) < 0.$$

This last inequality is true for $0 < k < 150$. Hence, the Bolts can generate more than \$1,000,000 in revenue for $\boxed{0 < k < 150}$.

4.39 Factoring the quadratic gives $(x + 77)(x - 27) \geq 0$. If $x \leq -77$, then $(x + 77)(x - 27) \geq 0$. If $-77 < x < 27$, then $(x + 77)(x - 27) < 0$. If $x \geq 27$, then $(x + 77)(x - 27) \geq 0$. Therefore, the solutions are $\boxed{x \leq -77 \text{ and } x \geq 27}$.

4.40 Since $P(x)$ is quadratic, we have $P(x) = ax^2 + bx + c$ for some constants a, b, and c. From $P(0) = -1$, $P(1) = 9$, $P(2) = 25$, we obtain the system of equations

$$c = -1,$$
$$a + b + c = 9,$$
$$4a + 2b + c = 25.$$

Since $c = -1$, the second and third equations become $a + b = 10$ and $4a + 2b = 26$, respectively. Dividing the second equation by 2, we get $2a + b = 13$. Subtracting the equation $a + b = 10$, we get $a = 3$, so $b = 10 - a = 7$. Therefore, $P(-1) = a - b + c = 3 - 7 - 1 = \boxed{-5}$.

4.41 If a and b are real numbers such that $a^b = 1$, then one of the following must hold: (1) $a = 1$, (2) $a = -1$ and b is an even integer, or (3) $b = 0$ and $a \neq 0$. We analyze each below:

(1) If $x^2 - 5x + 5 = 1$, then $(x - 4)(x - 1) = 0$, so $x = 1$ or $x = 4$.

(2) If $x^2 - 5x + 5 = -1$, then $(x - 2)(x - 3) = 0$, so $x = 2$ or $x = 3$. If $x = 2$, then $x^2 - 9x + 20 = 6$, and if $x = 3$, then $x^2 - 9x + 20 = 2$. Both exponents are even, so both are solutions.

(3) If $b = 0$, then $x^2 - 9x + 20 = 0$, so $(x - 4)(x - 5) = 0$. If $x = 4$, then from case (1), the base is 1. If $x = 5$, then $x^2 - 5x + 5 = 5 \neq 0$.

Therefore, the solutions are $x = \boxed{1, 2, 3, 4, \text{ and } 5}$.

4.42 Adding the first two equations, we get $x^2 + x = 6$, which simplifies to $x^2 + x - 6 = 0$, so $(x - 2)(x + 3) = 0$, which means $x = 2$ or $x = -3$.

If $x = -3$, then the second equation becomes $-3 - yz = -33$, so $yz = 30$. From the third equation, we have $z = 13 - y$. Substituting this expression for z into $yz = 30$, we get $y(13 - y) = 30$. This simplifies to $y^2 - 13y + 30 = 0$, so $(y - 10)(y - 3) = 0$, which means $y = 3$ or $y = 10$. Since $z = 13 - y$, we obtain the solutions $(x, y, z) = (-3, 3, 10)$ and $(-3, 10, 3)$.

If $x = 2$, then the second equation becomes $2 - yz = -33$, so $yz = 35$. Again from the third equation, we have $z = 13 - y$. Substituting this expression for z into $yz = 35$, we get $y(13 - y) = 35$, which simplifies to $y^2 - 13y + 35 = 0$. By the quadratic formula, the solutions to this quadratic are $y = \frac{13 \pm \sqrt{29}}{2}$. If $y = \frac{13 + \sqrt{29}}{2}$, then $z = 13 - y = \frac{13 - \sqrt{29}}{2}$, and vice versa. Thus, we obtain the solutions $(x, y, z) = (2, \frac{13 + \sqrt{29}}{2}, \frac{13 - \sqrt{29}}{2})$ and $(2, \frac{13 - \sqrt{29}}{2}, \frac{13 + \sqrt{29}}{2})$.

To summarize, the solutions are

$$(x, y, z) = \boxed{(-3, 3, 10), \ (-3, 10, 3), \ \left(2, \frac{13 + \sqrt{29}}{2}, \frac{13 - \sqrt{29}}{2}\right), \text{ and } \left(2, \frac{13 - \sqrt{29}}{2}, \frac{13 + \sqrt{29}}{2}\right)}.$$

4.43   By Vieta, the sum of the roots of the quadratic is $-(a - \frac{1}{a}) = -a + \frac{1}{a}$ and the product of the roots is -1. Aha! The two numbers $-a$ and $\frac{1}{a}$ have sum $-a + \frac{1}{a}$ and product -1. Testing, we find that both of these do indeed satisfy the equation, and the roots are $\boxed{-a \text{ and } \dfrac{1}{a}}$.

Challenge Problems

4.44   Cross-multiplying, we get

$$(m + 1)(x^2 - bx) = (m - 1)(ax - c)$$
$$\Rightarrow \qquad mx^2 - bmx + x^2 - bx = amx - cm - ax + c$$
$$\Rightarrow \quad (m + 1)x^2 + (-am - bm + a - b)x + (cm - c) = 0.$$

This equation is quadratic in x. If each root of the quadratic is the negative of the other, then their sum is 0. Hence, by Vieta's Formulas,

$$-\frac{-am - bm + a - b}{m + 1} = 0 \quad \Rightarrow \quad am + bm - a + b = 0 \quad \Rightarrow \quad (a + b)m = a - b \quad \Rightarrow \quad m = \boxed{\frac{a - b}{a + b}}.$$

4.45   Let $f(x) = \frac{1-x^2}{1+x^2}$. If the graph stays within $a < y \le b$, then the entire range of f must be within the interval $(a, b]$. So, let's find the range of f. The real number k is in the range of $f(x)$ if and only if

$$f(x) = \frac{1 - x^2}{1 + x^2} = k$$

for some real number x. Multiplying both sides by $1 + x^2$, we get

$$1 - x^2 = k(1 + x^2) \quad \Rightarrow \quad 1 - x^2 = kx^2 + k \quad \Rightarrow \quad (k + 1)x^2 = 1 - k \quad \Rightarrow \quad x^2 = \frac{1 - k}{1 + k}.$$

This equation has a solution in x if and only if $\frac{1-k}{1+k} \ge 0$. If $k < -1$, then $(1 - k)/(1 + k) < 0$. If $-1 < k \le 1$, then $(1 - k)/(1 + k) \ge 0$. If $k > 1$, then $(1 - k)/(1 + k) < 0$. (And of course, if $k = -1$, then $(1 - k)/(1 + k)$ is undefined.) Hence, $-1 < k \le 1$. Thus, we can take $\boxed{a = -1 \text{ and } b = 1}$.

4.46   By the quadratic formula, the roots of $z^2 + tz + 1 = 0$ in z are

$$z = \frac{-t \pm \sqrt{t^2 - 4}}{2}.$$

If $t = \pm 2$, then $t^2 - 4 = 0$, so $z = -t/2$, and $|z| = |t/2| = 1$. Otherwise, $|t| < 2$, so $t^2 - 4 < 0$, which means that the discriminant of the quadratic is negative. Hence, we can write

$$z = \frac{-t \pm \sqrt{t^2 - 4}}{2} = \frac{-t \pm i\sqrt{4 - t^2}}{2}.$$

Then the real part of z is $-t/2$ and the imaginary part is $\pm\sqrt{4 - t^2}/2$, so the magnitude of z is

$$|z| = \sqrt{\left(-\frac{t}{2}\right)^2 + \left(\pm\frac{\sqrt{4 - t^2}}{2}\right)^2} = \sqrt{\frac{t^2}{4} + \frac{4 - t^2}{4}} = \sqrt{1} = 1.$$

Thus, the magnitude of each root is always equal to 1.

Here is an alternative solution. Since the discriminant $t^2 - 4$ is negative, the roots of the quadratic form a complex conjugate pair, say r and $\bar{r}$. Then by Vieta's Formulas, the product of the roots is $r\bar{r} = 1$. But $r\bar{r} = |r|^2$, so $|r|^2 = 1$. Since $|r|$ is a nonnegative real number, $|r| = |\bar{r}| = 1$.

4.47 Suppose that both a and b are real numbers. Cross-multiplying, we get $(a + b)^2 = ab$. Expanding and simplifying gives $a^2 + ab + b^2 = 0$. We can view this as a quadratic in a. Then by the quadratic formula, we have

$$a = \frac{-b \pm \sqrt{b^2 - 4b^2}}{2} = \frac{-b \pm \sqrt{-3b^2}}{2}.$$

If b is nonzero, then $-3b^2 < 0$, which results in nonreal values for a. If $b = 0$, then $a = 0$, but then the expressions in the problem are undefined. Therefore, a and b cannot both be real numbers.

4.48 Let r and s be the roots of the quadratic. We can find the roots of the quadratic in terms of p by using the quadratic formula, or by completing the square. However, by Vieta's Formulas, $r + s = p$ and $rs = (p^2 - 1)/4$, so

$$(r - s)^2 = r^2 - 2rs + s^2 = r^2 + 2rs + s^2 - 4rs = (r + s)^2 - 4rs = p^2 - (p^2 - 1) = 1.$$

Therefore, the positive difference between r and s is $\boxed{1}$.

4.49 Since there is no obvious approach, we try to eliminate one of the variables. We try to do so in as nice a way as possible.

First, adding the equations, the xy terms are eliminated and we get $x^2 + y^2 = 10$. From $x^2 - xy = 6$, we have $xy = x^2 - 6$. Squaring, we get $x^2 y^2 = (x^2 - 6)^2$. Now we can substitute $y^2 = 10 - x^2$ to eliminate y, to get $x^2(10 - x^2) = (x^2 - 6)^2$. Expanding and rearranging gives $x^4 - 11x^2 + 18 = 0$, and factoring gives $(x^2 - 9)(x^2 - 2) = 0$. Hence, $x = 3, -3, \sqrt{2},$ or $-\sqrt{2}$. We can find the corresponding values of y by solving $x^2 - xy = 6$ for y to find $y = \frac{x^2 - 6}{x}$. Using this formula, we find that the solutions are $(x, y) = \boxed{(3, 1), (-3, -1), (\sqrt{2}, -2\sqrt{2}), \text{ and } (-\sqrt{2}, 2\sqrt{2})}$.

4.50 Since $f(x)$ is monic and the sum of the roots of $f(x)$ is a, by Vieta's Formulas, we have $f(x) = x^2 - ax + d_1$ for some constant d_1. Similarly, $g(x) = x^2 - bx + d_2$ and $h(x) = x^2 - cx + d_3$ for some constants d_2 and d_3. Then

$$f(x) + g(x) + h(x) = 3x^2 - (a + b + c)x + d_1 + d_2 + d_3.$$

Hence, again by Vieta's Formulas, the sum of the roots of $f(x) + g(x) + h(x) = 0$ is $\boxed{(a + b + c)/3}$.

4.51 Let $c = b - 1$, so $b = c + 1$ and we seek c/a. Then the given equation becomes

$$\begin{aligned} & a^2 + (c + 1)^2 + 2a + 1 = 2a(c + 1) + 2(c + 1) \\ \Rightarrow \quad & a^2 + c^2 + 2c + 1 + 2a + 1 = 2ac + 2a + 2c + 2 \\ \Rightarrow \quad & a^2 - 2ac + c^2 = 0. \end{aligned}$$

Factoring gives $(a - c)^2 = 0$, so $a = c$. Therefore, $(b - 1)/a = c/a = \boxed{1}$.

4.52 Note that $\frac{1}{r^2} + \frac{1}{s^2} = \frac{r^2 + s^2}{r^2 s^2} = \frac{(r + s)^2 - 2rs}{(rs)^2}$. By Vieta's Formulas, we have $r + s = 6/a$ and $rs = 12/a$. So,

$$\frac{(r + s)^2 - 2rs}{(rs)^2} = \frac{\left(\frac{6}{a}\right)^2 - 2\left(\frac{12}{a}\right)}{\left(\frac{12}{a}\right)^2} = \frac{\frac{36}{a^2} - \frac{24}{a}}{\frac{144}{a^2}} = \frac{36 - 24a}{144} = \boxed{\frac{3 - 2a}{12}}.$$

4.53 Since $(2 + (2 + (2 + x)^2)^2)^2$ is nonnegative, we can take the nonnegative square root of both sides, to get

$$2 + (2 + (2 + (2 + x)^2)^2)^2 = 123 \quad \Rightarrow \quad (2 + (2 + (2 + x)^2)^2)^2 = 121.$$

Again, since $(2 + (2 + x)^2)^2$ is nonnegative, we can take the nonnegative square root of both sides, to get

$$2 + (2 + (2 + x)^2)^2 = 11 \quad \Rightarrow \quad (2 + (2 + x)^2)^2 = 9.$$

Again, since $(2 + x)^2$ is nonnegative, we can take the nonnegative square root of both sides, to get

$$2 + (2 + x)^2 = 3 \quad \Rightarrow \quad (2 + x)^2 = 1.$$

Then $2 + x = \pm 1$, so the (real) solutions are $x = -3$ and -1, and their sum is $\boxed{-4}$.

4.54 Let the quadratic be $f(x) = ax^2 + bx + c$. We are given that the sum of the coefficients is 0, so $a + b + c = 0$. But this is also equal to $f(1)$, so $x = 1$ is one root of the quadratic. Then by Vieta's Formulas, the product of the roots is $\frac{c}{a}$, so the other root is $\frac{c}{a}$. Hence, both roots of the quadratic (namely 1 and $\frac{c}{a}$) are real.

4.55 Rearranging the second equation, we get $s^2 - 1 = 3rs - 4r^2$. Comparing this to the first equation, we see that $3rs - 4r^2 = 4r^2 + 4r$, so $8r^2 - 3rs + 4r = 0$, which means $r(8r - 3s + 4) = 0$. Therefore, either $r = 0$ or $8r - 3s + 4 = 0$.

If $r = 0$, then the first equation becomes $s^2 - 1 = 0$, so $s = \pm 1$.

If $8r - 3s + 4 = 0$, then $s = (8r + 4)/3$. Substituting this expression for s into the first equation, we get

$$\left(\frac{8r + 4}{3}\right)^2 - 1 = 4r^2 + 4r$$
$$\Rightarrow \quad (8r + 4)^2 - 9 = 36r^2 + 36r$$
$$\Rightarrow \quad 64r^2 + 64r + 16 - 9 = 36r^2 + 36r$$
$$\Rightarrow \quad 28r^2 + 28r + 7 = 0$$
$$\Rightarrow \quad 4r^2 + 4r + 1 = 0$$
$$\Rightarrow \quad (2r + 1)^2 = 0,$$

so $r = -1/2$, and then $s = (8r + 4)/3 = 0$.

Therefore, the solutions are $(r, s) = \boxed{(0, 1), (0, -1), \text{ and } (-1/2, 0)}$.

4.56 From our given equation, we have $rs + r + s + 1 = 13$, so $r + s = 12 - rs$. Suppose $c = r + s$, so that the coefficient of the linear term of a monic quadratic with roots r and s is $-c$, as desired. From $r + s = 12 - rs$, we have $rs = 12 - (r + s) = 12 - c$, so the constant term of a monic quadratic with roots r and s must be $12 - c$. Therefore, if $(r + 1)(s + 1) = 13$, then for some constant c, there must be a quadratic of the form $x^2 - cx + (12 - c)$ that has roots r and s.

4.57 Subtracting the equation $12 = y^2 + x$ from the equation $12 = x^2 + y$, we get

$$0 = x^2 - y^2 + y - x = (x + y)(x - y) - (x - y) = (x + y - 1)(x - y).$$

Therefore, either $x + y - 1 = 0$ or $x - y = 0$.

If $x + y - 1 = 0$, then $y = 1 - x$. Substituting this expression for y into $x^2 + y = 12$, we get

$$x^2 + 1 - x = 12 \quad \Rightarrow \quad x^2 - x - 11 = 0.$$

By the quadratic formula, we have $x = \dfrac{1 \pm \sqrt{1^2 - 4 \cdot (-11)}}{2} = \dfrac{1 \pm \sqrt{45}}{2} = \dfrac{1 \pm 3\sqrt{5}}{2}$. Since $y = 1 - x$, when $x = \dfrac{1 + 3\sqrt{5}}{2}$, we have $y = \dfrac{1 - 3\sqrt{5}}{2}$, and when $x = \dfrac{1 - 3\sqrt{5}}{2}$, we have $y = \dfrac{1 + 3\sqrt{5}}{2}$.

If $x = y$, then $x^2 + y = 12$ becomes $x^2 + x = 12$, from which we find $x = 3$ or $x = -4$. The corresponding values of y are $y = 3$ and $y = -4$.

Therefore, the solutions are

$$(x, y) = \boxed{(3, 3), \ (-4, -4), \ \left(\frac{1 + 3\sqrt{5}}{2}, \frac{1 - 3\sqrt{5}}{2}\right), \text{ and } \left(\frac{1 - 3\sqrt{5}}{2}, \frac{1 + 3\sqrt{5}}{2}\right)}.$$

4.58 Let $x = \frac{a+b}{a-b}$. Since $0 < a < b$, we have $x < 0$. Since we're given an equation involving the squares of a and b, we can try to square x:

$$x^2 = \left(\frac{a+b}{a-b}\right)^2 = \frac{a^2 + 2ab + b^2}{a^2 - 2ab + b^2}.$$

We can now substitute $a^2 + b^2 = 6ab$ to give $x^2 = \frac{a^2 + 2ab + b^2}{a^2 - 2ab + b^2} = \frac{6ab + 2ab}{6ab - 2ab} = \frac{8ab}{4ab} = 2$. We know that $x < 0$, so $x = \boxed{-\sqrt{2}}$.

4.59 Viewing the second equation as a quadratic in xy, and using the quadratic formula, we have

$$xy = \frac{13 \pm \sqrt{13^2 - 4 \cdot 2 \cdot 18}}{2 \cdot 2} = \frac{13 \pm \sqrt{25}}{4} = \frac{13 \pm 5}{4},$$

so $xy = 2$ or $xy = 9/2$. (We also could have factored to find $(xy - 2)(2xy - 9) = 0$.)

If $xy = 2$, then by Vieta's Formulas, x and y are the roots of the quadratic $t^2 - \frac{9}{2}t + 2 = 0$. Multiplying both sides by 2, and then factoring the quadratic, gives $(2t - 1)(t - 4) = 0$, so $t = 4$ or $t = 1/2$. Therefore, the solutions are $(x, y) = (4, 1/2)$ and $(1/2, 4)$.

If $xy = 9/2$, then by Vieta's Formulas, x and y are the roots of the quadratic $t^2 - \frac{9}{2}t + \frac{9}{2} = 0$. Again, we can multiply by 2 and factor to find $(2t - 3)(t - 3) = 0$, so $t = 3$ or $t = 3/2$. Therefore, the solutions are $(x, y) = (3, 3/2)$ and $(3/2, 3)$.

To summarize, the solutions are $(x, y) = \boxed{(4, 1/2), (1/2, 4), (3, 3/2), \text{ and } (3/2, 3)}$.

4.60 Let the roots be r_1 and r_2. Then by Vieta's Formulas, we have $r_1 + r_2 = m/2$ and $r_1 r_2 = -8/2 = -4$. Therefore,

$$(r_1 - r_2)^2 = (r_1 + r_2)^2 - 4r_1 r_2 = \frac{m^2}{4} + 16.$$

But from the given information, we have $(r_1 - r_2)^2 = (m - 1)^2 = m^2 - 2m + 1$, so $\frac{m^2}{4} + 16 = m^2 - 2m + 1$. Rearranging gives $3m^2 - 8m - 60 = 0$, and factoring gives $(m - 6)(3m + 10) = 0$. Therefore, the solutions are $m = \boxed{6 \text{ and } -10/3}$.

4.61 One approach is to multiply everything out, and solve. The algebra is somewhat lengthy, but in the end works out nicely.

When solving a complicated looking equation, one thing that is always good to try is to see if the equation has any obvious solutions. We see that $x = 0$ is one solution, because both terms $(x - a)/b$ and $a/(x - b)$ become $-a/b$, and both the other terms $(x - b)/a$ and $b/(x - a)$ become $-b/a$. This raises the question, are there any other values of x for which $(x - a)/b = a/(x - b)$?

Cross-multiplying, we obtain the equation $(x - a)(x - b) = ab$, so

$$x^2 - (a + b)x + ab = ab \quad \Rightarrow \quad x^2 - (a + b)x = 0 \quad \Rightarrow \quad x[x - (a + b)] = 0.$$

Hence, $(x - a)/b = a/(x - b)$ for both $x = 0$ and $x = a + b$. Note that for $x = a + b$, the two other terms in the original equation become $(x - b)/a = b/(x - a) = 1$. Therefore, $x = a + b$ is also a solution of the given equation.

However, not only does this give us another solution, it gives us a nice way of rewriting the given equation:

$$\frac{x - a}{b} - \frac{a}{x - b} = \frac{b}{x - a} - \frac{x - b}{a}.$$

We have just seen that both sides become 0 when $x = 0$ and when $x = a + b$, so we expect a factor of $x[x - (a + b)]$

in both sides. Multiplying both sides by $ab(x - a)(x - b)$, we get

$$
\begin{aligned}
& a(x-a)[(x-a)(x-b) - ab] = b(x-b)[ab - (x-a)(x-b)] \\
\Rightarrow \quad & a(x-a)[x^2 - (a+b)x + ab - ab] = b(x-b)[ab - x^2 + (a+b)x - ab] \\
\Rightarrow \quad & a(x-a)[x^2 - (a+b)x] = b(x-b)[-x^2 + (a+b)x] \\
\Rightarrow \quad & [a(x-a) + b(x-b)][x^2 - (a+b)x] = 0 \\
\Rightarrow \quad & [(a+b)x - (a^2 + b^2)]x[x - (a+b)] = 0.
\end{aligned}
$$

Therefore, the solutions are $x = \boxed{0,\, a + b,\, \text{and } (a^2 + b^2)/(a + b)}$.

4.62   Each equation contains two different variables, so there is not much we can do with each individual equation. However, by adding all the equations, we can group the quadratic and linear terms of each variable:

$$
x_1^2 - 2x_1 + x_2^2 - 4x_2 + x_3^2 - 6x_3 + x_4^2 - 8x_4 + x_5^2 - 10x_5 + 55 = 0.
$$

Completing the squares in each variable, we get

$$
(x_1 - 1)^2 + (x_2 - 2)^2 + (x_3 - 3)^2 + (x_4 - 4)^2 + (x_5 - 5)^2 - (1^2 + 2^2 + 3^2 + 4^2 + 5^2) + 55 = 0,
$$

which becomes $(x_1 - 1)^2 + (x_2 - 2)^2 + (x_3 - 3)^2 + (x_4 - 4)^2 + (x_5 - 5)^2 = 0$.

A sum of real squares is 0 if and only if each of the squares is equal to 0, so the only possible solution to the system is $(x_1, x_2, x_3, x_4, x_5) = \boxed{(1, 2, 3, 4, 5)}$. Checking, we find that this solution does satisfy all five equations.

CHAPTER 5

Conics

Exercises for Section 5.1

5.1.1 *Solution 1: Use the geometric description of a parabola.* The vertex is 4 above the directrix, so the focus is 4 above the vertex, at $(1, 7)$. Therefore, if (x, y) is on the parabola, it is equidistant from $(1, 7)$ and $y = -1$ (and is above the line $y = -1$). This gives us

$$\sqrt{(x - 1)^2 + (y - 7)^2} = y - (-1).$$

Squaring both sides gives $(x - 1)^2 + (y - 7)^2 = (y + 1)^2$, so $(x - 1)^2 = (y + 1)^2 - (y - 7)^2 = 16y - 48$, from which we find $\boxed{y = \frac{1}{16}(x - 1)^2 + 3}$.

Solution 2: Use the standard form. If a parabola has a horizontal directrix with vertex (h, k) and focus $(h, k + \frac{1}{4a})$, then the equation of the directrix is $y = k - \frac{1}{4a}$. We are given that $h = 1$ and $k = 3$, so from the equation of the directrix we have $3 - \frac{1}{4a} = -1$, which gives $a = \frac{1}{16}$. Therefore, the equation of the parabola is $\boxed{y = \frac{1}{16}(x - 1)^2 + 3}$.

5.1.2

(a) The focus and vertex are on a horizontal line, so the directrix is vertical. The focus is 1 unit to the right of the vertex, so the directrix is 1 unit to the left of the vertex. Therefore, the directrix is $\boxed{x = 1}$.

(b) The line through the focus and vertex is the axis of symmetry of the parabola, so the slope of the axis of symmetry is $(7 - 5)/(3 - 2) = 2$. The axis is perpendicular to the directrix, so the directrix has slope $-\frac{1}{2}$.

   Let P be the point where the axis and directrix meet, and let the focus and vertex be F and V, respectively. Points P, F, and V are all on the axis of symmetry, and by the definition of a parabola, we must have $PV = VF$. Because V is 1 unit to the left of and 2 units below F, point P must be 1 unit to the left of and 2 units below V. This means point P is $(1, 3)$. (Alternatively, we could note that V is the midpoint of $\overline{PF}$, and used the formula for the midpoint of a segment to find P.) Therefore, the directrix is the line through $(1, 3)$ with slope $-\frac{1}{2}$, which means it is the graph of the equation $y - 3 = -\frac{1}{2}(x - 1)$. Rearranging this equation gives $\boxed{x + 2y - 7 = 0}$.

5.1.3

(a) The equation is already in standard form for a parabola. Because x is squared and a is positive, the parabola opens upward. The lowest point on the graph is the vertex, and it occurs when the squared term is 0, which is when $x = 3$. This gives $y = 2$, so the vertex is $\boxed{(3, 2)}$. The axis of symmetry is the vertical line through the vertex, which is $\boxed{x = 3}$.

(b)   Completing the square in y, we get

$$x = 5y^2 - 20y + 23 = 5(y^2 - 4y + 4) + 3 = 5(y-2)^2 + 3.$$

Therefore, the parabola opens rightward. The leftmost point on the graph is the vertex, and it occurs when the squared term is 0, which is when $y = 2$. This gives us $x = 3$, so the vertex is $(h, k) = \boxed{(3, 2)}$. The axis of symmetry is the horizontal line through the vertex, which is $\boxed{y = 2}$.

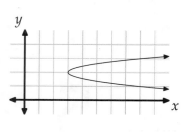

5.1.4

(a)   The directrix is vertical and the focus is to the left of the directrix. Any point (x, y) on the parabola is therefore also to the left of the directrix, and $4 - x$ units from the directrix. This distance must equal the distance between the point and the focus, so we have $4 - x = \sqrt{(x-2)^2 + (y-2)^2}$. Squaring both sides gives $(4-x)^2 = (x-2)^2 + (y-2)^2$. Solving for x gives $\boxed{x = -\frac{1}{4}(y-2)^2 + 3}$.

(b)   Because the axis is horizontal and the vertex is $(4, 6)$, the equation is $x = a(y-6)^2 + 4$ for some constant a. Because $(-2, 2)$ is on the parabola, we must have $-2 = a(2-6)^2 + 4$. Solving this equation gives $a = -3/8$, so our equation is $\boxed{x = -\frac{3}{8}(y-6)^2 + 4}$.

5.1.5   The horizontal line passing through the focus is given by $y = k + \frac{1}{4a}$, so to find the endpoints of the latus rectum, we solve the equation $k + \frac{1}{4a} = a(x-h)^2 + k$. Subtracting k from both sides, we get $a(x-h)^2 = \frac{1}{4a}$, so $(x-h)^2 = \frac{1}{4a^2}$. Hence, we have $x - h = \pm\frac{1}{2a}$, so $x = h \pm \frac{1}{2a}$. Therefore, the length of the latus rectum is $h + \frac{1}{2a} - (h - \frac{1}{2a}) = \frac{1}{a}$.

Exercises for Section 5.2

5.2.1   Let the equation of the parabola be $x = ay^2 + by + c$. Then from the given points, we obtain the system of equations

$$20 = 9a - 3b + c,$$
$$12 = a + b + c,$$
$$33 = 4a + 2b + c.$$

Subtracting the first equation from the second equation, and subtracting the second equation from the third, we get the system of equations $-8 = -8a + 4b$, $21 = 3a + b$. Multiplying the second equation by 4, we get $84 = 12a + 4b$. Subtracting the first equation, we get $92 = 20a$, so $a = 92/20 = 23/5$. Then $b = 21 - 3a = 36/5$, and $c = 12 - a - b = 1/5$. Therefore, $\boxed{x = \frac{23}{5}y^2 + \frac{36}{5}y + \frac{1}{5}}$.

5.2.2   Place the parabola in the Cartesian plane so that it passes through the points $(15, 0)$ and $(-15, 0)$ and opens downwards. The vertex of the parabola is at $(0, k)$, so the equation of the parabola is $y = ax^2 + k$.

Since the parabola passes through the point $(15, 0)$, we have $225a + k = 0$. We may place the fly at the point $(10, 25)$, so $100a + k = 25$. Subtracting this from $225a + k = 0$ gives $125a = -25$, so $a = -25/125 = -1/5$, and $k = -225a = 45$. Therefore, the center of the arch is $\boxed{45 \text{ meters}}$ high.

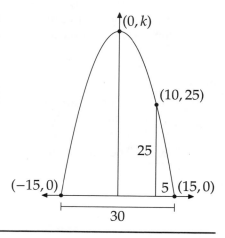

5.2.3 When the graph of $y = x^2 + 2x - 2$ is reflected over the line $y = x$, we obtain the graph of $x = y^2 + 2y - 2$. These graphs are shown at right.

The two graphs appear to intersect at four points, but to make the argument rigorous, we solve the equations algebraically.

The points of intersection satisfy both $y = x^2 + 2x - 2$ and $x = y^2 + 2y - 2$. Subtracting the first equation from the second, we get

$$x - y = y^2 + 2y - x^2 - 2x.$$

Moving all the terms to the left gives $x^2 - y^2 + 3x - 3y = 0$. Factoring the left side gives

$$x^2 - y^2 + 3x - 3y = (x - y)(x + y) + 3(x - y) = (x - y)(x + y + 3),$$

so we have $(x - y)(x + y + 3) = 0$. This means $x = y$ or $x + y + 3 = 0$.

If $x = y$, then the equation $y = x^2 + 2x - 2$ simplifies as $(x - 1)(x + 2) = 0$, so $x = 1$ or $x = -2$, which produces the two solutions $(x, y) = (1, 1)$ and $(-2, -2)$.

If $x + y + 3 = 0$, then $y = -x - 3$, and the equation $y = x^2 + 2x - 2$ simplifies as $x^2 + 3x + 1 = 0$. By the quadratic equation, we have $x = \frac{-3 \pm \sqrt{5}}{2}$, which produces the two solutions

$$(x, y) = \left(\frac{-3 + \sqrt{5}}{2}, \frac{-3 - \sqrt{5}}{2} \right) \quad \text{and} \quad \left(\frac{-3 - \sqrt{5}}{2}, \frac{-3 + \sqrt{5}}{2} \right).$$

Thus, the two graphs do intersect at $\boxed{\text{four}}$ points.

5.2.4 We seek the value of a such that the graphs of $y = x$ and $y = ax^2 + 6$ meet at exactly one point. In other words, we want the value of a such that there is only one pair (x, y) that satisfies both equations. Substituting $y = x$ from the linear equation into the quadratic gives $x = ax^2 + 6$, or $ax^2 - x + 6 = 0$. This equation has only one solution x if its discriminant equals 0. Therefore, we must have $(-1)^2 - 4a(6) = 0$, which gives us $a = \boxed{1/24}$.

Exercises for Section 5.3

5.3.1 Completing the square in $f(x)$, we get

$$f(x) = -2x^2 + 6x + 9 = -2(x^2 - 3x) + 9 = -2\left(x^2 - 3x + \frac{9}{4} \right) - (-2)\left(\frac{9}{4} \right) + 9 = -2\left(x - \frac{3}{2} \right)^2 + \frac{27}{2}.$$

The term $-2(x - 3/2)^2$ is always at most 0, and equal to 0 when $x = 3/2$. Hence, the maximum value of $f(x)$ is $\boxed{27/2}$.

5.3.2 Let the length of each additional interior fence be a, so the length of one side of the original rectangle is also a. Let the other dimension of the original rectangle be b. Then we must have $4a + 2b = 80$, so $b = (80 - 4a)/2 = 40 - 2a$, and the area of the big rectangle, in square yards, is $ab = a(40 - 2a) = -2a^2 + 40a$. Completing the square gives $-2a^2 + 40a = -2(a^2 - 20a) = -2(a^2 - 20a + 100) - (-2)(100) = -2(a - 10)^2 + 200$. Therefore, the maximum possible value is $\boxed{200}$, which occurs when $a = 10$.

5.3.3 Completing the square in x, we get

$$x^2 + px + q = x^2 + px + \frac{p^2}{4} + q - \frac{p^2}{4} = \left(x + \frac{p}{2} \right)^2 + q - \frac{p^2}{4}.$$

The term $(x + p/2)^2$ is always at least 0, and equal to 0 when $x = -p/2$. Hence, the minimum value of $x^2 + px + q$ is $q - p^2/4$. But we are given that this is equal to 0. Therefore, $q = \boxed{p^2/4}$.

5.3.4 Completing the square in both x and y, we find $x^2 - 4x + y^2 + 6y = (x - 2)^2 + (y + 3)^2 - 13$. Therefore, the coldest point in the plane is $\boxed{(2, -3)}$, and its temperature is $\boxed{-13}$.

5.3.5 Let l, w, and h denote the length, width, and height of the table, respectively. Then the top of the table consists of lw cubes, and the number of cubes in each leg (not counting the cube in top layer) is $h - 1$, so $lw + 4(h - 1) = 68$, which simplifies as $lw + 4h = 72$. The volume of the space between the table's top and the floor, excluding the legs, is
$$lw(h - 1) - 4(h - 1) = (lw - 4)(h - 1).$$
Since $lw = 72 - 4h$, this volume simplifies as
$$(lw - 4)(h - 1) = (68 - 4h)(h - 1) = 4(17 - h)(h - 1) = -4h^2 + 72h - 68.$$
Completing the square gives $-4h^2 + 72h - 68 = -4(h^2 - 18h) - 68 = -4(h^2 - 18h + 81) - (-4)(81) - 68 = -4(h - 9)^2 + 256$. Therefore, the maximum volume is $\boxed{256}$.

Exercises for Section 5.4

5.4.1 Completing the square in x and y, we get
$$4x^2 + 4y^2 + 12x + 16y + 9 = 4(x^2 + 3x) + 4(y^2 + 4y) + 9 = 4\left(x + \frac{3}{2}\right)^2 + 4(y + 2)^2 - 9 - 16 + 9,$$
so the original equation is equivalent to $4\left(x + \frac{3}{2}\right)^2 + 4(y + 2)^2 - 16 = 0$. Dividing by 4 and rearranging gives $\left(x + \frac{3}{2}\right)^2 + (y + 2)^2 = 4$. Therefore, the center of the circle is is $\boxed{(-3/2, -2)}$, and the radius is $\sqrt{4} = \boxed{2}$.

5.4.2 Since $(3, 2)$ and $(7, -4)$ are the endpoints of a diameter, the center of the circle is
$$\left(\frac{3 + 7}{2}, \frac{2 - 4}{2}\right) = (5, -1),$$
and the radius is $\sqrt{(5 - 3)^2 + (-1 - 2)^2} = \sqrt{13}$. Therefore, the equation of the circle is $\boxed{(x - 5)^2 + (y + 1)^2 = 13}$.

5.4.3 The endpoints of the common chord are the intersections of the two circles. Hence, we can find their coordinates by finding the ordered pairs (x, y) that satisfy both equations. Subtracting the second equation from the first, we get $6x - 2 = 4$, so $x = 1$. Substituting $x = 1$ in the equation $x^2 + y^2 = 4$, we get $y^2 = 3$, so $y = \pm\sqrt{3}$. Therefore, the length of the common chord is $\sqrt{3} - (-\sqrt{3}) = \boxed{2\sqrt{3}}$.

5.4.4 Place square $ABCD$ in the Cartesian plane, so that $A = (0, 0)$, $B = (4, 0)$, $C = (4, 4)$, and $D = (0, 4)$. Then $M = (2, 4)$. The equation of the circle with radius 2 and center M is given by
$$(x - 2)^2 + (y - 4)^2 = 4,$$
which expands as $x^2 + y^2 - 4x - 8y + 16 = 0$. The equation of the circle with radius 4 and center A is given by $x^2 + y^2 - 16 = 0$. Subtracting the first equation from the second gives $4x + 8y - 32 = 0$, so $y = (32 - 4x)/8 = (8 - x)/2$. Substituting this into the equation $x^2 + y^2 - 16 = 0$, the left side becomes

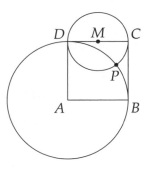

$$x^2 + \frac{(8 - x)^2}{4} - 16 = \frac{5x^2 - 16x}{4} = \frac{x(5x - 16)}{4},$$

so we have $x(5x - 16)/4 = 0$, which means $x = 0$ or $x = 16/5$. The solution $x = 0$ corresponds to the point D, so the solution $x = \boxed{16/5}$ corresponds to the point P. Since $\overline{AD}$ is on the y-axis, the x-coordinate of P is the desired distance from P to $\overline{AD}$.

Exercises for Section 5.5

5.5.1 The ellipse has center $(-2, 3)$, major axis (in the x direction) $10 = 2 \cdot 5$, and minor axis (in the y direction) $8 = 2 \cdot 4$, so the equation of the ellipse is

$$\boxed{\frac{(x + 2)^2}{5^2} + \frac{(y - 3)^2}{4^2} = 1}.$$

5.5.2 Completing the square in y, we get $x^2 + 4(y + 1)^2 = 1$, so we have $x^2 + \dfrac{(y + 1)^2}{(1/2)^2} = 1$.

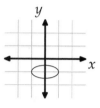

Both squared terms are 0 when $(x, y) = \boxed{(0, -1)}$, so this point is the center. The denominator of the x term, 1, is larger than the denominator of the y term, which is $1/4$. Therefore, the major axis is horizontal. Its endpoints are the points at which the x term equals 1 and the y term equals 0, so the denominator of the x term tells us that these endpoints are $\sqrt{1}$ to the left and right of the center. The major axis therefore has length $2\sqrt{1} = \boxed{2}$. Similarly, the minor axis is vertical, with length $2\sqrt{1/4} = \boxed{1}$. We find the foci by using the relationship

$$(\text{major axis length})^2 = (\text{minor axis length})^2 + (\text{distance between foci})^2,$$

which tells us that the foci are $\sqrt{3}$ apart. Since the major axis is horizontal, the foci are $\sqrt{3}/2$ to the right and left of the center, at $\boxed{(\sqrt{3}/2, -1) \text{ and } (-\sqrt{3}/2, -1)}$.

5.5.3 Completing the square in x and y, we get $(x + 2)^2 + 2(y - 3)^2 = 20$. Dividing both sides by 20, we get $\frac{(x+2)^2}{20} + \frac{(y-3)^2}{10} = 1$, so we have

$$\frac{(x + 2)^2}{(2\sqrt{5})^2} + \frac{(y - 3)^2}{(\sqrt{10})^2} = 1.$$

The squared terms are both 0 when $(x, y) = \boxed{(-2, 3)}$, so this point is the center. The denominator of the x term is larger than that of the y term, so the major axis is horizontal. Its endpoints are the points at which the x term equals 1 and the y term equals 0, so the denominator of the x term tells us that they are $2\sqrt{5}$ to the left and right of the center. The major axis therefore has length $2(2\sqrt{5}) = \boxed{4\sqrt{5}}$. From the denominator of the y term, we see that the endpoints of the minor axis are $\sqrt{10}$ above and below the center, so the minor axis has length $\boxed{2\sqrt{10}}$.

We find the foci by using the relationship

$$(\text{major axis length})^2 = (\text{minor axis length})^2 + (\text{distance between foci})^2,$$

which tells us that the foci are $\sqrt{10}$ from the center. Since the major axis is horizontal, the foci are to the left and right of the center, at $\boxed{(-2 - \sqrt{10}, 3) \text{ and } (-2 + \sqrt{10}, 3)}$.

5.5.4 Let $F_1 = (0, 0)$ and $F_2 = (14, 0)$. Completing the square in the equation for the parabola, we get $y = x^2 - 10x + 37 = (x - 5)^2 + 12$, so the vertex of the parabola is $V = (5, 12)$. To find the length of the major axis of the ellipse, we recall that for any point P on an ellipse with foci F_1 and F_2, the sum $PF_1 + PF_2$ equals the length of the major axis of the ellipse. Since V is on the ellipse with foci F_1 and F_2, the length of the major axis is $VF_1 + VF_2$. We have

$$VF_1 + VF_2 = \sqrt{5^2 + 12^2} + \sqrt{(14-5)^2 + 12^2} = \sqrt{169} + \sqrt{225} = 13 + 15 = 28,$$

so the major axis of the ellipse has length $\boxed{28}$.

5.5.5   We begin with a circle with radius b, which has area πb^2. Scaling the circle by a factor of a/b in the x direction produces an ellipse with axes $2a$ and $2b$.

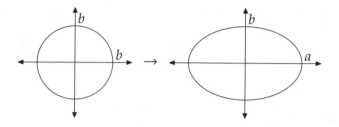

Since the area of the circle is πb^2, the area of the ellipse is $\pi b^2 \cdot (a/b) = ab\pi$, as desired.

Exercises for Section 5.6

5.6.1   Since $(5,0)$ is a vertex, we have $a = 5$. Since $(-13,0)$ is a focus, we have $c = 13 = \sqrt{a^2 + b^2}$. Therefore, $a^2 + b^2 = 169$, so $b^2 = 169 - a^2 = 144$, which implies $b = 12$. Therefore, the equation of the hyperbola is $\boxed{\dfrac{x^2}{5^2} - \dfrac{y^2}{12^2} = 1}$.

5.6.2

(a)   The hyperbola has center $(2,1)$. The vertices are at $(4,1)$ and $(0,1)$, so $a = 2$. Therefore, the equation of the hyperbola is of the form $\dfrac{(x-2)^2}{2^2} - \dfrac{(y-1)^2}{b^2} = 1$. Substituting $x = 6$ and $y = 7$, we get $\dfrac{16}{4} - \dfrac{36}{b^2} = 1$, so $b^2 = 12$, which implies $b = 2\sqrt{3}$. Therefore, the equation of the hyperbola is

$$\boxed{\dfrac{(x-2)^2}{2^2} - \dfrac{(y-1)^2}{(2\sqrt{3})^2} = 1}.$$

(b)   The graph appears to be a hyperbola with asymptotes $x = -1$ and $y = 4$. Assuming this is the case, the equation of the hyperbola has the form $(x+1)(y-4) = k$, for some nonzero constant k. Substituting $x = 0$ and $y = 6$, we get $1 \cdot 2 = k$, so $k = 2$. Therefore, the equation of the hyperbola is $\boxed{(x+1)(y-4) = 2}$.

5.6.3

(a)   The squared terms are 0 at $(x, y) = \boxed{(3,0)}$, so this point is the center of the hyperbola. The vertices are the points at which the negative term is 0, which occurs when $y = 0$. This gives us $(x-3)^2/9 = 1$, from which we have $x = 6$ and $x = 0$. Therefore, the vertices are $\boxed{(0,0) \text{ and } (6,0)}$. Finally, the asymptotes of the hyperbola are

$$\boxed{\dfrac{x-3}{3} = \pm y},$$

both of which make the left side of the original equation equal to 0. Using the asymptotes and vertices, we produce the graph at right above.

(b) Completing the square in x and y, we get $(x + 6)^2 - 3(y - 3)^2 = -12$. Dividing both sides by -12, we get $\dfrac{(y - 3)^2}{4} - \dfrac{(x + 6)^2}{12} = 1$, so

$$\frac{(y - 3)^2}{2^2} - \frac{(x + 6)^2}{(2\sqrt{3})^2} = 1.$$

The squared terms are 0 at $(x, y) = \boxed{(-6, 3)}$, so this point is the center of the hyperbola. The vertices are the points at which the negative term is 0, which occurs when $x = -6$. This gives us $(y - 3)^2/2^2 = 1$, from which we have $y = 5$ and $y = 1$. Therefore, the vertices are $\boxed{(-6, 5) \text{ and } (-6, 1)}$. Finally, the asymptotes of the hyperbola are

$$\boxed{\frac{y - 3}{2} = \pm\frac{x + 6}{2\sqrt{3}}},$$

both of which make the left side of the standard form equation above equal to 0. Using the asymptotes and vertices, we produce the graph at right above.

5.6.4 The asymptotes of the hyperbola are

$$\frac{x - h}{a} = \pm\frac{y - k}{b},$$

which have slopes b/a and $-b/a$. If the hyperbola is rectangular, then the asymptotes are perpendicular, meaning the product of their slopes is -1. Therefore, $-b^2/a^2 = -1$, or $a^2 = b^2$. Since a and b are positive, we have $a = b$, so $a/b = \boxed{1}$.

5.6.5 If a point on the line $y = ax$ has x-coordinate 0, then the y-coordinate is also 0. Substituting $x = y = 0$ into the equation for $\mathcal{H}$, we get $\frac{9}{a^2} - 0 = 1$, so $a^2 = 9$. Thus, the equation of $\mathcal{H}$ is

$$\frac{(x - 3)^2}{9} - \frac{y^2}{b^2} = 1.$$

Now we seek the intersections of $\mathcal{H}$ and the line $y = bx$. Substituting $y = bx$, we get $\frac{(x-3)^2}{9} - x^2 = 1$, which simplifies as $8x^2 + 6x = 0$, so $x = 0$ or $x = -3/4$. Therefore, the x-coordinates of the points of intersection of $\mathcal{H}$ and the line $y = bx$ are $\boxed{0 \text{ and } -3/4}$.

5.6.6 We could choose points (x, y) on the hyperbola and use a ton of algebra to find the foci. Or, we could be a little more crafty. First, we note that the hyperbola is the result of rotating the graph of some equation of the form $\frac{x^2}{a^2} - \frac{y^2}{b^2} = 1$ about the origin. Rotating a hyperbola won't change the angle between its asymptotes, so, because the graph of $xy = 1$ is a rectangular hyperbola, the graph of $\frac{x^2}{a^2} - \frac{y^2}{b^2} = 1$ must be a rectangular hyperbola. This means that we must have $a = b$.

The vertices of the graph of $xy = 1$ are $(1, 1)$ and $(-1, -1)$, so the distance between them is $2\sqrt{2}$. The distance between the vertices of the graph of $\frac{x^2}{a^2} - \frac{y^2}{b^2} = 1$ must therefore also be $2\sqrt{2}$. This means that $a = \sqrt{2}$. Since we have $a = b = \sqrt{2}$, we have $c = \sqrt{a^2 + b^2} = 2$, so the distance from the center of the hyperbola to a focus is 2. The center of the hyperbola is the origin, so the foci are on the line $x = y$, and are 2 units from the origin. Therefore, the foci are $\boxed{(\sqrt{2}, \sqrt{2}) \text{ and } (-\sqrt{2}, -\sqrt{2})}$.

Review Problems

5.35

(a) The coefficient of y^2 is 0 and the coefficient of x^2 is nonzero, so the graph is a $\boxed{\text{parabola}}$.

(b) If we divide both sides by 9, then the resulting equation is in the standard form of an $\boxed{\text{ellipse}}$.

(c) Completing the square in y, we get $x^2 + (y+9)^2 = 0$. The graph is a $\boxed{\text{point}}$, namely the point $(0,-9)$.

(d) The coefficients of both x^2 and y^2 are both positive, so we think the graph is an $\boxed{\text{ellipse}}$. To confirm that the graph is not empty and not just a point, we complete the square to find $(x+1)^2 + 4y^2 = 1$, so $(x+1)^2 + y^2/(1/4) = 1$, which is the standard form of an ellipse.

(e) The coefficients of both x^2 and y^2 are both positive, and are equal to each other, so the graph is a $\boxed{\text{circle}}$. To confirm the graph is not empty and not just a point, we divide both sides by 6 and rearrange to get $x^2 + 2x + y^2 - 3y = 0$. Completing the square gives $(x+1)^2 + (y - \frac{3}{2})^2 = \frac{13}{4}$, so the graph is indeed a circle.

(f) Rearranging, we get $x(y-2) = 4$. The graph is a $\boxed{\text{hyperbola}}$.

(g) Adding 1 to both sides, we get $(x+1)^2 = (y+2)^2$, so $x+1 = \pm(y+2)$. The graph is $\boxed{\text{two lines}}$.

(h) Rearranging the given equation, we get $x(x+1) = y$. The coefficient of y^2 is 0, so the graph is a $\boxed{\text{parabola except for the point } (-1,0)}$.

(i) Completing the square in y gives $x^2 - 4(y-1)^2 = -4$, so we have $\frac{(y-1)^2}{4} - x^2 = 1$. This equation is in the standard form of a $\boxed{\text{hyperbola}}$.

5.36 The equation has y equal to a quadratic in x, so its graph is a parabola. Completing the square, we get

$$y = 4x^2 - 12x + 8 = 4\left(x - \frac{3}{2}\right)^2 - 1.$$

The vertex is where the squared term is 0, which gives us a vertex of $\boxed{(3/2, -1)}$. Since the parabola opens upward (because x is squared and the coefficient of the squared term in standard form is positive), the axis of symmetry is the vertical line through the vertex, which is $\boxed{x = 3/2}$.

5.37 The equation is in the standard form of a horizontally-opening parabola. Since the coefficient of the squared term is positive, the parabola opens to the right. The squared term is 0 for $y = 4$, which gives us a vertex of $(x, y) = \boxed{(-1, 4)}$. The axis of symmetry is the horizontal line through the vertex, which is $\boxed{y = 4}$. The graph is at right.

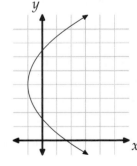

5.38 Completing the square, we get $x = 4y^2 + 6y + c = 4\left(y + \frac{3}{4}\right)^2 + c - \frac{9}{4}$. Thus, the vertex is $(c - 9/4, -3/4)$. If this point lies on the y-axis, then $c = \boxed{9/4}$.

5.39 Since the points $(1,5)$ and $(1,3)$ have the same x-coordinate, the parabola has a vertical axis, and thus a horizontal directrix. The focus is 2 below the vertex, so the directrix is 2 above the vertex, at $y = 7$. Therefore, if (x, y) is on the parabola, it is equidistant from $(1,3)$ and $y = 7$ (and (x, y) is below the line $y = 7$). This gives us $\sqrt{(x-1)^2 + (y-3)^2} = 7 - y$. Squaring both sides gives $(x-1)^2 + (y-3)^2 = (7-y)^2$, so $(x-1)^2 = (7-y)^2 - (y-3)^2 = (10 - 2y)(4) = 40 - 8y$. Solving for y gives $\boxed{y = -\frac{1}{8}(x-1)^2 + 5}$.

We could also have used the formulas in the text. The vertex gives us $h = 1$ and $k = 5$, and the focus gives us $k + \frac{1}{4a} = 3$. Solving for a, we find $a = -\frac{1}{8}$. Therefore, the equation of the parabola is $\boxed{y = -\frac{1}{8}(x-1)^2 + 5}$.

5.40

(a) The graph appears to be an upward-opening parabola with vertex $(2, 1)$, so the equation has the form $y = a(x-2)^2 + 1$. The graph passes through the point $(0, 3)$. Substituting $x = 0$ and $y = 3$, we get $3 = 4a + 1$, so $a = 1/2$. Therefore, the equation of the parabola is $\boxed{y = \frac{1}{2}(x-2)^2 + 1}$.

(b) The graph appears to be an ellipse with center $(3, 2)$. The major axis (in the y direction) has length $8 = 2 \cdot 4$, and the minor axis (in the x direction) has length $6 = 2 \cdot 3$, so the equation of the ellipse is

$$\boxed{\frac{(x-3)^2}{3^2} + \frac{(y-2)^2}{4^2} = 1}.$$

(c) The graph appears to be a circle with center $(-3, 2)$, so the equation has the form

$$(x+3)^2 + (y-2)^2 = r^2.$$

The graph passes through the origin. Substituting $x = 0$ and $y = 0$, we get $13 = r^2$. Therefore, the equation of the circle is $\boxed{(x+3)^2 + (y-2)^2 = 13}$.

(d) The graph appears to be a hyperbola with center $(-2, -1)$. The vertices are $(-2, 1)$ and $(-2, -3)$, so $a = 2$. The hyperbola opens upward and downward, so the equation has the form

$$\frac{(y+1)^2}{2^2} - \frac{(x+2)^2}{b^2} = 1.$$

Because the hyperbola passes through $(4, 3)$, we must have $\frac{(3+1)^2}{2^2} - \frac{(4+2)^2}{b^2} = 1$. Solving this equation for b gives $b = 2\sqrt{3}$, so the equation of our hyperbola is

$$\boxed{\frac{(y+1)^2}{2^2} - \frac{(x+2)^2}{(2\sqrt{3})^2} = 1}.$$

5.41 The vertex of the parabola $y = x^2 + a^2$ is $(0, a^2)$, so if this point lies on the line $y = x + a$, then $a^2 = a$. Rearranging and factoring this equation gives $a(a-1) = 0$, so $a = 0$ or $a = 1$. Thus, there are only $\boxed{2}$ possible values of a.

5.42

(a) Completing the square in y, we get $p(y) = 2y^2 - 4y + 19 = 2(y^2 - 2y) + 19 = 2(y^2 - 2y + 1) + 17 = 2(y-1)^2 + 17$. Therefore, the minimum value of $p(y)$ is $\boxed{17}$.

(b) Completing the square in r, we get $37 - 16r - r^2 = -(r^2 + 16r) + 37 = -(r+8)^2 + 101$. Therefore, the maximum value of $37 - 16r - r^2$ is $\boxed{101}$.

(c) Let $y = x^2$. Then $x^4 - 3x^2 - 13 = y^2 - 3y - 13$. Completing the square in y, we get

$$y^2 - 3y - 13 = \left(y - \frac{3}{2}\right)^2 - \frac{61}{4}.$$

Since $y = x^2$, the only restriction on y is that $y \geq 0$, so y may be equal to $3/2$. Therefore, the minimum value of $f(x)$ is $\boxed{-61/4}$.

(d) Completing the square in both x and y, we get

$$x^2 + 4y^2 - 6x + 4y + 5 = (x-3)^2 + 4\left(y + \frac{1}{2}\right)^2 - 5.$$

Hence, the minimum value is $\boxed{-5}$.

5.43 Suppose that GetThere Airlines increases their ticket price to $200 + 10n = 10(20 + n)$, in dollars. Then the number of tickets they sell is $40000 - 1000n = 1000(40 - n)$. Therefore, their total revenue is

$$10(20 + n) \cdot 1000(40 - n) = 10000(20 + n)(40 - n) = 10000(800 + 20n - n^2).$$

This is maximized when $n = -\frac{20}{2 \cdot (-1)} = 10$. Therefore, they should charge $200 + 10 \cdot 10 = \boxed{\$300}$ per ticket.

5.44 Let the length and width of the rectangle be x and y, respectively. Then the perimeter of the rectangle is $2x + 2y = 36$, so $x + y = 18$. Also, the length of the diagonal is $\sqrt{x^2 + y^2}$. Minimizing $\sqrt{x^2 + y^2}$ is equivalent to minimizing $x^2 + y^2$.

Substituting $y = 18 - x$, and completing the square, we get

$$x^2 + y^2 = x^2 + (18 - x)^2 = 2x^2 - 36x + 324 = 2(x - 9)^2 + 162,$$

so the minimum value of $x^2 + y^2$ is 162. Therefore, the minimum value of the length of the diagonal is $\sqrt{162} = \boxed{9\sqrt{2}}$.

5.45 Completing the square in both x and y, we get $(x - 8)^2 + (y - 16)^2 = 320$. Thus, the equation represents a circle with radius r, such that $r^2 = 320$. Therefore, the area of the circle is $\pi r^2 = \boxed{320\pi}$.

5.46 The equation $x^2 + y^2 = r$ represents a circle centered at the origin with radius $\sqrt{r}$. If the line $x + y = r$ intersects the circle at a point (a, b), then we have $a^2 + b^2 = r$ and $a + b = r$, so the point (b, a) is also on both the circle and the line. Therefore, the only way the line can only intersect the circle at exactly one point is if $a = b$, so the point of tangency must be $(r/2, r/2)$. (We can also see this from the fact that the radius to the point of tangency must be perpendicular to the tangent. The tangent has slope -1, so the radius to this point has slope 1).

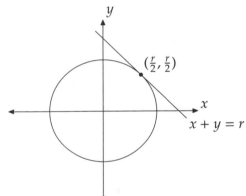

Since $(r/2, r/2)$ is on the circle, we have $\frac{r^2}{4} + \frac{r^2}{4} = r$, so $r^2 = 2r$. Dividing both sides by r (which must be nonzero), we find $r = \boxed{2}$.

5.47 The equation of the circle is of the form $(x - a)^2 + (y - b)^2 = r^2$. Substituting the values for the three vertices, we obtain the system of equations

$$(-2 - a)^2 + (5 - b)^2 = r^2,$$
$$(-4 - a)^2 + (-3 - b)^2 = r^2,$$
$$a^2 + (-3 - b)^2 = r^2.$$

These equations expand as

$$a^2 + 4a + b^2 - 10b + 29 = r^2,$$
$$a^2 + 8a + b^2 + 6b + 25 = r^2,$$
$$a^2 + b^2 + 6b + 9 = r^2.$$

Subtracting the third equation from the second equation, we get $8a + 16 = 0$, so $a = -16/8 = -2$. Subtracting the third equation from the first equation, we get $4a - 16b + 20 = 0$, so $b = (4a + 20)/16 = 12/16 = 3/4$. Finally, substituting into the third equation, we find

$$r^2 = (-2)^2 + \left(\frac{3}{4}\right)^2 + 6 \cdot \frac{3}{4} + 9 = \frac{289}{16}.$$

Therefore, the equation of the circumcircle is $\boxed{(x + 2)^2 + \left(y - \frac{3}{4}\right)^2 = \frac{289}{16}}$.

(Note: There are several other ways to do this problem. We could have found the perpendicular bisectors of two sides of the triangle. The center of the desired circle is the intersection of the perpendicular bisectors of the sides. We also could have noted that the triangle is isosceles to simplify our work somewhat.)

5.48 The major axis connects $(3, 6)$ and $(3, -2)$, so the center of the ellipse is $(3, 2)$, which means the equation in standard form is $\frac{(x-3)^2}{a^2} + \frac{(y-2)^2}{b^2} = 1$. The major axis is vertical, and its length is 8, so the denominator of the y term is $(8/2)^2 = 4^2$. The minor axis has length 6 and is horizontal, so the denominator of the x term is $(6/2)^2 = 3^2$. Therefore, desired equation is

$$\boxed{\frac{(x-3)^2}{3^2} + \frac{(y-2)^2}{4^2} = 1}.$$

5.49

(a) The center is where both squared terms are 0, which is $\boxed{(3,0)}$. The denominator of the x term is squared, so the major axis is horizontal. We find the endpoints of the axes by investigating where the squared terms equal 1. From $(x-3)^2/9 = 1$, we have $x - 3 = \pm 3$, which tells us that the endpoints of the major axis are 3 to the right and left of the center. So, the major axis has length $\boxed{6}$. Similarly, the length of the minor axis is $2\sqrt{1} = \boxed{2}$. Using the lengths of the axes, along with the center, we can produce the graph at right. From the relationship

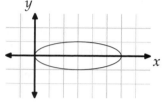

$$(\text{major axis length})^2 = (\text{minor axis length})^2 + (\text{distance between foci})^2,$$

we find that the foci are $4\sqrt{2}$ apart. Since the major axis is horizontal, they are $2\sqrt{2}$ to the left and right of the center, at $\boxed{(3 + 2\sqrt{2}, 0) \text{ and } (3 - 2\sqrt{2}, 0)}$.

(b) Completing the square in both x and y, we get $4(x+2)^2 + (y-3)^2 = 36$. Dividing both sides by 36, we get $\frac{(x+2)^2}{9} + \frac{(y-3)^2}{36} = 1$, so $\frac{(x+2)^2}{3^2} + \frac{(y-3)^2}{6^2} = 1$.

In this form, we see that the major axis is vertical because the denominator of the y term is larger than that of the x term. These denominators also tell us that the major axis has length $2 \cdot 6 = \boxed{12}$ and the minor axis has length $2 \cdot 3 = \boxed{6}$. The center is $\boxed{(-2, 3)}$. The relationship

$$(\text{major axis length})^2 = (\text{minor axis length})^2 + (\text{distance between foci})^2$$

tells us that the foci are $6\sqrt{3}$ apart. The major axis is vertical, so the foci are $3\sqrt{3}$ above and below the center, at $\boxed{(-2, 3 + 3\sqrt{3}) \text{ and } (-2, 3 - 3\sqrt{3})}$.

5.50 Let $F_1 = (4, 1)$. As a focus, F_1 lies on the major axis. Since F_1 has no coordinates in common with the point $(6, 5)$, the point $(6, 5)$ must be an endpoint of the minor axis, and the equation of this axis must be $x = 6$. Furthermore, the other focus F_2 is the reflection of F_1 over the minor axis, so F_2 must be at the point $(8, 1)$.

It then follows that the center of the ellipse is $(6, 1)$, and the length of the minor axis is $2 \cdot 4$. The distance between F_1 and $(6, 5)$ is

$$\sqrt{(4-6)^2 + (1-5)^2} = \sqrt{20} = 2\sqrt{5},$$

so the length of the major axis is $2 \cdot 2\sqrt{5}$. Therefore, the equation of the ellipse in this case is

$$\boxed{\frac{(x-6)^2}{(2\sqrt{5})^2} + \frac{(y-1)^2}{4^2} = 1}.$$

5.51

(a) We can rewrite the equation as

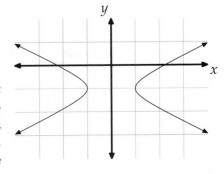

$$x^2 - \frac{(y+1)^2}{(1/2)^2} = 1,$$

which is the equation of a hyperbola. The squared terms are 0 at $(x,y) = \boxed{(0,-1)}$, so this point is the center of the hyperbola. The vertices are the points at which the negative term is 0, which occurs when $y = -1$. This gives us $x^2 = 1$, from which we have $x = 1$ and $x = -1$. Therefore, the vertices are $\boxed{(1,-1) \text{ and } (-1,-1)}$. Finally, the asymptotes of the hyperbola are $\boxed{x = \pm 2(y+1)}$, both of which make the left side of the standard form equation equal to 0. Using the asymptotes and vertices, we produce the graph at right above.

(b) Completing the square in both x and y, we get $2(x+2)^2 - (y-3)^2 = 8$. Dividing both sides by 8, we get $\frac{(x+2)^2}{4} - \frac{(y-3)^2}{8} = 1$, so

$$\frac{(x+2)^2}{2^2} - \frac{(y-3)^2}{(2\sqrt{2})^2} = 1,$$

which is the equation of a hyperbola. The squared terms are 0 at $(x,y) = \boxed{(-2,3)}$, so this point is the center of the hyperbola. The vertices are the points at which the negative term is 0, which occurs when $y = 3$. This gives us $(x+2)^2/2^2 = 1$, from which we have $x = 0$ and $x = -4$. Therefore, the vertices are $\boxed{(0,3) \text{ and } (-4,3)}$. Finally, the asymptotes of the hyperbola are

$$\boxed{\frac{x+2}{2} = \pm\frac{y-3}{2\sqrt{2}}},$$

both of which make the left side of the standard form equation equal to 0. Using the asymptotes and vertices, we produce the graph at right above.

5.52 Let $c = \sqrt{a^2 + b^2}$. One vertex of the hyperbola is $(h, k+a)$, and the closest focus is $(h, k+c)$, so the distance between these two points is $c - a$. The other vertex is at $(h, k-a)$, so the distance from the focus to this vertex is $a + c$. Therefore, the total distance is $(c-a) + (a+c) = 2c = \boxed{2\sqrt{a^2+b^2}}$.

5.53 The graph of (a) is a leftward-opening parabola that is entirely to the left of the y-axis, so it won't get anywhere near the target point. The graph of (c) is a circle with center $(-2, 0)$ and a small radius. It won't get anywhere near the target point either. The graph of (d) is an ellipse with center $(0, 2)$ and short axes, so this graph won't be close to the target point.

This leaves $\boxed{\text{(b)}}$ by process of elimination. The graph of (b) is a hyperbola, so we can't immediately eliminate it like we did the others. We can check that we are correct by first completing the square in x and y to get $16(x-3)^2 - 9(y+3)^2 = 144$. When $x = 265$, we get $9(y+3)^2 = 1098160$, which has solutions $y = 346.31043\ldots$ and $y = -352.31043\ldots$. So, the graph does indeed pass within 1 unit of $(265, 346)$.

Challenge Problems

5.54 Let $A = (a, 4a^2 + 7a - 1)$ and $B = (b, 4b^2 + 7b - 1)$. Then the midpoint of $\overline{AB}$ is

$$\left(\frac{a+b}{2}, \frac{4a^2 + 7a + 4b^2 + 7b - 2}{2}\right).$$

Since this midpoint is also the origin, we obtain the system of equations

$$a + b = 0,$$
$$4a^2 + 7a + 4b^2 + 7b - 2 = 0.$$

From the first equation, we have $b = -a$. Substituting into the second equation, we get $8a^2 = 2$, so $a^2 = 1/4$, which means $a = 1/2$ or $a = -1/2$.

These two solutions correspond to the two points A and B, so the points, in some order. Letting $x = 1/2$ in $y = 4x^2 + 7x - 1$ gives $y = 7/2$, and letting $x = -1/2$ gives $y = -7/2$. Therefore, the points A and B are $(1/2, 7/2)$ and $(-1/2, -7/2)$. The distance between these two points is

$$\sqrt{\left(\frac{1}{2} + \frac{1}{2}\right)^2 + \left(\frac{7}{2} + \frac{7}{2}\right)^2} = \sqrt{1^2 + 7^2} = \sqrt{50} = \boxed{5\sqrt{2}}.$$

5.55 Suppose that $a \geq b$. (The case $b > a$ is essentially the same.) Then one of the foci of the ellipse is $(c, 0)$, where $c = \sqrt{a^2 - b^2}$. The equation of the line passing through this focus and parallel to the minor axis is $x = c$. To find the intersection points of this line with the ellipse, we substitute $x = c$ into $1 = \frac{x^2}{a^2} + \frac{y^2}{b^2}$ to get

$$1 = \frac{c^2}{a^2} + \frac{y^2}{b^2} = \frac{a^2 - b^2}{a^2} + \frac{y^2}{b^2} = 1 - \frac{b^2}{a^2} + \frac{y^2}{b^2},$$

so

$$\frac{y^2}{b^2} = \frac{b^2}{a^2} \quad \Rightarrow \quad y^2 = \frac{b^4}{a^2}.$$

Therefore, $y = \pm b^2/a$. Hence, the length of the latus rectum is $\boxed{2b^2/a}$.

5.56 The equation of the original parabola may be expressed in the form $y = a(x - h)^2 + k$. Then the equation of its reflection in the x-axis is $y = -a(x - h)^2 - k$.

One of these parabolas is translated 5 units to the right, and the other is translated 5 units to the left. If the original parabola is translated 5 units to the right, and the reflected parabola is translated 5 units to the left, then $f(x) = a(x - h - 5)^2 + k$ and $g(x) = -a(x - h + 5)^2 - k$, so

$$(f + g)(x) = f(x) + g(x) = a(x - h - 5)^2 + k - a(x - h + 5)^2 - k = -20ax + 20ah,$$

which is the equation of a non-horizontal line. If the translations occur the other way, then $f(x) = a(x - h + 5)^2 + k$ and $g(x) = -a(x - h - 5)^2 - k$, so

$$(f + g)(x) = f(x) + g(x) = a(x - h + 5)^2 + k - a(x - h - 5)^2 - k = 20ax - 20ah,$$

which is again a non-horizontal line. In either case, the graph of $(f + g)(x)$ is a non-horizontal line, so the answer is $\boxed{(D)}$.

5.57 We can simplify the expression as

$$\frac{3x^2 + 9x + 17}{3x^2 + 9x + 7} = 1 + \frac{10}{3x^2 + 9x + 7}.$$

Thus, finding the maximum value of the given expression is equivalent to maximizing $1/(3x^2 + 9x + 7)$. If the denominator is always positive, then maximizing the fraction is equivalent finding the minimum value of $3x^2 + 9x + 7$. Completing the square gives $3x^2 + 9x + 7 = 3\left(x + \frac{3}{2}\right)^2 + \frac{1}{4}$, so $3x^2 + 9x + 7$ is indeed strictly positive. Furthermore, because the minimum of $3x^2 + 9x + 7$ is $1/4$, the maximum of the original expression is $1 + 10/(1/4) = \boxed{41}$.

5.58 From $f(x) + f(y) \leq 0$, we have $x^2 + 6x + y^2 + 6y + 2 \leq 0$. Completing the square in x and y, we get $(x+3)^2 + (y+3)^2 \leq 16$, so the set of the points that satisfy $f(x) + f(y) \leq 0$ is the disc with center $(-3, -3)$ and radius 4. (A disc consists of a circle and the interior of the circle.)

From $f(x) - f(y) \leq 0$, we have $x^2 + 6x - y^2 - 6y \leq 0$. Completing the square in x and y gives $(x+3)^2 - (y+3)^2 \leq 0$. Factoring the left side as a difference of squares gives $(x-y)(x+y+6) \leq 0$. Hence, either $x - y \leq 0$ and $x + y + 6 \geq 0$, or $x - y \geq 0$ and $x + y + 6 \leq 0$. The set of points (x, y) that satisfy $x - y \leq 0$ and $x + y + 6 \geq 0$ is the quadrant that is "above" the lines $y = x$ and $y = -x - 6$, and the set of points (x, y) that satisfy $x - y \geq 0$ and $x + y + 6 \leq 0$ is the quadrant that is "below" the lines $y = x$ and $y = -x - 6$.

We graph the circle and the quadrants below, with their intersection R shaded.

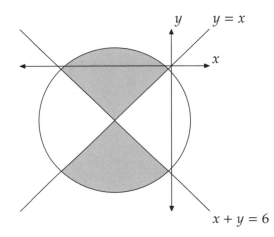

Thus, we see that R consists of two quarter-circles because the lines are perpendicular and intersect at $(-3, -3)$, which is the center of the circle. Hence, the area of R is $\frac{2}{4} \cdot \pi \cdot 4^2 = \boxed{8\pi}$.

5.59 Let $F = (3, -28)$, and let the two parabolas intersect at P and Q. Let P_x and P_y be the feet of the altitudes from P to the x-axis and y-axis, respectively.

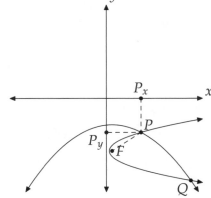

Since P lies on a parabola with focus F and the x-axis as directrix, we have $PF = PP_x$. But P also lies on a parabola with focus F and the y-axis as directrix, so $PF = PP_y$. Therefore, $PP_x = PP_y$, so the magnitudes of the coordinates of P are equal. Furthermore, the x-coordinate of P is positive and the y-coordinates of P is negative, so P lies on the line $y = -x$. By the same argument, Q also lies on the line $y = -x$. Therefore, the slope of $\overline{PQ}$ is $\boxed{-1}$.

5.60 The given equation can be rewritten as $x^2 + y^2 - x - y = 0$. Completing the square in both x and y, we get

$$\left(x - \frac{1}{2}\right)^2 + \left(y - \frac{1}{2}\right)^2 = \frac{1}{2}.$$

Hence, the point (x, y) lies on the circle with center $(1/2, 1/2)$ and radius $1/\sqrt{2}$. Therefore, the maximum value of

x is $1/2 + 1/\sqrt{2} = \boxed{(1+\sqrt{2})/2}$.

5.61 Place the circle in the Cartesian plane so that its equation is $x^2 + y^2 = r^2$, and let $A = (-r, 0)$ and $B = (r, 0)$. Let $P = (p, 0)$. We therefore have $PA^2 + PB^2 = (p + r)^2 + (p - r)^2 = 2p^2 + 2r^2$. Since the chord $\overline{CD}$ is parallel to $\overline{AB}$, the y-coordinates of C and D are equal, and the x-coordinates are opposites. Let $C = (c, d)$ and $D = (-c, d)$, so $c^2 + d^2 = r^2$.

Then

$$PC^2 + PD^2 = (p - c)^2 + d^2 + (p + c)^2 + d^2 = p^2 - 2pc + c^2 + d^2 + p^2 + 2pc + c^2 + d^2 = 2p^2 + 2c^2 + 2d^2 = 2p^2 + 2r^2,$$

which equals our earlier expression for $PA^2 + PB^2$.

5.62 Since $(2, 3)$ is the vertex of the parabola, its equation is given by $y = a(x - 2)^2 + 3$. The parabola has two x-intercepts, so $a < 0$. (If $a > 0$, then $y = a(x - 2)^2 + 3 \geq 3 > 0$ for all x.) The equation $a(x - 2)^2 + 3 = 0$ becomes $(x - 2)^2 = -3/a$, whose roots are $2 \pm \sqrt{-3/a}$.

The root $2 + \sqrt{-3/a}$ is always positive, and the root $2 - \sqrt{-3/a}$ is positive if and only if $\sqrt{-3/a} < 2$. Squaring both sides, we get $-3/a < 4$, which is equivalent to $a < -3/4$. Since $a = k/2000$, we have $k = 2000a < -1500$. Hence, the greatest possible integer value of k is $\boxed{-1501}$.

5.63 Let $P = (0, 1)$, and let the other two points be Q and R. By symmetry, $Q = (-u, v)$ and $R = (u, v)$ for some real numbers u and v.

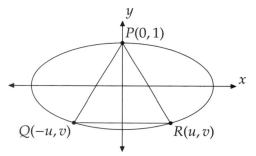

Then $QR = 2u$ and $PR^2 = u^2 + (v - 1)^2$. The triangle is equilateral, so we have $PR^2 = QR^2 = 4u^2$, which means we have $u^2 + (v - 1)^2 = 4u^2$. Solving for u^2 gives $u^2 = (v - 1)^2/3$. Also, since R lies on the ellipse, we have $u^2 + 4v^2 = 4$. Substituting $u^2 = (v - 1)^2/3$ and simplifying, we get $13v^2 - 2v - 11 = 0$. Factoring gives $(v - 1)(13v + 11) = 0$, so $v = 1$ or $v = -11/13$. Since $v < 1$, we have $v = -11/13$. Then

$$u^2 = 4 - 4v^2 = \frac{192}{169},$$

so the side of the triangle is $2u = 2\sqrt{\dfrac{192}{169}} = \boxed{\dfrac{16\sqrt{3}}{13}}$.

5.64 Completing the square in both x and y, we get $(x - 7)^2 + (y - 3)^2 = 64$. This is the equation of the circle with center $O = (7, 3)$ and radius 8.

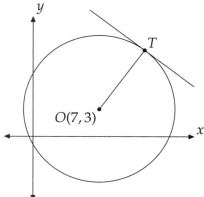

Given a constant c, the equation $3x + 4y = c$ implies $y = -3x/4 + c/4$, which is the equation of a line. As c varies, the position of the line varies. Specifically, when $c > 0$, increasing c moves the line away from the origin. However, varying c does not change the slope of the line. So, increasing c for $c > 0$ is like sliding the line away from the origin. The farthest we can make the line from the origin such that it still intersects the circle occurs when the line is tangent to the circle. Therefore, $3x + 4y = c$ is maximized when the line is tangent to the circle. Let the point of tangency be T.

Then radius $\overline{OT}$ is perpendicular to the line, so it has slope $4/3$. If point T is (x, y), we must therefore have $(x - 7)^2 + (y - 3)^2 = 64$ because T is on the circle, and $(y - 3)/(x - 7) = 4/3$ because $\overline{OT}$ has slope $4/3$. We therefore have $y - 3 = (4/3)(x - 7)$. Substituting this into the equation for the circle gives $(x - 7)^2 + \frac{16}{9}(x - 7)^2 = 64$, so $(x - 7)^2 = 64(9/25)$, which gives $x - 7 = \pm 24/5$. We therefore have $x = 59/5$ or $x = 11/5$. Since $y - 3 = (4/3)(x - 7)$ and we wish to maximize $3x + 4y$, we take the larger solution for x. This gives us $47/5$ for y, and $3x + 4y = \boxed{73}$.

5.65 Place the diagram in the Cartesian plane, so that $A = (-2, 0)$ and $B = (2, 0)$, which means the origin O is the midpoint of $\overline{AB}$. Note that triangle ABC is equilateral with side length 4. Therefore, triangle CAO is a $30°$-$60°$-$90°$ triangle. Hence, $OC/OA = \sqrt{3}$, so $OC = \sqrt{3} \cdot OA = 2\sqrt{3}$. Also, by symmetry, C lies on the y-axis, so place C at $(0, 2\sqrt{3})$.

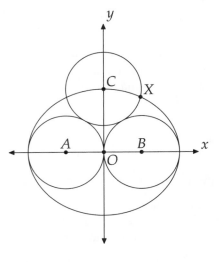

The locus of points X such that $AX + XB = AC + CB = 8$ is an ellipse with foci A and B. The minor axis has length $2OC = 4\sqrt{3}$, and the distance between the foci is $AB = 4$, so the major axis has length $\sqrt{(4\sqrt{3})^2 + 4^2} = 8$. Therefore, the equation of the ellipse is

$$\frac{x^2}{4^2} + \frac{y^2}{(2\sqrt{3})^2} = 1.$$

The equation of the circle with center C is $x^2 + (y - 2\sqrt{3})^2 = 4$, so $x^2 = 4 - (y - 2\sqrt{3})^2$. To find the coordinates of X, we substitute this into the equation for the ellipse, and we get

$$\frac{4 - (y - 2\sqrt{3})^2}{16} + \frac{y^2}{12} = 1,$$

which simplifies as $y^2 + 12\sqrt{3}y - 72 = 0$. By the quadratic formula, we have

$$y = \frac{-12\sqrt{3} \pm \sqrt{432 + 4 \cdot 72}}{2} = \frac{-12\sqrt{3} \pm \sqrt{720}}{2} = \frac{-12\sqrt{3} \pm 12\sqrt{5}}{2} = -6\sqrt{3} \pm 6\sqrt{5}.$$

Since X lies on the circle with center C, y must be positive, so $y = 6\sqrt{5} - 6\sqrt{3}$.

Since y is also the height of triangle AXB with respect to base $\overline{AB}$, the area of triangle AXB is

$$\frac{1}{2} \cdot 4 \cdot (6\sqrt{5} - 6\sqrt{3}) = \boxed{12\sqrt{5} - 12\sqrt{3}}.$$

5.66 The graph of f^{-1} is the reflection of the graph of f in the line $y = x$. Let P be on the graph of f and Q on the graph of f^{-1}. We claim that if PQ is minimized, then Q is the reflection of P in the line $y = x$. We prove this rigorously as follows.

Suppose that Q is not the reflection of P over the line $y = x$. Let P' be the reflection of P, and let Q' be the reflection of Q, so that P' is distinct from Q, and Q' is distinct from P. Also, P' lies on the graph of f^{-1}, and Q' lies on the graph of f. Let $\overline{PQ}$ and $\overline{P'Q'}$ intersect at X, so by symmetry, X lies on the line $y = x$.

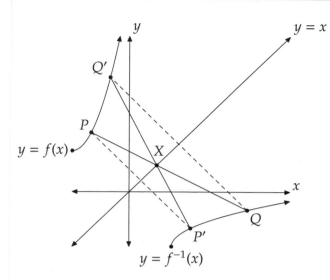

By the Triangle Inequality, we have $PP' < PX + XP'$ and $QQ' < QX + XQ'$, so $PP' + QQ' < PX + XQ + P'X + XQ' = PQ + P'Q'$. By symmetry, we have $P'Q' = PQ$, so $PP' + QQ' < 2PQ$. In other words, either PP' or QQ' (or both) is less than PQ, which means that PQ cannot be the shortest distance between a point on the graph of f and a point on the graph of f^{-1} if Q is not the reflection of P over the line $y = x$.

Therefore, when PQ is minimized, then Q is the reflection of P over the line $y = x$. In other words, $P = (a, f(a))$ and $Q = (f(a), a)$ for some real number a. Then

$$PQ = \sqrt{[a - f(a)]^2 + [f(a) - a]^2} = \sqrt{2[f(a) - a]^2} = \sqrt{2}|f(a) - a| = \sqrt{2}\left|(a + 3)^2 + \frac{9}{4} - a\right| = \sqrt{2}\left|a^2 + 5a + \frac{45}{4}\right|.$$

Completing the square gives $a^2 + 5a + \frac{45}{4} = \left(a + \frac{5}{2}\right)^2 + 5$, so the minimum value of PQ is $\boxed{5\sqrt{2}}$.

5.67 Completing the square in both x and y, the given equation becomes

$$\left[4\left(x + \frac{3}{4}\right)^2 + \frac{7}{4}\right]\left[4\left(y - \frac{3}{2}\right)^2 + 16\right] = 28.$$

However, we have $(x + 3/4)^2 \geq 0$ for all x and $(y - 3/2)^2 \geq 0$ for all y, so

$$\left[4\left(x + \frac{3}{4}\right)^2 + \frac{7}{4}\right]\left[4\left(y - \frac{3}{2}\right)^2 + 16\right] \geq \frac{7}{4} \cdot 16 = 28.$$

The only way we can have equality hold is if both $4\left(x + \frac{3}{4}\right)^2$ and $4\left(y - \frac{3}{2}\right)^2$ equal 0. For any other values of x and/or y, the product on the left side will be greater than 28. Therefore, the only values that satisfy the original equation are $(x, y) = \boxed{(-3/4, 3/2)}$.

5.68 First, we find the directrix. Since the given vertex is the midpoint of the origin and the given focus, by Problem 5.1, the directrix passes through the origin. Also, the directrix is perpendicular to the line joining the vertex and the focus. Therefore, the equation of the directrix is $y = -x$.

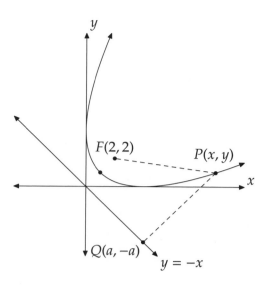

Now, let $P(x, y)$ be a point on the parabola. Let Q be the foot of the altitude from P to the directrix, so $Q = (a, -a)$ for some real number a, and PQ is the distance from P to the directrix. Since $\overline{PQ}$ is perpendicular to the directrix, it has slope 1. Hence, $(y + a)/(x - a) = 1$, which implies $y + a = x - a$, so $a = (x - y)/2$.

Finally, since P lies on the parabola, we have $PF = PQ$, so $PF^2 = PQ^2$. By the distance formula, we have $(x - 2)^2 + (y - 2)^2 = (x - a)^2 + (y + a)^2$. Substituting $a = (x - y)/2$ and simplifying, we find that the equation of the parabola is given by $\boxed{x^2 - 2xy + y^2 - 8x - 8y + 16 = 0}$.

5.69 If the graphs of $y = f(x)$ and $y = g(x)$ do not intersect, then the equation $f(x) = g(x)$ has no solutions in real numbers. In other words, the quadratic

$$f(x) - g(x) = x^2 + (2b - 2a)x + 1 - 2ab$$

has no real roots. Therefore, the discriminant

$$(2b - 2a)^2 - 4(1 - 2ab) = 4a^2 - 8ab + 4b^2 - 4 + 8ab = 4a^2 + 4b^2 - 4 = 4(a^2 + b^2 - 1)$$

is negative, which means $a^2 + b^2 < 1$. Thus, S is the interior of a circle with center $(0,0)$ and radius 1, which has area $\boxed{\pi}$.

5.70 Let $k = y/x$, so $y = kx$. Substituting into the given equation, we get

$$(3k^2 + k + 2)x^2 - (20k + 11)x + 40 = 0.$$

The discriminant of this quadratic is

$$(20k + 11)^2 - 4 \cdot (3k^2 + k + 2) \cdot 40 = -80k^2 + 280k - 199,$$

which must be nonnegative.

The solution to the inequality $-80k^2 + 280k - 199 \geq 0$ is of the form $x \in [k_1, k_2]$, where k_1 and k_2 are the roots of the corresponding quadratic equation $-80k^2 + 280k - 199 = 0$, so $a = k_1$ and $b = k_2$. Then by Vieta's Formulas, $a + b = k_1 + k_2 = 280/80 = \boxed{7/2}$.

CHAPTER 6

Polynomial Division

Exercises for Section 6.1

6.1.1 The degree of $g(x) + 3 = 2x^2 + 5x + 10$ and the degree of $g(x) + 9 = 2x^2 + 5x + 16$ are both 2, so the degree of their product $h(x)$ is $2 + 2 = \boxed{4}$.

6.1.2

(a) The degree of $f(x)$ is $\boxed{4}$.

(b) The degree of $g(x)$ is $\boxed{3}$.

(c) We have $f(g(x)) = f(2x^3 + 3x^2 + 4x + 5) = (2x^3 + 3x^2 + 4x + 5)^4 + (2x^3 + 3x^2 + 4x + 5)^2 - 2$. Without expanding, we can tell that the term of highest degree in $f(g(x))$ will be $(2x^3)^4 = 16x^{12}$, so the degree of $f(g(x))$ is $\boxed{12}$. In general, if $f(x)$ and $g(x)$ are nonzero polynomials, then the degree of $f(g(x))$ is equal to $(\deg f) \cdot (\deg g)$.

(d) First, we find the degree of $f^2(x)$. We have $f(f(x)) = (x^4 + x^2 - 2)^4 + (x^4 + x^2 - 2)^2 - 2$. The highest degree term comes from $(x^4)^4$ in the expansion of $(x^4 + x^2 - 2)^4$. So, the degree of $f(f(x))$ is $4 \cdot 4 = 16$. Similarly, when we find $f(f(f(x)))$, the highest degree term is the $(x^4)^{16}$ term that occurs when we expand the $(x^4 + x^2 - 2)^{16}$ term of $f^2(f(x))$. So, the degree of $f(f(f(x)))$ is $4 \cdot 16 = 64$. Similarly, the degree of $f^4(x)$ is $4 \cdot 64 = \boxed{256}$.

(e) For any polynomial $p(x)$, the sum of the coefficients of $p(x)$ is $p(1)$. Therefore, the sum of the coefficients of $g(f(x))$ is $g(f(1)) = g(0) = \boxed{5}$.

6.1.3 Setting $x = 1$, we get $a + b + c + d = (1^2 + 2 \cdot 1 - 8)(1 - 3) - (1 - 2)(1^2 + 5 \cdot 1 + 4) = \boxed{20}$.

6.1.4 Let

$$f(x) = a_m x^m + a_{m-1} x^{m-1} + \cdots + a_1 x + a_0,$$
$$g(x) = b_n x^n + b_{n-1} x^{n-1} + \cdots + b_1 x + b_0,$$

where $a_m \neq 0$ and $b_n \neq 0$.

Then

$$\begin{aligned}
f(g(x)) &= f(b_n x^n + b_{n-1} x^{n-1} + \cdots + b_1 x + b_0) \\
&= a_m (b_n x^n + b_{n-1} x^{n-1} + \cdots + b_1 x + b_0)^m \\
&\quad + a_{m-1} (b_n x^n + b_{n-1} x^{n-1} + \cdots + b_1 x + b_0)^{m-1} \\
&\quad + \cdots + a_1 (b_n x^n + b_{n-1} x^{n-1} + \cdots + b_1 x + b_0) + a_0.
\end{aligned}$$

In this polynomial, the term of greatest degree is $a_m b_n^m x^{mn}$. Its coefficient is nonzero because a_m and b_n are nonzero. Therefore, the degree of $f(g(x))$ is $\boxed{mn}$.

Exercises for Section 6.2

6.2.1

(a)

$$
\begin{array}{r}
x - 26 \\
x + 7\overline{\smash{\big)}\ x^2 - 19x + 17} \\
\underline{x^2 + 7x} \\
-26x + 17 \\
\underline{-26x - 182} \\
199
\end{array}
$$

The quotient is $\boxed{x - 26}$ and the remainder is $\boxed{199}$.

(b)

$$
\begin{array}{r}
3x^2 - 3x + 4 \\
x^2 + x + 4\overline{\smash{\big)}\ 3x^4 + 0x^3 + 13x^2 - 9x + 122} \\
\underline{3x^4 + 3x^3 + 12x^2} \\
-3x^3 + x^2 - 9x + 122 \\
\underline{-3x^3 - 3x^2 - 12x} \\
4x^2 + 3x + 122 \\
\underline{4x^2 + 4x + 16} \\
- x + 106
\end{array}
$$

The quotient is $\boxed{3x^2 - 3x + 4}$ and the remainder is $\boxed{-x + 106}$.

(c)

$$
\begin{array}{r}
2x^2 - 3x + \tfrac{11}{2} \\
2x + 3\overline{\smash{\big)}\ 4x^3 + 0x^2 + 2x - 1} \\
\underline{4x^3 + 6x^2} \\
-6x^2 + 2x - 1 \\
\underline{-6x^2 - 9x} \\
11x - 1 \\
\underline{11x + \tfrac{33}{2}} \\
-\tfrac{35}{2}
\end{array}
$$

The quotient is $\boxed{2x^2 - 3x + \tfrac{11}{2}}$ and the remainder is $\boxed{-\tfrac{35}{2}}$.

(d)

$$
\begin{array}{r}
19x^4 + 19x^3 + 19x^2 + 19x + 19 \\
x-1\overline{\smash{\big)}\ 19x^5 + 0x^4 + 0x^3 + 0x^2 + 0x + 0} \\
\underline{19x^5 - 19x^4} \\
19x^4 + 0x^3 + 0x^2 + 0x + 0 \\
\underline{19x^4 - 19x^3} \\
19x^3 + 0x^2 + 0x + 0 \\
\underline{19x^3 - 19x^2} \\
19x^2 + 0x + 0 \\
\underline{19x^2 - 19x} \\
19x + 0 \\
\underline{19x - 19} \\
19
\end{array}
$$

The quotient is $\boxed{19x^4 + 19x^3 + 19x^2 + 19x + 19}$ and the remainder is $\boxed{19}$.

6.2.2 We perform the long division:

$$
\begin{array}{r}
x^2 + (c+3)x + (3c+13) \\
x-3\overline{\smash{\big)}\ x^3 + cx^2 + 4x - 21} \\
\underline{x^3 - 3x^2} \\
(c+3)x^2 + 4x - 21 \\
\underline{(c+3)x^2 - (3c+9)x} \\
(3c+13)x - 21 \\
\underline{(3c+13)x - (9c+39)} \\
9c+18
\end{array}
$$

We see the remainder is $9c + 18$, which is 0 when $c = \boxed{-2}$.

6.2.3 We perform the long division:

$$
\begin{array}{r}
x + 0 \\
x^2+3x-2\overline{\smash{\big)}\ x^3 + 3x^2 + ax + 13} \\
\underline{x^3 + 3x^2 - 2x} \\
(a+2)x + 13
\end{array}
$$

We see the remainder is $(a + 2)x + 13$, which is a constant when $a = \boxed{-2}$.

6.2.4 Teresa incorrectly finds that $3x^4 + 2x^3 - 7x^2 + 4x - 1 = (x + 2)(3x^3 - 4x^2 + x + 2) + 5$. One quick way to check the calculation would be to set x to a specific value, and see if both sides agree. For example, setting $x = 0$, we get $-1 = 4 + 5$, which is not correct. Therefore, Teresa must have made a mistake. Note that letting $x = 0$ is essentially the same as just looking at the constant terms. The constant in the quartic is -1, but the constant on the right side is $2 \cdot 2 + 5 = 9$.

Exercises for Section 6.3

6.3.1 From the synthetic division at right, we find that the quotient is $x^2 + 7x + 16$ and the remainder is 67.

$$
\begin{array}{r|rrrr}
3 & 1 & 4 & -5 & 19 \\
  &   & 3 & 21 & 48 \\
\hline
  & 1 & 7 & 16 & 67
\end{array}
$$

6.3.2

(a) $(x + 3)(x^2 - 3x + 8) - 17 = x^3 - 3x^2 + 8x + 3x^2 - 9x + 24 - 17 = x^3 - x + 7.$

(b)

$$(3y + 2)\left(y^3 - 2y^2 + 3y - \frac{17}{3}\right) + \frac{40}{3} = 3y^4 - 6y^3 + 9y^2 - 17y + 2y^3 - 4y^2 + 6y - \frac{34}{3} + \frac{40}{3}$$

$$= 3y^4 - 4y^3 + 5y^2 - 11y + 2.$$

6.3.3 From the synthetic division at right, we see that the quotient is $\boxed{x^4 - 5x^3 + 2x^2 + x - 19}$ and the remainder is $\boxed{115}$. If you have a different answer, make sure you didn't overlook the fact that there is no x^4 term in $x^5 - 23x^3 + 11x^2 - 14x + 20$, so we need to remember to include a 0 in the top row of the synthetic division.

$$
\begin{array}{r|rrrrrr}
-5 & 1 & 0 & -23 & 11 & -14 & 20 \\
   &   & -5 & 25 & -10 & -5 & 95 \\
\hline
   & 1 & -5 & 2 & 1 & -19 & 115
\end{array}
$$

6.3.4 First, we need to make the coefficient of x in the divisor to be 1 to use synthetic division. We have

$$\frac{4x^4 - 10x^3 + 14x^2 + 7x - 19}{2x - 1} = \frac{1}{2} \cdot \frac{4x^4 - 10x^3 + 14x^2 + 7x - 19}{x - \frac{1}{2}}.$$

The synthetic division at right then gives

$$
\begin{array}{r|rrrrr}
\frac{1}{2} & 4 & -10 & 14 & 7 & -19 \\
 &   & 2 & -4 & 5 & 6 \\
\hline
 & 4 & -8 & 10 & 12 & -13
\end{array}
$$

$$\frac{4x^4 - 10x^3 + 14x^2 + 7x - 19}{2x - 1} = \frac{1}{2} \cdot \frac{4x^4 - 10x^3 + 14x^2 + 7x - 19}{x - \frac{1}{2}}$$

$$= \frac{1}{2}\left(4x^3 - 8x^2 + 10x + 12 + \frac{-13}{x - \frac{1}{2}}\right)$$

$$= 2x^3 - 4x^2 + 5x + 6 + \frac{-13}{2x - 1}.$$

Therefore, the quotient is $\boxed{2x^3 - 4x^2 + 5x + 6}$ and the remainder is $\boxed{-13}$.

6.3.5 First, the coefficient of z in the divisor must be 1 to use synthetic division. We have

$$\frac{3z^3 - 4z^2 - 14z + 3}{3z + 5} = \frac{1}{3} \cdot \frac{3z^3 - 4z^2 - 14z + 3}{z + \frac{5}{3}}.$$

Our synthetic division at right gives us

$$
\begin{array}{r|rrrr}
-\frac{5}{3} & 3 & -4 & -14 & 3 \\
 &   & -5 & 15 & -\frac{5}{3} \\
\hline
 & 3 & -9 & 1 & \frac{4}{3}
\end{array}
$$

$$\frac{3z^3 - 4z^2 - 14z + 3}{3z + 5} = \frac{1}{3} \cdot \frac{3z^3 - 4z^2 - 14z + 3}{z + \frac{5}{3}}$$

$$= \frac{1}{3}\left(3z^2 - 9z + 1 + \frac{4/3}{z + \frac{5}{3}}\right)$$

$$= z^2 - 3z + \frac{1}{3} + \frac{4/3}{3z + 5}.$$

So, the quotient is $\boxed{z^2 - 3z + \frac{1}{3}}$ and the remainder is $\boxed{\frac{4}{3}}$.

6.3.6 As usual, we first make the coefficient of x equal to 1:

$$\frac{x^4 + 3x^3 - x^2 + 7x - 1}{2 - x} = -\frac{x^4 + 3x^3 - x^2 + 7x - 1}{x - 2}.$$

Then, we perform the synthetic division to find

$$\begin{array}{r|rrrrr} 2 & 1 & 3 & -1 & 7 & -1 \\ & & 2 & 10 & 18 & 50 \\ \hline & 1 & 5 & 9 & 25 & 49 \end{array}$$

$$\begin{aligned} \frac{x^4 + 3x^3 - x^2 + 7x - 1}{2 - x} &= -\frac{x^4 + 3x^3 - x^2 + 7x - 1}{x - 2} \\ &= -\left(x^3 + 5x^2 + 9x + 25 + \frac{49}{x - 2}\right) \\ &= -x^3 - 5x^2 - 9x - 25 + \frac{49}{2 - x}. \end{aligned}$$

Therefore, the quotient is $\boxed{-x^3 - 5x^2 - 9x - 25}$ and the remainder is $\boxed{49}$.

6.3.7 The leading term of the quotient times the leading term of the divisor should give the leading term of the polynomial we are dividing. However, we get $(2x)(x^2) = 2x^3$ instead of x^3. Stan probably used 3 on the left of the division bracket of his synthetic division, and thereby mistakenly divided by $x - 3$ instead of by $2x - 3$.

Exercises for Section 6.4

6.4.1

(a) The remainder when $g(x)$ is divided by $x + 1$ is $g(-1) = 2 \cdot (-1)^6 - (-1)^4 + 4 \cdot (-1)^2 - 8 = \boxed{-3}$.

(b) The remainder when $g(x)$ is divided by $x - 1$ is $g(1) = 2 \cdot 1^6 - 1^4 + 4 \cdot 1^2 - 8 = \boxed{-3}$.

(c) The remainder when $g(x)$ is divided by $x + 3$ is $g(-3) = 2 \cdot (-3)^6 - (-3)^4 + 4 \cdot (-3)^2 - 8 = \boxed{1405}$.

(d) The remainder when $g(x)$ is divided by $x - 3$ is $g(3) = 2 \cdot 3^6 - 3^4 + 4 \cdot 3^2 - 8 = \boxed{1405}$.

(e) Since $x^2 + 4x + 3 = (x + 1)(x + 3)$ has degree 2, the remainder is of the form $ax + b$ for some constants a and b. Let $q(x)$ be the quotient, so $g(x) = (x + 1)(x + 3)q(x) + ax + b$.

Taking $x = -1$ and $x = -3$, respectively, we obtain the equations

$$\begin{aligned} g(-1) &= -a + b, \\ g(-3) &= -3a + b. \end{aligned}$$

We found $g(-1) = -3$ and $g(-3) = 1405$ above, so we have $-a + b = -3$ and $-3a + b = 1405$. Solving this system, we find $a = -704$ and $b = -707$. Therefore, the remainder is $\boxed{-704x - 707}$.

(f) Since $x^2 - 4x + 3 = (x - 1)(x - 3)$ has degree 2, the remainder is of the form $ax + b$ for some constants a and b. Let $q(x)$ be the quotient, so $g(x) = (x - 1)(x - 3)q(x) + ax + b$.

Taking $x = 1$ and $x = 3$, respectively, and using $g(1) = -3$ and $g(3) = 1405$ from above, we obtain the equations

$$\begin{aligned} -3 &= a + b, \\ 1405 &= 3a + b. \end{aligned}$$

Solving this system, we find $a = 704$ and $b = -707$. Therefore, the remainder is $\boxed{704x - 707}$.

6.4.2 Let $h(y) = y^2 + my + 2$. By the Remainder Theorem, $R_1 = h(1) = m + 3$ and $R_2 = h(-1) = -m + 3$. Since $R_1 = R_2$, we have $m + 3 = -m + 3$, so $m = \boxed{0}$.

6.4.3 Since $(x - 1)(x - 2)$ has degree 2, the remainder is of the form $ax + b$ for some constants a and b. Let $q(x)$ be the quotient, so $x^{100} = (x - 1)(x - 2)q(x) + ax + b$.

Letting $x = 1$ and $x = 2$, respectively, we obtain the equations $1 = a + b$ and $2^{100} = 2a + b$. Solving this system, we find $a = 2^{100} - 1$ and $b = 2 - 2^{100}$. Therefore, the remainder is $\boxed{(2^{100} - 1)x - 2^{100} + 2}$.

6.4.4 From the given information, we have $f(x) = d(x)q(x) + r(x)$. If $x = a$ is a root of $d(x)$, then $d(a) = 0$, so $f(a) = d(a)q(a) + r(a) = r(a)$, as desired.

6.4.5 The divisor has degree 3, so the remainder is of the form $ax^2 + bx + c$ for some constants a, b, and c. Let $q(x)$ be the quotient, so
$$x^{100} - 4x^{98} + 5x + 6 = (x^3 - 2x^2 - x + 2)q(x) + ax^2 + bx + c.$$
To eliminate the $q(x)$, we must find values of x such that $x^3 - 2x^2 - x + 2 = 0$. We notice the similarity between the first two coefficients and the next two. Hoping to exploit this, we factor x^2 out of the first two terms, and we have $x^3 - 2x^2 - x + 2 = x^2(x - 2) - x + 2$. Aha! We can factor more:
$$x^2(x - 2) - x + 2 = x^2(x - 2) - 1(x - 2) = (x^2 - 1)(x - 2) = (x - 1)(x + 1)(x - 2).$$
Therefore, we have
$$x^{100} - 4x^{98} + 5x + 6 = (x - 2)(x - 1)(x + 1)q(x) + ax^2 + bx + c.$$
Taking $x = 2$, $x = 1$, and $x = -1$, respectively, we obtain the equations
$$2^{100} - 4 \cdot 2^{98} + 5 \cdot 2 + 6 = 16 = 4a + 2b + c,$$
$$1^{100} - 4 \cdot 1^{98} + 5 \cdot 1 + 6 = 8 = a + b + c,$$
$$(-1)^{100} - 4 \cdot (-1)^{98} + 5 \cdot (-1) + 6 = -2 = a - b + c.$$
Subtracting the third equation from the second equation, we get $10 = 2b$, so $b = 5$. Then the first and second equations become $6 = 4a + c$ and $3 = a + c$.

Solving this system, we get $a = 1$ and $c = 2$. Therefore, the remainder is $ax^2 + bx + c = \boxed{x^2 + 5x + 2}$.

Review Problems

6.20 First, we write $f(x) + cg(x)$ as a polynomial:
$$f(x) + cg(x) = x^2 + 3x + 3 + c(2x^2 - x + 2) = (2c + 1)x^2 + (3 - c)x + (3 + 2c) = 0.$$
This becomes a binomial if one coefficient is zero, and the other two coefficients are nonzero. Solving each of $2c + 1 = 0$, $3 - c = 0$, and $3 + 2c = 0$ gives the values $c = \boxed{-1/2, 3, \text{ and } -3/2}$.

6.21

(a)
```
-4 | 9    1     0    -12     21
   |     -36   140  -560   2288
   -----------------------------
     9   -35   140  -572 | 2309
```

The quotient is $\boxed{9x^3 - 35x^2 + 140x - 572}$ and the remainder is $\boxed{2309}$.

(b)
$$\begin{array}{r|rrrr} 3 & 1 & -3 & -9 & 27 \\ & & 3 & 0 & -27 \\ \hline & 1 & 0 & -9 & 0 \end{array}$$

The quotient is $\boxed{x^2 - 9}$ and the remainder is $\boxed{0}$.

(c) Note that all of the exponents of y are even, so let $x = y^2$. Then we are dividing $2x^3 + 3x^2 + 4x + 5$ by $x - 1$.

$$\begin{array}{r|rrrr} 1 & 2 & 3 & 4 & 5 \\ & & 2 & 5 & 9 \\ \hline & 2 & 5 & 9 & 14 \end{array}$$

The quotient is $2x^2 + 5x + 9$ and the remainder is 14. In terms of y, the quotient is $\boxed{2y^4 + 5y^2 + 9}$ and the remainder is $\boxed{14}$. (We could also have used long division.)

(d) In order to use synthetic division, we first make the coefficient of the linear term of the divisor equal to 1. We're also careful to note that we are dividing by $2x + 5$, not $5x + 2$:

$$\frac{8x^3 + 16x^2 - 7x + 4}{2x + 5} = \frac{1}{2} \cdot \frac{8x^3 + 16x^2 - 7x + 4}{x + \frac{5}{2}}.$$

Our synthetic division is shown at right. We find

$$\begin{array}{r|rrrr} -\frac{5}{2} & 8 & 16 & -7 & 4 \\ & & -20 & 10 & -\frac{15}{2} \\ \hline & 8 & -4 & 3 & -\frac{7}{2} \end{array}$$

$$\begin{aligned} \frac{8x^3 + 16x^2 - 7x + 4}{2x + 5} &= \frac{1}{2} \cdot \frac{8x^3 + 16x^2 - 7x + 4}{x + \frac{5}{2}} \\ &= \frac{1}{2}\left(8x^2 - 4x + 3 + \frac{-7/2}{x + \frac{5}{2}}\right) \\ &= 4x^2 - 2x + \frac{3}{2} + \frac{-7/2}{2x + 5}. \end{aligned}$$

So, our quotient is $\boxed{4x^2 - 2x + \frac{3}{2}}$ and the remainder is $\boxed{-\frac{7}{2}}$.

(e) In order to use synthetic division, we first make the coefficient of the linear term of the divisor equal to 1:

$$\frac{3r^4 + 16r^3 - 5r + 19}{3r - 2} = \frac{1}{3} \cdot \frac{3r^4 + 16r^3 - 5r + 19}{r - \frac{2}{3}}.$$

We then perform the synthetic division shown at right. We have

$$\begin{array}{r|rrrrr} \frac{2}{3} & 3 & 16 & 0 & -5 & 19 \\ & & 2 & 12 & 8 & 2 \\ \hline & 3 & 18 & 12 & 3 & 21 \end{array}$$

$$\begin{aligned} \frac{3r^4 + 16r^3 - 5r + 19}{3r - 2} &= \frac{1}{3} \cdot \frac{3r^4 + 16r^3 - 5r + 19}{r - \frac{2}{3}} \\ &= \frac{1}{3}\left(3r^3 + 18r^2 + 12r + 3 + \frac{21}{r - \frac{2}{3}}\right) \\ &= r^3 + 6r^2 + 4r + 1 + \frac{21}{3r - 2}. \end{aligned}$$

The quotient is $\boxed{r^3 + 6r^2 + 4r + 1}$ and the remainder is $\boxed{21}$.

(f) Since we are dividing by a quadratic, we perform long division:

$$
\begin{array}{r}
t^2 + 3t + 0 \\
t^2 - 3t + 8\,\overline{\smash{\big)}\,t^4 + 0t^3 - t^2 + 3t - 7} \\
\underline{t^4 - 3t^3 + 8t^2} \\
3t^3 - 9t^2 + 3t - 7 \\
\underline{3t^3 - 9t^2 + 24t} \\
- 21t - 7
\end{array}
$$

The quotient is $\boxed{t^2 + 3t}$ and the remainder is $\boxed{-21t - 7}$.

6.22   The sum of the coefficients of $P(x)$ is $P(1)$. Setting $x = 1$, we get

$$1 + 23 - 18 - 24 + 108 = (1 - 3 - 2 + 9)P(1) \quad \Rightarrow \quad P(1) = \boxed{18}.$$

6.23   By the Remainder Theorem, the remainder when $f(x)$ is divided by $x - a$ is $f(a)$, and the remainder when $f(x)$ is divided by $x + a$ is $f(-a)$, so these are equal if and only if $f(a) = f(-a)$. In general, this is $\boxed{\text{not true}}$.

However, this turns out to be the case in Exercise 6.4.1 because every term in $g(x) = 2x^6 - x^4 + 4x^2 - 8$ has an even exponent. Hence, $g(a) = g(-a)$ for all real numbers a.

6.24   If we divide the polynomial $f(x)$ by the polynomial $d(x)$ and find the quotient $q(x)$ and remainder $r(x)$, then we have $f(x) = d(x) \cdot q(x) + r(x)$, where the degree of $r(x)$ (if $r(x)$ is nonzero) is less than the degree of $d(x)$. Therefore, the degree of $r(x)$ is less than the degree of $d(x) \cdot q(x)$. So, the degree of $d(x) \cdot q(x) + r(x)$ equals the degree of $d(x) \cdot q(x)$. The degree of $f(x)$ must equal the degree of $d(x) \cdot q(x) + r(x)$, so $\deg f = \deg(d \cdot q) = \deg d + \deg q$. Therefore, if $\deg f = 3$ and $\deg d = 1$, we must have $\deg q = 2$.

However, Beth divided a cubic by a linear expression and came up with a linear quotient. Therefore, she knows she made a mistake, because the quotient must be quadratic.

6.25   Let $f(x) = x^{13} + 1$. Then by the Remainder Theorem, the remainder when $f(x) = x^{13} + 1$ is divided by $x - 1$ is $f(1) = 1^{13} + 1 = 1 + 1 = \boxed{2}$.

6.26   From the given information, we have $f(x) = d(x)q(x) + r(x)$. Since the degree of $f(x)$ is 12 and the degree of $r(x)$ is 2, the degree of $d(x)q(x)$ is 12, so $\deg d + \deg q = 12$. However, we also know that in a division, the degree of the remainder $r(x)$ is less than the degree of the divisor $d(x)$, so $\deg d \geq 3$. Therefore, $\boxed{0 \leq \deg q \leq 9}$.

6.27   Since we are dividing by a cubic, we perform long division:

$$
\begin{array}{r}
x + 1 \\
x^3 + 3x^2 + 9x + 3\,\overline{\smash{\big)}\,x^4 + 4x^3 + 6px^2 + 4qx + r} \\
\underline{x^4 + 3x^3 + 9x^2 + 3x} \\
x^3 + (6p - 9)x^2 + (4q - 3)x + r \\
\underline{x^3 + 3x^2 + 9x + 3} \\
(6p - 12)x^2 + (4q - 12)x + (r - 3)
\end{array}
$$

Thus, the quotient is $x + 1$ and the remainder is $(6p - 12)x^2 + (4q - 12)x + (r - 3)$. Hence, for the remainder to be 0, we require $6p - 12 = 4q - 12 = r - 3 = 0$, which implies $p = 2$, $q = 3$, and $r = 3$. Therefore, $(p + q)r = \boxed{15}$.

6.28 Since $x^2 - 1 = (x - 1)(x + 1)$ has degree 2, the remainder is of the form $ax + b$ for some constants a and b. Let $q(x)$ be the quotient, so $13x^6 + 3x^4 + 9x^3 + 2x^2 + 17 = (x - 1)(x + 1)q(x) + ax + b$.

Taking $x = 1$ and $x = -1$, respectively, we obtain the equations $44 = a + b$ and $26 = -a + b$. Solving this system, we find $a = 9$ and $b = 35$. Therefore, the remainder is $\boxed{9x + 35}$.

6.29 Since $(x - 1)(x - 2)(x - 3)$ has degree 3, the remainder is of the form $ax^2 + bx + c$ for some constants a, b, and c. Let $q(x)$ be the quotient, so

$$f(x) = (x - 1)(x - 2)(x - 3)q(x) + ax^2 + bx + c.$$

Taking $x = 1$, $x = 2$, and $x = 3$, respectively, we obtain the equations

$$f(1) = 2 = a + b + c,$$
$$f(2) = 3 = 4a + 2b + c,$$
$$f(3) = 5 = 9a + 3b + c.$$

Taking the difference of the first two equations to eliminate c, we get $3a + b = 1$. Similarly, taking the difference of the the last two equations, we get $5a + b = 2$. Taking the difference of these equations, we get $2a = 1$, so $a = 1/2$. Then it quickly follows that $b = -1/2$ and $c = 2$. Therefore, the remainder is $ax^2 + bx + c = \boxed{x^2/2 - x/2 + 2}$.

6.30 Let $q(x)$ be the quotient when $f(x)$ is divided by $x - a$, so $f(x) = (x - a)q(x) + r$. Since c is nonzero, we can rewrite this as

$$f(x) = c(x - a) \cdot \frac{q(x)}{c} + r.$$

Thus, the remainder when $f(x)$ is divided by $c(x - a)$ is also r, with quotient $q(x)/c$.

Challenge Problems

6.31 Because the roots of $f(x)$ are solutions of the equation $g(x) = 0$, we guess that $f(x)$ is a factor of $g(x)$. In other words, we guess that we can write $g(x) = f(x) \cdot q(x)$ for some polynomial $q(x)$. If this is the case, then any solution to the equation $f(x) = 0$ is also a solution to $g(x) = 0$.

Dividing $g(x)$ by $f(x)$, we find that $3x^3 - 5x^2 - 3x + 12 = (x^2 - 3x + 3)(3x + 4)$. Therefore, the third solution of $g(x) = 0$ is $\boxed{-4/3}$.

6.32 Since the remainder when $f(x)$ is divided by $(x - a)(x - b)(x - c)$ is $px + q$, we can write

$$f(x) = (x - a)(x - b)(x - c)g(x) + px + q$$

for some polynomial $g(x)$. Taking $x = a$, $x = b$, $x = c$, respectively, we obtain the equations

$$f(a) = pa + q,$$
$$f(b) = pb + q,$$
$$f(c) = pc + q.$$

Then

$$(b - c)f(a) + (c - a)f(b) + (a - b)f(c) = (b - c)(pa + q) + (c - a)(pb + q) + (a - b)(pc + q)$$
$$= pab - pac + bq - cq + pbc - pab + cq - aq + pac - pbc + aq - bq$$
$$= 0.$$

6.33 Since $x^2 - 3x + 2$ factors as $(x - 1)(x - 2)$, we can divide by $x^2 - 3x + 2$ in two steps using synthetic division as follows. First, we divide by $x - 1$, as shown at right. We find

$$
\begin{array}{r|rrrrr}
1 & 1 & -3 & 4 & 11 & -9 \\
  &   & 1 & -2 & 2 & 13 \\
\hline
  & 1 & -2 & 2 & 13 & 4
\end{array}
$$

$$x^4 - 3x^3 + 4x^2 + 11x - 9 = (x - 1)(x^3 - 2x^2 + 2x + 13) + 4.$$

Then we divide the quotient $x^3 - 2x^2 + 2x + 13$ by $x - 2$, as shown at right. Hence, $x^3 - 2x^2 + 2x + 13 = (x - 2)(x^2 + 2) + 17$. Substituting this expression into our first equation, we get

$$
\begin{array}{r|rrrr}
2 & 1 & -2 & 2 & 13 \\
  &   & 2 & 0 & 4 \\
\hline
  & 1 & 0 & 2 & 17
\end{array}
$$

$$
\begin{aligned}
x^4 - 3x^3 + 4x^2 + 11x - 9 &= (x - 1)(x^3 - 2x^2 + 2x + 13) + 4 \\
&= (x - 1)[(x - 2)(x^2 + 2) + 17] + 4 \\
&= (x - 1)(x - 2)(x^2 + 2) + 17(x - 1) + 4 \\
&= (x - 1)(x - 2)(x^2 + 2) + 17x - 13.
\end{aligned}
$$

The quotient is $x^2 + 2$ and the remainder is $17x - 13$.

6.34 By the Remainder Theorem, we have $P(1) = 5$ and $P(2) = -4$. Let

$$Q(x) = x^{81} + Lx^{57} + Gx^{41} + Hx^{19} + Kx + R = P(x) - (2x + 1) + (Kx + R).$$

Since $Q(x)$ is divisible by $(x - 1)(x - 2)$, we have $Q(1) = Q(2) = 0$. But from the above equation, we have $Q(1) = P(1) - 3 + K + R = K + R + 2$, and $Q(2) = P(2) - 5 + 2K + R = 2K + R - 9$, which gives us the system of equations $K + R + 2 = 0, 2K + R - 9 = 0$. Solving this system gives $(K, R) = \boxed{(11, -13)}$.

6.35 Since $\deg P(x) = 3$ and $\deg R(x) = 3$, we have

$$\deg P(Q(x)) = \deg(P(x)R(x)) = \deg P(x) + \deg R(x) = 3 + 3 = 6.$$

But $\deg P(Q(x)) = \deg P(x) \cdot \deg Q(x) = 3 \deg Q(x)$, so $\deg Q(x) = 2$. Then $Q(x)$ is quadratic, so $Q(x) = ax^2 + bx + c$ for some constants a, b, and c, where $a \neq 0$.

Now we look at the condition $P(Q(x)) = P(x) \cdot R(x)$. Taking $x = 1$, this equation becomes $P(Q(1)) = P(1) \cdot R(1) = 0$, since $P(1) = 0$. Similarly, $P(Q(2)) = P(Q(3)) = 0$. Hence, for such a polynomial $R(x)$ to exist, the polynomial $Q(x)$ must satisfy $P(Q(1)) = P(Q(2)) = P(Q(3)) = 0$.

Conversely, if $P(Q(1)) = P(Q(2)) = P(Q(3)) = 0$, then by the Factor Theorem, $P(Q(x))$ is divisible by $x - 1$, $x - 2$, and $x - 3$, so

$$P(Q(x)) = (x - 1)(x - 2)(x - 3)R(x) = P(x) \cdot R(x)$$

for some polynomial $R(x)$. Hence, these two conditions are equivalent, that is, $P(Q(x)) = P(x) \cdot R(x)$ for some polynomial $R(x)$ if and only if $P(Q(1)) = P(Q(2)) = P(Q(3)) = 0$.

Furthermore, since $P(x) = (x - 1)(x - 2)(x - 3)$, we have $P(Q(1)) = P(Q(2)) = P(Q(3)) = 0$ if and only if each of $Q(1)$, $Q(2)$, and $Q(3)$ is equal to one of 1, 2, or 3. Now that we have established the possible values of $Q(1)$, $Q(2)$, and $Q(3)$, we can solve for the coefficients of $Q(x) = ax^2 + bx + c$. Taking $x = 1$, $x = 2$, and $x = 3$, respectively, we obtain the system of equations

$$
\begin{aligned}
Q(1) &= a + b + c, \\
Q(2) &= 4a + 2b + c, \\
Q(3) &= 9a + 3b + c.
\end{aligned}
$$

Taking the difference of the first two equations, we get $3a + b = Q(2) - Q(1)$, and taking the difference of the last two equations, we get $5a + b = Q(3) - Q(2)$. Taking the difference of these two equations, we get

$2a = Q(1) - 2Q(2) + Q(3)$, so $a = \frac{Q(1)-2Q(2)+Q(3)}{2}$. Then, we have

$$b = Q(2) - Q(1) - 3a = Q(2) - Q(1) - \frac{3Q(3) - 6Q(2) + 3Q(1)}{2} = \frac{-5Q(1) + 8Q(2) - 3Q(3)}{2},$$

and

$$c = Q(1) - a - b = Q(1) - \frac{Q(1) - 2Q(2) + Q(3)}{2} - \frac{-5Q(1) + 8Q(2) - 3Q(3)}{2} = 3Q(1) - 3Q(2) + Q(3).$$

Thus, there exists a solution (a, b, c) for any triple of values $(Q(1), Q(2), Q(3))$. However, the polynomial $Q(x) = ax^2 + bx + c$ is quadratic if and only if $a \neq 0$, which means $Q(1) - 2Q(2) + Q(3) \neq 0$, or

$$Q(2) \neq \frac{Q(1) + Q(3)}{2}.$$

In other words, $Q(2)$ cannot be the average of $Q(1)$ and $Q(3)$.

Since each of $Q(1)$, $Q(2)$, and $Q(3)$ must be one of 1, 2, or 3, there are $3^3 = 27$ possible triples $(Q(1), Q(2), Q(3))$. Of these 27 triples, $Q(2)$ is the average of $Q(1)$ and $Q(3)$ in the following five triples:

$$(1, 1, 1), \ (2, 2, 2), \ (3, 3, 3), \ (1, 2, 3), \ (3, 2, 1).$$

Therefore, there are $27 - 5 = \boxed{22}$ polynomials $Q(x)$ that satisfy the problem.

6.36 Since $x^3 + x = x(x^2 + 1) = x(x - i)(x + i)$ has degree 3, the remainder is of the form $ax^2 + bx + c$ for some constants a, b, and c. Let $q(x)$ be the quotient, so

$$x^{81} + x^{49} + x^{25} + x^9 + x = x(x - i)(x + i)q(x) + ax^2 + bx + c.$$

Taking $x = 0$, we get $c = 0$. Taking $x = i$, we get $5i = ai^2 + bi + c = -a + bi$ (because $c = 0$). Similarly, taking $x = -i$, we get $-5i = -a - bi$. Adding $5i = -a + bi$ and $-5i = -a - bi$ gives $a = 0$, from which we find $b = 5$. Therefore, the remainder is $ax^2 + bx + c = \boxed{5x}$.

6.37 Since $f(x)$ has degree 5, we have $f(x) = ax^5 + bx^4 + cx^3 + dx^2 + ex + f$ for some constants a, b, c, d, e, and f. Furthermore, since $f(x)$ is divisible by x^3, the coefficients d, e, and f are all 0, so $f(x) = ax^5 + bx^4 + cx^3$.

We also know that $f(x) - 1$ is divisible by $(x - 1)^3$. Hence, by the Remainder Theorem, $f(1) = 1$, so $a + b + c = 1$. Unfortunately, this is all that the Remainder Theorem can tell us, because all the roots of $(x - 1)^3$ are the same.

To divide $f(x) - 1$ by the expression $(x - 1)^3$, we can divide by $x - 1$ three times in a row, each time making sure that the remainder is 0. To divide $f(x) - 1 = ax^5 + bx^4 + cx^3 - 1$ by $x - 1$, we can use synthetic division:

$$
\begin{array}{c|cccccc}
1 & a & b & c & 0 & 0 & -1 \\
  &   & a & a+b & a+b+c & a+b+c & a+b+c \\
\hline
  & a & a+b & a+b+c & a+b+c & a+b+c & a+b+c-1 \\
\end{array}
$$

Hence, the quotient is $ax^4 + (a + b)x^3 + (a + b + c)x^2 + (a + b + c)x + (a + b + c)$ and the remainder is $a + b + c - 1 = 0$. Therefore, $a + b + c = 1$, which we already derived above. To make things easier, we use $a + b + c = 1$ to write the quotient as $ax^4 + (a + b)x^3 + x^2 + x + 1$.

We then divide by $x - 1$ again:

$$
\begin{array}{c|ccccc}
1 & a & a+b & 1 & 1 & 1 \\
  &   & a & 2a+b & 2a+b+1 & 2a+b+2 \\
\hline
  & a & 2a+b & 2a+b+1 & 2a+b+2 & 2a+b+3 \\
\end{array}
$$

Hence, the quotient is $ax^3 + (2a + b)x^2 + (2a + b + 1)x + (2a + b + 2)$ and the remainder is $2a + b + 3 = 0$. Therefore, $2a + b = -3$, so we can write the quotient as $ax^3 - 3x^2 - 2x - 1$. Finally, we require this expression to be divisible by $x - 1$. The polynomial $ax^3 - 3x^2 - 2x - 1$ is divisible by $x - 1$ if the remainder is 0 when we divide by $x - 1$. However, by the Remainder Theorem, this remainder is what results when we let $x = 1$ in $ax^3 - 3x^2 - 2x - 1$. Therefore, we must have $a - 3 - 2 - 1 = 0$, so $a = 6$. Since $2a + b = -3$, we have $b = -15$. And finally, since $a + b + c = 1$, we have $c = 10$. Therefore, $f(x) = \boxed{6x^5 - 15x^4 + 10x^3}$.

Another approach is to use a substitution. Let $y = x - 1$, so $x = y + 1$. Then, $f(x) - 1$ is divisible by $(x - 1)^3$ if and only if $f(y + 1) - 1$ is divisible by y^3. But

$$
\begin{aligned}
f(y + 1) - 1 &= a(y + 1)^5 + b(y + 1)^4 + c(y + 1)^3 - 1 \\
&= a(y^5 + 5y^4 + 10y^3 + 10y^2 + 5y + 1) + b(y^4 + 4y^3 + 6y^2 + 4y + 1) + c(y^3 + 3y^2 + 3y + 1) - 1 \\
&= ay^5 + (5a + b)y^4 + (10a + 4b + c)y^3 + (10a + 6b + 3c)y^2 + (5a + 4b + 3c)y + (a + b + c - 1).
\end{aligned}
$$

This is divisible by y^3 if and only if the coefficients of y^2, y, and 1 are 0, which gives us the system of equations

$$
\begin{aligned}
10a + 6b + 3c &= 0, \\
5a + 4b + 3c &= 0, \\
a + b + c &= 1.
\end{aligned}
$$

We can then solve this system to find a, b, and c.

CHAPTER 7

Polynomial Roots Part I

Exercises for Section 7.1

7.1.1 If $x + 2$ is a factor of $x^4 - 2x^3 + ax^2 - ax + 7$, then this polynomial is equal to 0 when $x = -2$. Hence, substituting $x = -2$, we get $(-2)^4 - 2(-2)^3 + a(-2)^2 - a(-2) + 7 = 6a + 39$, so $a = -39/6 = \boxed{-13/2}$.

7.1.2 By the Factor Theorem, a polynomial $p(x)$ is divisible by $x - 1$ if and only if $p(1) = 0$. But $p(1)$ is also equal to the sum of the coefficients. For the polynomial given in the problem, the sum of the coefficients is $1 + 6 - 7 + 2 - 2 = 0$. Therefore, it is divisible by $x - 1$.

7.1.3 Because $-5/3$ is a root of $f(x)$, we know that $x + \frac{5}{3}$ is a factor of $f(x)$. Dividing $x + \frac{5}{3}$ into the polynomial $f(x) = 6x^4 + 19x^3 - 36x^2 - 49x + 60$, we get

$$f(x) = \left(x + \frac{5}{3}\right)(6x^3 + 9x^2 - 51x + 36).$$

All the coefficients of the cubic on the right are divisible by 3, so we can factor 3 out to find the factorization $f(x) = 3(x + \frac{5}{3})(2x^3 + 3x^2 - 17x + 12)$. To find more roots of $f(x)$, we must find more roots of $2x^3 + 3x^2 - 17x + 12$, which we'll call $g(x)$. Fortunately, there's an easy root to find—the sum of the coefficients of $g(x)$ is 0, so $g(1) = 0$. Therefore, $x - 1$ is a factor of $g(x)$, which means it is also a factor of $f(x)$. Dividing $g(x)$ by $x - 1$ gives

$$f(x) = 3\left(x + \frac{5}{3}\right)g(x) = 3\left(x + \frac{5}{3}\right)(x - 1)(2x^2 + 5x - 12).$$

Factoring the quadratic gives us $f(x) = 3(x + \frac{5}{3})(x - 1)(2x - 3)(x + 4)$. Therefore, the roots of $f(x)$ are $\boxed{-4, -5/3, 1, 3/2}$.

7.1.4 It is not necessarily true that $q(x) = 0$ if $f(x) = 0$. For example, let $q(x)$ be a nonzero constant c. Then $f(x) = (x - a)q(x) = c(x - a)$. We see that $f(a) = c(a - a) = 0$, but $q(a) = c \neq 0$.

7.1.5 Since $p(x)$ is divisible by $x - 3$, we have $p(3) = 0$. Substituting $x = 3$, we get $p(3) = 3 \cdot 3^3 - 20 \cdot 3^2 + 3k + 12 = 0$, so $3k - 87 = 0$, which gives $k = 29$.

Dividing $3x^3 - 20x^2 + 29x + 12$ by $x - 3$, we get $3x^2 - 11x - 4$. Factoring this quadratic, we find $p(x) = (x - 3)(3x^2 - 11x - 4) = \boxed{(x - 3)(3x + 1)(x - 4)}$.

Exercises for Section 7.2

7.2.1

(a) First, we look for integer roots. If there are any integer roots, then they must be among the factors of 30: 1, 2, 3, 5, 6, and their negatives, and so on. Checking these values, we find $f(2) = 2^3 - 4 \cdot 2^2 - 11 \cdot 2 + 30 = 0$, so by the Factor Theorem, $x - 2$ is a factor of $f(x)$. Dividing $f(x)$ by $x - 2$, we get

$$f(x) = x^3 - 4x^2 - 11x + 30 = (x - 2)(x^2 - 2x - 15).$$

The quadratic $x^2 - 2x - 15$ factors as $(x - 5)(x + 3)$. Therefore, the roots are $x = \boxed{2, 5, \text{ and } -3}$.

(b) We look for integer roots among the factors of 150: 1, 2, 3, 5, 6, and their negatives, and so on. Checking these values, we find $g(2) = 2^4 + 5 \cdot 2^3 - 19 \cdot 2^2 - 65 \cdot 2 + 150 = 0$, so by the Factor Theorem, $t - 2$ is a factor. (We could also have used synthetic division to determine that $t - 2$ is a factor.) Dividing $g(t)$ by $t - 2$, we get

$$g(t) = t^4 + 5t^3 - 19t^2 - 65t + 150 = (t - 2)(t^3 + 7t^2 - 5t - 75).$$

We continue to look for integer roots, this time among the factors of 75: 1, 3, 5, 15, and their negatives, and so on. Checking these values, we find that for $t = 3$, we have $3^3 + 7 \cdot 3^2 - 5 \cdot 3 - 75 = 0$, so by the Factor Theorem, $t - 3$ is a factor. (We could also have used synthetic division to determine that $t - 3$ is a factor.) Dividing by $t - 3$, we get

$$g(t) = (t - 2)(t^3 + 7t^2 - 5t - 75) = (t - 2)(t - 3)(t^2 + 10t + 25) = (t - 2)(t - 3)(t + 5)^2.$$

Therefore, the roots are $t = \boxed{2, 3, -5, \text{ and } -5}$.

(c) We look for integer roots among the factors of 36: 1, 2, 3, 4, 6, and their negatives, and so on. Checking these values, we find $f(1) = -1 - 12 + 6 + 64 - 93 + 36 = 0$, so by the Factor Theorem, $x - 1$ is a factor of $f(x)$. In fact, we find that we may divide by $x - 1$ three times:

$$-x^5 - 12x^4 + 6x^3 + 64x^2 - 93x + 36 = (x - 1)(-x^4 - 13x^3 - 7x^2 + 57x - 36),$$
$$= (x - 1)^2(-x^3 - 14x^2 - 21x + 36),$$
$$= (x - 1)^3(-x^2 - 15x - 36).$$

Then the quadratic $-x^2 - 15x - 36 = -(x^2 + 15x + 36)$ factors as $-(x + 3)(x + 12)$. Therefore, the roots are $x = \boxed{1, 1, 1, -3, \text{ and } -12}$.

(d) We look for integer roots among the factors 24: 1, 2, 3, 4, 6, and their negatives, and so on. To find an integer root, we must also check the negative values, and we find $h(-2) = 6 \cdot (-2)^3 - 5 \cdot (-2)^2 - 22 \cdot (-2) + 24 = 0$, so by the Factor Theorem, $y + 2$ is a factor of $h(y)$. Dividing $h(y)$ by $y + 2$, we get

$$h(y) = 6y^3 - 5y^2 - 22y + 24 = (y + 2)(6y^2 - 17y + 12).$$

Then the quadratic $6y^2 - 17y + 12$ factors as $(2y - 3)(3y - 4)$. Therefore, the roots are $y = \boxed{3/2, 4/3, \text{ and } -2}$.

7.2.2

(a) Moving all the terms to one side, we obtain the inequality $t^3 + 10t^2 + 17t - 28 > 0$. Factoring, the inequality becomes $(t + 7)(t + 4)(t - 1) > 0$.

We list the signs of the factors in the following table:

	$t + 7$	$t + 4$	$t - 1$	$(t + 7)(t + 4)(t - 1)$
$t < -7$	$-$	$-$	$-$	$-$
$-7 < t < -4$	$+$	$-$	$-$	$+$
$-4 < t < 1$	$+$	$+$	$-$	$-$
$t > 1$	$+$	$+$	$+$	$+$

Therefore, the solution is $t \in \boxed{(-7, -4) \cup (1, +\infty)}$.

(b) Moving all the terms to one side, we obtain the inequality $r^4 - 6r^3 - 12r^2 + 70r + 75 \geq 0$. Factoring, the inequality becomes $(r+3)(r+1)(r-5)^2 \geq 0$.

We list the signs of the factors in the following table:

	$r+3$	$r+1$	$(r-5)^2$	$(r+3)(r+1)(r-5)^2$
$r < -3$	$-$	$-$	$+$	$+$
$-3 < r < -1$	$+$	$-$	$+$	$-$
$-1 < r < 5$	$+$	$+$	$+$	$+$
$r > 5$	$+$	$+$	$+$	$+$

Since the inequality is nonstrict (meaning that equality is allowed), we include the roots. Therefore, the solution is $r \in \boxed{(-\infty, -3] \cup [-1, +\infty)}$.

(c) The inequality factors as $(r+1)(r-2)(r^2+2r+9) \leq 0$. Completing the square in the quadratic factor, we get $r^2 + 2r + 9 = (r+1)^2 + 8 \geq 8$, so the quadratic factor is always positive.

We list the signs of the factors in the following table:

	$r+1$	$r-2$	r^2+2r+9	$(r+1)(r-2)(r^2+2r+9)$
$r < -1$	$-$	$-$	$+$	$+$
$-1 < r < 2$	$+$	$-$	$+$	$-$
$r > 2$	$+$	$+$	$+$	$+$

Since the inequality is nonstrict, we include the roots. Therefore, the solution is $r \in \boxed{[-1, 2]}$.

7.2.3 $\boxed{\text{No!}}$ While it is true that no other divisors of 16 are roots of $f(x)$, it is possible that 2 is a multiple root of $f(x)$. To test if this is the case, we first divide $f(x)$ by $x - 2$ to find $f(x) = (x-2)(x^3 - 6x^2 + 12x - 8)$. We find that 2 is a root of $x^3 - 6x^2 + 12x - 8$, so we again divide by $x - 2$ to find $f(x) = (x-2)^2(x^2 - 4x + 4) = (x-2)^4$. All the roots of $f(x)$ are integers, and all four are 2.

7.2.4 We will prove that $f(x)$ has no integer roots by showing that $f(x)$ is an odd integer for all integers x. Let $f(x) = a_n x^n + a_{n-1} x^{n-1} + \cdots + a_1 x + a_0$. Because all the a_i are integers, every term of $f(2) = a_n(2^n) + a_{n-1}(2^{n-1}) + \cdots + a_1(2) + a_0$ except the last term must be even. Therefore, $f(2)$ is an even number plus a_0. Since $f(2) = 3$, we know a_0 must be odd. Furthermore, if k is even, then all of the terms of $f(k) = a_n k^n + a_{n-1} k^{n-1} + \cdots + a_1 k + a_0$ except the last are even, so $f(k)$ is odd.

Next, we note that because $f(7) = a_n(7)^n + a_{n-1}(7)^{n-1} + \cdots + a_1(7) + a_0$, and because both $f(7)$ and a_0 are odd, the sum $a_n(7)^n + a_{n-1}(7)^{n-1} + \cdots + a_1(7)$ is even. For any odd integer k, the parity (oddness or evenness) of each term of $f(k)$ matches the parity of the corresponding term of $f(7)$ above. (For example, the term $a_n k^n$ is odd if $a_n(7^n)$ is odd, and $a_n k^n$ is even if $a_n(7^n)$ is even.) Therefore, the parity of $f(k)$ is the same as the parity of $f(7)$. So, because $f(7)$ is odd, $f(k)$ must also be odd.

Since $f(x)$ is odd for all even integers x and for all odd integers x, there are no integer roots of $f(x)$.

Exercises for Section 7.3

7.3.1

(a) By the Rational Root Theorem, if p/q is a root of $g(y)$, then p divides 10 and q divides 12. Note that $g(0) = 10 > 0$, and $g(1) = -15 < 0$, so $g(y)$ must have a root between 0 and 1. (We don't know yet if this root is rational, but it does give us a place to start looking.)

Checking the rational numbers of the form p/q, where p divides 10 and q divides 12, and where p/q is between 0 and 1, we find that the cubic is 0 for $y = 1/2$, so by the Factor Theorem, $2y - 1$ is a factor of $g(y)$.

Dividing by $2y - 1$, we get

$$12y^3 - 28y^2 - 9y + 10 = (2y - 1)(6y^2 - 11y - 10).$$

Then the quadratic $6y^2 - 11y - 10$ factors as $(2y - 5)(3y + 2)$. Therefore, the roots are $r = \boxed{1/2,\ 5/2,\ \text{and } -2/3}$.

(b) We first note that all the coefficients are positive, so $f(x) > 0$ for all $x > 0$. Because $f(0) = 2$ and $f(-1) = -12$, we know that there is a root between 0 and -1. We check for rational roots between 0 and -1 by trying negatives of fractions in which the denominator is a divisor of 45 and the numerator is a divisor of 2. We get a little lucky, because the simplest such fraction to test is $-1/3$, and it works. We therefore divide $f(x)$ by $x + \frac{1}{3}$, then factor the resulting quadratic to find:

$$f(x) = \left(x + \frac{1}{3}\right)(45x^2 + 33x + 6) = 3\left(x + \frac{1}{3}\right)(15x^2 + 11x + 2) = 3\left(x + \frac{1}{3}\right)(3x + 1)(5x + 2).$$

Therefore, the roots are $\boxed{-1/3, -1/3,\ \text{and } -2/5}$.

7.3.2 We see why both $g(0)$ and $g(1)$ have the same sign by looking at the factorization of $g(x)$:

$$g(x) = 30\left(x + \frac{3}{2}\right)(x - 5)\left(x - \frac{1}{3}\right)\left(x - \frac{3}{5}\right).$$

When $x = 0$, the first binomial is positive and the rest are negative, so $g(0) < 0$. When $x = 1$, the binomial $(x - 5)$ is negative and the other three are positive, so $g(1) < 0$ also. The reason these two end up being the same sign despite the fact that $g(x)$ has roots between 0 and 1 is that $g(x)$ has *two* roots between 0 and 1. Both factors change from negative to positive as we go from $x = 0$ to $x = 1$, leaving the sign of $g(x)$ the same for both $x = 0$ and $x = 1$.

7.3.3 The Rational Root Theorem only applies when all of the coefficients are integers. The polynomial $f(x) = x^3 + \frac{3}{4}x^2 - 4x - 3$ has a non-integer coefficient, namely 3/4, so the Rational Root Theorem does not apply. To apply the Rational Root Theorem, we would have to multiply the polynomial by 4, to obtain a polynomial with integer coefficients.

7.3.4 To solve the inequality $3(x - \frac{1}{3})(x + 2)(2x - 3)^2 > 0$ graphically, we consider the behavior of the function $f(x) = 3(x - \frac{1}{3})(x + 2)(2x - 3)^2$. Note that the roots of $f(x)$ are $x = -2$, 1/3, and 3/2, so the graph of $y = f(x)$ can only cross the x-axis at these points.

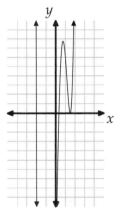

When x is much less than 0, all four linear factors of $f(x)$ are negative, so $f(x)$ is positive. Therefore, the graph is above the x-axis when x is much less than 0. As we increase x towards $x = 0$, the graph must stay above the x-axis until the first point where it crosses the x-axis, which is $(-2, 0)$, corresponding to the smallest root of $f(x)$. Since the factor $x + 2$ appears only once in $f(x)$, we know that $f(x)$ changes sign as x goes from slightly less than -2 to slightly more than -2. In other words, the graph then goes below the x-axis.

The graph must then stay below the x-axis until the next time it intersects the x-axis, which is at $(1/3, 0)$, corresponding to the next highest root of $f(x)$. As with -2, we know that $f(x)$ changes sign when x goes from slightly below 1/3 to slightly above 1/3, so the graph crosses the x-axis at $(1/3, 0)$, and is then above the x-axis until the next time it reaches the x-axis.

The next point at which the graph meets the x-axis is $(3/2, 0)$, corresponding to the last root of $f(x)$. However, the factor $(2x - 3)$ appears twice (is squared), so we see that the sign of $f(x)$ does not change as x goes from slightly less than 3/2 to slightly more than 3/2. In other words, the graph touches the x-axis at $(3/2, 0)$, but does not cross it. So, the graph remains above the x-axis for $x > 3/2$. Thereafter, the graph does not meet the x-axis again, because there are no more roots of $f(x)$.

Combining all those values of x for which $f(x)$ is above the x-axis, we see that $f(x) = 3(x - \frac{1}{3})(x + 2)(2x - 3)^2 > 0$ when $x \in (-\infty, -2) \cup (\frac{1}{3}, \frac{3}{2}) \cup (\frac{3}{2}, +\infty)$.

7.3.5

(a) We can rearrange the inequality as $24x^3 + 26x^2 - 21x - 9 \geq 0$.

First, we look for integer roots of the left side, so we can factor the polynomial. If there are any integer roots, then they must be among the factors of 9. Checking these values, we find that there are no integer roots, so we turn our attention to rational roots. By the Rational Root Theorem, if p/q is a rational root of the left side, where p/q is in lowest terms, then p divides 9 and q divides 24.

Checking these values, we find that the left side becomes 0 for $x = 3/4$, so by the Factor Theorem, $4x - 3$ is a factor of $24x^3 + 26x^2 - 21x - 9$. Dividing $24x^3 + 26x^2 - 21x - 9$ by $4x - 3$, we get

$$24x^3 + 26x^2 - 21x - 9 = (4x - 3)(6x^2 + 11x + 3).$$

Then the quadratic $6x^2 + 11x + 3$ factors as $(2x + 3)(3x + 1)$, so the inequality becomes

$$(2x + 3)(3x + 1)(4x - 3) \geq 0.$$

To solve the inequality, we list the signs of the factors in the following table:

	$2x + 3$	$3x + 1$	$4x - 3$	$(2x + 3)(3x + 1)(4x - 3)$
$x < -3/2$	$-$	$-$	$-$	$-$
$-3/2 < x < -1/3$	$+$	$-$	$-$	$+$
$-1/3 < x < 3/4$	$+$	$+$	$-$	$-$
$x > 3/4$	$+$	$+$	$+$	$+$

And of course, we have $(2x + 3)(3x + 1)(4x - 3) = 0$ for $x = -3/2, -1/3,$ and $3/4$. Therefore, the solution is $x \in \boxed{[-3/2, -1/3] \cup [3/4, +\infty)}$.

(b) First, we look for integer roots of the left side, so we can factor the polynomial. If there are any integer roots, then they must be among the factors of 12. Checking these values, we find the quartic is 0 for $s = -3$, so by the Factor Theorem, $s + 3$ is a factor. Dividing by $s + 3$, we get

$$6s^4 + 13s^3 - 2s^2 + 35s - 12 = (s + 3)(6s^3 - 5s^2 + 13s - 4).$$

If we look for integer roots of the cubic, which must be among the factors of 4, we find there are none, so we then turn to rational roots. If a/b is a rational root (in lowest terms), then by the Rational Root Theorem, a must divide 4 and b must divide 6. Checking these values, we find the cubic is 0 for $s = 1/3$, so by the Factor Theorem, $3s - 1$ is a factor. Dividing by $3s - 1$, we get

$$6s^3 - 5s^2 + 13s - 4 = (3s - 1)(2s^2 - s + 4).$$

The discriminant of the quadratic $2s^2 - s + 4$ is $1^2 - 4 \cdot 2 \cdot 4 = -31$, which is negative, so it has no real roots. Furthermore, the coefficient of s^2 is positive, so $2s^2 - s + 4 > 0$ for all real numbers s. (Alternatively, we might have noted that the minimum value of the quadratic is $4 - (-1)^2/(4 \cdot 2) = 31/8$, so the quadratic cannot be negative.)

To solve the inequality, we list the signs of the factors in the following table:

	$s + 3$	$3s - 1$	$2s^2 - s + 4$	$(s + 3)(3s - 1)(2s^2 - s + 4)$
$s < -3$	$-$	$-$	$+$	$+$
$-3 < s < 1/3$	$+$	$-$	$+$	$-$
$s > 1/3$	$+$	$+$	$+$	$+$

Therefore, the solution is $\boxed{s \in (-3, 1/3)}$.

Exercises for Section 7.4

7.4.1

(a) We begin by dividing $6r^3 + 31r^2 + 34r - 15$ by $r - 1$ using the synthetic division at right. We find that $6r^3 + 31r^2 + 34r - 15 = (r - 1)(6r^2 + 37r + 71) + 56$. Since the remainder 56 is positive and all the coefficients in the quotient $6r^2 + 37r + 71$ are nonnegative, the cubic has no roots greater than 1.

$$\begin{array}{r|rrrr} 1 & 6 & 31 & 34 & -15 \\ & & 6 & 37 & 71 \\ \hline & 6 & 37 & 71 & 56 \end{array}$$

Looking for integer roots less than 1, which must be among the factors of 15, we find that the cubic is 0 for $r = -3$, so by the Factor Theorem, $r + 3$ is a factor. Dividing by $r + 3$, we get

$$6r^3 + 31r^2 + 34r - 15 = (r + 3)(6r^2 + 13r - 5).$$

Then the quadratic $6r^2 + 13r - 5$ factors as $(3r - 1)(2r + 5)$. Therefore, the solutions are $r = \boxed{-3, 1/3, \text{ and } -5/2}$.

(b) We observe that all the even powers of r have positive coefficients, and all the odd powers of r have negative coefficients. Therefore, all the real roots of the quartic must be positive.

Looking for (positive) integer roots, which must be among the factors of 9, we find that the quartic is 0 for $r = 3$, so by the Factor Theorem, $r - 3$ is a factor. In fact, we find that we may divide by $r - 3$ twice:

$$4r^4 - 28r^3 + 61r^2 - 42r + 9 = (r - 3)(4r^3 - 16r^2 + 13r - 3) = (r - 3)^2(4r^2 - 4r + 1).$$

Then the quadratic $4r^2 - 4r + 1$ factors as $(2r - 1)^2$. Therefore, the solutions are $r = \boxed{3 \text{ and } 1/2}$. (Note that the roots are 3, 3, 1/2, and 1/2.)

7.4.2 $\boxed{\text{No}}$, it is not necessarily true that all the other roots are either greater than 1 or less than 0. For example, let $f(x) = (2x - 1)(3x - 1)(20x - 3)$. Then $f(0) = (-1)(-1)(-3) = -3$, and $f(1) = (2 - 1)(3 - 1)(20 - 3) = 34$. Furthermore, the roots of $f(x)$ are 1/2, 1/3, and 3/20, all of which are between 0 and 1. The fact that $f(0) = -3$ and $f(1) = 34$ does tell us that there is a root between 0 and 1, but there could be more than one such root.

7.4.3 Suppose that we have a polynomial in x, where some of the coefficients are positive and others are 0. If we substitute a positive value for x, then we obtain a sum where each term is either positive or 0, and at least one term is positive, so the sum must be positive. Therefore, the polynomial $\boxed{\text{cannot have a positive root}}$. If all of the coefficients are negative or 0, then we can multiply the polynomial by -1 to obtain a polynomial in which all the coefficients are positive or 0.

7.4.4 We first rearrange the equation as $x^3 = -(a_2 x^2 + a_1 x + a_0)$. So, to make x as large as possible, we want $a_2 x^2 + a_1 x + a_0$ to be as small as possible. We expect that we can make this expression as small as possible by making the a_i as small as possible. This suggests that $f(x) = x^3 - 2x^2 - 2x - 2$ is the desired polynomial with the greatest root r. (We will justify this claim below.) To find bounds on r, we substitute the endpoints of the intervals given in the choices. Since $f(5/2) = -31/8$, $f(3) = 1$, and $f(7/2) = 75/8$, we conclude that $5/2 < r < 3$, so the answer is $\boxed{\text{(D)}}$.

Now, let's take a look at why the polynomial with the greatest root is $f(x) = x^3 - 2x^2 - 2x - 2$. We let $g(x) = x^3 + a_2 x^2 + a_1 x + a_0$, where $|a_i| \le 2$ for all i, and at least one of the a_i is not equal to -2. Furthermore, let r be the greatest root of $f(x)$, as before. Because r is the greatest root of $f(x)$, and the leading coefficient of $f(x)$ is positive, we know that $f(x) > 0$ for all $x > r$. (Otherwise, the graph of f goes below the x-axis to the right of $x = r$. This graph must then cross the x-axis again to the right of $x = r$, because $f(x)$ is positive for very large values of x.)

Let t be the greatest root of $g(x)$, so we have $g(t) = t^3 + a_2 t^2 + a_1 t + a_0 = 0$. Therefore, we have $f(t) - g(t) = (-2 - a_2)t^2 + (-2 - a_1)t + (-2 - a_0)$. Because $a_0 \ge -2$, $a_1 \ge -2$, and $a_2 \ge -2$, if $t > 0$, then we must have $f(t) - g(t) \le 0$, which means $f(t) \le g(t)$. However, we cannot have $f(t) \le 0$ if $t > r$. Therefore, we must have either $t < 0$ (in which case $t < r$), or we must have $t \le r$. This tells us that our $f(x)$ does indeed produce our desired maximal root.

Exercises for Section 7.5

7.5.1

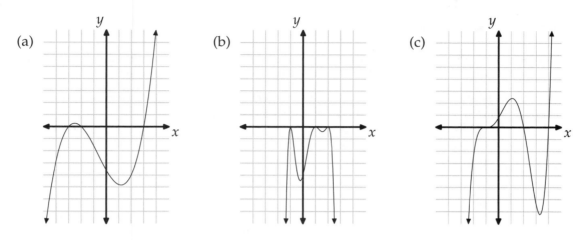

7.5.2

(a) The quartic factors as $x^4 - 8x^2 + 16 = (x^2 - 4)^2 = [(x - 2)(x + 2)]^2 = (x - 2)^2(x + 2)^2$, so the roots are $x = \boxed{2, 2, -2, \text{ and } -2}$.

(b) Since $x^4 - 8x^2 + 16 = (x^2 - 4)^2 \geq 0$ for all real x, the inequality $x^4 - 8x^2 + 16 < 0$ has $\boxed{\text{no solutions}}$.

(c)

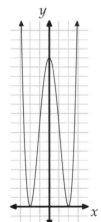

7.5.3

(a) Suppose that -4 is a double root of $f(x)$. Then $f(x) = (x + 4)^2 g(x)$ for some polynomial $g(x)$ that does not have -4 as a root. Therefore, $g(x)$ has the same sign (positive or negative) for values of x that are close to -4. Also, $(x + 4)^2 > 0$ for all $x \neq -4$, so $f(x) = (x + 4)^2 g(x)$ has the same sign for values of x that are slightly less than -4 and for values of x that are slightly greater than -4. Graphically speaking, the graph of $f(x)$ would be tangent to the x-axis at $x = -4$.

However, the graph shows that $f(x)$ is positive for values of x that are slightly less than -4, and $f(x)$ is negative for values of x that are slightly greater than -4. Therefore, -4 cannot be a double root of $f(x)$.

(b) We know that $x = -4, -3, 0$, and 2 are all roots $f(x) = 0$. Therefore, the degree of $f(x)$ is at least $\boxed{4}$. (The degree of $f(x)$ could be greater than 4, because there may be nonreal roots.)

(c) Let $\deg f(x) = 4$. By the Factor Theorem, $x + 4$, $x + 3$, x, and $x - 2$ are all factors of $f(x)$, so

$$f(x) = (x + 4)(x + 3)x(x - 2)q(x)$$

for some polynomial $q(x)$. Taking the degree of both sides, we get $4 = \deg f = 4 + \deg q$, so $\deg q = 0$, which means that $q(x)$ is a constant, say c.

Since $f(-2) = 4$, we let $x = -2$ and we have $4 = (-2+4)(-2+3)(-2)(-2-2)c$, so $c = 1/4$. Therefore, we have $f(x) = \boxed{\dfrac{1}{4}(x+4)(x+3)x(x-2)}$.

7.5.4 If the graphs of $y = x$ and $y = f(x)$ intersect at $x = r$, then r is a root of the polynomial $f(x) - x = 0$. Therefore, if the graphs intersect at seven different points, then the polynomial $f(x) - x$ would have seven real roots.

But the degree of $f(x) - x$ is 6, so by the Fundamental Theorem of Algebra, the polynomial $f(x) - x$ has at most 6 real roots. Therefore, the two graphs cannot intersect at seven different points.

Exercises for Section 7.6

7.6.1

(a) Let $g(x) = f(x) - 1$, so $g(0) = g(1) = g(2) = g(3) = 0$ and $g(4) = -1$, and $\deg g = 4$. Then $g(x) = cx(x-1)(x-2)(x-3)$ for some constant c. Taking $x = 4$, we get $g(4) = c \cdot 4 \cdot 3 \cdot 2 \cdot 1 = 24c$. Since $g(4) = -1$, we have $c = -1/24$. Therefore, $g(x) = -\frac{1}{24}x(x-1)(x-2)(x-3)$, so $f(5) = g(5)+1 = -\frac{1}{24} \cdot 5 \cdot 4 \cdot 3 \cdot 2 + 1 = -5+1 = -4$.

(b) Let $g(x) = f(x) - 1$, so $g(0) = g(1) = g(2) = \cdots = g(n-1) = 0$ and $g(n) = -1$, and $\deg g = n$. Then

$$g(x) = cx(x-1)(x-2)\cdots[x-(n-1)]$$

for some constant c. Taking $x = n$, we get $g(n) = cn \cdot (n-1) \cdots 1 = cn!$. Since $g(n) = -1$, we have $c = -1/n!$. Therefore, $g(x) = -\frac{1}{n!}x(x-1)\cdots[x-(n-1)]$, so

$$f(n+1) = g(n+1) + 1 = -\frac{1}{n!} \cdot (n+1) \cdot n \cdots 2 + 1 = -\frac{1}{n!} \cdot (n+1)! + 1 = -(n+1) + 1 = -n.$$

7.6.2 Note that $0 = 1^3 - 1$, $7 = 2^3 - 1$, $26 = 3^3 - 1$, and $63 = 4^3 - 1$. So, $p(x) = x^3 - 1$ is one possible polynomial of degree 3. By the Identity Theorem, there is at most one polynomial of degree at most 3 that satisfies these four conditions, so the only such polynomial is $\boxed{x^3 - 1}$.

7.6.3 By the Remainder Theorem, we have $p(a) = a$, $p(b) = b$, and $p(c) = c$.

When $p(x)$ is divided by $(x-a)(x-b)(x-c)$, since $(x-a)(x-b)(x-c)$ has degree 3, the remainder has degree at most 2, or the remainder is 0. Let $q(x)$ be the quotient, so

$$p(x) = (x-a)(x-b)(x-c)q(x) + r(x).$$

Taking $x = a$, $x = b$, and $x = c$, respectively, we obtain the equations $p(a) = r(a)$, $p(b) = r(b)$, and $p(c) = r(c)$. Using our values of $p(a)$, $p(b)$, and $p(c)$ above, we have $r(x) = x$ for $x = a, b$, and c. Therefore, we have found 3 different roots of the polynomial $r(x) - x$. But $r(x) - x$ has degree at most 2 or is 0, so we must have $r(x) = x$ for all x. Therefore, the remainder is $\boxed{x}$.

7.6.4 Let $f(x) = ax^3 + bx^2 + cx + d$, a polynomial of degree at most 3. Then the given equations can be rewritten as $f(1) = 1$, $f(2) = 16$, $f(3) = 81$, and $f(4) = 256$. Note that $1 = 1^4$, $16 = 2^4$, $81 = 3^4$, and $256 = 4^4$, so we have 4 values of x such that $f(x) = x^4$. Therefore, the polynomial $x^4 - f(x)$ has roots 1, 2, 3, and 4. Because the degree of $f(x)$ is at most 3, the degree of $x^4 - f(x)$ is 4. Since 1, 2, 3, and 4 are the roots of this quartic, we have

$$x^4 - f(x) = k(x-1)(x-2)(x-3)(x-4),$$

so

$$f(x) = x^4 - k(x-1)(x-2)(x-3)(x-4).$$

The degree of the left side is at most 3, so the degree of the right is also at most 3, which means the coefficient of x^4 is 0. The coefficient of x^4 in the right side is $1 - k$, so $k = 1$. Therefore,

$$f(x) = x^4 - (x-1)(x-2)(x-3)(x-4) = 10x^3 - 35x^2 + 50x - 24.$$

Hence, the solution to the system is $(a, b, c, d) = \boxed{(10, -35, 50, -24)}$.

7.6.5 We have $P(x) = \frac{1}{x}$ for $n = 1, 2, 3, \ldots, 1997$, so it appears that we have 1997 roots of $P(x) - \frac{1}{x}$. Unfortunately, this expression is not a polynomial, so we can't use our polynomial tools on it. We get around this difficulty by noting that when n is nonzero, $P(n) = \frac{1}{n}$ also means $nP(n) = 1$. Since this equation holds for $n = 1, 2, 3, \ldots, 1997$, we have 1997 roots of $xP(x) - 1$.

Let $Q(x) = xP(x) - 1$. Since $\deg P = 1996$, we have $\deg Q = 1997$. Because $Q(1) = Q(2) = \cdots = Q(1997) = 0$, we have

$$Q(x) = (x-1)(x-2)\cdots(x-1997)R(x)$$

for some polynomial $R(x)$. Taking the degree of both sides, we get $\deg Q = 1997 + \deg R = 1997$, so $\deg R = 0$, which means that $R(x)$ is a constant.

Letting $R(x) = c$ and $x = 0$ gives $Q(0) = (-1)(-2)\cdots(-1997)c = -1997!c$. Since $Q(0) = 0P(0) - 1 = -1$, we have $c = \frac{1}{1997!}$. Therefore, we have $Q(x) = \frac{1}{1997!}(x-1)(x-2)\cdots(x-1997)$, so $Q(1998) = \frac{1}{1997!} \cdot 1997 \cdot 1996 \cdots 1 = 1$. But $Q(1998) = 1998P(1998) - 1$, so $1998P(1998) - 1 = 1$. Therefore, $P(1998) = \frac{2}{1998} = \boxed{\frac{1}{999}}$.

Review Problems

7.30 Let $f(x) = x^3 + 2kx^2 + k^2x + k - 4$. If $x - 2$ is a factor of $f(x)$, then by the Factor Theorem, $f(2) = 0$. Since $f(2) = 2^3 + 2k \cdot 2^2 + k^2 \cdot 2 + k - 4 = 2k^2 + 9k + 4 = (k+4)(2k+1)$, the values of k such that $f(2) = 0$ are $k = \boxed{-4 \text{ and } -1/2}$.

7.31 Expanding the equation $f(f(x)) = f(x)$, we get $(x^2 + 4x)^2 + 4(x^2 + 4x) = x^2 + 4x$. We can expand both sides, but if we keep the appearances of $x^2 + 4x$ intact, it makes the equation much easier to solve. Subtracting $x^2 + 4x$ from both sides, we get $(x^2 + 4x)^2 + 3(x^2 + 4x) = 0$. Simplifying the left side gives

$$(x^2 + 4x)^2 + 3(x^2 + 4x) = (x^2 + 4x)[(x^2 + 4x) + 3] = x(x+4)(x+3)(x+1).$$

Therefore, the solutions are $x = \boxed{0, -4, -3, \text{ and } -1}$.

7.32 The Rational Root Theorem states how to find all rational roots of a polynomial with integer coefficients. However, a polynomial may also have irrational roots, which the Rational Root Theorem will not be able to determine. Indeed, it turns out that the root of $2x^3 - x^2 - 5x + 3 = 0$ that is between 0 and 1 is $x = (-1 + \sqrt{5})/2 \approx 0.618$, which is irrational.

7.33 The integer root of $4x^3 - 41x^2 + 10x - 1989 = 0$ must be a factor of 1989, whose prime factorization is $3^2 \cdot 13 \cdot 17$. Checking these values, we find that the integer root is $x = \boxed{13}$.

7.34

(a) The quartic $f(x) = (x-1)^2(x-2)(x-3) = x^4 - 7x^3 + 17x^2 - 17x + 6$ satisfies all of the given conditions, and so demonstrates that it is possible for 3 to be a root of $f(x)$.

(b) Suppose that 3 is a root of $f(x)$. Then, we must have $f(x) = (x-3)g(x)$, where $g(x)$ has integer coefficients. Since the constant term of $f(x)$ is 6, the constant term of $g(x)$ must be $6/(-3) = -2$. In order for 3 to be a double root of $f(x)$, it must also be a root of $g(x)$. But 3 does not divide 2, so by the Rational Root Theorem, 3 cannot be a root of $g(x)$. Therefore, 3 cannot be a double root of $f(x)$.

7.35 First, we give an example to show that the Rational Root Theorem can fail if all the coefficients are rational, but not integers. Consider the polynomial

$$x^2 - \frac{5}{2}x + 1 = 0.$$

If the Rational Root Theorem applied to this polynomial, then the only possible rational roots would be ± 1. But

$$x^2 - \frac{5}{2}x + 1 = \frac{2x^2 - 5x + 2}{2} = \frac{(x-2)(2x-1)}{2},$$

so this polynomial does have rational roots not equal to 1 or -1, namely 2 and 1/2.

To see why the Rational Root Theorem fails, we look at the proof in the text. We showed that if p/q is a root of $f(x) = a_n x^n + a_{n-1} x^{n-1} + \cdots + a_1 x + a_0$, where p/q is in lowest terms, then

$$\frac{a_n p^n}{q} = -a_{n-1} p^{n-1} - a_{n-2} p^{n-2} q - \cdots - a_1 p q^{n-2} - a_0 q^{n-1}.$$

To show that q divides a_n, we used the fact that the right side is an integer. However, if the coefficients a_i are not necessarily integers, then the right side is not necessarily an integer either. Thus, we require that all the coefficients a_i be integers.

7.36

(a) Moving all the terms to one side, we get $2x^3 + 19x^2 + 52x + 35 = 0$. Note that since all the coefficients are positive, any real roots must be negative. First, we look for integer roots. If there are any integer roots, then they must be among the (negative) factors of 35. We quickly see that -1 is a root, and dividing by $x + 1$ gives $2x^3 + 19x^2 + 52x + 35 = (x+1)(2x^2 + 17x + 35)$. Factoring the quadratic gives us $2x^3 + 19x^2 + 52x + 35 = (x+1)(x+5)(2x+7)$, so the solutions are $x = \boxed{-1, -5, \text{ and } -7/2}$.

(b) Moving all the terms to one side, we get $16y^3 + 48y^2 - 61y + 12 = 0$. If there are any integer roots, then they must be among the factors of 12. Checking these values, we find that the cubic is 0 for $y = -4$, so by the Factor Theorem, $y + 4$ is a factor. Dividing by $y + 4$, we get

$$16y^3 + 48y^2 - 61y + 12 = (y+4)(16y^2 - 16y + 3).$$

Then the quadratic $16y^2 - 16y + 3$ factors as $(4y-1)(4y-3)$. Therefore, the solutions are $y = \boxed{1/4, 3/4, \text{ and } -4}$.

7.37 Let the four consecutive positive integers be n, $n + 1$, $n + 2$, and $n + 3$. Then the sum of three of the numbers n^3, $(n+1)^3$, $(n+2)^3$, and $(n+3)^3$ is equal to the fourth remaining number. Since all of these numbers are positive, the greatest number must be equal to the sum of the other three. In other words, $n^3 + (n+1)^3 + (n+2)^3 = (n+3)^3$, which simplifies to $2(n^3 - 6n - 9) = 0$. Since n is a positive integer, n is a factor of 9. Checking these values, we find that the only solution is $n = 3$, so the only set is $\boxed{\{3, 4, 5, 6\}}$.

7.38

(a) First, we look for integer roots. We start with the easy ones to check, 0, 1, and -1. Fortunately, $f(-1) = 0$, so $r + 1$ is a factor of $f(r)$. Dividing $f(r)$ by $r + 1$ gives us

$$6r^3 - 31r^2 - 31r + 6 = (r+1)(6r^2 - 37r + 6).$$

Then the quadratic $6r^2 - 37r + 6$ factors as $(6r - 1)(r - 6)$. Therefore, the roots are $r = \boxed{6, 1/6, \text{ and } -1}$.

(b) Note that all the coefficients of $g(t)$ are positive, so if there are any real roots, they must be negative. First, we look for integer roots. If there are any integer roots, then they must be among the (negative) factors of 14. We find that $g(-1) = 0$, and dividing by $t + 1$ gives us

$$g(t) = (t+1)(6t^3 + 35t^2 + 53t + 14).$$

We continue our hunt for roots with the cubic. We find that $t = -2$ is a root, and dividing the cubic by $t + 2$ gives us $g(t) = (t+1)(t+2)(6t^2 + 23t + 7)$. Then the quadratic $6t^2 + 23t + 7$ factors as $(3t+1)(2t+7)$. Therefore, the roots are $r = \boxed{-1, -2, -1/3, \text{ and } -7/2}$.

(c) We have $h(0) = 30$, $h(1) = 15$, and $h(2) = -8$, so there is a root between 1 and 2. If the root is rational, its numerator divides 30 and its denominator divides 12. Fortunately, the simplest choice, 3/2, works. Dividing $h(t)$ by $t - \frac{3}{2}$ gives us

$$h(t) = \left(t - \frac{3}{2}\right)(12t^2 - 22t - 20) = 2\left(t - \frac{3}{2}\right)(6t^2 - 11t - 10).$$

The quadratic $6t^2 - 11t - 10$ factors as $(2t-5)(3t+2)$, so our roots are $\boxed{-2/3, 3/2, \text{ and } 5/2}$.

7.39 Let $r = a/b$. Then the calculator will let Kai play his game when $30r^3 + 11r^2 = 59r + 12$, which simplifies to $30r^3 + 11r^2 - 59r - 12 = 0$. By the Rational Root Theorem, a is a factor of 12 and b is a factor of 30. Checking these values, we find that $r = 4/3$ is a root, so $\boxed{a = 4 \text{ and } b = 3}$.

7.40

(a) Moving all the terms to one side, the inequality becomes $6t^5 + 2t^4 \geq 0$, so $2t^4(3t+1) \geq 0$. Since $t^4 \geq 0$ for all t, this inequality holds if and only if $3t + 1 \geq 0$. Therefore, the solution is $t \in \boxed{[-1/3, +\infty)}$.

(b) First, we look for integer roots of the polynomial on the left side. We quickly find that $x = 1$ is a root, and factoring out $x - 1$ gives us $(x-1)(2x^3 + 15x^2 + 24x - 16) \geq 0$. If there are any integer roots of the cubic, then they must be among the factors of 16. Checking these values, we find that $x = -4$ is a root, so $x + 4$ is a factor. Factoring out $x + 4$ gives us $(x-1)(x+4)(2x^2 + 7x - 4) \geq 0$. The quadratic $2x^2 + 7x - 4$ factors as $(x+4)(2x-1)$, so the the inequality becomes $(x-1)(x+4)^2(2x-1) \geq 0$.

Since $(x+4)^2 \geq 0$ for all x, the given inequality holds if and only if $(x-1)(2x-1) \geq 0$. Therefore, the solution is $x \in \boxed{(-\infty, 1/2] \cup [1, +\infty)}$.

7.41

(a) It is possible that $f(x)$ has negative roots if the nonzero coefficients of $f(x)$ alternate in sign. For example, let $f(x) = x^2 - 1$. The nonzero coefficients alternate in sign, but $x^2 - 1 = (x-1)(x+1)$, so $x = -1$ is a negative root.

(b) The nonzero condition is important, because it ensures that all the terms with even degree have positive coefficients, and all the terms with odd degree have negative coefficients, or vice versa.

Hence, we can amend the statement to allow some coefficients to be equal to zero as follows. If all the terms with even degree have nonnegative coefficients, and all the terms with odd degree have nonpositive coefficients (or vice versa), then the polynomial does not have any negative roots.

7.42 Every intersection of the graphs of $y = p(x)$ and $y = q(x)$ corresponds to a solution of $p(x) = q(x)$, or $p(x) - q(x) = 0$. The degrees of $p(x)$ and $q(x)$ are both 4, but both leading coefficients are 1, so $p(x) - q(x)$ has degree at most 3. Therefore, the equation $p(x) - q(x) = 0$ has at most 3 solutions, so the graphs of $p(x)$ and $q(x)$ intersect in at most $\boxed{3}$ points.

7.43

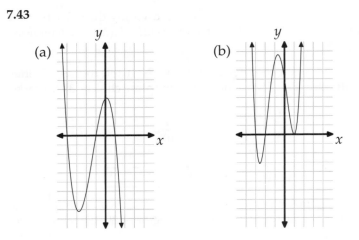

(a)

(b)

7.44 It is clear from the graph that y cannot be a function of x. Instead, we can try to express x as a function of y; more specifically we let $x = p(y)$, where $p(y)$ is a polynomial in y.

From the y-intercepts, we have $p(3) = p(-2) = p(-4) = 0$, so we can let $p(y) = c(y - 3)(y + 2)(y + 4)$ for some constant c. From the x-intercept, we have $p(0) = 3$, so $c \cdot (-3) \cdot 2 \cdot 4 = 3$, which means $c = -1/8$. Therefore,

$$x = \boxed{-\frac{1}{8}(y - 3)(y + 2)(y + 4)}.$$

7.45 If the line is vertical, then it is of the form $x = c$ for some constant c. Then the line $x = c$ intersects the graph of $y = f(x)$ at the one point $(c, f(c))$.

Otherwise, the line is of the form $y = ax + b$ for some constants a and b. Then every intersection of the graph of $y = f(x)$ and the line $y = ax + b$ corresponds to a solution of the equation $f(x) = ax + b$, which must also be a solution of $f(x) - ax - b = 0$. The polynomial $f(x) - ax - b$ has degree n, so it has at most n roots. Therefore, the number of intersections is at most n.

7.46 Let $g(x) = f(x) + x^2$. Then $g(x)$ is a monic quartic polynomial, and $g(-1) = g(2) = g(-3) = g(4) = 0$. Therefore, $g(x) = (x+1)(x-2)(x+3)(x-4)$, so $f(x) = (x+1)(x-2)(x+3)(x-4) - x^2$. Hence, $f(1) = (1+1)(1-2)(1+3)(1-4) - 1^2 = \boxed{23}$.

Challenge Problems

7.47 To help us solve this problem, we graph the function

$$f(x) = \frac{2(x - 1)^2(2x + 3)(x + 5)^3}{100}$$

at right. (We have scaled the function by a factor of 100 to make the behavior of the function easier to see.)

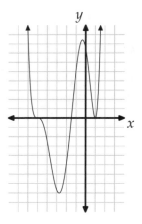

Then, the given equation becomes

$$\frac{2(x - 1)^2(2x + 3)(x + 5)^3}{100} = \frac{0.001}{100} = 10^{-5},$$

and all solutions correspond to the points where the graph of $y = f(x)$ intersects the line $y = 10^{-5}$. Since 10^{-5} is such a small number, all these points of intersection must be close to where the graph of $y = f(x)$ crosses the x-axis, which occurs at $x = -5$, $x = -3/2$, and $x = 1$.

If $x < -5$, then $f(x) > 0$, so there is one solution to $f(x) = 10^{-5}$ where x is slightly less than -5. Next, $f(x) < 0$ if x is slightly less than $-3/2$, and $f(x) > 0$ if x is slightly greater than $-3/2$, so there is another solution to $f(x) = 10^{-5}$ where x is slightly greater than $-3/2$.

Finally, we consider the behavior of $f(x)$ around $x = 1$. We note that $f(1) = 0$, but that $f(x) > 0$ for x both slightly less than 1 and slightly greater than 1, so around $x = 1$, the graph of $y = f(x)$ intersects the line $y = 10^{-5}$ twice, once on either side of $x = 1$.

Therefore, there are $\boxed{4}$ solutions to the equation $2(x - 1)^2(2x + 3)(x + 5)^3 = 0.001$.

7.48   The easiest way to ensure that both equations $y = f(x)$ and $x = f(y)$ are satisfied is to have $y = x$.

The equation $f(x) = x$ becomes $x^3 - x^2 - 13x + 24 = x$, which simplifies as $x^3 - x^2 - 14x + 24 = 0$. First, we look for integer roots. If there are any integer roots, then they must be among the factors of 24. Checking these values, we find that the cubic becomes 0 for $x = 2, 3$, and -4. This gives us the three solutions $\boxed{(2,2), (3,3), \text{ and } (-4,-4)}$. (Note: There are other ordered pairs that satisfy the problem, but they are much, much harder to find.)

7.49   Since $g(x)$ is divisible by $x - 4$, by the Factor Theorem, we have $g(4) = 0$. Since

$$g(4) = 4^3 - 4^2 - (m^2 + m) \cdot 4 + 2m^2 + 4m + 2 = 50 - 2m^2,$$

we have $50 - 2m^2 = 0$, so $m = \pm 5$. We check both solutions.

If $m = 5$, then $g(x) = x^3 - x^2 - 30x + 72 = (x - 4)(x^2 + 3x - 18) = (x - 4)(x + 6)(x - 3)$, and all the roots are integers.

If $m = -5$, then $g(x) = x^3 - x^2 - 20x + 32 = (x - 4)(x^2 + 3x - 8)$, but $x^2 + 3x - 8$ does not have integer roots.

Therefore, the only solution is $m = \boxed{5}$.

7.50   By the Remainder Theorem, the remainder when $f(x)$ is divided by $x - a$ is $f(a)$, and the remainder when $f(x)$ is divided by $x + a$ is $f(-a)$. Therefore, $f(a) = f(-a)$ for all real numbers a.

Let $f(x) = c_n x^n + c_{n-1} x^{n-1} + \cdots + c_1 x + c_0$. Then, we have

$$f(-x) = c_n(-x)^n + c_{n-1}(-x)^{n-1} + \cdots - c_1 x + c_0 = (-1)^n c_n x^n + (-1)^{n-1} c_{n-1} x^{n-1} + \cdots - c_1 x + c_0.$$

By the Identity Theorem, $f(x)$ and $f(-x)$ are the same polynomial because $f(a) = f(-a)$ for more than n values of a. Therefore, $c_k = (-1)^k c_k$ for all $0 \le k \le n$. If k is even, then this equation becomes $c_k = c_k$. But if k is odd, then this equation becomes $-c_k = c_k$, so $c_k = 0$. Therefore, the polynomial $f(x)$ only contains terms of even degree.

7.51   Let $P(x) = a_{n-1}x^{n-1} + a_{n-2}x^{n-2} + \cdots + a_1 x + a_0$. We are given that $P(0) = 0$, so $a_0 = 0$. Then the equation $54x^n + P(x) = 315$ becomes $54x^n + a_{n-1}x^{n-1} + a_{n-2}x^{n-2} + \cdots + a_1 x - 315 = 0$.

By the Rational Root Theorem, for any rational root a/b (in lowest terms), a divides $315 = 3^2 \cdot 5 \cdot 7$ and b divides $54 = 2 \cdot 3^3$. Therefore,

$$\frac{a}{b} = \pm \frac{3^p \cdot 5^q \cdot 7^r}{2^s \cdot 3^t} = \pm 2^{-s} \cdot 3^{p-t} \cdot 5^q \cdot 7^r,$$

where $0 \le p \le 2, 0 \le q \le 1, 0 \le r \le 1, 0 \le s \le 1$, and $0 \le t \le 3$. Then $-3 \le p - t \le 2$.

There are 2 possible choices for $-s$, the exponent of 2. There are 2 possible choices for q, the exponent of 5. There are 2 possible choices for r, the exponent of 7. Finally, there are 6 possible choices for $p - t$, the exponent of 3. And each rational root may be positive or negative, so the number of possible rational roots is $2 \cdot 2 \cdot 2 \cdot 6 \cdot 2 = \boxed{96}$.

For a complete solution, we should show that for any such root a/b, there does exist such a polynomial $P(x)$. Since a divides 315, $c = 315/a$ is an integer, and since b divides 54, $d = 54/b$ is also an integer. Consider the polynomial

$$(bx - a)(dx^{n-1} + c),$$

which clearly has $x = a/b$ as a root. Expanding, we get

$$bdx^n - adx^{n-1} + bcx - ac = 54x^n - adx^{n-1} + bcx - 315.$$

We want $P(x)$ to satisfy $54x^n + P(x) - 315 = 0$, so take

$$P(x) = -adx^{n-1} + bcx.$$

Then clearly $P(x)$ has degree $n - 1$ and $P(0) = 0$, as desired.

7.52 Let $F(x) = P(x) - 10x$. Then $F(1) = F(2) = F(3) = 0$, so by the Factor Theorem,

$$F(x) = (x - 1)(x - 2)(x - 3)Q(x)$$

for some polynomial $Q(x)$. Taking the degree of both sides, we get $\deg F = 4 = 3 + \deg Q$, so $\deg Q = 1$. Furthermore, $P(x)$ is monic, so $F(x)$ is monic. Therefore, $Q(x) = x - r$ for some real number r. Hence, $F(x) = (x-1)(x-2)(x-3)(x-r)$.

Then, we have $P(x) = F(x) + 10x$, so

$$
\begin{aligned}
\frac{P(12) + P(-8)}{10} &= \frac{F(12) + 10 \cdot 12 + F(-8) + 10 \cdot (-8)}{10} \\
&= \frac{(12-1)(12-2)(12-3)(12-r) + (-8-1)(-8-2)(-8-3)(-8-r) + 40}{10} \\
&= \frac{11880 - 990r + 990r + 7920 + 40}{10} \\
&= \boxed{1984}.
\end{aligned}
$$

7.53 Suppose that $f(x)$ has rational roots, say r and s. Since $f(x)$ is monic, r and s must be integers. Also, by Vieta's Formulas, we have $r + s = a - n$ and $rs = a$. We can then express n in terms of r and s as follows:

$$n = a - r - s = rs - r - s.$$

Adding 1 to both sides, we get $n + 1 = rs - r - s + 1 = (r - 1)(s - 1)$. Thus, whether such roots r and s exists depends on the prime factorization of $n + 1$. We note that since $rs = a$ is positive, r and s have the same sign. And since $r + s = a - n$ is negative, both r and s must be negative.

If $n + 1$ is a prime, say p, then there are two cases: $r - 1$ and $s - 1$ are equal to 1 and p, or -1 and $-p$, in some order. We found earlier that r and s must be negative, so neither of $r - 1$ and $s - 1$ can be 1 or -1. Therefore, no such roots r and s can exist if $n + 1$ is a prime.

If $n + 1$ is composite, then we can write $n + 1 = bc$, where b and c are positive integers such that $1 < b < n + 1$ and $1 < c < n + 1$. Let $r = -b + 1$ and $s = -c + 1$, so $a = (-b + 1)(-c + 1) = bc - b - c + 1$ and $n = bc - 1$. In this case, $a \leq n$, so these values of r and s work.

Hence, for a particular value of n, $f(x)$ has rational roots for some value of a if and only if $n + 1$ is composite. Since $2 < n + 1 \leq 50$, and there are 14 primes in this range, there are $\boxed{14}$ values of n for which $f(x)$ has irrational roots for all possible values of a.

7.54 Let the quadratic be $f(x) = ax^2 + bx + c$, where a, b, and c are all odd. By the Rational Root Theorem, if p/q is a root of the quadratic, where p and q are relatively prime, then $p \mid c$ and $q \mid a$. Since a and c are odd, both p and q must be odd. We have

$$f\left(\frac{p}{q}\right) = a \cdot \frac{p^2}{q^2} + b \cdot \frac{p}{q} + c = \frac{ap^2 + bpq + cq^2}{q^2}.$$

But if a, b, c, p, and q are all odd, then $ap^2 + bpq + cq^2$ is odd, which means we cannot have $f(p/q) = 0$. Therefore, the quadratic $f(x)$ cannot have any rational roots.

7.55 If $f(x_1) = f(x_2)$ for some real numbers x_1 and x_2 in the interval $[1, 5]$, then by the graph of $y = f(x)$, which is symmetric around $x = 3$, either $x_1 = x_2$ or $x_1 - 3 = 3 - x_2$. The latter gives us $x_1 + x_2 = 2 \cdot 3 = 6$. Therefore, if $f(x) = f(f(x))$, either $x = f(x)$ or $x + f(x) = 6$.

The solutions of $x = f(x)$ correspond to the points where the graph of $y = f(x)$ intersects the graph of $y = x$. There are two such intersections; one in the interval $[1, 3]$, and one in the interval $[3, 5]$.

The solutions of $x + f(x) = 6$ correspond to the points where the graph of $y = f(x)$ intersects the graph of $y = 6 - x$. Again, there are two such intersections; one in the interval $[1, 3]$, and one in the interval $[3, 5]$.

Therefore, the equation $f(x) = f(f(x))$ has $\boxed{4}$ solutions.

7.56 Consider the polynomial

$$f(x) = 10^{10}(x - 9)(x - 8) \cdots (x - 1)(x)(x + 1)(x + 2) \cdots (x + 9).$$

The graph of this polynomial passes through $(c, 0)$ for all integers c with $-10 < c < 10$. For each such integer c, the value of $f(x)$ goes from negative to positive, or vice versa, as x goes from just below c to just above c. The factor 10^{10} ensures that the magnitude of $f(x)$ is very large for all $x = c \pm 0.1$. Combining these observations, we see that the graph of $f(x)$ stays within $1/3$ of the graph of $x = c$ for all integers c with $-10 < c < 10$, and that $f(x)$ ranges from -10 to 10 for values very close to c for each such integer c. In other words, it passes within $1/3$ of each point (a, b), where a and b are single digit integers.

7.57 Let $g(x) = f(x) - 5$, so $g(p) = g(q) = g(r) = g(s) = 0$. Then by the Factor Theorem, we know that $(x - p)(x - q)(x - r)(x - s)$ is a factor of $g(x)$, so $g(x) = (x - p)(x - q)(x - r)(x - s)h(x)$ for some polynomial $h(x)$ with integer coefficients.

Suppose that there is an integer k such that $f(k) = 8$. Then

$$(k - p)(k - q)(k - r)(k - s)h(k) = g(k) = f(k) - 5 = 3.$$

Since each of the factors $k - p, k - q, k - r$, and $k - s$ is an integer, each must have absolute value 1 or 3. Furthermore, at most one can have absolute value 3, which means that at least three of them have absolute value 1. In other words, at least three of them are 1 or -1, which means that two of the factors are equal. However, since p, q, r, and s are all distinct, we cannot have two of the factors be the same. Therefore, there is no integer k such that $f(k) = 8$.

7.58 Expanding both sides gives $a^3 + a^2b^2 + ab + b^3 = a^3 - 3a^2b + 3ab^2 - b^3$. Rearranging gives $2b^3 + a^2b^2 - 3ab^2 + 3a^2b + ab = 0$. Because b is nonzero, we can divide by b to find $2b^2 + a^2b - 3ab + 3a^2 + a = 0$. We can view this as a quadratic in b by writing $2b^2 + (a^2 - 3a)b + (3a^2 + a) = 0$. If this quadratic has an integer root, then the discriminant must be a perfect square. The discriminant of the quadratic is

$$(a^2 - 3a)^2 - 4(2)(3a^2 + a) = a^4 - 6a^3 - 15a^2 - 8a = a(a + 1)^2(a - 8).$$

Since $(a + 1)^2$ is a perfect square, the discriminant is a perfect square if and only if $a(a - 8)$ is a perfect square.

We note that $a(a - 8) < 0$ if $0 < a < 8$, so we cannot have a in this range. Next, we note that $a(a - 8) = 0$ for $a = 0$ and $a = 8$. We are given that a is nonzero, and $a = 8$ gives the solution $(a, b) = (8, -10)$.

Next, we note that $a(a - 8) < a(a - 8) + 16 = a^2 - 8a + 16 = (a - 4)^2$, so $a(a - 8)$ is always 16 less than the perfect square $(a - 4)^2$. Since the positive difference between two consecutive squares, n^2 and $(n - 1)^2$, is $n^2 - (n - 1)^2 = 2n - 1$, and $2n - 1 > 16$ for $n > 8.5$, we know that all squares greater than 64 are more than 16 larger than the next smaller square. Hence, if $a(a - 8)$ is a perfect square, then $(a - 4)^2$ must be no greater than 64. This means that $-8 \le a - 4 \le 8$, so $-4 \le a \le 12$. We know that we cannot have $0 < a < 8$, so we must only test $-4, -3, -2, -1, 9, 10, 11$, and 12. (We have already taken care of $a = 8$.) Testing these, we find the only additional integer solutions are $(a, b) = (-1, -1)$, $(9, -6)$, and $(9, -21)$. Therefore, the solutions are $(a, b) = \boxed{(-1, -1), (9, -6), (8, -10), \text{ and } (9, -21)}$.

CHAPTER 8

Polynomial Roots Part II

Exercises for Section 8.1

8.1.1 First, we look for integer roots. If there are any integer roots, then they must be among the factors of 4. We find that the polynomial is 0 for $x = 2$, so by the Factor Theorem, $x - 2$ is a factor. Dividing by $x - 2$, we get $x^3 - 7x^2 + 12x - 4 = (x - 2)(x^2 - 5x + 2)$. Therefore, the roots are

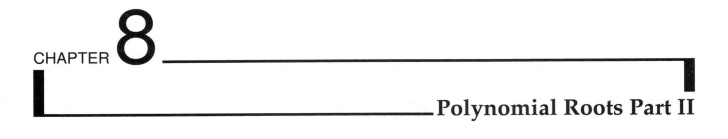

$$x = \boxed{2, \quad \frac{5 + \sqrt{17}}{2}, \quad \text{and} \quad \frac{5 - \sqrt{17}}{2}}.$$

8.1.2 If $2 - 3\sqrt{2}$ is the root of a polynomial with rational coefficients, then so is $2 + 3\sqrt{2}$. Their sum is 4 and their product is $(2 - 3\sqrt{2})(2 + 3\sqrt{2}) = 4 - 18 = -14$, so the quadratic with roots $2 - 3\sqrt{2}$ and $2 + 3\sqrt{2}$ is $x^2 - 4x - 14$.

If $1 - \sqrt{3}$ is the root of a polynomial with rational coefficients, then so is $1 + \sqrt{3}$. Their sum is 2 and their product is $(1 - \sqrt{3})(1 + \sqrt{3}) = 1 - 3 = -2$, so the quadratic with roots $1 - \sqrt{3}$ and $1 + \sqrt{3}$ is $x^2 - 2x - 2$.

Therefore, the desired quartic is $(x^2 - 4x - 14)(x^2 - 2x - 2) = \boxed{x^4 - 6x^3 - 8x^2 + 36x + 28}$.

8.1.3 The number $\sqrt[3]{4}$ is a root of $x^3 - 4 = 0$. By the Rational Root Theorem, if this polynomial has a rational root, then it must be an integer (since the polynomial $x^3 - 4$ is monic), and it must divide 4. Checking these values, we find that no factor of 4 is a root of $x^3 - 4$. Therefore, $\sqrt[3]{4}$ is irrational.

8.1.4

(a) Let $y = x^2$. Then $x^4 - 3x^2 + 2 = y^2 - 3y + 2 = (y - 1)(y - 2) = (x^2 - 1)(x^2 - 2) = \boxed{(x - 1)(x + 1)(x^2 - 2)}$.

(b) Using real numbers, we can factor $x^2 - 2$ to get $\boxed{(x - 1)(x + 1)(x - \sqrt{2})(x + \sqrt{2})}$.

8.1.5 Let $p(x) = (x - 4)^3 - 7$, so $p(x)$ has rational coefficients. Then $p(4 + \sqrt[3]{7}) = (4 + \sqrt[3]{7} - 4)^3 - 7 = 0$, but $p(4 - \sqrt[3]{7}) = (4 - \sqrt[3]{7} - 4)^3 - 7 = -14$. Thus, $p(4 - \sqrt[3]{7})$ is not necessarily 0.

8.1.6 Let $y = \sqrt{\sqrt{2} + \sqrt{3}}$. Then $y^2 = \sqrt{2} + \sqrt{3}$. In the text, we found that $\sqrt{2} + \sqrt{3}$ is a root of $x^4 - 10x^2 + 1$. Since y^2 is a root of $x^4 - 10x^2 + 1$, letting $x = y^2$ reveals that y is a root of the polynomial $\boxed{y^8 - 10y^4 + 1}$.

Exercises for Section 8.2

8.2.1   Note that all the coefficients are positive, so if there are any real roots, then they must be negative. Checking negative integers that divide 18, we find $g(-1) = 6$ and $g(-2) = -8$, so there's a root between -1 and -2. If a/b is a root of $g(x)$ (in lowest terms), then by the Rational Root Theorem, a must divide 18 and b must divide 2. Our only option between -1 and -2 is $-3/2$, so we try that. We find $g(-3/2) = 0$, so by the Factor Theorem, $2x + 3$ is a factor of $g(x)$. Dividing by $2x + 3$, we find $2x^3 + 5x^2 + 15x + 18 = (2x + 3)(x^2 + x + 6)$. Therefore, the roots are

$$x = \boxed{-\frac{3}{2}, \quad \frac{-1 + i\sqrt{23}}{2}, \quad \text{and} \quad \frac{-1 - i\sqrt{23}}{2}}.$$

8.2.2   First, we look for integer roots. If there are any integer roots, then they must be among the factors of 36. Checking these values, we find $f(3) = f(-4) = 0$, so by the Factor Theorem, $(t - 3)(t + 4)$ is a factor of $f(t)$. Dividing by $(t - 3)(t + 4)$, we find

$$2t^4 - 23t^2 + 27t - 36 = (t - 3)(t + 4)(2t^2 - 2t + 3).$$

Therefore, the roots are $t = \boxed{3, \quad -4, \quad \dfrac{1 + i\sqrt{5}}{2}, \quad \text{and} \quad \dfrac{1 - i\sqrt{5}}{2}}$.

8.2.3   The coefficients of g are all real, so nonreal solutions are present in conjugate pairs. Since $2 + i$ and $3 + i$ are roots of g, then so are $\boxed{2 - i}$ and $\boxed{3 - i}$. That gives a total of 4 roots, which is the maximum possible, so there are no other roots of g.

8.2.4   Since the nonreal roots of a polynomial with real coefficients come in conjugate pairs, the number of nonreal roots of the polynomial is even. Since the polynomial is of degree 4, the total number of roots is 4, so the number of real roots must also be even. We are given that the polynomial has at least one real root, so the polynomial must have at least two real roots.

8.2.5   We seek complex numbers z such that $z^4 = -1$, or $z^4 + 1 = 0$. Multiplying both sides by 4, we get $4z^4 + 4 = 0$. Thus, we can transform this equation to $x^4 + 4 = 0$ via the substitution $x = z\sqrt[4]{4} = z\sqrt[4]{2^2} = z\sqrt{2}$. The roots of $x^4 + 4 = 0$ are $1 + i$, $1 - i$, $-1 + i$, and $-1 - i$, so, from $z = x/\sqrt{2}$, we find that the roots of $z^4 + 1 = 0$ are

$$z = \boxed{\frac{1 + i}{\sqrt{2}}, \quad \frac{1 - i}{\sqrt{2}}, \quad \frac{-1 + i}{\sqrt{2}}, \quad \text{and} \quad \frac{-1 - i}{\sqrt{2}}}.$$

8.2.6   Let the integer roots be r and s. We can then write the polynomial in the form

$$(x - r)(x - s)(x^2 + ax + b),$$

for some real numbers a and b. This expands as

$$x^4 + (a - r - s)x^3 + (b - ar - as + rs)x^2 + (ars - br - bs)x + brs.$$

We know that all the coefficients are integers. The coefficient of x^3 is $a - r - s$, and r and s are both integers, so a is an integer. Then the coefficient of x^2 is $b - ar - as + rs$, and ar, as, and rs are all integers, so b is also an integer.

Each of the roots listed is nonreal, so each must be a root of the quadratic $x^2 + ax + b$, if it is to be a root at all. Furthermore, the other root must be the conjugate. For example, if one root is $\frac{1 + i\sqrt{11}}{2}$, then the other root must be $\frac{1 - i\sqrt{11}}{2}$. The sum of these roots is 1, and their product is

$$\frac{1 + i\sqrt{11}}{2} \cdot \frac{1 - i\sqrt{11}}{2} = \frac{1 + 11}{4} = 3.$$

Hence, $\frac{1+i\sqrt{11}}{2}$ and $\frac{1-i\sqrt{11}}{2}$ are the roots of $x^2 - x + 3 = 0$. This shows $\frac{1+i\sqrt{11}}{2}$ can be a root of the original polynomial, so it must be the answer. We check the other choices to be sure:

Root	Quadratic
$\frac{1+i\sqrt{11}}{2}$	$x^2 - x + 3$
$\frac{1+i}{2}$	$x^2 - x + \frac{1}{2}$
$\frac{1}{2} + i$	$x^2 - x + \frac{5}{4}$
$1 + \frac{i}{2}$	$x^2 - 2x + \frac{5}{4}$
$\frac{1+i\sqrt{13}}{2}$	$x^2 - x + \frac{7}{2}$

The only quadratic of the form $x^2 + ax + b$ where a and b are integers is $x^2 - x + 3$, so the only possible root is $\boxed{\dfrac{1 + i\sqrt{11}}{2}}$.

8.2.7 Note that all the coefficients are positive, so if there are any real roots, then they must be negative. First, we look for integer roots (which must be negative factors of 5), and find that there are none. Since the quartic is monic, all rational roots must be integers, so there are no rational roots either.

We know that we can factor the quartic into linear and/or quadratic factors with real coefficients, but we also know that there are no rational roots (and hence, we can't easily find linear factors). So, next we try to factor the quartic as the product of two quadratics. Let $x^4 + 5x^2 + 4x + 5 = (x^2 + ax + b)(x^2 + cx + d)$, where a, b, c, and d are integers. Expanding the right side, we get

$$x^4 + 5x^2 + 4x + 5 = x^4 + (a + c)x^3 + (b + d + ac)x^2 + (ad + bc)x + bd.$$

Equating the corresponding coefficients on both sides, we obtain the system of equations

$$a + c = 0,$$
$$b + d + ac = 5,$$
$$ad + bc = 4,$$
$$bd = 5.$$

Now, we have a number game much like factoring a quadratic. From the first equation, we have $c = -a$. The equation $bd = 5$ suggests that we try the values $b = 1$ and $d = 5$. Then from the second equation, we have $a = 1$ and $c = -1$. These values satisfy the third equation. Hence, we have the factorization

$$x^4 + 5x^2 + 4x + 5 = (x^2 + x + 1)(x^2 - x + 5).$$

Therefore, the roots are $x = \boxed{\dfrac{-1 + i\sqrt{3}}{2}, \quad \dfrac{-1 - i\sqrt{3}}{2}, \quad \dfrac{1 + i\sqrt{19}}{2}, \quad \text{and} \quad \dfrac{1 - i\sqrt{19}}{2}}$.

Exercises for Section 8.3

8.3.1 By Vieta's Formulas, the product of the roots is $-(-6)/3 = \boxed{2}$.

8.3.2 Combining the fractions, we get $\frac{1}{ab} + \frac{1}{bc} + \frac{1}{ca} = \frac{a+b+c}{abc}$. By Vieta's Formulas, $a + b + c = 5$ and $abc = 19$. Therefore, $(a + b + c)/(abc) = \boxed{5/19}$.

8.3.3 We present two ways to find t_1 and t_2.

Method 1: Vieta's Formulas. Since $x + 2$ and $x - 1$ are factors, 1 and -2 are roots. Let the third root be r. By Vieta's Formulas, we have $1 + (-2) + r = 0$, so $r = 1$. Therefore, we have $x^3 - t_1x + t_2 = (x + 2)(x - 1)^2 = x^3 - 3x + 2$, so $t_1 = 3$ and $t_2 = 2$.

Method 2: Substitute the roots into the polynomial. Since $x + 2$ is a factor, we have $(-2)^3 - t_1(-2) + t_2 = 0$, so $2t_1 + t_2 - 8 = 0$. Also, since $x - 1$ is a factor, we have $1 - t_1 + t_2 = 0$. Solving this system of two linear equations, we get $t_1 = 3$ and $t_2 = 2$.

Therefore, $x^2 - t_1x + t_2 = x^2 - 3x + 2 = (x - 1)(x - 2)$, whose roots are $\boxed{1 \text{ and } 2}$.

8.3.4 Since the coefficients of the given cubic are real, the conjugate of $1 + i$, namely $1 - i$, must be another root. Let the third root be r. Then by Vieta's Formulas, we have

$$-4 = (1 + i)(1 - i) + r(1 + i) + r(1 - i) = 2r + 2,$$

so $r = -3$. Therefore, the other two roots are $\boxed{1 - i \text{ and } -3}$.

8.3.5 *Solution 1: Vieta.* The expansion of $(2 + r)(2 + s)(2 + t)(2 + u)$ is the sum of the 16 possible products formed by multiplying one term from each of the four factors. We therefore have

$$(2 + r)(2 + s)(2 + t)(2 + u) = 16 + 8(r + s + t + u) + 4(rs + rt + ru + st + su + tu)$$
$$+ 2(rst + rsu + rtu + stu) + rstu.$$

We evaluate each of the expressions on the right with Vieta's Formulas, and find

$$(2 + r)(2 + s)(2 + t)(2 + u) = 16 + 8\left(\frac{-1}{3}\right) + 4\left(\frac{2}{3}\right) + 2\left(-\frac{7}{3}\right) + \frac{2}{3} = \boxed{\frac{52}{3}}.$$

Solution 2: Recognize the desired expression. Because r, s, t and u are the roots of $f(x)$, and the leading coefficient of $f(x)$ is 3, we have
$$f(x) = 3x^4 - x^3 + 2x^2 + 7x + 2 = 3(x - r)(x - s)(x - t)(x - u).$$

At first, it seems like we are stuck, since our desired factors are of the form $2 + r$, not $2 - r$. However, if we let $x = -2$, each factor becomes the negative of a factor in our desired product:

$$f(-2) = 3(-2 - r)(-2 - s)(-2 - t)(-2 - u) = 3(-1)^4(2 + r)(2 + s)(2 + t)(2 + u) = 3(2 + r)(2 + s)(2 + t)(2 + u).$$

Therefore, we have $3(2 + r)(2 + s)(2 + t)(2 + u) = f(-2) = 52$, so $(2 + r)(2 + s)(2 + t)(2 + u) = \boxed{52/3}$.

8.3.6 Let the roots be a, a, and b. Then, Vieta tells us $2a + b = -(-9)/2 = 9/2$ and $a^2 + ab + ab = 12/2 = 6$. Simplifying the latter equation gives $a(a + 2b) = 6$. Solving $2a + b = 9/2$ for b gives $b = 9/2 - 2a$. Substituting this into $a(a + 2b) = 6$ gives $a(9 - 3a) = 6$. Dividing by 3 and rearranging gives us $a^2 - 3a + 2 = 0$, so $a = 1$ or $a = 2$.

If $a = 1$ is a root of $2x^3 - 9x^2 + 12x - k$, then we have $2 - 9 + 12 - k = 0$, so $k = 5$. If $a = 2$ is a root, then $2 \cdot 8 - 9 \cdot 4 + 12 \cdot 2 - k = 0$, so $k = 4$. Checking our answers, we find

$$2x^3 - 9x^2 + 12x - 5 = (x - 1)^2(2x - 5), \text{ and}$$
$$2x^3 - 9x^2 + 12x - 4 = (x - 2)^2(2x - 1).$$

Therefore, the desired values of k are $\boxed{4 \text{ and } 5}$.

8.3.7 Let w and z be the two roots that sum to $3 + 4i$. The coefficients of the quartic are real, and we are given that all the roots are nonreal, so the roots must come in conjugate pairs. Furthermore, w and z cannot be conjugates of each other, because if they were, their sum would be a real number, but $3 + 4i$ is not real. Therefore, the other two roots of the quartic are $\overline{w}$ and $\overline{z}$. From here, we offer two solutions:

Solution 1: Build the quartic. We are told that the product of the second pair of roots is $\overline{w} \cdot \overline{z} = 13 + i$. But $\overline{w} \cdot \overline{z} = \overline{wz}$, so $\overline{wz} = 13 + i$. Therefore, we have $wz = \overline{(\overline{wz})} = \overline{13 + i} = 13 - i$. We now know the sum and the product of w and z, so we can use Vieta's Formulas to determine that w and z are the roots of the quadratic $x^2 - (3 + 4i)x + (13 - i) = 0$. Similarly, $\overline{w} + \overline{z} = \overline{w + z} = 3 - 4i$ and $\overline{w} \cdot \overline{z} = 13 + i$, so $\overline{w}$ and $\overline{z}$ are the roots of the quadratic $x^2 - (3 - 4i)x + (13 + i) = 0$.

Therefore, the original quartic is the product of these two quadratics:

$$x^4 + ax^3 + bx^2 + cx + d = [x^2 - (3 + 4i)x + (13 - i)][x^2 - (3 - 4i)x + (13 + i)].$$

We seek b, the coefficient of x^2, which we can calculate as

$$13 + i + (3 + 4i)(3 - 4i) + 13 - i = \boxed{51}.$$

Solution 2: Find b with Vieta. Vieta's Formulas tell us that b equals the sum of the six possible products of pairs of the roots. So, using our sums and products of pairs of the roots found earlier, we have

$$b = w \cdot z + w \cdot \overline{w} + w \cdot \overline{z} + \overline{w} \cdot z + \overline{w} \cdot \overline{z} + z \cdot \overline{z} = w \cdot z + \overline{w} \cdot \overline{z} + w(\overline{w} + \overline{z}) + z(\overline{w} + \overline{z})$$
$$= w \cdot z + \overline{w} \cdot \overline{z} + (w + z)(\overline{w} + \overline{z}) = (13 - i) + (13 + i) + (3 + 4i)(3 - 4i) = 51.$$

A little factoring can go a long way!

Exercises for Section 8.4

8.4.1 Since t is a root of $x^3 - x + 2$, we have $t^3 - t + 2 = 0$, so $t^3 = t - 2$. Therefore,

$$t^6 - t^2 + 4t = (t^3)^2 - t^2 + 4t = (t - 2)^2 - t^2 + 4t = t^2 - 4t + 4 - t^2 + 4t = \boxed{4}.$$

8.4.2

(a) $(r + 3)(r + 4) = r^2 + 7r + 12 = (r^2 + 7r + 17) - 5 = \boxed{-5}$.

(b) $(r - 2)(r + 9) = r^2 + 7r - 18 = (r^2 + 7r + 17) - 35 = \boxed{-35}$.

(c) We can find the product $(r - 1)(r + 2)(r + 8)(r + 5)$ by first finding the products $(r - 1)(r + 8)$ and $(r + 2)(r + 5)$. We have

$$(r - 1)(r + 8) = r^2 + 7r - 8 = (r^2 + 7r + 17) - 25 = -25,$$
$$(r + 2)(r + 5) = r^2 + 7r + 10 = (r^2 + 7r + 17) - 7 = -7,$$

so $(r - 1)(r + 2)(r + 8)(r + 5) = (-25) \cdot (-7) = \boxed{175}$.

8.4.3 Let $z_1, z_2, \ldots, z_{20}$ be the 20 roots of $z^{20} - 19z^2 + 2 = 0$. Then $z_i^{20} - 19z_i^2 + 2 = 0$ for all $1 \le i \le 20$, and adding all 20 of these equations gives

$$z_1^{20} + z_2^{20} + \cdots + z_{20}^{20} = 19(z_1^2 + z_2^2 + \cdots + z_{20}^2) - 40.$$

By Vieta's Formulas, $z_1 + z_2 + \cdots + z_{20} = 0$ and $z_1z_2 + z_1z_3 + \cdots + z_{19}z_{20} = 0$. Squaring $z_1 + z_2 + \cdots + z_{20}$, we get

$$(z_1 + z_2 + \cdots + z_{20})^2 = z_1^2 + z_2^2 + \cdots + z_{20}^2 + 2(z_1z_2 + z_1z_3 + \cdots + z_{19}z_{20}),$$

so $z_1^2 + z_2^2 + \cdots + z_{20}^2 = 0$. Therefore, $z_1^{20} + z_2^{20} + \cdots + z_{20}^{20} = \boxed{-40}$.

8.4.4 By Vieta's Formulas, we have $a + b = 3c$ and $c + d = 3a$. Subtracting the first equation from the second gives $d - b = 4(a - c)$.

Since a is a root of $x^2 - 3cx - 8d$, we have $a^2 - 3ac - 8d = 0$. Since c is a root of $x^2 - 3ax - 8b$, we have $c^2 - 3ac - 8b = 0$. Subtracting the second equation from the first gives $a^2 - c^2 - 8d + 8b = 0$. Rearranging this equation gives $a^2 - c^2 = 8d - 8b$, and factoring gives $(a - c)(a + c) = 8(d - b)$. Since $d - b = 4(a - c)$, we have $(a - c)(a + c) = 32(a - c)$. Since $a \ne c$, we can divide both sides by $a - c$ to get $a + c = 32$. Therefore, $a + b + c + d = 3c + 3a = 3(a + c) = 3 \cdot 32 = \boxed{96}$.

Review Problems

8.28

(a) First, we look for integer roots. If there are any integer roots, then they must be among the factors of 21. Checking these values, we find $f(-3) = 0$, so by the Factor Theorem, $r + 3$ is a factor of $f(r)$. Dividing by $r + 3$, we get $2r^3 + 7r^2 - 4r - 21 = (r + 3)(2r^2 + r - 7)$. Therefore, the roots are

$$r = \boxed{-3, \quad \frac{-1 + \sqrt{57}}{4}, \quad \text{and} \quad \frac{-1 - \sqrt{57}}{4}}.$$

(b) First, we look for integer roots. If there are any integer roots, then they must be among the factors of 24. Checking these values, we find $g(-2) = 0$. Dividing $g(s)$ by $s + 2$ gives us $g(s) = (s + 2)(s^2 - 2s + 12)$. Applying the quadratic formula gives us $1 \pm i\sqrt{11}$ as the roots of the quadratic factor. So, the roots of $g(s)$ are $\boxed{-2, 1 + i\sqrt{11}, \text{ and } 1 - i\sqrt{11}}$.

8.29 Since the coefficients of the polynomial are real, the complex conjugate of $-3 + 5i$, namely $-3 - 5i$, must also be a root. The sum of $-3 + 5i$ and $-3 - 5i$ is -6, and their product is $(-3 + 5i)(-3 - 5i) = 9 + 25 = 34$. Therefore, the quadratic whose roots are $-3 + 5i$ and $-3 - 5i$ is $x^2 + 6x + 34$.

Now, we could divide the quartic by the quadratic to get the other two roots. However, by Vieta's Formulas, the sum of all four roots of the quartic is -9, and their product is -136. The sum of the roots of the quadratic is -6, and their product is 34. Hence, the sum of the other two roots is $-9 - (-6) = -3$, and their product is $-136/34 = -4$. Therefore, the other two roots are the roots of the quadratic $x^2 + 3x - 4 = (x - 1)(x + 4)$. Therefore, the roots of the quartic other than $-3 + 5i$ are $\boxed{-3 - 5i, 1, \text{ and } -4}$.

8.30

(a) We are given that the polynomial has five distinct roots, but a polynomial of degree 6 has six roots (counting multiplicity), so one of the roots must be repeated.

(b) Since the coefficients of the polynomial are real, the nonreal roots must come in conjugate pairs. The nonreal roots $2 - i$ and $2 + i$ already form a conjugate pair, as do the nonreal roots $1 - i$ and $1 + i$. We also know that 1 is a root, so the sixth and final root must be real. Therefore, the double root is $\boxed{1}$.

8.31 We have $r_1 i = -b + ai$ and $r_2 i = b + ai$. Since the polynomial has real coefficients, the complex conjugate of $r_1 i = -b + ai$, namely $-b - ai$, must also be a root. Since a and b are nonzero, $-b - ai$ cannot be equal to $r_2 i = b + ai$. Similarly, the conjugate of $r_2 i = b + ai$, namely $b - ai$, must also be a root. We have shown that the polynomial must have at least four roots. Furthermore, the polynomial

$$(x + b - ai)(x + b + ai)(x - b - ai)(x - b + ai) = [(x + b)^2 + a^2][(x - b)^2 + a^2]$$

has real coefficients, and its roots are $-b + ai$, $-b - ai$, $b + ai$, and $b - ai$. Therefore, the minimum degree of the polynomial is $\boxed{4}$.

8.32 Since the coefficients of the polynomial are rational, the radical conjugate of $2 + \sqrt{3}$, namely $2 - \sqrt{3}$, must also be a root. The sum of $2 + \sqrt{3}$ and $2 - \sqrt{3}$ is 4. By Vieta's Formulas, the sum of all three roots is 11, so the third root is 7. Therefore, the other two roots are $\boxed{2 - \sqrt{3} \text{ and } 7}$.

8.33 Since the coefficients of the polynomial are real, the complex conjugate of $3 - 2i\sqrt{5}$, namely $3 + 2i\sqrt{5}$, is also a root. The sum of $3 - 2i\sqrt{5}$ and $3 + 2i\sqrt{5}$ is 6, and their product is $9 + 20 = 29$. Therefore, $3 - 2i\sqrt{5}$ and $3 + 2i\sqrt{5}$ are the roots of $x^2 - 6x + 29$.

Since the coefficients of the polynomial are rational, the radical conjugate of $-1 + \sqrt{3}$, namely $-1 - \sqrt{3}$, is also a root. The sum of $-1 + \sqrt{3}$ and $-1 - \sqrt{3}$ is -2, and their product is $1 - 3 = -2$. Therefore, $-1 + \sqrt{3}$ and $-1 - \sqrt{3}$ are the roots of $x^2 + 2x - 2$.

Therefore, the polynomial we seek is $(x^2 - 6x + 29)(x^2 + 2x - 2) = \boxed{x^4 - 4x^3 + 15x^2 + 70x - 58}$.

8.34 The number $\sqrt{c}$ is a root of $x^2 - c = 0$. Since this polynomial is monic, any rational root must also be an integer. However, since c is not a perfect square, $\sqrt{c}$ is not an integer. Therefore, $\sqrt{c}$ is irrational.

8.35 Let $x = \sqrt{2 - \sqrt{2}}$. Then $x^2 = 2 - \sqrt{2}$, so $x^2 - 2 = -\sqrt{2}$. Squaring both sides, we get $x^4 - 4x^2 + 4 = 2$, so $x^4 - 4x^2 + 2 = 0$. Thus, $\sqrt{2 - \sqrt{2}}$ is a root of $x^4 - 4x^2 + 2$.

By the Rational Root Theorem, if $x^4 - 4x^2 + 2$ has a rational root, then it must be an integer (since $x^4 - 4x^2 + 2$ is monic), and it must divide 2. Checking these values, we find that none of the factors of 2 is a root of $x^4 - 4x^2 + 2$. Therefore, $\sqrt{2 - \sqrt{2}}$ is irrational.

8.36 First, we write $8 - 2y + 4y^3 - y^4$ as $-(y^4 - 4y^3 + 2y - 8)$.

(a) We look for integer roots, which must be among the factors of 8. We find the quartic is 0 for $y = 4$, so by the Factor Theorem, $y - 4$ is factor. Dividing by $y - 4$, we find
$$-(y^4 - 4y^3 + 2y - 8) = -(y - 4)(y^3 + 2).$$
The cubic $y^3 + 2$ cannot be factored further over the integers because no factor of 2 is a root of $y^3 + 2$, so the complete factorization over the integers is $\boxed{-(y - 4)(y^3 + 2)}$.

(b) The cubic $y^3 + 2$ has the real root $y = -\sqrt[3]{2}$, so by the sum of cubes factorization (or by synthetic division), we have
$$y^3 + 2 = (y + \sqrt[3]{2})(y^2 - \sqrt[3]{2}y + \sqrt[3]{4}).$$
The roots of the quadratic are not real, so we cannot factor any further over the reals. Therefore, the complete factorization over the real numbers is $\boxed{-(y - 4)(y + \sqrt[3]{2})(y^2 - \sqrt[3]{2}y + \sqrt[3]{4})}$.

(c) By the quadratic formula, the roots of $y^2 - \sqrt[3]{2}y + \sqrt[3]{4} = 0$ are
$$y = \frac{\sqrt[3]{2} \pm \sqrt{\sqrt[3]{4} - 4\sqrt[3]{4}}}{2} = \frac{\sqrt[3]{2} \pm i\sqrt{3}\sqrt{\sqrt[3]{4}}}{2} = \frac{\sqrt[3]{2} \pm i\sqrt[3]{2}\sqrt{3}}{2} = \frac{\sqrt[3]{2} \pm i\sqrt[6]{108}}{2}.$$
Therefore, the complete factorization over the complex numbers is
$$\boxed{-(y - 4)(y + \sqrt[3]{2})\left(y - \frac{\sqrt[3]{2} + i\sqrt[6]{108}}{2}\right)\left(y - \frac{\sqrt[3]{2} - i\sqrt[6]{108}}{2}\right)}.$$

8.37

(a) Let $p(x) = x^3 + ax^2 + bx + c$. (If the coefficient of x^3 is not 1, then we can divide $p(x)$ by this coefficient. This does not affect the roots of the cubic.) To show that $p(x)$ has a root, it suffices to show that $p(x) > 0$ for some value of x, and $p(x) < 0$ for some other value of x.

Consider the behavior of $p(x)$ as x increases. As x becomes very large, the term x^3 grows much faster than the other terms, so it dominates them. And since x^3 is positive for $x > 0$, $p(x)$ must also be positive for sufficiently large values of x. Hence, $p(x) > 0$ for some value of x.

Similarly, the term x^3 dominates as x becomes much less than 0. And since x^3 is negative for $x < 0$, $p(x)$ must also be negative for sufficiently low values of x. Hence, $p(x) < 0$ for some value of x. Therefore, $p(x) = 0$ for some value of x in between, so $p(x)$ has a root.

(b) By the Fundamental Theorem of Algebra, $p(x)$ has three roots, say r_1, r_2, and r_3. Consider root r_1. If r_1 is real, then we are done. Otherwise, r_1 is nonreal, so let $r_1 = u + vi$, where $v \neq 0$. Since the coefficients of $p(x)$ are real, another root, say r_2, must be its complex conjugate, so $r_2 = u - vi$. Then by Vieta's Formulas, $r_1 + r_2 + r_3 = (u + vi) + (u - vi) + r_3 = 2u + r_3$ is a real number. Since u is real, the root r_3 is real.

8.38

(a) Since $-\sqrt{2}$ is the radical conjugate of $\sqrt{2}$, if $\sqrt{2}$ is a root of $f(x)$, then $-\sqrt{2}$ is also a root of $f(x)$.

(b) Let $f(x) = x^3 - 2$. Then $\sqrt[3]{2}$ is a root of $f(x)$, but $-\sqrt[3]{2}$ is not a root of $f(x)$. Therefore, in general, if $\sqrt[3]{2}$ is a root of $f(x)$, then $-\sqrt[3]{2}$ is not necessarily a root of $f(x)$.

8.39 By Vieta's Formulas, we have

$$p + q + r + s = 5,$$
$$pq + pr + ps + qr + qs + rs = 2,$$
$$pqr + pqs + prs + qrs = -7,$$
$$pqrs = -11.$$

(a) $pqr + pqs + prs + qrs = \boxed{-7}$.

(b) $\frac{1}{p} + \frac{1}{q} + \frac{1}{r} + \frac{1}{s} = \frac{pqr+pqs+prs+qrs}{pqrs} = \frac{-7}{-11} = \boxed{\frac{7}{11}}$.

(c) $p^2 + q^2 + r^2 + s^2 = (p + q + r + s)^2 - 2(pq + pr + ps + qr + qs + rs) = 5^2 - 2 \cdot 2 = \boxed{21}$.

(d) $p^2qrs + pq^2rs + pqr^2s + pqrs^2 = pqrs(p + q + r + s) = (-11) \cdot 5 = \boxed{-55}$.

8.40 Multiplying both sides by $2004x$, we obtain the quadratic

$$2003x^2 + 2004x + 2004 = 0.$$

Let the roots be r and s. Then by Vieta's Formulas, $r + s = -2004/2003$ and $rs = 2004/2003$. Therefore, the sum of the reciprocals of the roots is

$$\frac{1}{r} + \frac{1}{s} = \frac{r + s}{rs} = \frac{-2004/2003}{2004/2003} = \boxed{-1}.$$

We can also find the sum of the reciprocals of the roots via a substitution. Let $x = 1/y$. Then the given equation becomes

$$\frac{2003}{2004} \cdot \frac{1}{y} + 1 + y = 0.$$

Multiplying both sides by $2004y$, we get $2004y^2 + 2004y + 2003 = 0$. This quadratic has roots $1/r$ and $1/s$. Therefore, $1/r + 1/s = -2004/2004 = -1$, as above.

8.41 We are given that $a + bi$ and $a - bi$ are two of the roots of $x^3 + qx + r$. Let the third root be c. Then by Vieta's Formulas, the sum of the roots is $0 = (a + bi) + (a - bi) + c = 2a + c$, so $c = -2a$.

Therefore, $q = (a + bi)(a - bi) + (a + bi)(-2a) + (a - bi)(-2a) = \boxed{-3a^2 + b^2}$.

8.42 We see that $3 - 2i$ is a root of $p(x)$. Since $p(x)$ has real coefficients, the conjugate of $3 - 2i$, namely $3 + 2i$, is also a root of $p(x)$. The sum of $3 - 2i$ and $3 + 2i$ is 6, and their product is $9 + 4 = 13$, so the quadratic with roots $3 - 2i$ and $3 + 2i$ is $x^2 - 6x + 13 = 0$.

Let the third root of $p(x)$ be r. Since $p(0)$ equals the constant term, and $p(x)$ is monic, Vieta's Formulas tell us that $(3 - 2i)(3 + 2i)r = 52$, so $r = 4$. Hence, $p(x) = (x^2 - 6x + 13)(x - 4) = \boxed{x^3 - 10x^2 + 37x - 52}$.

8.43 By Vieta's Formulas, we have $r + s + t = 4$ and $rs + rt + st = 5$. Then the length of an interior diagonal of a box with sides r, s, and t is

$$\sqrt{r^2 + s^2 + t^2} = \sqrt{(r + s + t)^2 - 2(rs + rt + st)} = \sqrt{4^2 - 2 \cdot 5} = \boxed{\sqrt{6}}.$$

8.44   Since $x^3 + 4x = 8$, we have $x^3 = 8 - 4x$. Then

$$x^7 + 64x^2 = x \cdot (x^3)^2 + 64x^2 = x(8 - 4x)^2 + 64x^2 = x(16x^2 - 64x + 64) + 64x^2 = 16x^3 - 64x^2 + 64x + 64x^2$$
$$= 16x^3 + 64x = 16(x^3 + 4x) = 16 \cdot 8 = \boxed{128}.$$

Challenge Problems

8.45   Since $r^2 - 2r + 3 = 0$, we have $r^3 - 2r^2 + 3r = 0$ (or $r^3 = 2r^2 - 3r$), and $r^4 - 2r^3 + 3r^2 = 0$ (or $r^4 = 2r^3 - 3r^2$). Then

$$r^4 - 4r^3 + 2r^2 + 4r + 3 = (2r^3 - 3r^2) - 4r^3 + 2r^2 + 4r + 3 = -2r^3 - r^2 + 4r + 3$$
$$= -2(2r^2 - 3r) - r^2 + 4r + 3 = -4r^2 + 6r - r^2 + 4r + 3$$
$$= -5r^2 + 10r + 3 = -5(r^2 - 2r) + 3 = (-5) \cdot (-3) + 3 = \boxed{18}.$$

8.46   Let the roots be $r_1, r_2, \ldots, r_n$. By Vieta's Formulas, we have the equations $r_1 + r_2 + \cdots + r_n = -a_1$ and $r_1 r_2 + r_1 r_3 + \cdots + r_{n-1} r_n = a_2$.

Then the sum of the squares of the roots is

$$r_1^2 + r_2^2 + \cdots + r_n^2 = (r_1 + r_2 + \cdots + r_n)^2 - 2(r_1 r_2 + r_1 r_3 + \cdots + r_{n-1} r_n) = \boxed{a_1^2 - 2a_2}.$$

8.47   Let $Q(x) = P(x) - 4$. Then $Q(\sqrt[3]{7} + \sqrt[3]{49}) = P(\sqrt[3]{7} + \sqrt[3]{49}) - 4 = 0$, so $\sqrt[3]{7} + \sqrt[3]{49}$ is a root of $Q(x)$. Thus, we seek a monic polynomial $Q(x)$ that has $\sqrt[3]{7} + \sqrt[3]{49}$ as a root.

To find such a polynomial, let $x = \sqrt[3]{7} + \sqrt[3]{49}$. Then

$$x^3 = 7 + 3\sqrt[3]{7^2}\sqrt[3]{49} + 3\sqrt[3]{7}\sqrt[3]{49^2} + 49 = 56 + 3 \cdot 7\sqrt[3]{7} + 3 \cdot 7\sqrt[3]{49} = 56 + 21x.$$

Thus, $\sqrt[3]{7} + \sqrt[3]{49}$ is a root of the cubic $x^3 - 21x - 56$. Now we must prove that $\sqrt[3]{7} + \sqrt[3]{49}$ is not a root of a polynomial with rational coefficients and smaller degree.

First, we show that $\sqrt[3]{7} + \sqrt[3]{49}$ is irrational. Because it is a root of $x^3 - 21x - 56$, if $\sqrt[3]{7} + \sqrt[3]{49}$ were rational, then by the Rational Root Theorem, it would be an integer, and a factor of 56. However, we find that none of the factors of 56 is a root of $x^3 - 21x - 56$, so $\sqrt[3]{7} + \sqrt[3]{49}$ is irrational. Therefore, $\sqrt[3]{7} + \sqrt[3]{49}$ is not the root of a linear polynomial with rational coefficients.

Now, suppose that $\sqrt[3]{7} + \sqrt[3]{49}$ is the root of a quadratic with rational coefficients, say $ax^2 + bx + c$. Dividing $x^3 - 21x - 56$ by $ax^2 + bx + c$, we obtain a quotient $S(x)$ and a remainder $R(x)$, with rational coefficients:

$$x^3 - 21x - 56 = S(x)(ax^2 + bx + c) + R(x).$$

Taking $x = \sqrt[3]{7} + \sqrt[3]{49}$, we get $R(\sqrt[3]{7} + \sqrt[3]{49}) = 0$ (since $\sqrt[3]{7} + \sqrt[3]{49}$ is a root of both $x^3 - 21x - 56$ and $ax^2 + bx + c$). However, the remainder $R(x)$ has degree at most 1, and as we showed above, $\sqrt[3]{7} + \sqrt[3]{49}$ is not the root of a linear polynomial with rational coefficients, so $R(x) = 0$.

Therefore, $ax^2 + bx + c$ divides $x^3 - 21x - 56$, and $S(x)$ is linear. Since $S(x)$ has rational coefficients, $S(x)$ has a rational root r. Then r is a rational root of $x^3 - 21x - 56$, but as we also showed above, $x^3 - 21x - 56$ has no rational roots, which gives us a contradiction.

Hence, $x^3 - 21x - 56$ is the polynomial of smallest degree with rational coefficients that has $\sqrt[3]{7} + \sqrt[3]{49}$ as a root, so the polynomial $P(x)$ we seek is $Q(x) + 4 = x^3 - 21x - 56 + 4 = \boxed{x^3 - 21x - 52}$.

8.48 To deal with the appearances of i in the coefficients, let $y = ix$. Then the given cubic becomes

$$iy^3 - 8y^2 - 22iy + 21 = i(ix)^3 - 8(ix)^2 - 22i(ix) + 21 = i(-ix^3) - 8(-x^2) - 22i(ix) + 21 = x^3 + 8x^2 + 22x + 21.$$

We look for roots of this new polynomial. First, we look for integer roots. If there are any integer roots, then they must be among the factors of 21. Checking these values, we find the cubic becomes 0 for $x = -3$. So by the Factor Theorem, $x + 3$ is a factor of the cubic. Dividing by $x + 3$, we find

$$x^3 + 8x^2 + 22x + 21 = (x + 3)(x^2 + 5x + 7).$$

The roots of the quadratic are $\frac{-5 \pm i\sqrt{3}}{2}$. Since $y = ix$, the solutions of the original cubic are

$$y = \boxed{-3i, \quad \frac{\sqrt{3} - 5i}{2}, \quad \text{and} \quad \frac{-\sqrt{3} - 5i}{2}}.$$

8.49 By Vieta's Formulas, we have $x_1 + x_2 = a + d$ and $x_1 x_2 = ad - bc$, so $x_1^3 x_2^3 = (ad - bc)^3$. Also,

$$\begin{aligned}
x_1^3 + x_2^3 &= (x_1 + x_2)(x_1^2 - x_1 x_2 + x_2^2) = (x_1 + x_2)[(x_1 + x_2)^2 - 3x_1 x_2] \\
&= (a + d)[(a + d)^2 - 3(ad - bc)] = (a + d)(a^2 + 2ad + d^2 - 3ad + 3bc) \\
&= (a + d)(a^2 - ad + d^2 + 3bc) = a^3 + d^3 + 3abc + 3bcd.
\end{aligned}$$

Hence, x_1^3 and x_2^3 are the roots of the equation $y^2 - (a^3 + d^3 + 3abc + 3bcd)y + (ad - bc)^3 = 0$.

8.50 By Vieta's Formulas, we have $r_1 + r_2 + r_3 = 0$, $r_1 r_2 + r_1 r_3 + r_2 r_3 = -1$, and $r_1 r_2 r_3 = 1$. Then

$$\begin{aligned}
r_1(r_2 - r_3)^2 + r_2(r_3 - r_1)^2 + r_3(r_1 - r_2)^2 &= r_1(r_2^2 - 2r_2 r_3 + r_3^2) + r_2(r_1^2 - 2r_1 r_3 + r_3^2) + r_3(r_1^2 - 2r_1 r_2 + r_2^2) \\
&= r_1^2 r_2 + r_1 r_2^2 + r_1^2 r_3 + r_1 r_3^2 + r_2^2 r_3 + r_2 r_3^2 - 6r_1 r_2 r_3 \\
&= (r_1 + r_2)r_1 r_2 + (r_1 + r_3)r_1 r_3 + (r_2 + r_3)r_2 r_3 - 6r_1 r_2 r_3.
\end{aligned}$$

Because $r_1 + r_2 + r_3 = 0$, we have $r_1 + r_2 = -r_3$, $r_1 + r_3 = -r_2$, and $r_2 + r_3 = -r_1$, so

$$\begin{aligned}
r_1(r_2 - r_3)^2 + r_2(r_3 - r_1)^2 + r_3(r_1 - r_2)^2 &= (r_1 + r_2)r_1 r_2 + (r_1 + r_3)r_1 r_3 + (r_2 + r_3)r_2 r_3 - 6r_1 r_2 r_3 \\
&= (-r_3)r_1 r_2 + (-r_2)r_1 r_3 + (-r_1)r_2 r_3 - 6r_1 r_2 r_3 \\
&= -9r_1 r_2 r_3 \\
&= \boxed{-9}.
\end{aligned}$$

8.51 If $\sqrt[4]{2}$ is a root of $f(x)$, then $-\sqrt[4]{2}$ is a also a root of $f(x)$.

To prove this, we will prove that if $\sqrt[4]{2}$ is a root of $f(x)$, and $f(x)$ has rational coefficients, then $f(x)$ is divisible by $x^4 - 2$. We begin by assuming that all the coefficients of $f(x)$ are integers. (If they are not, then we can multiply $f(x)$ by all the denominators of the coefficients to obtain integer coefficients.)

Suppose dividing $f(x)$ by $x^4 - 2$ gives a quotient $q(x)$ and remainder $r(x)$, so we have

$$f(x) = (x^4 - 2)q(x) + r(x),$$

where the degree of $r(x)$ is at most 3. Then $f(x)$ is divisible by $x^4 - 2$ if and only if $r(x)$ is the zero polynomial. Let $r(x) = ax^3 + bx^2 + cx + d$, where a, b, c, and d are integers, so $r(x)$ is the zero polynomial if and only if $a = b = c = d = 0$. Substituting $x = \sqrt[4]{2}$ into the equation

$$f(x) = (x^4 - 2)q(x) + r(x) = (x^4 - 2)q(x) + ax^3 + bx^2 + cx + d,$$

we get $f(\sqrt[4]{2}) = r(\sqrt[4]{2}) = a\sqrt[4]{8} + b\sqrt{2} + c\sqrt[4]{2} + d = 0$, since $\sqrt[4]{2}$ is a root of $f(x)$.

For the sake of contradiction, suppose that not all of a, b, c, and d are equal to 0. We may assume that not all of a, b, c, and d are even; otherwise, we simply keep dividing this equation by 2, until one of the four numbers is odd.

To make this equation easier to work with, we move the fourth roots to one side, to get

$$a\sqrt[4]{8} + c\sqrt[4]{2} = -b\sqrt{2} - d.$$

Squaring both sides, we get $2a^2\sqrt{2} + 4ac + c^2\sqrt{2} = 2b^2 + 2bd\sqrt{2} + d^2$. Since $\sqrt{2}$ is irrational, it follows that $2a^2 + c^2 = 2bd$ and $4ac = 2b^2 + d^2$.

Then $c^2 = 2bd - 2a^2 = 2(bd - a^2)$, so c is even. Let $c = 2c_1$. Also, $d^2 = 4ac - 2b^2 = 2(2ac - b^2)$, so d is even. Let $d = 2d_1$.

Then from $2a^2 + c^2 = 2bd$, we have $2a^2 + 4c_1^2 = 4bd_1$, so $a^2 + 2c_1^2 = 2bd_1$, or $a^2 = 2bd_1 - 2c_1^2 = 2(bd_1 - c_1^2)$, so a is even. From $4ac = 2b^2 + d^2$, we have $8ac_1 = 2b^2 + 4d_1^2$, so $4ac_1 = b^2 + 2d_1^2$, or $b^2 = 4ac_1 - 2d_1^2 = 2(2ac_1 - d_1^2)$, so b is even. This contradicts the fact that not all of a, b, c, and d are even. Therefore, the only integers a, b, c, and d such that $a\sqrt[4]{8} + b\sqrt{2} + c\sqrt[4]{2} + d = 0$ are $a = b = c = d = 0$. Hence, $r(x) = ax^3 + bx^2 + cx + d = 0$, so $f(x) = (x^4 - 2)q(x)$. Then $f(-\sqrt[4]{2}) = 0$, so $-\sqrt[4]{2}$ is also a root of $f(x)$, as desired.

8.52 Since r, s, and t are the roots of $x^3 + 9x^2 - 9x - 8$, we have

$$x^3 + 9x^2 - 9x - 8 = (x - r)(x - s)(x - t). \qquad (*)$$

By Vieta's Formulas, we have $r + s + t = -9$. Then $r + s = -9 - t$, $s + t = -9 - r$, and $t + r = -9 - s$, so

$$(r + s)(s + t)(t + r) = (-9 - t)(-9 - r)(-9 - s).$$

Substituting $x = -9$ into $(*)$, we get $(-9 - r)(-9 - s)(-9 - t) = (-9)^3 + 9 \cdot (-9)^2 - 9 \cdot (-9) - 8 = \boxed{73}$.

8.53 By Vieta's Formulas, we have $m + n + p = 0$. Therefore, by the Factor Theorem, one of the roots of the quadratic $mx^2 + nx + p = 0$ is $x = 1$. The product of the roots of this quadratic is p/m, so this is the other root of the quadratic. This root is rational, since p and m are rational.

8.54 Since p, q, and r are the roots of $x^3 - x^2 + x - 2$, we have

$$p^3 - p^2 + p - 2 = 0,$$
$$q^3 - q^2 + q - 2 = 0,$$
$$r^3 - r^2 + r - 2 = 0.$$

Adding these equations, we get $p^3 + q^3 + r^3 = (p^2 + q^2 + r^2) - (p + q + r) + 6$. By Vieta's Formulas, we have $p + q + r = 1$ and $pq + pr + qr = 1$. To find $p^2 + q^2 + r^2$, we square $p + q + r$ to get

$$(p + q + r)^2 = p^2 + q^2 + r^2 + 2(pq + pr + qr).$$

Therefore, $p^2 + q^2 + r^2 = (p + q + r)^2 - 2(pq + pr + qr) = 1 - 2 = -1$. So, we have

$$p^3 + q^3 + r^3 = (p^2 + q^2 + r^2) - (p + q + r) + 6 = -1 - 1 + 6 = \boxed{4}.$$

8.55 The left side is an alternating sum of the elementary symmetric sums of the n values $n/1$, $n/2$, $\ldots$, n/n. Thus, we can model the left side with a polynomial with these values as roots.

Let

$$f(x) = \left(x - \frac{n}{1}\right)\left(x - \frac{n}{2}\right)\cdots\left(x - \frac{n}{n}\right)$$
$$= x^n - \left(\frac{n}{1} + \frac{n}{2} + \cdots + \frac{n}{n}\right)x^{n-1} + \left(\frac{n^2}{1 \cdot 2} + \frac{n^2}{1 \cdot 3} + \cdots + \frac{n^2}{(n-1)n}\right)x^{n-2} - \cdots + (-1)^n\frac{n^n}{n!}.$$

Taking $x = 1$, we get

$$1 - \left(\frac{n}{1} + \frac{n}{2} + \cdots + \frac{n}{n}\right) + \left(\frac{n^2}{1 \cdot 2} + \frac{n^2}{1 \cdot 3} + \cdots + \frac{n^2}{(n-1)n}\right) - \cdots + (-1)^n \frac{n^n}{n!} = 0,$$

since $n/n = 1$ is a root of $f(x)$. Therefore,

$$\left(\frac{n}{1} + \frac{n}{2} + \cdots + \frac{n}{n}\right) - \left(\frac{n^2}{1 \cdot 2} + \frac{n^2}{1 \cdot 3} + \cdots + \frac{n^2}{(n-1)n}\right) + \cdots + (-1)^{n-1} \frac{n^n}{n!} = 1,$$

as desired.

8.56 Let the equation of the line be $y = mx + b$. Then the x_i are the roots of $2x^4 + 7x^3 + 3x - 5 = mx + b$, or $2x^4 + 7x^3 + (3 - m)x - (b + 5) = 0$. Then by Vieta's Formulas, we have

$$\frac{x_1 + x_2 + x_3 + x_4}{4} = -\frac{7/2}{4} = \boxed{-\frac{7}{8}}.$$

Thus, the sum is independent of the line.

8.57 Let $r_1, r_2, \ldots, r_n$ be the roots of $a_n x^n + a_{n-1} x^{n-1} + a_{n-2} x^{n-2} + \cdots + a_0$. Then

$$a_n r_i^n + a_{n-1} r_i^{n-1} + a_{n-2} r_i^{n-2} + \cdots + a_0 = 0$$

for all $1 \le i \le n$. Multiplying both sides by r_i^{k-n}, we get

$$a_n r_i^k + a_{n-1} r_i^{k-1} + a_{n-2} r_i^{k-2} + \cdots + a_0 r_i^{k-n} = 0.$$

We have n such equations, one for each root. Adding all of them gives

$$a_n t_k + a_{n-1} t_{k-1} + a_{n-2} t_{k-2} + \cdots + a_0 t_{k-n} = 0.$$

8.58 Let $f(x) = (x - \alpha)(x - \beta)(x - \gamma) = x^3 - x - 1$. By Vieta's Formulas, we have $\alpha + \beta + \gamma = 0$, $\alpha\beta + \alpha\gamma + \beta\gamma = -1$, and $\alpha\beta\gamma = 1$. Then

$$\frac{1+\alpha}{1-\alpha} + \frac{1+\beta}{1-\beta} + \frac{1+\gamma}{1-\gamma} = \frac{(1+\alpha)(1-\beta)(1-\gamma) + (1-\alpha)(1+\beta)(1-\gamma) + (1-\alpha)(1-\beta)(1+\gamma)}{(1-\alpha)(1-\beta)(1-\gamma)}.$$

The denominator is equal to $(1 - \alpha)(1 - \beta)(1 - \gamma) = f(1) = -1$, and the numerator is equal to

$(1 + \alpha - \beta - \gamma - \alpha\beta - \alpha\gamma + \beta\gamma + \alpha\beta\gamma)$
$\quad + (1 - \alpha + \beta - \gamma - \alpha\beta + \alpha\gamma - \beta\gamma + \alpha\beta\gamma)$
$\quad + (1 - \alpha - \beta + \gamma + \alpha\beta - \alpha\gamma - \beta\gamma + \alpha\beta\gamma) = 3 - (\alpha + \beta + \gamma) - (\alpha\beta + \alpha\gamma + \beta\gamma) + 3\alpha\beta\gamma = 3 - 0 - (-1) + 3 \cdot 1 = 7,$

so $\dfrac{1+\alpha}{1-\alpha} + \dfrac{1+\beta}{1-\beta} + \dfrac{1+\gamma}{1-\gamma} = \dfrac{7}{-1} = \boxed{-7}$.

Alternatively, let $y = \frac{1+x}{1-x}$. Solving for x, we get $x = \frac{y-1}{y+1}$. Substituting into the cubic $x^3 - x - 1 = 0$, we get

$$\left(\frac{y-1}{y+1}\right)^3 - \frac{y-1}{y+1} - 1 = 0,$$

which simplifies as $y^3 + 7y^2 - y + 1 = 0$. Since α, β, and γ are the roots of $x^3 - x - 1$, and $y = \frac{1+x}{1-x}$, the roots of $y^3 + 7y^2 - y + 1$ are $\frac{1+\alpha}{1-\alpha}$, $\frac{1+\beta}{1-\beta}$, and $\frac{1+\gamma}{1-\gamma}$. Therefore, we have

$$\frac{1+\alpha}{1-\alpha} + \frac{1+\beta}{1-\beta} + \frac{1+\gamma}{1-\gamma} = -7.$$

8.59   Let r_1, r_2, and r_3 be the roots of $x^3 + 1 = 0$. We see that $x^3 + 1 = (x + 1)(x^2 - x + 1)$. The roots of the quadratic $x^2 - x + 1 = 0$ are distinct and nonreal, which makes them complicated to work with, so we alter our approach accordingly.

Since $x^3 + 1 = (x - r_1)(x - r_2)(x - r_3)$ has degree 3, the remainder is of the form $s(x) = ax^2 + bx + c$ for some constants a, b, and c. Let $q(x)$ be the quotient, so

$$x^{81} + x^{48} + 2x^{27} + x^6 + 3 = (x - r_1)(x - r_2)(x - r_3)q(x) + ax^2 + bx + c.$$

Taking $x = r_1$, we get $r_1^{81} + r_1^{48} + 2r_1^{27} + r_1^6 + 3 = ar_1^2 + br_1 + c$. Since r_1 is a root of $x^3 + 1 = 0$, we know that $r_1^3 = -1$. Hence,

$$r_1^{81} + r_1^{48} + 2r_1^{27} + r_1^6 + 3 = (r_1^3)^{27} + (r_1^3)^{16} + 2(r_1^3)^9 + (r_1^3)^2 + 3 = (-1)^{27} + (-1)^{16} + 2 \cdot (-1)^9 + (-1)^2 + 3 = 2.$$

Hence, $ar_1^2 + br_1 + c = 2$, so $s(r_1) = 2$. Note that in our calculation, we have only used the fact that $r_1^3 = -1$. The other roots r_2 and r_3 also have this property, so this same process will give us $s(r_2) = 2$ and $s(r_3) = 2$. Therefore, because the r_i are distinct, we have found three roots to the polynomial $s(x) - 2$. But the degree of $s(x) - 2$ is no greater than 2, so we must have $s(x) = 2$ for all x. Therefore, the remainder is $\boxed{2}$.

8.60   By Vieta's Formulas, we have $a + b + c = 9$, $ab + ac + bc = 11$, and $abc = 1$, so $\sqrt{abc} = 1$. (Note that the roots a, b, and c are positive.)

We have $s^2 = a + b + c + 2\sqrt{ab} + 2\sqrt{ac} + 2\sqrt{bc} = 9 + 2(\sqrt{ab} + \sqrt{ac} + \sqrt{bc})$, so $s^2 - 9 = 2(\sqrt{ab} + \sqrt{ac} + \sqrt{bc})$. Squaring, we get

$$s^4 - 18s^2 + 81 = 4(ab + ac + bc + 2\sqrt{ab}\sqrt{ac} + 2\sqrt{ab}\sqrt{bc} + 2\sqrt{ac}\sqrt{bc})$$
$$= 4[ab + ac + bc + 2\sqrt{abc}(\sqrt{a} + \sqrt{b} + \sqrt{c})] = 4(11 + 2s) = 44 + 8s,$$

so $s^4 - 18s^2 - 8s + 37 = 0$. Therefore, $s^4 - 18s^2 - 8s = \boxed{-37}$.

8.61   By Vieta's Formulas, we have $\beta_1 + \beta_2 + \beta_3 = 0$, $\beta_1\beta_2 + \beta_1\beta_3 + \beta_2\beta_3 = -3$, and $\beta_1\beta_2\beta_3 = 1$. Therefore, for all i,

$$(\alpha_i + \beta_1)(\alpha_i + \beta_2)(\alpha_i + \beta_3) = \alpha_i^3 + (\beta_1 + \beta_2 + \beta_3)\alpha_i^2 + (\beta_1\beta_2 + \beta_1\beta_3 + \beta_2\beta_3)\alpha_i + \beta_1\beta_2\beta_3 = \alpha_i^3 - 3\alpha_i + 1.$$

Because α_1 and α_2 are roots of $x^2 - 5x - 2$, we have $\alpha_i^2 = 5\alpha_i + 2$ for $i = 1, 2$. Therefore, we have

$$\alpha_i^3 = 5\alpha_i^2 + 2\alpha_i = 5(5\alpha_i + 2) + 2\alpha_i = 27\alpha_i + 10.$$

Therefore, $\alpha_i^3 - 3\alpha_i + 1 = 24\alpha_i + 11$. Vieta's Formulas also give us $\alpha_1 + \alpha_2 = 5$ and $\alpha_1\alpha_2 = -2$, so

$$(\alpha_1 + \beta_1)(\alpha_1 + \beta_2)(\alpha_1 + \beta_3)(\alpha_2 + \beta_1)(\alpha_2 + \beta_2)(\alpha_2 + \beta_3) = (24\alpha_1 + 11)(24\alpha_2 + 11) = 576\alpha_1\alpha_2 + 264(\alpha_1 + \alpha_2) + 121$$
$$= 576 \cdot (-2) + 264 \cdot 5 + 121 = \boxed{289}.$$

8.62   We tackle a more general result, by finding the average of the roots of $p(q(x))$ in terms of the average of the roots of $q(x)$, where $p(x)$ and $q(x)$ are arbitrary polynomials with degree greater than 1.

Let $p(x) = p_n x^n + p_{n-1} x^{n-1} + \cdots + p_1 x + p_0$ and $q(x) = q_m x^m + q_{m-1} x^{m-1} + \cdots + q_1 x + q_0$. Then

$$p(q(x)) = p(q_m x^m + q_{m-1} x^{m-1} + \cdots + q_1 x + q_0)$$
$$= p_n(q_m x^m + q_{m-1} x^{m-1} + \cdots + q_1 x + q_0)^n$$
$$+ p_{n-1}(q_m x^m + q_{m-1} x^{m-1} + \cdots + q_1 x + q_0)^{n-1}$$
$$+ \cdots$$
$$+ p_1(q_m x^m + q_{m-1} x^{m-1} + \cdots + q_1 x + q_0) + p_0.$$

This equation tells us a number of things. First, when expanded, the term of highest degree is $p_n q_m^n x^{mn}$, so the degree of $p(q(x))$ is mn. Hence, from Vieta's Formulas, to find the sum of the roots of $p(q(x)) = 0$, we must find the coefficients of x^{mn} and x^{mn-1} in $p(q(x))$.

Next, we look at how the terms of degree mn and $mn - 1$ can arise in $p(q(x))$. Let $0 \le t \le n$. Then the term of highest degree in

$$p_t(q_m x^m + q_{m-1} x^{m-1} + \cdots + q_1 x + q_0)^t$$

is $p_t(q_m x^m)^t = p_t q_m^t x^{tm}$. If $t \le n - 1$, then $tm \le mn - m < mn - 1$, so the only term in $p(q(x))$ that can produce terms with degree mn and $mn - 1$ is $p_n(q_m x^m + q_{m-1} x^{m-1} + \cdots + q_1 x + q_0)^n$.

As we saw above, the only term of degree mn is $p_n q_m^n x^{mn}$. Now we must find the coefficient of x^{mn-1} in

$$p_n(q_m x^m + q_{m-1} x^{m-1} + \cdots + q_1 x + q_0)^n$$
$$= p_n(q_m x^m + q_{m-1} x^{m-1} + \cdots + q_1 x + q_0)(q_m x^m + q_{m-1} x^{m-1} + \cdots + q_1 x + q_0) \cdots (q_m x^m + q_{m-1} x^{m-1} + \cdots + q_1 x + q_0).$$

We have n factors of $q_m x^m + q_{m-1} x^{m-1} + \cdots + q_1 x + q_0$. The only way we can obtain a term of degree $mn - 1$ is if the term $q_m x^m$ is chosen from $n - 1$ of the factors, and the term $q_{m-1} x^{m-1}$ is chosen from the remaining factor. There are n factors from which to choose $q_{m-1} x^{m-1}$, so the term of degree $mn - 1$ in the expansion is

$$n p_n q_m^{n-1} q_{m-1} x^{mn-1}.$$

Therefore, by Vieta's Formulas, the sum of the roots of $p(q(x))$ is

$$-\frac{n p_n q_m^{n-1} q_{m-1}}{p_n q_m^n} = -\frac{n q_{m-1}}{q_m},$$

which means the average of the mn roots of $p(q(x))$ is

$$-\frac{n q_{m-1}}{mn q_m} = -\frac{q_{m-1}}{m q_m}.$$

However, this is the same as the average of the roots of $q(x)$. Therefore, in general, the average of the roots of $p(q(x))$ is equal to the average of the roots of $q(x)$.

Since $g_{k+1} = f(g_k(x))$, it follows that $r_{k+1} = r_k$ for all $k \ge 1$. Therefore, since $r_{19} = 89$, we have $r_{89} = \boxed{89}$.

8.63   One way to show that $\alpha^2 - 2\alpha - 14$ is a root of the cubic is to find the third root in terms of α, then show that the three roots satisfy Vieta's Formulas for the polynomial $x^3 - 21x + 35$. The sum of the roots of this polynomial is 0, so if α and $\alpha^2 + 2\alpha - 14$ are roots, then $-(\alpha + \alpha^2 + 2\alpha - 14) = -\alpha^2 - 3\alpha + 14$ must also be a root. Letting $\beta = \alpha^2 + 2\alpha - 14$ and $\gamma = -\alpha^2 - 3\alpha + 14$, we must check the other symmetric sums of α, β, and γ, and see if they match what Vieta's Formulas give us.

We first find $\beta\gamma$, since we'll need this for both $\alpha\beta + \alpha\gamma + \beta\gamma$ and $\alpha\beta\gamma$. We have

$$\beta\gamma = (\alpha^2 + 2\alpha - 14)(-\alpha^2 - 3\alpha + 14) = -\alpha^4 - 5\alpha^3 + 22\alpha^2 + 70\alpha - 196.$$

Uh-oh, we'll need to deal with α^4 and α^3. Fortunately, we know that α is a root of $x^3 - 21x + 35 = 0$, so $\alpha^3 - 21\alpha + 35 = 0$, which means $\alpha^3 = 21\alpha - 35$. Multiplying by α gives $\alpha^4 = 21\alpha^2 - 35\alpha$. Now, we have

$$\beta\gamma = -\alpha^4 - 5\alpha^3 + 22\alpha^2 + 70\alpha - 196 = -(21\alpha^2 - 35\alpha) - 5(21\alpha - 35) + 22\alpha^2 + 70\alpha - 196$$
$$= -21\alpha^2 + 35\alpha - 105\alpha + 175 + 22\alpha^2 + 70\alpha - 196 = \alpha^2 - 21,$$

so $\alpha\beta + \alpha\gamma + \beta\gamma = \alpha(\beta + \gamma) + \beta\gamma = \alpha \cdot (-\alpha) + \alpha^2 - 21 = -21$ and $\alpha\beta\gamma = \alpha(\alpha^2 - 21) = \alpha^3 - 21\alpha = -35$.

Thus, α, β, and γ are the roots of $x^3 - 21x + 35 = 0$. In particular, $\beta = \alpha^2 + 2\alpha - 14$ is a root.

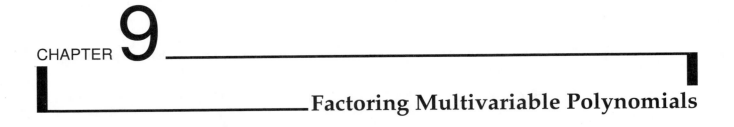

CHAPTER **9**

Factoring Multivariable Polynomials

Exercises for Section 9.1

9.1.1

(a) The first two terms factor as $xy + 3x = x(y + 3)$. The last two terms factor as $-4y - 12 = -4(y + 3)$. Therefore,
$$xy + 3x - 4y - 12 = x(y + 3) - 4(y + 3) = \boxed{(x - 4)(y + 3)}.$$

(b) The first two terms factor as $2ab - 14a = 2a(b - 7)$. The last two terms factor as $5b - 35 = 5(b - 7)$. Therefore,
$$2ab - 14a + 5b - 35 = 2a(b - 7) + 5(b - 7) = \boxed{(2a + 5)(b - 7)}.$$

9.1.2   We begin by multiplying through by all the denominators in the equation to get $7y + 7x = xy$. Rearranging this equation gives us $xy - 7x - 7y = 0$. We apply Simon's Favorite Factoring Trick by adding 49 to both sides to give $xy - 7x - 7y + 49 = 49$, which we can factor as $(x - 7)(y - 7) = 49$. Since x and y are integers, we need $x - 7$ and $y - 7$ to be integers that multiply to 49. The pairs of integers that multiply to 49 are

$$49 = (1)(49) = (-1)(-49) = (7)(7) = (-7)(-7).$$

If $x - 7 = 1$ and $y - 7 = 49$, we have $(x, y) = (8, 56)$. Similarly, $(x, y) = (56, 8)$ must also be a solution. From $49 = (-1)(-49)$, we have the solutions $(x, y) = (6, -42)$ and $(x, y) = (-42, 6)$. From $49 = (7)(7)$, we have $(x, y) = (14, 14)$, and from $49 = (-7)(-7)$, we get $(x, y) = (0, 0)$, which we must discard because we cannot divide by 0. Our solutions are $\boxed{(14, 14); (8, 56); (56, 8); (6, -42); \text{ and } (-42, 6)}$.

9.1.3   The first two terms factor as $x^3 + 2xy^2 = x(x^2 + 2y^2)$, and the last two terms factor as $5x^2 + 10y^2 = 5(x^2 + 2y^2)$, so $x^3 + 2xy^2 + 5x^2 + 10y^2 = x(x^2 + 2y^2) + 5(x^2 + 2y^2) = \boxed{(x + 5)(x^2 + 2y^2)}$.

9.1.4   Moving all the terms to one side, we get $4x^2y^2 + 36x^2 - y^2 - 9 = 0$. The first two terms factor as $4x^2y^2 + 36x^2 = 4x^2(y^2 + 9)$, so
$$4x^2y^2 + 36x^2 - y^2 - 9 = 4x^2(y^2 + 9) - (y^2 + 9) = (4x^2 - 1)(y^2 + 9),$$
which gives us $(4x^2 - 1)(y^2 + 9) = 0$.

The second factor $y^2 + 9$ can never be 0, so $4x^2 - 1 = 0$, or $x = \pm 1/2$. Therefore, the solutions are of the form $\boxed{(1/2, y) \text{ and } (-1/2, y), \text{ where } y \text{ is any real number}}$.

9.1.5   Adding 1 to both sides, we find $xy + x + y + 1 = 55$, so $(x + 1)(y + 1) = 55$. The prime factorization of 55 is $5 \cdot 11$. Since x and y are positive integers, both $x + 1$ and $y + 1$ are at least 2. Hence, $x + 1$ and $y + 1$ are equal to 5 and 11 in some order. Then $(x + 1) + (y + 1) = 5 + 11 = 16$, so $x + y = 16 - 2 = \boxed{14}$.

9.1.6   The first two terms factor as $8 - 2y = 2(4 - y)$, and the last two terms factor as $4y^3 - y^4 = y^3(4 - y)$, so
$$8 - 2y + 4y^3 - y^4 = 2(4 - y) + y^3(4 - y) = (4 - y)(y^3 + 2).$$

Therefore, the real roots are $y = \boxed{4 \text{ and } -\sqrt[3]{2}}$.

9.1.7 Looking for a way to group terms so that they factor, we find that $x^2y + xy^2 = xy(x + y)$ and $x^2 + 2xy + y^2 = (x + y)^2$. Hence,

$$x^2y + xy^2 + x^2 + 2xy + y^2 + x + y = xy(x + y) + (x + y)^2 + (x + y) = (x + y)(xy + x + y + 1) = \boxed{(x + y)(x + 1)(y + 1)}.$$

9.1.8 Using Simon's Favorite Factoring Trick on the first equation, we find $xy + 2x + 3y + 6 = 4 + 6$, so $(x+3)(y+2) = 10$. Similarly, $yz - y + 2z - 2 = 5 - 2$, so $(y + 2)(z - 1) = 3$, and $xz - x + 3z - 3 = 33 - 3$, so $(x + 3)(z - 1) = 30$.

To make these equations easier to work with, let $a = x + 3$, $b = y + 2$, and $c = z - 1$. Then these equations become

$$ab = 10,$$
$$bc = 3,$$
$$ac = 30.$$

Multiplying all three equations, we get $a^2b^2c^2 = 900$, so $abc = 30$ or $abc = -30$.

If $abc = 30$, then $a = 30/(bc) = 10$, $b = 30/(ac) = 1$, and $c = 30/(ab) = 3$, so $x = 7$, $y = -1$, and $z = 4$.

If $abc = -30$, then $a = -30/(bc) = -10$, $b = -30/(ac) = -1$, and $c = 30/(ab) = -3$, so $x = -13$, $y = -3$, and $z = -2$.

Therefore, the solutions are $(x, y, z) = \boxed{(7, -1, 4) \text{ and } (-13, -3, -2)}$.

Exercises for Section 9.2

9.2.1 Using difference of squares, we have $x^6 - 1 = (x^3)^2 - 1 = (x^3 - 1)(x^3 + 1)$. Using difference of cubes, we have $x^3 - 1 = (x-1)(x^2+x+1)$. The roots of the quadratic are $\frac{-1 \pm i\sqrt{3}}{2}$. Using sum of cubes, we have $x^3 + 1 = (x+1)(x^2 - x + 1)$. The roots of the quadratic are $\frac{1 \pm i\sqrt{3}}{2}$.

Therefore, the roots of $x^6 - 1$ are

$$\boxed{1, \quad -1, \quad \frac{-1 + i\sqrt{3}}{2}, \quad \frac{-1 - i\sqrt{3}}{2}, \quad \frac{1 + i\sqrt{3}}{2}, \quad \text{and} \quad \frac{1 - i\sqrt{3}}{2}}.$$

9.2.2 Adding the two equations, we get $2xy + 16 = x^2 + y^2$, so $x^2 - 2xy + y^2 = 16$, which means $(x - y)^2 = 16$. Hence, $x - y = 4$ or $x - y = -4$. Subtracting the two original equations, we get $x^2 - y^2 = -2$, so $(x + y)(x - y) = -2$.

If $x - y = 4$, then $x + y = -2/4 = -1/2$, so $x = 7/4$ and $y = -9/4$. If $x - y = -4$, then $x + y = -2/(-4) = 1/2$, so $x = -7/4$ and $y = 9/4$. Therefore, the solutions are $(x, y) = \boxed{(7/4, -9/4) \text{ and } (-7/4, 9/4)}$.

9.2.3 The expression $2^{27} + 3^{27}$ can be factored many ways. For example, factoring as a sum of 27th powers, we have $2^{27} + 3^{27} = (2 + 3)(2^{26} - 2^{25} \cdot 3 + 2^{24} \cdot 3^2 - \cdots + 3^{26})$.

Factoring as a sum of 9th powers, we have

$$2^{27} + 3^{27} = 8^9 + 27^9 = (8 + 27)(8^8 - 8^7 \cdot 27 + 8^6 \cdot 27^2 - \cdots + 27^8).$$

Since $8 + 27 = 35$, $2^{27} + 3^{27}$ is divisible by 35.

9.2.4 A direct substitution for x or y would result in a 9th degree polynomial, so we look for another way to work with the equations. Adding and subtracting the equations, we get $x^3 + y^3 = 9x + 9y$, and $x^3 - y^3 = 7x - 7y$.

These equations can be factored as follows:

$$(x + y)(x^2 - xy + y^2) = 9(x + y),$$
$$(x - y)(x^2 + xy + y^2) = 7(x - y).$$

Since $|x| \neq |y|$, we have $x + y \neq 0$ and $x - y \neq 0$. Hence, we can divide the first equation by $x + y$ and the second by $x - y$ to get

$$x^2 - xy + y^2 = 9,$$
$$x^2 + xy + y^2 = 7.$$

Adding these equations gives us $2x^2 + 2y^2 = 16$, so $x^2 + y^2 = \boxed{8}$.

9.2.5 Multiplying both sides by $ac(a + c)$ to clear the fractions, we get

$$ac = a(a + c) + c(a + c) = a^2 + 2ac + c^2,$$

so $a^2 + ac + c^2 = 0$. Multiplying by $a - c$ gives $(a - c)(a^2 + ac + c^2) = 0$, so $a^3 - c^3 = 0$, and we have $(a/c)^3 = a^3/c^3 = \boxed{1}$.

9.2.6 We can write 7,999,999,999 as the difference of two cubes:

$$7{,}999{,}999{,}999 = 8{,}000{,}000{,}000 - 1 = 2000^3 - 1 = (2000 - 1)(2000^2 + 2000 + 1) = 1999 \cdot 4002001.$$

We are given that 7,999,999,999 has at most two prime factors, so they must be 1999 and 4002001. Therefore, the answer is $\boxed{4002001}$.

9.2.7 Multiplying both sides by $x - 1$ to clear the fraction, and using difference of cubes twice, we find

$$10 = (x - 1)(x^2 + x + 1)(x^6 + x^3 + 1) = (x^3 - 1)(x^6 + x^3 + 1) = x^9 - 1,$$

so $x^9 = 11$. Therefore, $x = \boxed{\sqrt[9]{11}}$.

9.2.8 We can try to show that $n^4 - 20n^2 + 4$ is composite for all integers n by trying to factor the expression. This expression is a quartic in n, but it is also a quadratic in n^2, so we may complete the square by finding a square of a binomial with n^4 and $20n^2$ as terms in the expansion of the square:

$$n^4 - 20n^2 + 4 = (n^2 - 10)^2 - 100 + 4 = (n^2 - 10)^2 - 96.$$

This expression does not factor over the integers, so we instead try to complete the square by finding a square of a binomial with n^4 and 4 as terms in the expansion of the square:

$$n^4 - 20n^2 + 4 = n^4 - 4n^2 + 4 - 16n^2 = (n^2 - 2)^2 - (4n)^2.$$

Thus, we have a difference of squares, which factors as

$$(n^2 - 2)^2 - (4n)^2 = (n^2 - 4n - 2)(n^2 + 4n - 2).$$

To show that this expression is always composite, it suffices to show that both factors are always at least 2. For $n \geq 5$, we have $n^2 - 4n - 2 = (n - 2)^2 - 6 \geq 3^2 - 6 = 3$, and $n^2 + 4n - 2 = (n + 2)^2 - 6 \geq 7^2 - 6 = 43$, so both factors are greater than 2. Therefore, $n^4 - 20n^2 + 4$ is composite for all integers $n > 4$.

Exercises for Section 9.3

9.3.1 Let $f(a, b, c) = (a - b)^3 + (b - c)^3 + (c - a)^3$. Then $f(a, b, a) = (a - b)^3 + (b - a)^3 + (a - a)^3 = 0$, so $f(a, b, c) = 0$ for $a = c$. This means $f(a, b, c)$ is divisible by $c - a$.

Similarly, we have $f(a, b, b) = (a-b)^3 + (b-b)^3 + (b-a)^3 = 0$, so $f(a, b, c)$ is divisible by $c - b$. Finally, letting $b = a$, we have $f(a, a, c) = (a-a)^3 + (a-c)^3 + (c-a)^3 = 0$, so $f(a, b, c)$ is divisible by $a - b$.

Therefore, $f(a, b, c)$ is divisible by $(a-b)(c-a)(c-b)$. Since both $f(a, b, c)$ and $(a-b)(c-a)(c-b)$ have degree 3, we have

$$f(a, b, c) = (a-b)^3 + (b-c)^3 + (c-a)^3 = k(a-b)(c-a)(c-b)$$

for some constant k.

Taking $a = -1$, $b = 0$, and $c = 1$, we get $(-1)^3 + (-1)^3 + 2^3 = k(-1)(2)(1)$, or $6 = -2k$, so $k = -3$. Therefore,

$$(a-b)^3 + (b-c)^3 + (c-a)^3 = -3(a-b)(c-a)(c-b) = 3(a-b)(b-c)(c-a).$$

9.3.2 Let the left side be $f(a, b, c, d)$. We show that a is a factor of $f(a, b, c, d)$ by letting $a = 0$. This gives us

$$\begin{aligned} f(0, b, c, d) &= (b+c+d)^4 + (b-c-d)^4 + (-b+c-d)^4 + (-b-c+d)^4 \\ &\quad - (b+c-d)^4 - (b-c+d)^4 - (-b+c+d)^4 - (b+c+d)^4 \\ &= (b+c+d)^4 + (b-c-d)^4 + (b-c+d)^4 + (b+c-d)^4 \\ &\quad - (b+c-d)^4 - (b-c+d)^4 - (b-c-d)^4 - (b+c+d)^4 = 0, \end{aligned}$$

so by the Factor Theorem, $f(a, b, c, d)$ is divisible by a.

Similarly, it can be shown that $f(a, b, c, d)$ is divisible by b, c, and d, so $f(a, b, c, d)$ is divisible by $abcd$. Since $f(a, b, c, d)$ has degree 4, and the polynomial $abcd$ has degree 4, we have

$$\begin{aligned} (a+b+c+d)^4 &+ (a+b-c-d)^4 + (a-b+c-d)^4 + (a-b-c+d)^4 \\ &- (a+b+c-d)^4 - (a+b-c+d)^4 - (a-b+c+d)^4 - (-a+b+c+d)^4 = kabcd \end{aligned}$$

for some constant k. To find this constant, let $a = b = c = d = 1$. Then the equation above becomes

$$4^4 + 0^4 + 0^4 + 0^4 - 2^4 - 2^4 - 2^4 - 2^4 = k,$$

so $k = 192$. Hence, the expression is equal to $192abcd$.

9.3.3 Let $f(a, b, c) = abc - ab - bc - ca + a + b + c - 1$. To find a factor of $f(a, b, c)$, we look for values of a that may make all the terms cancel. For example, if $a = 1$, then the term abc would cancel with bc, so we try taking $a = 1$:

$$f(1, b, c) = bc - b - bc - c + 1 + b + c - 1 = 0.$$

Hence, $f(a, b, c)$ is divisible by $a - 1$.

Similarly,

$$\begin{aligned} f(a, 1, c) &= ac - a - c - ca + a + 1 + c - 1 = 0, \quad \text{and} \\ f(a, b, 1) &= ab - ab - b - a + a + b + 1 - 1 = 0, \end{aligned}$$

so $f(a, b, c)$ is also divisible by $b - 1$ and $c - 1$. Therefore, $f(a, b, c)$ is divisible by $(a-1)(b-1)(c-1)$.

Since $f(a, b, c) = abc - ab - bc - ca + a + b + c - 1$ has degree 3 and $(a-1)(b-1)(c-1)$ has degree 3,

$$f(a, b, c) = abc - ab - bc - ca + a + b + c - 1 = k(a-1)(b-1)(c-1)$$

for some constant k. To find this constant, set $a = b = c = 0$. Then the equation above becomes $-1 = k(-1)(-1)(-1)$, so $k = 1$. Therefore, $abc - ab - bc - ca + a + b + c - 1 = \boxed{(a-1)(b-1)(c-1)}$.

9.3.4 Multiplying both sides by abc, we see that we must show that

$$bc(b-c) + ac(c-a) + ab(a-b) = (c-b)(a-c)(b-a).$$

When $a = b$ the left side is

$$bc(b - c) + bc(c - b) + b^2(b - b) = 0,$$

so $b - a$ is a factor of the left side. Similarly, $c - b$ and $a - c$ are also factors, so we have

$$bc(b - c) + ac(c - a) + ab(a - b) = k(c - b)(a - c)(b - a)$$

for some constant k. Letting $a = 1$, $b = 2$, and $c = 3$, we find that $k = 1$, so we have

$$bc(b - c) + ac(c - a) + ab(a - b) = (c - b)(a - c)(b - a).$$

Dividing by abc gives the desired result.

9.3.5 From the given information, we have $\frac{1}{x} + \frac{1}{y} + \frac{1}{z} = \frac{1}{x+y+z}$. Multiplying both sides by $xyz(x + y + z)$ to clear fractions, we get

$$yz(x + y + z) + xz(x + y + z) + xy(x + y + z) = xyz,$$

so we have

$$yz(x + y + z) + xz(x + y + z) + xy(x + y + z) - xyz = 0.$$

We hope to simplify the left side, so we let it be $f(x, y, z)$ and try to factor. We find that if $x = -y$, we have

$$f(-y, y, z) = yz(z) + (-y)(z)(z) + (-y)(y)(z) - (-y)yz = 0,$$

so $x + y$ is a factor of $f(x, y, z)$. Similarly, $y + z$ and $z + x$ are also factors of $f(x, y, z)$, and because f has degree 3, we have

$$yz(x + y + z) + xz(x + y + z) + xy(x + y + z) - xyz = k(x + y)(x + z)(y + z)$$

for some constant k. Letting $x = y = z = 1$ gives us $k = 1$. Hence, the equation

$$\frac{1}{x} + \frac{1}{y} + \frac{1}{z} = \frac{1}{x + y + z}$$

holds if and only if $(x + y)(x + z)(y + z) = 0$, for all x, y, and z such that all the fractions are defined.

If $x + y = 0$, then $y = -x$, and since n is an odd integer, we have $x^n + y^n + z^n = x^n + (-x)^n + z^n = z^n$, and $(x + y + z)^n = z^n$. Hence, $x^n + y^n + z^n = (x + y + z)^n$. The same holds if $x + z = 0$ or $y + z = 0$.

9.3.6 Let $f(x, y) = (x+y)^7 - (x^7 + y^7)$. Then $f(0, y) = y^7 - y^7 = 0$, so $f(x, y)$ is divisible by x. Also, $f(x, 0) = x^7 - x^7 = 0$, so $f(x, y)$ is divisible by y. Finally, taking $y = -x$, we get $f(x, -x) = (x - x)^7 - [x^7 + (-x)^7] = 0$, so $f(x, y)$ is divisible by $x + y$.

Expanding $(x + y)^7 - (x^7 + y^7)$, we get

$$f(x, y) = 7x^6y + 21x^5y^2 + 35x^4y^3 + 35x^3y^4 + 21x^2y^5 + 7xy^6.$$

Factoring out $xy(x + y)$, we get $f(x, y) = xy(x + y)(7x^4 + 14x^3y + 21x^2y^2 + 14xy^3 + 7y^4)$.

We observe that we can take out a further factor of 7, to get $7xy(x + y)(x^4 + 2x^3y + 3x^2y^2 + 2xy^3 + y^4)$. Finally, as seen in Problem 9.15, this expression factors as $\boxed{7xy(x + y)(x^2 + xy + y^2)^2}$.

Review Problems

9.22 Using difference of squares, we have $x^2 - y^2 = (x + y)(x - y) = (2 \cdot 2001^{1002})(-2 \cdot 2001^{-1002}) = \boxed{-4}$.

9.23

(a) Using difference of squares twice, we have $16y^4 - 1 = (4y^2 - 1)(4y^2 + 1) = \boxed{(2y - 1)(2y + 1)(4y^2 + 1)}$.

(b) Using sum of fifth powers, we have

$$243x^5 + 32z^{10} = (3x)^5 + (2z^2)^5 = \boxed{(3x + 2z^2)(81x^4 - 54x^3z^2 + 36x^2z^4 - 24xz^6 + 16z^8)}.$$

9.24 The first two terms factor as $x^7 - x^4 = x^4(x^3 - 1)$, so

$$x^7 - x^4 - x^3 + 1 = x^4(x^3 - 1) - (x^3 - 1) = (x^4 - 1)(x^3 - 1).$$

Then $x^4 - 1$ factors as $(x^2 - 1)(x^2 + 1) = (x - 1)(x + 1)(x^2 + 1)$, and $x^3 - 1 = (x - 1)(x^2 + x + 1)$, so

$$(x^4 - 1)(x^3 - 1) = (x - 1)^2(x + 1)(x^2 + 1)(x^2 + x + 1).$$

Hence, the roots are $\boxed{1, \quad 1, \quad -1, \quad i, \quad -i, \quad \dfrac{-1 + i\sqrt{3}}{2}, \quad \text{and} \quad \dfrac{-1 - i\sqrt{3}}{2}}$.

9.25 The expression $x^8 - 1$ factors as

$$x^8 - 1 = (x^4 - 1)(x^4 + 1) = (x^2 - 1)(x^2 + 1)(x^4 + 1) = (x - 1)(x + 1)(x^2 + 1)(x^4 + 1).$$

Therefore, the roots of $x^8 - 1$ are

$$\boxed{1, \quad -1, \quad i, \quad -i, \quad \dfrac{1 + i}{\sqrt{2}}, \quad \dfrac{1 - i}{\sqrt{2}}, \quad \dfrac{-1 + i}{\sqrt{2}}, \quad \text{and} \quad \dfrac{-1 - i}{\sqrt{2}}}.$$

(We found the roots of $x^4 + 1 = 0$ in Problem 8.2.5.)

9.26 Using difference of cubes, we have

$$19x^3 = a^3 - b^3 = (a - b)(a^2 + ab + b^2) = x(a^2 + ab + b^2).$$

Since $a \neq b$, we have $x \neq 0$, so we can divide both sides of $19x^3 = x(a^2 + ab + b^2)$ by x to get $a^2 + ab + b^2 = 19x^2$.

Substituting $b = a - x$, we get $a^2 + a(a - x) + (a - x)^2 = 19x^2$. Expanding the left side and rearranging gives $3a^2 - 3ax - 18x^2 = 0$. Factoring gives us $3(a - 3x)(a + 2x) = 0$. Therefore, we have $a = \boxed{3x \text{ or } -2x}$.

9.27 By difference of cubes, we have $x^3 - y^3 = (x - y)(x^2 + xy + y^2) = 2(x^2 + xy + y^2)$, so the equation $x^3 - y^3 = 98$ becomes $2(x^2 + xy + y^2) = 98$. Dividing both sides by 2 and substituting $y = x - 2$, we get

$$x^2 + x(x - 2) + (x - 2)^2 = 49,$$

Simplifying the equation gives us $3(x - 5)(x + 3) = 0$. Then $x = 5$ or $x = -3$, and $y = 3$ or $y = -5$, respectively. Therefore, the solutions are $(x, y) = \boxed{(5, 3) \text{ and } (-3, -5)}$.

9.28 Let the numbers be x and y. These numbers are Simonish if $x + y = xy$. Equivalently, they satisfy $xy - x - y + 1 = 1$, or $(x - 1)(y - 1) = 1$. Hence, the numbers x and y are Simonish if and only if $x - 1$ and $y - 1$ are reciprocals. In other words, all Simonish pairs (x, y) are of the form $x = 1 + u$ and $y = 1 + v$, where u and v are reciprocals.

9.29 Moving all the terms to one side, we get $6xy - 3x - 2y + 1 = 0$, so

$$(3x - 1)(2y - 1) = 0.$$

Thus, the graph consists of the lines $x = 1/3$ and $y = 1/2$.

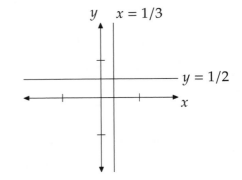

9.30

(a) The first three terms form the square $p^2 - 2p + 1 = (p - 1)^2$, and the last three terms form the square $-q^2 - 2qr - r^2 = -(q + r)^2$, so by difference of squares, we have

$$p^2 - 2p + 1 - q^2 - 2qr - r^2 = (p - 1)^2 - (q + r)^2 = \boxed{(p + q + r - 1)(p - q - r - 1)}.$$

(b) By difference of squares, we have

$$4x^2y^2 - (x^2 - z^2 + y^2)^2 = (2xy)^2 - (x^2 - z^2 + y^2)^2 = (2xy + x^2 - z^2 + y^2)(2xy - x^2 + z^2 - y^2).$$

We also have $2xy + x^2 - z^2 + y^2 = (x + y)^2 - z^2 = (x + y + z)(x + y - z)$ and $2xy - x^2 + z^2 - y^2 = z^2 - (x - y)^2 = (z + x - y)(z - x + y)$.

Hence,

$$4x^2y^2 - (x^2 - z^2 + y^2)^2 = (x + y + z)(x + y - z)(z + x - y)(z - x + y)$$
$$= \boxed{(x + y + z)(-x + y + z)(x - y + z)(x + y - z)}.$$

(c) By difference of squares, we have

$$(a^2 - b^2)(b^2 - c^2) + (a - b)(b - c) = (a - b)(a + b)(b - c)(b + c) + (a - b)(b - c) = \boxed{(a - b)(b - c)[(a + b)(b + c) + 1]}.$$

(d) Looking at all the terms of degree 2, we see that $x^2 + 5xy + 6y^2 = (x + 2y)(x + 3y)$. Hence,

$$x^2 + 5xy + 6y^2 + x + 3y = (x + 2y)(x + 3y) + (x + 3y) = \boxed{(x + 3y)(x + 2y + 1)}.$$

9.31 Moving all the terms to one side, we get $6cd + 4c - 3d - 79 = 0$. We'd like to use Simon's Favorite Factoring Trick. To figure out how, we start by factoring $2c$ out of the first two terms, which gives $2c(3d + 2) - 3d - 79 = 0$. We see now that adding $-2 + 2$ gives us another factor of $(3d + 2)$:

$$2c(3d + 2) - 3d - 2 + 2 - 79 = 2c(3d + 2) - 1(3d + 2) - 77 = (2c - 1)(3d + 2) - 77.$$

So, we have $(2c - 1)(3d + 2) - 77 = 0$, which gives us $(2c - 1)(3d + 2) = 77$. Hence, $3d + 2$ must be a factor of 77.

The only factors of 77 that are of the form $3d + 2$ for some integer d are 11, 77, -1, and -7, which lead to the values $d = 3, 25, -1$, and -3, respectively. Then the corresponding values of $2c - 1$ are 7, 1, -77, and -11, which lead to the values $c = 4, 1, -38$, and -5, respectively.

Therefore, the solutions are $(c, d) = \boxed{(4, 3), (1, 25), (-38, -1) \text{ and } (-5, -3)}$.

Here's a look at another way to apply Simon's Favorite Factoring Trick. We divide $6cd + 4c - 3d - 79 = 0$ by 6 to set the coefficient of cd to be 1:

$$cd + \frac{2}{3}c - \frac{1}{2}d - \frac{79}{6} = 0.$$

This becomes

$$\left(c - \frac{1}{2}\right)\left(d + \frac{2}{3}\right) + \frac{1}{2} \cdot \frac{2}{3} - \frac{79}{6} = 0.$$

Multiplying both sides by 6 and simplifying, we get $(2c - 1)(3d + 2) = 77$, as before.

9.32

(a) The first two terms factor as $9a^5 - 4a^3 = a^3(9a^2 - 4)$, and the last two terms factor as $-81a^2 + 36 = -9(9a^2 - 4)$, so $9a^5 - 4a^3 - 81a^2 + 36 = a^3(9a^2 - 4) - 9(9a^2 - 4) = (9a^2 - 4)(a^3 - 9) = \boxed{(3a + 2)(3a - 2)(a^3 - 9)}$. (If $a^3 = 9$, then $a = \sqrt[3]{9}$, which is irrational, so $a^3 - 9$ does not factor.)

(b) Same as part (a).

(c) Over the real numbers, we can factor $a^3 - 9 = (a - \sqrt[3]{9})(a^2 + \sqrt[3]{9}a + 3\sqrt[3]{3})$. If $a^3 - 9 = 0$, then the only real root is $a = \sqrt[3]{9}$, so the quadratic has no real roots. Hence, the desired factorization is

$$9a^5 - 4a^3 - 81a^2 + 36 = \boxed{(3a + 2)(3a - 2)(a - \sqrt[3]{9})(a^2 + \sqrt[3]{9}a + 3\sqrt[3]{3})}.$$

(d) From the quadratic formula, the roots of $a^2 + \sqrt[3]{9}a + 3\sqrt[3]{3}$ are

$$a = \frac{-\sqrt[3]{9} \pm \sqrt{\sqrt[3]{81} - 12\sqrt[3]{3}}}{2} = \frac{-\sqrt[3]{9} \pm \sqrt{-9\sqrt[3]{3}}}{2} = \frac{-3^{2/3} \pm 3^{7/6}i}{2}.$$

Thus, the given expression factors over the complex numbers as

$$\boxed{(3a + 2)(3a - 2)(a - \sqrt[3]{9})\left(a - \frac{-3^{2/3} + 3^{7/6}i}{2}\right)\left(a - \frac{-3^{2/3} - 3^{7/6}i}{2}\right)}.$$

9.33 Moving all the terms to one side, we get $a^2 - b^2 + ac - bc = 0$, so $(a - b)(a + b) + c(a - b) = 0$, which means $(a - b)(a + b + c) = 0$. Since $a \neq b$, we may divide by $a - b$ to get $a + b + c = 0$. Therefore, $a = -(b + c)$.

9.34 Note that the given expression is linear in a, so we can group the terms accordingly:

$$(bc + be + cd + de)a + bcf + bef + cdf + def.$$

We also note that all the terms that do not contain a factor of a contain a factor of f, so this expression becomes

$$(bc + be + cd + de)a + (bc + be + cd + de)f = (bc + be + cd + de)(a + f).$$

Finally, $bc + be + cd + de = b(c + e) + d(c + e) = (b + d)(c + e)$, so $(bc + be + cd + de)(a + f) = \boxed{(a + f)(b + d)(c + e)}$.

9.35

(a) Applying a difference of squares twice, we find $x^{12} - y^{12} = (x^6 - y^6)(x^6 + y^6) = (x^3 - y^3)(x^3 + y^3)(x^6 + y^6)$.
Now, we have several sums and differences of cubes we can factor:

$$\begin{aligned} x^3 - y^3 &= (x - y)(x^2 + xy + y^2), \\ x^3 + y^3 &= (x + y)(x^2 - xy + y^2), \\ x^6 + y^6 &= (x^2 + y^2)(x^4 - x^2y^2 + y^4). \end{aligned}$$

Hence, $x^{12} - y^{12} = \boxed{(x - y)(x + y)(x^2 + y^2)(x^2 - xy + y^2)(x^2 + xy + y^2)(x^4 - x^2y^2 + y^4)}$.

(b) The only factor from part (a) that can be further factored over the real numbers is $x^4 - x^2y^2 + y^4$. To do so, we can use difference of squares:

$$x^4 - x^2y^2 + y^4 = x^4 + 2x^2y^2 + y^4 - 3x^2y^2 = (x^2 + y^2)^2 - (\sqrt{3}xy)^2 = (x^2 - \sqrt{3}xy + y^2)(x^2 + \sqrt{3}xy + y^2).$$

The discriminants of both quadratics is $(\pm\sqrt{3}y)^2 - 4y^2 = -y^2$, which is negative for all nonzero real y, so neither factors over the real numbers. Hence, the complete factorization over the real numbers is given by

$$x^{12} - y^{12} = \boxed{(x - y)(x + y)(x^2 + y^2)(x^2 - xy + y^2)(x^2 + xy + y^2)(x^2 - \sqrt{3}xy + y^2)(x^2 + \sqrt{3}xy + y^2)}.$$

9.36 Putting the expression over the common denominator $(a - b)(b - c)(c - a)$, and taking care to get the signs correct, we get

$$\frac{b + c}{(a - b)(a - c)} + \frac{a + c}{(b - c)(b - a)} + \frac{a + b}{(c - a)(c - b)} = \frac{-(b + c)(b - c) + (a + c)(a - c) - (a + b)(a - b)}{(a - b)(b - c)(c - a)}$$

$$= \frac{-b^2 + c^2 + a^2 - c^2 - a^2 + b^2}{(a - b)(b - c)(c - a)}$$

$$= 0.$$

9.37 We attempt to write $x^4 + 4y^4$ as a difference of squares that we can factor. We therefore seek a binomial that, when squared, gives us terms of x^4 and $4y^4$. Since $(x^2 + 2y^2)^2$ does so, we produce our difference of squares by adding $4x^2y^2 - 4x^2y^2$ to $x^4 + 4y^4$:

$$x^4 + 4y^4 = x^4 + 4x^2y^2 + 4y^4 - 4x^2y^2 = (x^2 + 2y^2)^2 - (2xy)^2 = \boxed{(x^2 + 2xy + 2y^2)(x^2 - 2xy + 2y^2)}.$$

9.38 Let $f(x) = (b - c)^3(x - a)^3 + (c - a)^3(x - b)^3 + (a - b)^3(x - c)^3$. Then

$$f(a) = (b - c)^3(a - a)^3 + (c - a)^3(a - b)^3 + (a - b)^3(a - c)^3 = 0,$$

so $f(x)$ is divisible by $x - a$. Similarly, $f(b) = f(c) = 0$, so $f(x)$ is divisible by $x - b$ and $x - c$. Therefore, $f(x)$ is divisible by $(x - a)(x - b)(x - c)$.

Since $f(x)$ has degree 3 (as a polynomial in x), and $(x-a)(x-b)(x-c)$ has degree 3, we have $f(x) = k(x-a)(x-b)(x-c)$ for some value k that does not depend on x (but k may depend on a, b, and c). The coefficient of x^3 in $f(x)$ is $(a - b)^3 + (b - c)^3 + (c - a)^3$, so

$$f(x) = [(a - b)^3 + (b - c)^3 + (c - a)^3](x - a)(x - b)(x - c).$$

By Problem 9.3.1, we have $(a - b)^3 + (b - c)^3 + (c - a)^3 = 3(a - b)(b - c)(c - a)$. Therefore, the given expression factors as $\boxed{3(a - b)(b - c)(c - a)(x - a)(x - b)(x - c)}$.

Note that we also could have let $g(x, a, b, c) = (b - c)^3(x - a)^3 + (c - a)^3(x - b)^3 + (a - b)^3(x - c)^3$. Just as $f(a) = 0$ above, we have $g(a, a, b, c) = 0$, so $x - a$ is a factor. Similarly, setting any two of the inputs to g equal makes $g(x, a, b, c) = 0$. Since the degree of $g(x, a, b, c)$ is at most 6, $g(x, a, b, c) = k'(a - b)(b - c)(c - a)(x - a)(x - b)(x - c)$ for some constant k'. Choosing $(x, a, b, c) = (0, 1, 2, 3)$ gives us $k = 3$, and we have the factorization found earlier.

9.39 Multiplying both sides by $(a + b)(a + c)(b + c)$, we get

$$2a(a + c)(b + c) + 2b(a + b)(a + c) + 2c(a + b)(b + c) + (b - c)(c - a)(a - b) = 3(a + b)(a + c)(b + c).$$

To prove this identity, let $f(a, b, c)$ denote the left side of this equation. Then

$$f(-b, b, c) = 2(-b)(-b + c)(b + c) + 2b(-b + b)(-b + c) + 2c(-b + b)(b + c) + (b - c)(c + b)(-b - b)$$

$$= -2b(-b + c)(b + c) + (b - c)(c + b)(-2b) = 0,$$

so $f(a,b,c)$ is divisible by $a+b$. Similarly, $f(a,b,c)$ is also divisible by $a+c$ and $b+c$. Therefore, $f(a,b,c)$ is divisible by $(a+b)(a+c)(b+c)$.

Since $f(a,b,c)$ has degree 3, and $(a+b)(a+c)(b+c)$ has degree 3, we have

$$f(a,b,c) = 2a(a+c)(b+c) + 2b(a+b)(a+c) + 2c(a+b)(b+c) + (b-c)(c-a)(a-b)$$
$$= k(a+b)(a+c)(b+c)$$

for some constant k. To find this constant k, we take $a=b=1$ and $c=0$. Then the equation above becomes

$$2 \cdot 1 \cdot 1 \cdot 1 + 2 \cdot 1 \cdot 2 \cdot 1 + 2 \cdot 0 \cdot 2 \cdot 1 + 1 \cdot (-1) \cdot 0 = 2k,$$

so $k = 3$. Hence,

$$2a(a+c)(b+c) + 2b(a+b)(a+c) + 2c(a+b)(b+c) + (b-c)(c-a)(a-b) = 3(a+b)(a+c)(b+c),$$

and dividing by $(a+b)(a+c)(b+c)$ gives the desired result.

Challenge Problems

9.40 Taking the difference of the two equations, we get $x^2 - y^2 = 33y - 33x$, which factors as $(x-y)(x+y) = 33(y-x)$. Since $x \neq y$, we can divide both sides by $x-y$ to get $x+y = -33$. If we square this equation, we get $x^2 + 2xy + y^2 = 33^2$, which gives us a term of xy.

From the given equations, we have $x^2 = 33y + 907$ and $y^2 = 33x + 907$. Substituting these expressions into $x^2 + 2xy + y^2 = 33^2$, we get $33y + 907 + 2xy + 33x + 907 = 33^2$. Therefore,

$$2xy = 33^2 - 33(x+y) - 2 \cdot 907 = 33^2 + 33^2 - 2 \cdot 907 = 364,$$

so $xy = 364/2 = \boxed{182}$.

9.41 Since $x^3 - 1 = 0$, we have $(x-1)(x^2 + x + 1) = 0$. Since $x \neq 1$, we may divide by $x - 1$ to get $x^2 + x + 1 = 0$. Then $1 - x + x^2 = (1 + x + x^2) - 2x = -2x$, and $1 + x - x^2 = (1 + x + x^2) - 2x^2 = -2x^2$, so

$$(1 - x + x^2)(1 + x - x^2) = (-2x)(-2x^2) = 4x^3 = \boxed{4}.$$

9.42 The first equation factors as $x(1 - y^3) = 7$, so $x(1 - y)(1 + y + y^2) = 7$, and the second equation factors as $xy(y-1) = 3$. Hence, dividing the first equation by the second equation, we get $-\frac{1+y+y^2}{y} = \frac{7}{3}$, which simplifies as $3y^2 + 10y + 3 = 0$. Factoring gives $(y+3)(3y+1) = 0$, which means $y = -3$ or $y = -1/3$. Then $x = 1/4$ or $x = 27/4$, respectively, so the solutions are $(x,y) = \boxed{(1/4, -3) \text{ and } (27/4, -1/3)}$.

9.43 Multiplying by xy to clear the fractions, the equation becomes $x^2 + x^2 y = y^2 + xy^2$. Moving all the terms to one side, we get $x^2 y - xy^2 + x^2 - y^2 = 0$. Factoring the left side gives us

$$x^2 y - xy^2 + x^2 - y^2 = xy(x - y) + (x + y)(x - y) = (x - y)(xy + x + y),$$

so $(x-y)(xy + x + y) = 0$. Since $x \neq y$, we may divide by $x - y$ to get $xy + x + y = 0$, so $xy = -(x+y)$. Then, we have

$$\frac{1}{x} + \frac{1}{y} = \frac{x+y}{xy} = \boxed{-1}.$$

9.44 The first thing we note is the identity

$$(x-1)g(x) = (x-1)(x^5 + x^4 + x^3 + x^2 + x + 1) = x^6 - 1.$$

Therefore, if a polynomial is divisible by $x^6 - 1$, then it is divisible by $g(x)$.

We want to find the remainder of $g(x^{12}) = x^{60} + x^{48} + x^{36} + x^{24} + x^{12} + 1$ when divided by $g(x)$. We see that $x^{60} - 1 = (x^6)^{10} - 1 = (x^6 - 1)(x^{54} + x^{48} + x^{42} + \cdots + 1)$, so $x^{60} - 1$ is divisible by $x^6 - 1$.

Similarly,

$$x^{48} - 1 = (x^6 - 1)(x^{42} + x^{36} + x^{30} + \cdots + 1),$$
$$x^{36} - 1 = (x^6 - 1)(x^{30} + x^{24} + x^{18} + \cdots + 1),$$
$$x^{24} - 1 = (x^6 - 1)(x^{18} + x^{12} + x^6 + 1),$$
$$x^{12} - 1 = (x^6 - 1)(x^6 + 1),$$

so

$$(x^{60} - 1) + (x^{48} - 1) + (x^{36} - 1) + (x^{24} - 1) + (x^{12} - 1) = x^{60} + x^{48} + x^{36} + x^{24} + x^{12} - 5 = g(x^{12}) - 6$$

is divisible by $x^6 - 1$, which means it is divisible by $g(x)$. Therefore, the remainder of $g(x^{12})$ when divided by $g(x)$ is $\boxed{6}$.

9.45 We note that both $x^{73} - 1$ and $x^{37} - 1$ are divisible by $x - 1$, so we can rewrite the given polynomial as

$$37x^{73} - 73x^{37} + 36 = 37(x^{73} - 1) + 37 - 73(x^{37} - 1) - 73 + 36$$
$$= 37(x^{73} - 1) - 73(x^{37} - 1)$$
$$= 37(x - 1)(x^{72} + x^{71} + \cdots + 1) - 73(x - 1)(x^{36} + x^{35} + \cdots + 1).$$

Thus, the quotient is $Q(x) = 37(x^{72} + x^{71} + \cdots + 1) - 73(x^{36} + x^{35} + \cdots + 1)$, and the sum of the coefficients of $Q(x)$ is $37 \cdot 73 - 73 \cdot 37 = \boxed{0}$.

9.46 We are given that $x + \frac{1}{x} = 1$. Multiplying both sides by x, we get $x^2 + 1 = x$, so $x^2 - x + 1 = 0$. This reminds us of the sum of cubes identity. Indeed, $x^3 + 1 = (x + 1)(x^2 - x + 1) = 0$, so $x^3 = -1$. Squaring, we get $x^6 = 1$. Therefore, we have $x^7 + \dfrac{1}{x^7} = x + \dfrac{1}{x} = \boxed{1}$.

9.47

(a) Let $f(a) = a^3(c - b) + b^3(a - c) + c^3(b - a) = (c - b)a^3 + (b^3 - c^3)a + bc^3 - b^3c$, which is a cubic in a. Then

$$f(b) = b^3(c - b) + b^3(b - c) + c^3(b - b) = 0,$$

and

$$f(c) = c^3(c - b) + b^3(c - c) + c^3(b - c) = 0,$$

so b and c are roots of $f(a)$. Then by the Factor Theorem, $f(a)$ is divisible by $(a - b)(a - c)$.

Since the coefficient of a^2 in $f(a)$ is 0, by Vieta's Formulas, the sum of the roots is 0. Hence, the third root must be $-b - c$, so by the Factor Theorem, $f(a)$ is divisible by $a + b + c$.

Finally, the coefficient of a^3 is $c - b$, so the given expression factors as

$$a^3(c - b) + b^3(a - c) + c^3(b - a) = (c - b)(a - b)(a - c)(a + b + c) = \boxed{(a - b)(b - c)(c - a)(a + b + c)}.$$

(b) Let $f(a) = a^4(c - b) + b^4(a - c) + c^4(b - a)$. Then

$$f(b) = b^4(c - b) + b^4(b - c) + c^4(b - b) = 0,$$

and

$$f(c) = c^4(c - b) + b^4(c - c) + c^4(b - c) = 0,$$

so b and c are roots of $f(a)$. Then by the Factor Theorem, $f(a)$ is divisible by $(a - b)(a - c)$.

Writing $f(a)$ as a polynomial in a, we have

$$f(a) = (c - b)a^4 + (b^4 - c^4)a + bc^4 - b^4c = (c - b)a^4 + (b^4 - c^4)a + bc(c^3 - b^3)$$
$$= (c - b)a^4 + (b - c)(b^3 + b^2c + bc^2 + c^3)a + bc(c - b)(c^2 + cb + b^2).$$

Thus, we can take out a factor of $b - c$:

$$f(a) = (b - c)[-a^4 + (b^3 + b^2c + bc^2 + c^3)a - bc(b^2 + bc + c^2)]$$
$$= -(b - c)[a^4 - (b^3 + b^2c + bc^2 + c^3)a + bc(b^2 + bc + c^2)].$$

We know that $a - b$ and $a - c$ are also factors, so

$$a^4 - (b^3 + b^2c + bc^2 + c^3)a + bc(b^2 + bc + c^2) = (a - b)(a - c)g(a)$$

for some quadratic $g(a)$.

We can find $g(a)$ by using long division. Alternatively, we can use Vieta's Formulas to find $g(a)$. For example, Vieta's Formulas tell us that sum of the roots of

$$a^4 - (b^3 + b^2c + bc^2 + c^3)a + bc(b^2 + bc + c^2) = 0$$

is equal to 0, which means the sum of the roots of $(a - b)(a - c)g(a) = 0$ is also equal to 0. Therefore, the sum of the roots of $g(a) = 0$ is $-(b + c)$.

Also, the product of the roots of the quartic is $bc(b^2 + bc + c^2)$, so the product of the roots of $g(a) = 0$ is $b^2 + bc + c^2$. Therefore,

$$g(a) = a^2 + (b + c)a + (b^2 + bc + c^2) = a^2 + b^2 + c^2 + ab + ac + bc.$$

Hence, we have

$$a^4(c - b) + b^4(a - c) + c^4(b - a) = -(b - c)(a - b)(a - c)(a^2 + b^2 + c^2 + ab + ac + bc)$$
$$= \boxed{(a - b)(b - c)(c - a)(a^2 + b^2 + c^2 + ab + ac + bc)}.$$

9.48 Adding the two equations, we get $x^3 + y^3 = 16x + 16y$, which factors as

$$(x + y)(x^2 - xy + y^2) = 16(x + y).$$

Since $|x| \neq |y|$, we have $x + y \neq 0$, so we can divide both sides by $x + y$ to get $x^2 - xy + y^2 = 16$.

Similarly, we can subtract the two equations to get $x^3 - y^3 = 10x - 10y$, which factors as

$$(x - y)(x^2 + xy + y^2) = 10(x - y).$$

Since $|x| \neq |y|$, we have $x - y \neq 0$, so we can divide both sides by $x - y$, which gives us $x^2 + xy + y^2 = 10$.

Adding the equations $x^2 - xy + y^2 = 16$ and $x^2 + xy + y^2 = 10$, we get $2x^2 + 2y^2 = 26$, so $x^2 + y^2 = 13$. Subtracting the first equation from the second, we get $2xy = -6$, so $xy = -3$. Therefore,

$$(x^2 - y^2)^2 = (x - y)^2(x + y)^2 = (x^2 - 2xy + y^2)(x^2 + 2xy + y^2) = (13 + 2 \cdot 3)(13 - 2 \cdot 3) = 19 \cdot 7 = \boxed{133}.$$

9.49 We can write

$$x^n - n(x - 1) - 1 = (x^n - 1) - n(x - 1) = (x - 1)(x^{n-1} + x^{n-2} + \cdots + x + 1) - n(x - 1) = (x - 1)(x^{n-1} + x^{n-2} + \cdots + x - n + 1).$$

Let $f(x) = x^{n-1} + x^{n-2} + \cdots + x - n + 1$. Then

$$f(1) = \underbrace{1 + 1 + \cdots + 1}_{(n-1)\ 1's} - n + 1 = (n-1) - n + 1 = 0,$$

so by the Factor Theorem, $f(x)$ is divisible by $x - 1$. Therefore, $x^n - n(x-1) - 1$ is divisible by $(x-1)^2$.

9.50 Since $z^7 = 1$, we have $z^{700} = 1$. Then, we have

$$z^{100} + \frac{1}{z^{100}} + z^{300} + \frac{1}{z^{300}} + z^{500} + \frac{1}{z^{500}} = z^{100} + \frac{z^{700}}{z^{100}} + z^{300} + \frac{z^{700}}{z^{300}} + z^{500} + \frac{z^{700}}{z^{500}}$$
$$= z^{100} + z^{600} + z^{300} + z^{400} + z^{500} + z^{200}$$
$$= z^{600} + z^{500} + z^{400} + z^{300} + z^{200} + z^{100}.$$

We then have $z^{600} + z^{500} + z^{400} + z^{300} + z^{200} + z^{100} + 1 = \frac{z^{700} - 1}{z^{100} - 1} = \frac{1 - 1}{z^{100} - 1} = 0$. Therefore,

$$z^{100} + \frac{1}{z^{100}} + z^{300} + \frac{1}{z^{300}} + z^{500} + \frac{1}{z^{500}} = z^{600} + z^{500} + z^{400} + z^{300} + z^{200} + z^{100} = \boxed{-1}.$$

We must make sure that $z^{100} \neq 1$ in order to rewrite $z^{600} + z^{500} + z^{400} + z^{300} + z^{200} + z^{100} + 1$ as $(z^{700} - 1)/(z - 1)$. We show that $z^{100} \neq 1$ by contradiction. If $z^{100} = 1$, then $\frac{z^{100}}{(z^7)^{14}} = \frac{z^{100}}{z^{98}} = z^2 = 1$, and then $\frac{z^7}{(z^2)^3} = z = 1$. But we must have $z \neq 1$, so we have a contradiction. Therefore, $z^{100} \neq 1$, and our solution above is valid.

9.51 Using difference of sixth powers, we have $x^5 + x^4 y + x^3 y^2 + x^2 y^3 + xy^4 + y^5 = \frac{x^6 - y^6}{x - y}$. Then the numerator factors as

$$x^6 - y^6 = (x^3 - y^3)(x^3 + y^3) = (x - y)(x^2 + xy + y^2)(x + y)(x^2 - xy + y^2).$$

Therefore, $\dfrac{x^6 - y^6}{x - y} = \boxed{(x + y)(x^2 + xy + y^2)(x^2 - xy + y^2)}$.

9.52 We wish to factor a sum of several powers. It is difficult to find such a factorization because the exponents do not relate to each other in any obvious way, so we write each term in terms of its prime factorization:

$$6^6 + 8^4 + 27^4 = 2^6 \cdot 3^6 + 2^{12} + 3^{12}.$$

Let $x = 2$ and $y = 3$, so $2^6 \cdot 3^6 + 2^{12} + 3^{12} = x^{12} + x^6 y^6 + y^{12}$. We can then use difference of cubes to write

$$x^{12} + x^6 y^6 + y^{12} = (x^6)^2 + x^6 y^6 + (y^6)^2 = \frac{x^{18} - y^{18}}{x^6 - y^6}.$$

Then, we can use difference of squares: $\dfrac{x^{18} - y^{18}}{x^6 - y^6} = \dfrac{(x^9 + y^9)(x^9 - y^9)}{(x^3 + y^3)(x^3 - y^3)} = \dfrac{x^9 + y^9}{x^3 + y^3} \cdot \dfrac{x^9 - y^9}{x^3 - y^3}$. Finally, we can again use sum and difference of cubes, to find

$$\frac{x^9 + y^9}{x^3 + y^3} \cdot \frac{x^9 - y^9}{x^3 - y^3} = (x^6 - x^3 y^3 + y^6)(x^6 + x^3 y^3 + y^6) = (2^6 - 2^3 \cdot 3^3 + 3^6)(2^6 + 2^3 \cdot 3^3 + 3^6) = \boxed{577 \cdot 1009}.$$

Note that we could have gotten here a little faster with

$$x^{12} + x^6 y^6 + y^{12} = x^{12} + 2x^6 y^6 + y^{12} - x^6 y^6 = (x^6 + y^6)^2 - (x^3 y^3)^2 = (x^6 + y^6 + x^3 y^3)(x^6 + y^6 - x^3 y^3).$$

9.53 Let $a = \sqrt[3]{2}$, and let $x = \sqrt{\sqrt[3]{4} - 1} + \sqrt{\sqrt[3]{16} - \sqrt[3]{4}} = \sqrt{\sqrt[3]{4} - 1} + \sqrt{2\sqrt[3]{2} - \sqrt[3]{4}} = \sqrt{a^2 - 1} + \sqrt{2a - a^2}$. Then

$$x^2 = a^2 - 1 + 2\sqrt{a^2 - 1}\sqrt{2a - a^2} + 2a - a^2 = 2a - 1 + 2\sqrt{(a^2-1)(2a-a^2)} = 2a - 1 + 2\sqrt{-a^4 + 2a^3 + a^2 - 2a}.$$

Since $a^3 = 2$, we have

$$-a^4 + 2a^3 + a^2 - 2a = -2a + 4 + a^2 - 2a = a^2 - 4a + 4 = (2 - a)^2.$$

Since $a = \sqrt[3]{2} < 2$, we have $\sqrt{(2-a)^2} = 2 - a$, so $x^2 = 2a - 1 + 2(2 - a) = 3$. Since $x > 0$, we have $x = \sqrt{3}$, so $k = \boxed{3}$.

9.54 In the absence of any obvious factors, we can try substituting specific values, and then looking for a general pattern. For example, take $q = 1$ and $r = 2$. Then

$$p^3 + q^3 + r^3 - 3pqr = p^3 - 6p + 9.$$

We find that this factors as $(p + 3)(p^2 - 3p + 3)$. As another example, take $q = 3$ and $r = 5$. Then

$$p^3 + q^3 + r^3 - 3pqr = p^3 - 45p + 152.$$

We find that this factors as $(p + 8)(p^2 - 8p + 19)$. These examples suggest that $p + q + r$ is always a factor. To test this, we let $f(p, q, r) = p^3 + q^3 + r^3 - 3pqr$, then find $f(p, q, -p - q)$. We have

$$f(p, q, -p - q) = p^3 + q^3 + (-p - q)^3 - 3pq(-p - q) = p^3 + q^3 - (p + q)^3 + 3pq(p + q)$$
$$= p^3 + q^3 - (p^3 + 3p^2 q + 3pq^2 + q^3) + 3p^2 q + 3pq^2 = 0,$$

so $p + q + r$ is indeed a factor of $f(p, q, r)$. We therefore have

$$p^3 + q^3 + r^3 - 3pqr = (p + q + r)g(p, q, r).$$

To produce p^3, q^3, and r^3 on the right, we must have p^2, q^2, and r^2 in $g(p, q, r)$. We therefore have

$$p^3 + q^3 + r^3 - 3pqr = (p + q + r)(p^2 + q^2 + r^2 + h(p, q, r)).$$

Expanding the right side, we have

$$p^3 + q^3 + r^3 - 3pqr = p^3 + q^3 + r^3 + p^2(q + r) + q^2(p + r) + r^2(q + p) + (p + q + r)h(p, q, r),$$

so

$$-3pqr = p^2(q + r) + q^2(p + r) + r^2(q + p) + (p + q + r)h(p, q, r).$$

To get rid of the terms with squares in them, we include $-pq - qr - rp$ in $h(p, q, r)$, and try

$$-3pqr = p^2(q + r) + q^2(p + r) + r^2(q + p) + (p + q + r)(-pq - qr - rp + k(p, q, r)).$$

Expanding the last product on the right, we find that all the terms on the right cancel except for the terms in $-3pqr + (p + q + r)k(p, q, r)$. Since this must equal $-3pqr$, we have $k(p, q, r) = 0$, and we have our factorization:

$$p^3 + q^3 + r^3 - 3pqr = \boxed{(p + q + r)(p^2 + q^2 + r^2 - pq - pr - qr)}.$$

9.55 Let $x = 1/3$. Then the product can be rewritten as $P = (1 + x)(1 + x^2)(1 + x^4) \cdots (1 + x^{2^{100}})$. We multiply both sides by $1 - x$ to use difference of squares, and the product collapses:

$$(1 - x)P = (1 - x)(1 + x)(1 + x^2)(1 + x^4) \cdots (1 + x^{2^{100}})$$
$$= (1 - x^2)(1 + x^2)(1 + x^4) \cdots (1 + x^{2^{100}})$$
$$= (1 - x^4)(1 + x^4) \cdots (1 + x^{2^{100}})$$
$$= \cdots$$
$$= (1 - x^{2^{100}})(1 + x^{2^{100}})$$
$$= 1 - x^{2^{101}}.$$

Therefore, $P = \dfrac{1 - x^{2^{101}}}{1 - x} = \dfrac{1 - \frac{1}{3^{2^{101}}}}{1 - \frac{1}{3}} = \boxed{\dfrac{3}{2}\left(1 - \dfrac{1}{3^{2^{101}}}\right)}$.

9.56   Writing the given expression as a polynomial in x, we get

$$-x^3 + (y + z)x^2 + (y^2 - 2yz + z^2)x - y^3 + y^2 z + yz^2 - z^3.$$

The coefficient of x is $y^2 - 2yz + z^2 = (y - z)^2$, and the constant coefficient is

$$-y^3 + y^2 z + yz^2 - z^3 = -y^2(y - z) + z^2(y - z) = -(y - z)(y^2 - z^2) = -(y + z)(y - z)^2.$$

Thus, the first two terms factor as $-x^3 + (y + z)x^2 = x^2(-x + y + z)$, and the last two terms factor as $(y - z)^2 x - (y + z)(y - z)^2 = (y - z)^2(x - y - z)$, so

$$\begin{aligned}
-x^3 + (y + z)x^2 + (y^2 - 2yz + z^2)x - y^3 + y^2 z + yz^2 - z^3 &= x^2(-x + y + z) + (y - z)^2(x - y - z) \\
&= (-x + y + z)[x^2 - (y - z)^2] \\
&= (-x + y + z)[x - (y - z)][x + (y - z)] \\
&= \boxed{(-x + y + z)(x - y + z)(x + y - z)}.
\end{aligned}$$

9.57   We first write the expression with a common denominator, and we have

$$\frac{-a^3(b + c)(b - c) - b^3(c + a)(c - a) - c^3(a + b)(a - b)}{(a - b)(b - c)(c - a)}.$$

Next, we try factoring the numerator. We let

$$f(a, b, c) = -a^3(b + c)(b - c) - b^3(c + a)(c - a) - c^3(a + b)(a - b).$$

We hope to cancel factors in the numerator with those in the denominator, so we check whether $a - b$, $c - a$ and $b - c$ are factors. We find

$$f(b, b, c) = -b^3(b + c)(b - c) - b^3(c + b)(c - b) - c^3(b + b)(b - b) = 0,$$

so $f(a, b, c)$ is divisible by $a - b$. Similarly, we have $f(c, b, c) = f(a, c, c) = 0$, so $a - c$ and $b - c$ are factors. Therefore, we have

$$f(a, b, c) = (a - b)(b - c)(c - a)g(a, b, c)$$

for some polynomial $g(a, b, c)$. Since $\deg f = 5$ and the degree of $(a - b)(b - c)(c - a)$ is 3, we must have $\deg g = 2$.

Unfortunately, it's not so easy to find $g(a, b, c)$. We note that $f(a, b, c)$ is cubic in a, and that two of the factors we have found are linear in a. Therefore, we consider viewing $f(a, b, c)$ as a polynomial $h(a)$ in a, and treat b and c as constants. We have already found two roots of $h(a)$, namely b and c, so the remaining factor $g(a, b, c)$ must be linear in a. Therefore, we can write

$$h(a) = (a - b)(a - c)(b - c)(Pa + Q)$$

for some constants P and Q (where P and Q may be expressed in terms of b and c).

The leading coefficient in

$$h(a) = -a^3(b + c)(b - c) - b^3(c + a)(c - a) - c^3(a + b)(a - b)$$

is $-(b + c)(b - c)$, and the leading coefficient in $(a - b)(a - c)(b - c)(Pa + Q)$ is $(b - c)P$, so $P = -(b + c)$.

The constant coefficient in

$$h(a) = -a^3(b + c)(b - c) - b^3(c + a)(c - a) - c^3(a + b)(a - b)$$

is $h(0) = -b^3c^2 + b^2c^3 = b^2c^2(c-b)$, and the constant coefficient in $(a-b)(a-c)(b-c)(Pa+Q)$ is $bc(b-c)Q$, so $Q = -bc$. Hence,

$$
\begin{aligned}
h(a) &= -a^3(b+c)(b-c) - b^3(c+a)(c-a) - c^3(a+b)(a-b) \\
&= (a-b)(a-c)(b-c)(Pa+Q) \\
&= (a-b)(a-c)(b-c)[-(b+c)a-bc] \\
&= (a-b)(a-c)(b-c)(-ab-ac-bc) \\
&= (a-b)(b-c)(c-a)(ab+ac+bc),
\end{aligned}
$$

so we have

$$
\begin{aligned}
\frac{a^3(b+c)}{(a-b)(a-c)} + \frac{b^3(c+a)}{(b-c)(b-a)} + \frac{c^3(a+b)}{(c-a)(c-b)} &= \frac{-a^3(b+c)(b-c) - b^3(c+a)(c-a) - c^3(a+b)(a-b)}{(a-b)(b-c)(c-a)} \\
&= \frac{(a-b)(b-c)(c-a)(ab+ac+bc)}{(a-b)(b-c)(c-a)} = \boxed{ab+ac+bc}.
\end{aligned}
$$

9.58 By the Rational Root Theorem, the only possible rational roots are 1 and -1, and neither of these is a root of the quartic. Furthermore, there is no obvious way to group the terms to create a common factor.

Previously, when we encountered such a quartic, we were able to find a factorization by a creative manipulation of the terms and applying difference of squares. We can use the same approach here.

We can produce an x^4 by expanding a square of the form $(x^2+a)^2$. Then to match terms with degree 2 and lower, we can use a square of the form $b(x+c)^2$. Thus, we will attempt to find constants a, b, and c such that $x^4 - 4x - 1 = (x^2+a)^2 - b(x+c)^2$. Expanding, this becomes $x^4 - 4x - 1 = x^4 + (2a-b)x^2 - 2bcx + a^2 - bc^2$.

Equating the corresponding coefficients of the two sides of this equation, we obtain the system of equations $2a - b = 0$, $2bc = 4$, and $a^2 - bc^2 = -1$. Since we are only interested in finding some solution (a,b,c) rather than all solutions, we will use trial and error.

For example, from $2a - b = 0$, we get $b = 2a$, and from $2bc = 4$, we get $bc = 2$. If $a = 1$, then $b = 2$, and $c = 1$. Then $a^2 - bc^2 = -1$, so all equations are satisfied, which means we can write

$$x^4 - 4x - 1 = (x^2+1)^2 - 2(x+1)^2 = (x^2+1)^2 - [\sqrt{2}(x+1)]^2 = (x^2 - \sqrt{2}x + 1 - \sqrt{2})(x^2 + \sqrt{2}x + 1 + \sqrt{2}).$$

Using the quadratic formula, we find the roots are

$$\boxed{x = \frac{\sqrt{2} + \sqrt{-2 + 4\sqrt{2}}}{2}, \quad \frac{\sqrt{2} - \sqrt{-2 + 4\sqrt{2}}}{2}, \quad \frac{-\sqrt{2} + i\sqrt{2 + 4\sqrt{2}}}{2}, \quad \text{and} \quad \frac{-\sqrt{2} - i\sqrt{2 + 4\sqrt{2}}}{2}.}$$

9.59 Grouping the terms with even degree, we find

$$2000x^6 - 2 = 2(1000x^6 - 1) = 2(10x^2 - 1)(100x^4 + 10x^2 + 1) = (20x^2 - 2)(100x^4 + 10x^2 + 1),$$

and grouping the terms with odd degree, we find $100x^5 + 10x^3 + x = x(100x^4 + 10x^2 + 1)$. Hence, we have

$$2000x^6 + 100x^5 + 10x^3 + x - 2 = (20x^2 + x - 2)(100x^4 + 10x^2 + 1).$$

But $100x^4 + 10x^2 + 1 \geq 1$ for all real numbers x, so the real roots of the given polynomial must be the roots of the quadratic $20x^2 + x - 2 = 0$, which are

$$\boxed{x = \frac{-1 \pm \sqrt{161}}{40}.}$$

CHAPTER **10**

Sequences and Series

Exercises for Section 10.1

10.1.1   Let the first term be a and the common difference be d. We have $a + 3d = 203$ and $a + 12d = 167$. Subtracting the first equation from the second gives $9d = -36$, so $d = -4$. (Alternatively, we could note that we take 9 steps from the fourth term to the thirteenth term, and sequence goes down 36 over that time, so each step must be -4.)

From $d = -4$, we find that $a = 203 - 3d = 215$. The nth term of the sequence is therefore $215 - 4(n - 1)$. We seek the smallest value of n such that $215 - 4(n - 1) < 0$. Rearranging gives $4n > 219$, so $n > 54.75$. Therefore, the smallest n is $\boxed{55}$.

10.1.2

(a)   In five terms, the sequence goes from 10 to 4. In other words, it goes down 6 in five terms. Because the sequence is an arithmetic sequence, this is true for any consecutive five terms. In going from the tenth term to the twentieth terms, it must therefore go down 6 twice, from 4 to $4 - 2 \cdot 6 = \boxed{-8}$.

(b)   Let the first term of the arithmetic sequence be a, and let the common difference be d. Then $a + 4d = 10$ and $a + 9d = 4$. Subtracting the first equation from the second, we get $5d = -6$, so $d = -1.2$. Then $a = 14.8$. Therefore, the nth term of the arithmetic sequence is $a + (n - 1)d = \boxed{16 - 1.2n}$.

10.1.3   If b is the average of a and c, then $b = (a + c)/2$. Then $b - a = \frac{a+c}{2} - a = \frac{c-a}{2}$ and $c - b = c - \frac{a+c}{2} = \frac{c-a}{2}$. Hence, $b - a = c - b$, so a, b, and c are in arithmetic progression.

10.1.4   Let a and b be two consecutive terms in the original sequence. Then these terms become $a + k$ and $b + k$ in the new sequence. The difference between these two consecutive terms is then $(b + k) - (a + k) = b - a$, so the difference between consecutive terms is preserved.

Since the original sequence is arithmetic, all the differences between consecutive terms are equal. Therefore, all the differences between consecutive terms in the new sequence are also equal, so it is arithmetic.

10.1.5   You may have found the answer by simply experimenting with the numbers and finding patterns. Below, we prove that the solution is unique. (Finding the answer with trial and error is considerably easier than the solution below, which proves this solution is unique. So, this problem is not nearly as difficult as the solution below makes it look!)

Let the original arithmetic sequence be $a_1, a_2, a_3, a_4, a_5, a_6$, with common difference d. The student omitted one of these terms and miscopied another, so of the five terms the student wrote down, exactly four are correct. We separate into the following five cases.

Case 1: The term 113 is a miscopy, so the terms 137, 149, 155, and 173 are correct.

The common difference d must be no greater than $155 - 149 = 6$, so $d \le 6$. However, $173 - 137 = 36$. The greatest difference between any two terms in the original arithmetic sequence is $a_6 - a_1 = 5d \le 30$. Hence, this case is not possible.

Case 2: The term 137 is a miscopy, so the terms 113, 149, 155, and 173 are correct.

As in Case 1, d must divide $155 - 149 = 6$, so $d \le 6$. However, $173 - 113 = 60$, and again as in Case 1, the greatest difference between any two terms in the original arithmetic sequence is $5d \le 30$. Hence, this case is not possible.

Case 3: The term 149 is a miscopy, so the terms 113, 137, 155, and 173 are correct.

The common difference d must divide $137 - 113 = 24$, and it must divide $155 - 137 = 18$. Therefore, d must divide gcd$\{18, 24\} = 6$, so $d \le 6$. However, $173 - 113 = 60$, and $5d \le 30$, so this case is not possible.

Case 4: The term 155 is a miscopy, so the terms 113, 137, 149, and 173 are correct.

The common difference d must be no greater than $149 - 137 = 12$, so $d \le 12$. Also, $173 - 113 = 60 = 5 \cdot 12$, which means $d = 12$. Hence, the original arithmetic sequence consists of the terms 113, 125, 137, 149, 161, and 173.

Case 5: The term 173 is a miscopy, so the terms 113, 137, 149, and 155 are correct.

The common difference d must be no more than $155 - 149 = 6$, so $d \le 6$. However, $155 - 113 = 42$, and again as in Case 1, the greatest difference between any two terms in the original arithmetic sequence is $5d \le 30$. Hence, this case is not possible.

Therefore, original arithmetic sequence must have been the one found in Case 4, namely

$$\boxed{113, 125, 137, 149, 161, 173}.$$

10.1.6 Let the four roots be a, $a + d$, $a + 2d$, and $a + 3d$. By Vieta's Formulas, the sum of the roots is 0, so $a + (a + d) + (a + 2d) + (a + 3d) = 0$, so $4a + 6d = 0$, which means $d = -2a/3$. Then the four roots become a, $a/3$, $-a/3$, and $-a$.

Again by Vieta's Formulas, we have

$$-(3k + 4) = a \cdot \frac{a}{3} + a \cdot \left(-\frac{a}{3}\right) + a \cdot (-a) + \frac{a}{3} \cdot \left(-\frac{a}{3}\right) + \frac{a}{3} \cdot (-a) + \left(-\frac{a}{3}\right) \cdot (-a) = -\frac{10}{9}a^2,$$

and $k^2 = a \cdot \frac{a}{3} \cdot \left(-\frac{a}{3}\right) \cdot (-a) = \frac{a^4}{9}$.

From $-10a^2/9 = -(3k + 4)$, we have $10a^2 = 9(3k + 4)$. From $k^2 = a^4/9$, we have $a^4 = 9k^2$. Taking the square root, we get $a^2 = \pm 3k$.

If $a^2 = 3k$, then $30k = 9(3k + 4)$. Solving for k, we get $k = 12$, and the given quartic becomes

$$x^4 - 40x^2 + 144 = (x^2 - 4)(x^2 - 36) = (x + 6)(x + 2)(x - 2)(x - 6).$$

The roots $-6, -2, 2,$ and 6 are in arithmetic sequence.

If $a^2 = -3k$, then $-30k = 9(3k + 4)$. Solving for k, we get $k = -12/19$, and the given quartic becomes

$$x^4 - \frac{40}{19}x^2 + \frac{144}{361} = \frac{361x^4 - 760x^2 + 144}{361} = \frac{(19x^2 - 36)(19x^2 - 4)}{361} = \frac{(\sqrt{19}x + 6)(\sqrt{19}x + 2)(\sqrt{19}x - 2)(\sqrt{19}x - 6)}{361}.$$

The roots $-\frac{6}{\sqrt{19}}, -\frac{2}{\sqrt{19}}, \frac{2}{\sqrt{19}},$ and $\frac{6}{\sqrt{19}}$ are in arithmetic sequence.

Therefore, the solutions are $k = \boxed{12 \text{ and } -12/19}$.

Exercises for Section 10.2

10.2.1 Let n be the number of terms in the sequence. The first term is 639 and the common difference is -8, so we have $639 + (n-1)(-8) = -97$, so $n = 93$. Therefore, the sum of the series is

$$\frac{93[2 \cdot 639 + (93-1) \cdot (-8)]}{2} = \boxed{25203}.$$

We could have also used the fact that the first term is 639, the last term is -97, and that there are 93 terms, which tells us that the sum is equal to $\frac{93}{2}(639 - 97) = 25203$.

10.2.2 The ages of Mrs. White's 13 grandchildren form an arithmetic sequence with first term 5 and common difference 2. Then the sum of their ages is $\frac{13[2 \cdot 5 + (13-1) \cdot 2]}{2} = \boxed{221}$.

10.2.3

(a) The n smallest positive odd integers form an arithmetic sequence with first term 1 and common difference 2. Then the sum of these n numbers is $\frac{n[2 \cdot 1 + (n-1) \cdot 2]}{2} = \boxed{n^2}$.

(b) The n smallest positive odd integers that are greater than 100 form an arithmetic sequence with first term 101 and common difference 2. Then the sum of these n numbers is $\frac{n[2 \cdot 101 + (n-1) \cdot 2]}{2} = \boxed{n^2 + 100n}$.

We could also have used part (a). Since there are 50 positive odd integers less than 100, the sum of the first n positive odd integers greater than 100 equals the sum of the first $n + 50$ positive odd integers minus the sum of the first 50 positive odd integers. By part (a), this equals $(n + 50)^2 - 50^2 = n^2 + 100n$.

10.2.4 There are $b - a + 1$ integers from a to b inclusive. Therefore, our sum is an arithmetic series of $b - a + 1$ terms with first term a and last term b, so its sum is

$$a + (a + 1) + (a + 2) + \cdots + b = \boxed{\frac{(b - a + 1)(a + b)}{2}}.$$

10.2.5 Let $c_n = a_n + b_n$ for all $n \geq 1$. First, we note that $c_i - c_{i-1} = (a_i + b_i) - (a_{i-1} + b_{i-1}) = (a_i - a_{i-1}) + (b_i - b_{i-1})$. Because the terms a_i and b_i form arithmetic sequences, $a_i - a_{i-1}$ is a constant, and $b_i - b_{i-1}$ is a constant. Therefore, $c_i - c_{i-1}$ is constant for all $i > 1$, which means that $c_1, c_2, \ldots$ is an arithmetic sequence. Furthermore, $c_1 = a_1 + b_1 = 25 + 75 = 100$ and $c_{100} = 100$, so $c_n = 100$ for all $n \geq 1$. Therefore, the sum of the 100 terms $c_1, c_2, \ldots, c_{100}$ is $100 \cdot 100 = \boxed{10000}$.

10.2.6 Let the arithmetic sequence be $a_1, a_2, \ldots$. We are given that for any positive integer n,

$$a_1 + a_2 + \cdots + a_{n-1} + a_n = 3n^2 + 2n.$$

We'd like to isolate a_n, so we find an expression for the first $n - 1$ terms of the left side by using the given formula to determine that

$$a_1 + a_2 + \cdots + a_{n-1} = 3(n-1)^2 + 2(n-1) = 3n^2 - 4n + 1.$$

Subtracting the second equation from the first equation, we get $a_n = \boxed{6n - 1}$.

10.2.7 The sum of the first $3n$ positive integers is $3n(3n + 1)/2$, and the sum of the first n positive integers is $n(n + 1)/2$, so

$$\frac{3n(3n + 1)}{2} - \frac{n(n + 1)}{2} = 150,$$

which gives us $4n^2 + n = 150$. Rearranging and factoring gives $(n - 6)(4n + 25) = 0$. Therefore, $n = 6$, and the sum of the first $4n = 24$ positive integers is $24 \cdot 25/2 = \boxed{300}$.

10.2.8 Let a be the first term, and let d be the common difference. Using the formula for an arithmetic series, we get

$$\frac{r^2}{s^2} = \frac{\frac{r[2a + (r-1)d]}{2}}{\frac{s[2a + (s-1)d]}{2}} = \frac{r[2a + (r - 1)d]}{s[2a + (s - 1)d]}$$

so

$$\frac{r}{s} = \frac{2a + (r-1)d}{2a + (s-1)d}.$$

Cross-multiplying, this becomes $2ar + (s-1)rd = 2as + (r-1)sd$. Moving all the terms to one side, we get $2ar - 2as - rd + sd = 0$. Factoring the left side gives

$$2ar - 2as - rd + sd = 2a(r-s) - d(r-s) = (2a-d)(r-s),$$

so we have $(2a-d)(r-s) = 0$. Since this is true for all r and s, it is true when $r \neq s$, so $d = 2a$.

Then the 8th term is $a + (8-1)d = a + 7 \cdot 2a = 15a$, and the 23rd term is $a + (23-1)d = a + 22 \cdot 2a = 45a$, so the ratio is $(15a)/(45a) = \boxed{1/3}$.

Exercises for Section 10.3

10.3.1

(a) Let the geometric sequence have first term a and common ratio r. Then $ar = 6$ and $ar^5 = 3/8$. Dividing the second equation by the first gives $r^4 = \frac{1}{16}$. Since the sequence contains positive numbers, we have $r = \sqrt[4]{1/16} = \boxed{1/2}$.

(b) From part (a), we have $a = 6/r = 12$, so the mth term of the sequence is $12(1/2)^{m-1} = \boxed{24/2^m}$.

(c) The 1977th term of the sequence is $\dfrac{12}{2^{1976}} = \dfrac{3}{2^{1974}}$, so $a = \boxed{1974}$.

10.3.2   Let a be the number of people on the Steering Committee. Then the number of emails sent on day n is $3^n a$. We are given that $3^5 a - 3^4 a = 972$, so $162a = 972$, which gives us $a = 972/162 = \boxed{6}$.

10.3.3   The first term of the arithmetic progression is 9, so the next two terms are equal to $d + 9$ and $2d + 9$ for some real number d. Then the numbers $9, d + 11, 2d + 29$ form a geometric progression, so

$$\frac{d+11}{9} = \frac{2d+29}{d+11},$$

because each of these ratios equals the common ratio of the geometric progression. Cross-multiplying gives $9(2d + 29) = (d + 11)^2$, which simplifies as $d^2 + 4d - 140 = 0$, so $(d-10)(d+14) = 0$. Therefore, $d = 10$ or $d = -14$.

If we take $d = -14$, then we obtain the arithmetic progression $9, -5, -19$. Adding 2 to the second term and 20 to the third term, we obtain the geometric progression $9, -3, 1$. Thus, the smallest possible value for the third term in the geometric progression is $\boxed{1}$.

10.3.4

(a) If x, y, and z are in geometric progression, then $y = rx$ and $z = r^2 x$ for some real number r. Then $y^2 = r^2 x^2$ and $xz = r^2 x^2$, so $y^2 = xz$. Conversely, if $y^2 = xz$, then dividing by zy gives us $x/y = y/z$, which means that x, y, z is a geometric progression.

(b) In general, it is not true that if x, y, and z are in geometric progression, then $y = \sqrt{xz}$. For example, take $x = 1$, $y = -2$, and $z = 4$. Then x, y, and z are in geometric progression (with common ratio -2), but $\sqrt{xz} = 2 \neq y$.

10.3.5   Let the roots be a, ar, and ar^2. Then by Vieta's Formulas, we have

$$a + ar + ar^2 = \frac{19}{2},$$

$$a \cdot ar \cdot ar^2 = \frac{54}{2} = 27,$$

so we have $2a(r^2 + r + 1) = 19$ and $a^3r^3 = 27$. From the latter equation, we have $ar = \sqrt[3]{27} = 3$, so $a = 3/r$. Substituting this into $2a(r^2 + r + 1) = 19$, we get

$$\frac{6(r^2 + r + 1)}{r} = 19,$$

so $6(r^2 + r + 1) = 19r$. This simplifies as $6r^2 - 13r + 6 = 0$, so $(2r - 3)(3r - 2) = 0$, which means $r = 3/2$ or $r = 2/3$.

If $r = 3/2$, then $a = 3/(3/2) = 2$, and the roots are 2, 3, and 9/2. If $r = 2/3$, then $a = 3/(2/3) = 9/2$, and the roots are once again 9/2, 3, and 2. Now that we have the roots, we can use Vieta's Formulas to find k. We have

$$\frac{k}{2} = 2 \cdot 3 + 3 \cdot \frac{9}{2} + 2 \cdot \frac{9}{2} = \frac{57}{2},$$

so $k = \boxed{57}$.

10.3.6   Let the three numbers in geometric progression be a, ar, and ar^2, so $a + ar + ar^2 = 39$ and $a^2 + a^2r^2 + a^2r^4 = 741$. Factoring the left sides of these equations gives $a(r^2 + r + 1) = 39$ and $a^2(r^4 + r^2 + 1) = 741$.

We recognize that $r^4 + r^2 + 1$ is a factor of $r^6 - 1$ from the difference of cubes factorization of $r^6 - 1$:

$$r^4 + r^2 + 1 = \frac{(r^2 - 1)(r^4 + r^2 + 1)}{r^2 - 1} = \frac{r^6 - 1}{r^2 - 1}.$$

Then using difference of squares and difference of cubes, we find

$$r^4 + r^2 + 1 = \frac{r^6 - 1}{r^2 - 1} = \frac{(r^3 - 1)(r^3 + 1)}{(r - 1)(r + 1)} = \frac{r^3 - 1}{r - 1} \cdot \frac{r^3 + 1}{r + 1} = (r^2 + r + 1)(r^2 - r + 1).$$

Hence, the equation $a^2(r^4 + r^2 + 1) = 741$ becomes $a^2(r^2 + r + 1)(r^2 - r + 1) = 741$. Dividing by the equation $a(r^2 + r + 1) = 39$, we get $a(r^2 - r + 1) = \frac{741}{39} = 19$. Subtracting this from the equation $a(r^2 + r + 1) = 39$, we get $a(r^2 + r + 1) - a(r^2 - r + 1) = 20$, so $ar = 10$, which means $a = 10/r$.

Substituting $a = 10/r$ into the equation $a(r^2 + r + 1) = 39$, we get $\frac{10}{r}(r^2 + r + 1) = 39$, which simplifies as $10r^2 - 29r + 10 = 0$, so $(2r - 5)(5r - 2) = 0$, which gives $r = 5/2$ or $r = 2/5$.

If $r = 5/2$, then $a = 10/(5/2) = 4$, and the numbers are $a = 4$, $ar = 10$, and $ar^2 = 25$. If $r = 2/5$, then $a = 10/(2/5) = 25$, and the numbers are $a = 25$, $ar = 10$, and $ar^2 = 4$, which are the same numbers as before. Therefore, the three numbers are $\boxed{4, 10, \text{ and } 25}$.

Exercises for Section 10.4

10.4.1

(a)   Let n be the number of terms in the series. The first term is 2 and the common ratio is 4, so $2 \cdot 4^{n-1} = 2048$, or $4^{n-1} = 1024 = 4^5$. Then $n = 6$, so the sum of the series is $\frac{2(4^6 - 1)}{4 - 1} = \boxed{2730}$.

(b)   Let n be the number of terms in the series. The first term is 54 and the common ratio is $-1/3$, so

$$54 \cdot \left(-\frac{1}{3}\right)^{n-1} = \frac{2}{243} \quad \Rightarrow \quad \left(-\frac{1}{3}\right)^{n-1} = \frac{1}{6561} = \frac{1}{3^8}.$$

Then $n = 9$, so the sum of the series is $\dfrac{54[(-1/3)^9 - 1]}{(-1/3) - 1} = \boxed{\dfrac{9842}{243}}$.

(c)   The given series is the sum of two geometric series: $1 + 3 + 9 + \cdots + 729$ and $2 + 6 + 18 + \cdots + 1458$. We can find the sum of each series individually, but we can also observe that each term in the second series is double the respective term in the first series, so that the sum of the given series is also equal to three times the first series, which is $3 + 9 + 27 + \cdots + 2187$.

Let n be the number of terms in this series. The first term is 3 and the common ratio is 3, so $3 \cdot 3^{n-1} = 2187$, from which we find $n = 7$. Therefore, the sum of the series is $\frac{3(3^7 - 1)}{3 - 1} = \boxed{3279}$.

10.4.2

(a) The first term is $1/6$ and the common ratio is $1/6$, so the sum of the series is $\dfrac{1/6}{1 - \frac{1}{6}} = \boxed{\dfrac{1}{5}}$.

(b) The first term is 192 and the common ratio is $144/192 = 3/4$, so the sum of the series is $\dfrac{192}{1 - \frac{3}{4}} = \boxed{768}$.

(c) The first term is 2 and the common ratio is $-1/\sqrt{2}$, so the sum of the series is

$$\frac{2}{1 - \left(-\frac{1}{\sqrt{2}}\right)} \frac{1 + \frac{1}{\sqrt{2}}}{1 + \frac{1}{\sqrt{2}}} = \frac{2}{1 + \frac{1}{\sqrt{2}}} \cdot \frac{\sqrt{2}}{\sqrt{2}} = \frac{2\sqrt{2}}{1 + \sqrt{2}} = \frac{2\sqrt{2}(1 - \sqrt{2})}{(1 + \sqrt{2})(1 - \sqrt{2})} = \boxed{4 - 2\sqrt{2}}.$$

10.4.3 The given series can be expressed as the sum of two infinite geometric series:

$$\frac{1}{7} + \frac{2}{7^2} + \frac{1}{7^3} + \frac{2}{7^4} + \cdots = \left(\frac{1}{7} + \frac{1}{7^3} + \cdots\right) + \left(\frac{2}{7^2} + \frac{2}{7^4} + \cdots\right) = \frac{1/7}{1 - 1/7^2} + \frac{2/7^2}{1 - 1/7^2} = \frac{7}{48} + \frac{1}{24} = \boxed{\frac{3}{16}}.$$

10.4.4 For $|x| < 1$, the sum of the geometric series $1 + x + x^2 + \cdots$ is $1/(1 - x)$, so we have $1/(1 - x) = 4$. Solving this equation gives $x = \boxed{3/4}$.

10.4.5 We may compute each of the following sums as a geometric series:

$$a^8 + a^4 + 1 = \frac{a^{12} - 1}{a^4 - 1},$$
$$a^3 + a^2 + a + 1 = \frac{a^4 - 1}{a - 1},$$
$$a^9 + a^6 + a^3 + 1 = \frac{a^{12} - 1}{a^3 - 1}.$$

Hence,

$$\frac{(a^8 + a^4 + 1)(a^3 + a^2 + a + 1)}{a^9 + a^6 + a^3 + 1} = \frac{\frac{a^{12} - 1}{a^4 - 1} \cdot \frac{a^4 - 1}{a - 1}}{\frac{a^{12} - 1}{a^3 - 1}} = \frac{a^3 - 1}{a - 1} = a^2 + a + 1,$$

so our equation is now $a^2 + a + 1 = 21$. Rearranging and factoring gives $(a - 4)(a + 5) = 0$, so our solutions are $\boxed{a = 4 \text{ and } a = -5}$.

10.4.6 Let the first term and common ratio of the geometric series be a and r, respectively, and let the number of terms be n. Then the sum of the geometric series is $\frac{a(r^n - 1)}{r - 1}$.

If r were nonnegative, then every term in the series would be nonnegative, so the sum of the series would be nonnegative. But we are given that the sum of the series is negative, so r cannot be nonnegative.

Since r is negative, $r - 1$ is negative. Because a is positive, and we must have $a(r^n - 1)/(r - 1)$ be negative, we know that $r^n - 1$ must be positive. If n were odd, then $r^n - 1$ would be negative, so n must be even, as desired.

10.4.7 For a positive integer n, let $S_n = 1 + 5 + 5^2 + \cdots + 5^n$. We claim that $100S_n + 25 = 5^{n+3}$.

By the formula for the sum of a geometric series, we have $S_n = \frac{5^{n+1} - 1}{5 - 1} = \frac{1}{4}(5^{n+1} - 1)$. Then $100S_n + 25 = 25(5^{n+1} - 1) + 25 = 5^{n+3}$, as claimed.

Exercises for Section 10.5

10.5.1 Your answers may vary from those below, but still be correct, for the reasons described in the text.

(a) We have an arithmetic sequence with first term -91 and common difference 10. Let n be the number of terms, so $-91 + (n-1)10 = 1999$, or $n = 210$. Hence, we may write the arithmetic sequence as $\boxed{\{-91 + 10(k-1)\}_{k=1}^{210}}$. This can be simplified as $\{10k - 101\}_{k=1}^{210}$. Another possible answer is $\{10k - 91\}_{k=0}^{209}$.

(b) We have a geometric sequence with first term $1/8$ and common ratio 4. Let n be the number of terms, so $(1/8) \cdot 4^{n-1} = 8192$, or $4^{n-1} = 65536 = 4^8$, so $n = 9$. Hence, we may write the geometric sequence as $\boxed{\{4^{k-1}/8\}_{k=1}^{9}}$. This can be simplified as $\{4^k/32\}_{k=1}^{9}$, or $\{2^{2k-5}\}_{k=1}^{9}$.

(c) We have an infinite arithmetic sequence with first term 15 and common difference -4. Hence, we may write the arithmetic sequence as $\boxed{\{15 - 4(k-1)\}_{k=1}^{\infty}}$. This can be simplified as $\{-4k + 19\}_{k=1}^{\infty}$.

(d) We have an infinite geometric sequence with first term 54 and common ratio $2/3$. Hence, we may write the geometric sequence as $\boxed{\{54(2/3)^{k-1}\}_{k=1}^{\infty}}$. This can be simplified as $\{81(2/3)^k\}_{k=1}^{\infty}$.

10.5.2 Your answers may vary from those below, but still be correct, for the reasons described in the text.

(a) We can write the series as $\boxed{\displaystyle\sum_{k=1}^{7} 3}$.

(b) We have an arithmetic series with first term $\frac{1}{2}$, common difference $\frac{1}{2}$, and a total of six terms. Then the kth term is $\frac{1}{2} + \frac{1}{2}(k-1) = \frac{k}{2}$, so we can write the series as $\boxed{\displaystyle\sum_{k=1}^{6} \frac{k}{2}}$.

(c) We have an arithmetic series with first term n, common difference -1, and a total of n terms. Then the kth term is $n + (-1) \cdot (k-1) = n - k + 1$, so we can write the series as $\boxed{\displaystyle\sum_{k=1}^{n} (n - k + 1)}$. Another possible answer is $\displaystyle\sum_{k=1}^{n} k$.

(d) We have a series with n terms, and the kth term is given by $k/(k+1)$. Hence, we can write the series as $\boxed{\displaystyle\sum_{k=1}^{n} \frac{k}{k+1}}$.

(e) We have an infinite series, and the kth term is given by $x^k/k!$, where k starts at 1. Hence, we can write the series as $\boxed{\displaystyle\sum_{k=1}^{\infty} \frac{x^k}{k!}}$.

(f) We have an infinite series, and the kth term is given by $(-1)^k x^k$, where k starts at 0. Hence, we can write the series as $\boxed{\displaystyle\sum_{k=0}^{\infty} (-1)^k x^k}$.

10.5.3

(a) The sum begins with $i = -3$, so to start at 5, we can take $j = i + 8$. Then $i = j - 8$, so

$$\sum_{i=-3}^{71} \frac{2^i}{i+5} = \boxed{\sum_{j=5}^{79} \frac{2^{j-8}}{j-3}}.$$

(b) The sum begins with $i = 7$, so to start at 5, we can take $j = i - 2$. Then $i = j + 2$, so

$$\sum_{i=7}^{\infty} \frac{1}{i^2} = \boxed{\sum_{j=5}^{\infty} \frac{1}{(j+2)^2}}.$$

10.5.4

(a) $\displaystyle\sum_{i=1}^{15}(3i - 4) = 3\sum_{i=1}^{15} i - \sum_{i=1}^{15} 4 = 3 \cdot \frac{15 \cdot 16}{2} - 15 \cdot 4 = \boxed{300}.$

(b) $\displaystyle\sum_{i=1}^{n}(2 - i) = \sum_{i=1}^{n} 2 - \sum_{i=1}^{n} i = 2n - \frac{n(n+1)}{2} = \boxed{-\frac{n(n-3)}{2}}.$

(c) The series is an infinite geometric series with first term $4/2 = 2$ and common ratio $1/2$. Therefore, the sum of the series is $\frac{2}{1-1/2} = \boxed{4}$.

(d) Writing it out, the series becomes $\displaystyle\sum_{k=0}^{n} x^k = 1 + x + \cdots + x^n = \boxed{\frac{x^{n+1} - 1}{x - 1}}.$

(e) $\displaystyle\sum_{n=1}^{6}(3^n - 3n) = \sum_{n=1}^{6} 3^n - 3\sum_{n=1}^{6} n = 3 + 3^2 + \cdots + 3^6 - 3(1 + 2 + \cdots + 6) = \frac{3(3^6 - 1)}{3 - 1} - 3 \cdot \frac{6 \cdot 7}{2} = \boxed{1029}.$

(f) Writing it out, the series becomes $-1 + 4 - 9 + 16 - \cdots - 9801 + 10000$. We may compute this series several ways; one way is to simply add up all 100 terms. Another way is to pair the terms and use difference of squares:

$$-1 + 4 - 9 + 16 - 25 + 36 - \cdots - 9801 + 10000$$
$$= (2^2 - 1^2) + (4^2 - 3^2) + (6^2 - 5^2) + \cdots + (100^2 - 99^2)$$
$$= (2+1)(2-1) + (4+3)(4-3) + (6+5)(6-5) + \cdots + (100+99)(100-99)$$
$$= (2+1) + (4+3) + (6+5) + \cdots + (100+99)$$
$$= 1 + 2 + 3 + 4 + 5 + 6 + \cdots + 99 + 100 = \frac{100 \cdot 101}{2} = \boxed{5050}.$$

10.5.5 $\displaystyle\sum_{n=1}^{99}(107x_n - 74y_n) = 107\sum_{n=1}^{99} x_n - 74\sum_{n=1}^{99} y_n = 107 \cdot 7 - 74 \cdot 17 = \boxed{-509}.$

10.5.6 We can express the series as the sum of two infinite geometric series:

$$\sum_{k=0}^{\infty} \frac{3^k + 6^k}{9^k} = \sum_{k=0}^{\infty} \frac{3^k}{9^k} + \sum_{k=0}^{\infty} \frac{6^k}{9^k} = \sum_{k=0}^{\infty} \left(\frac{1}{3}\right)^k + \sum_{k=0}^{\infty} \left(\frac{2}{3}\right)^k = \frac{1}{1 - 1/3} + \frac{1}{1 - 2/3} = \frac{3}{2} + 3 = \boxed{\frac{9}{2}}.$$

10.5.7 We may express the product as

$$\prod_{k=0}^{350}[k^3 - (350 - k)] = \prod_{k=0}^{350}(k^3 + k - 350).$$

To try to simplify this further, we see if $k^3 + k - 350$ can be factored. First, we search for integer roots, which must be among the factors of 350. Checking these values, we find that the cubic is 0 for $k = 7$. Since one of the factors is 0, then the whole product must also be $\boxed{0}$.

Exercises for Section 10.6

10.6.1

(a) First, we split the sum into two parts: $\displaystyle\sum_{i=1}^{5}\sum_{j=1}^{5}(i+j) = \sum_{i=1}^{5}\sum_{j=1}^{5}i + \sum_{i=1}^{5}\sum_{j=1}^{5}j.$

The first sum is

$$\sum_{i=1}^{5}\sum_{j=1}^{5}i = \sum_{i=1}^{5}5i = 5\sum_{i=1}^{5}i = 5(1+2+3+4+5) = 75,$$

and the second sum is

$$\sum_{i=1}^{5}\sum_{j=1}^{5}j = \sum_{i=1}^{5}(1+2+3+4+5) = \sum_{i=1}^{5}15 = 5\cdot 15 = 75.$$

Therefore, $\displaystyle\sum_{i=1}^{5}\sum_{j=1}^{5}(i+j) = 75 + 75 = 150.$

(b) First, we observe that $3^{i+j} = 3^i \cdot 3^j$. The inner sum only depends on j, so we can write

$$\sum_{i=1}^{4}\sum_{j=1}^{7}3^{i+j} = \sum_{i=1}^{4}\sum_{j=1}^{7}3^i \cdot 3^j = \sum_{i=1}^{4}3^i\sum_{j=1}^{7}3^j.$$

Then

$$\sum_{i=1}^{4}3^i\sum_{j=1}^{7}3^j = \sum_{i=1}^{4}3^i(3+3^2+3^3+3^4+3^5+3^6+3^7)$$

$$= (3+3^2+3^3+3^4)(3+3^2+3^3+3^4+3^5+3^6+3^7) = 3\cdot\frac{3^4-1}{3-1}\cdot 3\cdot\frac{3^7-1}{3-1} = \boxed{393480}.$$

(c) The (third) inner sum only depends on k, so we can write

$$\sum_{i=1}^{4}\sum_{j=1}^{5}\sum_{k=1}^{6}ijk = \sum_{i=1}^{4}\sum_{j=1}^{5}ij\sum_{k=1}^{6}k = \sum_{i=1}^{4}\sum_{j=1}^{5}ij(1+2+3+4+5+6) = 21\sum_{i=1}^{4}\sum_{j=1}^{5}ij.$$

Then in this sum, the inner sum only depends on j, so we can write

$$21\sum_{i=1}^{4}\sum_{j=1}^{5}ij = 21\sum_{i=1}^{4}i\sum_{j=1}^{5}j = 21\sum_{i=1}^{4}i(1+2+3+4+5) = 21\cdot 15\sum_{i=1}^{4}i = 315(1+2+3+4) = \boxed{3150}.$$

10.6.2

(a)   The inner sum only depends on j, so we can write

$$\sum_{i=1}^{5}\sum_{j=1}^{i} ij = \sum_{i=1}^{5} i \sum_{j=1}^{i} j = \sum_{i=1}^{5} i(1+2+\cdots+i) = \sum_{i=1}^{5} i \cdot \frac{i(i+1)}{2} = \frac{1}{2}\sum_{i=1}^{5}(i^3+i^2) = \frac{1}{2}(2+12+36+80+150) = \boxed{140}.$$

(b)   We recognize the inner sum as a geometric series:

$$\sum_{i=0}^{5}\sum_{j=0}^{i} 2^j = \sum_{i=0}^{5}(1+2+\cdots+2^i) = \sum_{i=0}^{5}(2^{i+1}-1).$$

We can then split this sum into two parts:

$$\sum_{i=0}^{5}(2^{i+1}-1) = \sum_{i=0}^{5}2^{i+1} - \sum_{i=0}^{5}1 = 2\sum_{i=0}^{5}2^i - 6 = 2(1+2+\cdots+32)-6 = 2(64-1)-6 = \boxed{120}.$$

10.6.3   First, we tackle the inner sum:

$$\sum_{j=i}^{12}(2j-7) = 2\sum_{j=i}^{12} j - \sum_{j=i}^{12} 7.$$

The first sum is the sum of the first 12 integers minus the sum of the first $i-1$ integers, and the latter is the result of adding 7 exactly $13-i$ times. (Make sure you see why it's $13-i$, not $12-i$.) We therefore have:

$$2\sum_{j=i}^{12} j - \sum_{j=i}^{12} 7 = 2\left(\frac{13\cdot12}{2}-\frac{i(i-1)}{2}\right)-7(13-i) = 65+8i-i^2.$$

Hence, $\sum_{i=1}^{5}\sum_{j=i}^{1}(2j-7) = \sum_{i=1}^{5}(65+8i-i^2) = 5(65)+\sum_{i=1}^{5}(8i-i^2) = 325+7+12+15+16+15 = \boxed{390}$.

10.6.4   We investigate the sum on the left side by writing it out explicitly:

$$\sum_{i=1}^{n}\sum_{j=1}^{i} x_j = \sum_{i=1}^{n}(x_1+x_2+\cdots+x_i)$$
$$= x_1$$
$$+ x_1+x_2$$
$$+ x_1+x_2+x_3$$
$$+\cdots$$
$$+ x_1+x_2+x_3+\cdots+x_n$$
$$= nx_1+(n-1)x_2+(n-2)x_3+\cdots+x_n.$$

The right side is equal to $\sum_{i=1}^{n}(n+1-i)x_i = nx_1+(n-1)x_2+(n-2)x_3+\cdots+x_n$. Thus, the two sides are equal.

Review Problems

10.41   Let the common difference be d. Then the tenth term is $4+9d = 85$, so $d = 9$. Hence, the sum of the first fifty terms is $\frac{50\cdot[2\cdot4+(50-1)\cdot9]}{2} = \boxed{11225}$.

10.42

(a) We have an arithmetic series with first term 12 and common difference 3. Let n be the number of terms. Then $1002 = 12 + 3(n-1)$, so $n = 331$. Hence, the sum of the series is $\frac{331}{2}(12 + 1002) = \boxed{167817}$.

(b) $\displaystyle\sum_{n=1}^{40}(-3n + 4) = -3\sum_{n=1}^{40} n + \sum_{n=1}^{40} 4 = -3 \cdot \frac{40 \cdot 41}{2} + 40 \cdot 4 = \boxed{-2300}$.

(c) We have a geometric series with first term 2 and common ratio 1/3. Hence, the sum of the series is

$$\frac{2}{1 - 1/3} = \boxed{3}.$$

10.43

(a) The first term is 1 and the common ratio is 1/4, so the sum of the series is $\dfrac{1}{1 - 1/4} = \boxed{\dfrac{4}{3}}$.

(b) The first term is -3 and the common ratio is -2, so the number of terms of the sequence, n, satisfies $-3(-2)^{n-1} = -768$. Therefore, we have $(-2)^{n-1} = 256 = (-2)^8$, so $n - 1 = 8$, which means $n = 9$. So, the sum of the series is $-3((-2)^9 - 1)/(-2 - 1) = \boxed{-513}$.

(c) The first term is $-\sqrt{6}$ and the common ratio is $-1/\sqrt{3}$, so the sum of the series is

$$\frac{-\sqrt{6}}{1 + \frac{1}{\sqrt{3}}} = \frac{-\sqrt{6}}{1 + \frac{1}{\sqrt{3}}} \cdot \frac{\sqrt{3}}{\sqrt{3}} = -\frac{3\sqrt{2}}{1 + \sqrt{3}} = -\frac{3\sqrt{2}(1 - \sqrt{3})}{(1 + \sqrt{3})(1 - \sqrt{3})} = \boxed{\frac{3\sqrt{2} - 3\sqrt{6}}{2}}.$$

10.44

(a) $\boxed{\displaystyle\sum_{k=1}^{n} \sqrt{k}}$.

(b) We have an arithmetic series with first term 1 and common difference 3. Let n be the number of terms, so $1 + 3(n - 1) = 100$, or $n = 34$. Then we can write the series as

$$\sum_{k=1}^{34}[1 + 3(k - 1)] = \boxed{\sum_{k=1}^{34}(3k - 2)}.$$

(c) $\boxed{\displaystyle\prod_{k=5}^{12} k}$.

(d) $\boxed{\displaystyle\prod_{k=1}^{n}(x - x_k)}$.

10.45 Since $x, y, 9$ is an arithmetic sequence, we have $y - x = 9 - y$, so $x + 9 = 2y$. Since $3, x, y$ is a geometric sequence, we have $x/3 = y/x$, so $x^2 = 3y$. Substituting $x = 2y - 9$, we get $(2y - 9)^2 = 3y$, which simplifies as $4y^2 - 39y + 81 = 0$, so $(y - 3)(4y - 27) = 0$, which means $y = 3$ or $y = 27/4$.

If $y = 3$, then $x = 2 \cdot 3 - 9 = -3$. If $y = 27/4$, then $x = 2 \cdot 27/4 - 9 = 9/2$. Therefore, the solutions are $(x, y) = \boxed{(-3, 3) \text{ and } (9/2, 27/4)}$.

10.46 We can write the sum as

$$\frac{3}{2} + \frac{5}{4} + \frac{9}{8} + \frac{17}{16} + \frac{33}{32} + \frac{65}{64} - 7 = \left(1 + \frac{1}{2}\right) + \left(1 + \frac{1}{4}\right) + \left(1 + \frac{1}{8}\right) + \left(1 + \frac{1}{16}\right) + \left(1 + \frac{1}{32}\right) + \left(1 + \frac{1}{64} - 7\right)$$

$$= \frac{1}{2} + \frac{1}{4} + \frac{1}{8} + \frac{1}{16} + \frac{1}{32} + \frac{1}{64} - 1 = \frac{\frac{1}{2}[(\frac{1}{2})^6 - 1]}{\frac{1}{2} - 1} - 1 = \boxed{-\frac{1}{64}}.$$

10.47 Let the three roots be $r - d$, r, and $r + d$, for some complex numbers r and d. Then by Vieta's Formulas, we have

$$(r - d) + r + (r + d) = 6 \qquad \text{and} \qquad (r - d)r + (r - d)(r + d) + r(r + d) = 21,$$

from which we get $3r = 6$ and $3r^2 - d^2 = 21$. From $3r = 6$, we have $r = 2$. Substituting this into $3r^2 - d^2 = 21$, we have $12 - d^2 = 21$, so $d^2 = -9$. Then $d = \pm 3i$. Therefore, the roots of the cubic are $2 - 3i$, 2, and $2 + 3i$, so

$$a = -(2 - 3i) \cdot (2) \cdot (2 + 3i) = -2(2 - 3i)(2 + 3i) = -2(4 + 9) = \boxed{-26}.$$

10.48 Let the first term and common ratio of the infinite geometric progression be a and r, respectively. Then the sum of all the terms is

$$\frac{a}{1 - r} = 6,$$

so $a = 6(1 - r)$. We are also given that the sum of the first two terms is $a + ar = 9/2$, so $a(1 + r) = 9/2$. Substituting $a = 6(1 - r)$ into this equation gives $6(1 - r)(1 + r) = 9/2$, so $1 - r^2 = 3/4$. Therefore, we have $r = \pm 1/2$.

If $r = 1/2$, then $a = 6(1 - 1/2) = 3$, and if $r = -1/2$, then $a = 6(1 + 1/2) = 9$. Therefore, the possible first terms of the progression are $\boxed{3 \text{ and } 9}$.

10.49 We can define variables and do the calculations, but it will be easier to use the fact that the sum of an arithmetic series is the average of the first and last terms times the total number of terms.

The odd numbered terms form an arithmetic sequence, and there are $(47 + 1)/2 = 24$ odd numbered terms, so

$$24 \cdot \frac{t_1 + t_{47}}{2} = 1272,$$

which means $t_1 + t_{47} = 106$. Therefore, the sum of all 47 terms is $47 \cdot \frac{t_1 + t_{47}}{2} = 47 \cdot \frac{106}{2} = \boxed{2491}$.

10.50 Let $f(x) = ax + b$. Then for any nonnegative integer n, the difference between consecutive terms $f(n + 1)$ and $f(n)$ is

$$f(n + 1) - f(n) = a(n + 1) + b - (an + b) = a,$$

which is a constant. Therefore, the sequence $f(0), f(1), f(2), \dots$ is an arithmetic sequence.

10.51 Let the first term and common difference of the arithmetic sequence be a and d, respectively. Then

$$a + (p - 1)d = q,$$
$$a + (q - 1)d = p.$$

Subtracting the first equation from the second equation, we get $(q - p)d = p - q$. Since $p \neq q$, we can divide both sides by $q - p$ to get $d = -1$. Then $a = p + q - 1$. Therefore, the $(p + q)^{\text{th}}$ term is $a + (p + q - 1)d = \boxed{0}$.

10.52

(a) Let d be the common difference of the arithmetic sequence. Then the second, tenth, and thirty-fourth terms of the sequence are $1 + d$, $1 + 9d$, and $1 + 33d$, respectively. Therefore, $(1 + 9d)/(1 + d) = (1 + 33d)/(1 + 9d)$, so $(1 + 9d)^2 = (1 + d)(1 + 33d)$. This simplifies as $48d^2 - 16d = 0$, so $d = 0$ or $d = 1/3$. But we are given that $d \neq 0$, so $d = 1/3$.

Therefore, the arithmetic sequence is $\boxed{1, 4/3, 5/3, 2, \dots}$.

(b) The geometric sequence in part (a) begins 4/3, 4, 12, so the fourth term is 36. Then $1 + (n-1)/3 = 36$, so $n = \boxed{106}$.

10.53 We can express the given series as the sum of two infinite geometric series:

$$1 + \frac{1}{2} + \frac{1}{10} + \frac{1}{20} + \frac{1}{100} + \cdots = \left(1 + \frac{1}{10} + \frac{1}{100} + \cdots\right) + \left(\frac{1}{2} + \frac{1}{20} + \frac{1}{200} + \cdots\right)$$

$$= \left(1 + \frac{1}{10} + \frac{1}{100} + \cdots\right) + \frac{1}{2}\left(1 + \frac{1}{10} + \frac{1}{100} + \cdots\right)$$

$$= \frac{3}{2} \cdot \frac{1}{1 - 1/10} = \boxed{\frac{5}{3}}.$$

10.54 Both series in the numerator and denominator contain n terms. Using the formula for the sum of an arithmetic series in terms of the first term, last term, and number of terms, we get

$$\frac{115}{116} = \frac{1 + 3 + 5 + \cdots + (2n-1)}{2 + 4 + 6 + \cdots + 2n} = \frac{n \cdot \frac{1+(2n-1)}{2}}{n \cdot \frac{2+2n}{2}} = \frac{2n}{2(n+1)} = \frac{n}{n+1},$$

so $n = \boxed{115}$.

10.55 We are given that

$$18 = ar + ar^2 + ar^3 + \cdots + ar^{10} = ar(1 + r + r^2 + \cdots + r^9)$$

and

$$6 = \frac{1}{ar} + \frac{1}{ar^2} + \cdots + \frac{1}{ar^{10}} = \frac{1 + r + r^2 + \cdots + r^9}{ar^{10}}.$$

Dividing the first equation by the second equation, we get $a^2 r^{11} = 18/6 = 3$.

Therefore, the product of the ten numbers is

$$ar \cdot ar^2 \cdot ar^3 \cdots ar^{10} = a^{10} r^{10 \cdot 11/2} = a^{10} r^{55} = (a^2 r^{11})^5 = 3^5 = \boxed{243}.$$

10.56 The first geometric series is equal to $a + ar + ar^2 + \cdots = \frac{a}{1-r}$, and the second geometric series is equal to $a - ar + ar^2 - \cdots = \frac{a}{1+r}$, so their product is

$$\frac{a}{1-r} \cdot \frac{a}{1+r} = \frac{a^2}{1-r^2}.$$

The third geometric series is equal to $a^2 + a^2 r^2 + a^2 r^4 + \cdots = \frac{a^2}{1-r^2}$, so it is equal to the product of the first two geometric series.

10.57 Let the first term and common ratio of the first infinite geometric series be a and r, respectively. Then $a + ar + ar^2 + \cdots = \frac{a}{1-r} = 15$, and $a^2 + a^2 r^2 + a^2 r^4 + \cdots = \frac{a^2}{1-r^2} = 45$.

From the first equation, we have $a = 15(1-r)$. Substituting into the second equation, we get

$$45 = \frac{a^2}{1-r^2} = \frac{15^2(1-r)^2}{(1+r)(1-r)} = \frac{225(1-r)}{1+r}.$$

Solving $45 = 225(1-r)/(1+r)$ for r, we get $r = 2/3$, so $a = 15(1 - 2/3) = \boxed{5}$.

10.58 The sum of the first k terms is

$$g(1) + g(2) + \cdots + g(k-1) + g(k) = \frac{k}{3k-2}.$$

To find $g(k)$, we first find the sum of the first $k-1$ terms:

$$g(1) + g(2) + \cdots + g(k-1) = \frac{k-1}{3(k-1)-2} = \frac{k-1}{3k-5}.$$

Subtracting the second equation from the first equation, we find

$$g(k) = \frac{k}{3k-2} - \frac{k-1}{3k-5} = \boxed{-\frac{2}{(3k-2)(3k-5)}}.$$

10.59

(a) We note that the first sum goes from $i = -4$ to 19, and the second sum goes from $k = -2$ to 21. Hence, we set $k = i+2$, so $i = k-2$. Then the first sum becomes

$$\sum_{i=-4}^{19}(2-i^2) = \sum_{k=-2}^{21}[2-(k-2)^2] = \sum_{k=-2}^{21}(-2+4k-k^2),$$

which is the second sum, so they are equal.

(b) We write the first few terms of each sum:

$$\sum_{k=0}^{\infty}\frac{1}{2^k+4} = \frac{1}{2^0+4} + \frac{1}{2^1+4} + \frac{1}{2^2+4} + \frac{1}{2^3+4} + \cdots = \frac{1}{5} + \frac{1}{6} + \frac{1}{8} + \frac{1}{12} + \cdots,$$

$$\sum_{k=2}^{\infty}\frac{1}{2^k+1} = \frac{1}{2^2+1} + \frac{1}{2^3+1} + \frac{1}{2^4+1} + \frac{1}{2^5+1} + \cdots = \frac{1}{5} + \frac{1}{9} + \frac{1}{17} + \frac{1}{33} + \cdots.$$

The first terms are the same, but each remaining term of the first series is greater than the corresponding term of the second series. To see why, we note that the kth term of the first series is $\frac{1}{2^{k-1}+4}$, while the kth term of the second is $\frac{1}{2^{k+1}+1}$, which equals $\frac{1}{4\cdot2^{k-1}+1}$. For $k > 1$, we have $(4\cdot2^{k-1}+1) - (2^{k-1}+4) = 3\cdot2^{k-1} - 3 > 0$, so $4\cdot2^{k-1}+1 > 2^{k-1}+4$, which tells us that $\frac{1}{2^{k+1}+4} > \frac{1}{4\cdot2^{k-1}+1}$. Therefore, the terms of the series are not all the same, and their sums are not equal.

(c) To see if the the two sums are equal, we write out the terms. The first sum is

$$\sum_{j=1}^{20}\frac{j}{22-j} = \frac{1}{21} + \frac{2}{20} + \cdots + \frac{19}{3} + \frac{20}{2},$$

and the second sum is

$$\sum_{k=1}^{20}\frac{21-k}{k+1} = \frac{20}{2} + \frac{19}{3} + \cdots + \frac{2}{20} + \frac{1}{21}.$$

The terms in the two sums are the same, so they are equal.

10.60 The first term of the infinite geometric series $1/c + x/c^2 + x^2/c^3 + \cdots$ is $1/c$, and the common ratio is x/c, so its sum is

$$\frac{1/c}{1-x/c} = \frac{1}{c-x},$$

as desired.

10.61

(a) We can begin by expanding the inner sum:

$$\sum_{j=1}^{4}(i-j)^2 = \sum_{j=1}^{4}(i^2 - 2ij + j^2) = \sum_{j=1}^{4}i^2 - 2i\sum_{j=1}^{4}j + \sum_{j=1}^{4}j^2$$

$$= 4i^2 - 2i(1+2+3+4) + (1+4+9+16) = 4i^2 - 20i + 30.$$

Therefore, $\displaystyle\sum_{i=1}^{4}\sum_{j=1}^{4}(i-j)^2 = \sum_{i=1}^{4}(4i^2 - 20i + 30) = 14 + 6 + 6 + 14 = \boxed{40}$.

Alternatively, we could have written out all the terms:

$$\sum_{i=1}^{4}\sum_{j=1}^{4}(i-j)^2 = \sum_{i=1}^{4}[(i-1)^2 + (i-2)^2 + (i-3)^2 + (i-4)^2]$$

$$= 0^2 + 1^2 + 2^2 + 3^2$$
$$+ 1^2 + 0^2 + 1^2 + 2^2$$
$$+ 2^2 + 1^2 + 0^2 + 1^2$$
$$+ 3^2 + 2^2 + 1^2 + 0^2.$$

The square 0^2 appears 4 times, 1^2 appears 6 times, 2^2 appears 4 times, and 3^2 appears 2 times. Therefore, the sum is $4 \cdot 0^2 + 6 \cdot 1^2 + 4 \cdot 2^2 + 2 \cdot 3^2 = 40$.

(b) We have

$$\sum_{i=1}^{5}\sum_{j=-3}^{5}j2^i = \sum_{i=1}^{5}[(-3) + (-2) + (-1) + 0 + 1 + 2 + 3 + 4 + 5]2^i$$

$$= \sum_{i=1}^{5}9 \cdot 2^i = 9(2 + 4 + \cdots + 2^5) = 18(1 + 2 + \cdots + 2^4) = 18 \cdot 31 = \boxed{558}.$$

10.62

(a) We have $\displaystyle\sum_{j=0}^{5}\sum_{k=0}^{j}3 = \sum_{j=0}^{5}3(j+1) = 3\sum_{j=0}^{5}(j+1) = 3(1+2+3+4+5+6) = 3 \cdot 21 = \boxed{63}$.

(b) We can write $\displaystyle\sum_{i=1}^{8}\sum_{j=1}^{2i}\frac{2j}{i} = 2\sum_{i=1}^{8}\frac{1}{i}\sum_{j=1}^{2i}j$. The inner sum is $\displaystyle\sum_{j=1}^{2i}j = 1+2+\cdots+2i = \frac{2i(2i+1)}{2} = 2i^2 + i$. Therefore,

$$2\sum_{i=1}^{8}\frac{1}{i}\sum_{j=1}^{2i}j = 2\sum_{i=1}^{8}\frac{2i^2+i}{i} = 2\sum_{i=1}^{8}(2i+1) = 2\sum_{i=1}^{8}(2i) + 2\sum_{i=1}^{8}1 = 4\sum_{i=1}^{8}i + 2(8) = 4 \cdot \frac{8(9)}{2} + 16 = \boxed{160}.$$

10.63 We have

$$\sum_{j=0}^{k}\sum_{i=0}^{k}2j = 2\sum_{j=0}^{k}j\sum_{i=0}^{k}1 = 2\sum_{j=0}^{k}j(k+1) = 2(k+1)\sum_{j=0}^{k}j = 2(k+1)(1+2+\cdots+k) = 2(k+1) \cdot \frac{k(k+1)}{2} = \boxed{k(k+1)^2}.$$

10.64

(a) The reciprocals of the terms of the harmonic sequence form an arithmetic sequence. The first two terms of the arithmetic sequence are $\frac{1}{3}$ and $\frac{1}{4}$, so the common difference is $\frac{1}{4} - \frac{1}{3} = -\frac{1}{12}$. Therefore, the first four terms of the arithmetic sequence are $\frac{1}{3}, \frac{1}{4}, \frac{1}{4} - \frac{1}{12} = \frac{1}{6}$, and $\frac{1}{6} - \frac{1}{12} = \frac{1}{12}$. Then the first four terms of the harmonic sequence are 3, 4, 6, and 12, so $S_4 = 3 + 4 + 6 + 12 = \boxed{25}$.

(b) The reciprocals of the terms of the harmonic sequence form an arithmetic sequence in which the first term is 3 and the fifth term is 1. Therefore, if the common difference is d, then $4d = 1 - 3 = -2$, or $d = -\frac{1}{2}$. Hence, the terms in the arithmetic sequence are $3, \frac{5}{2}, 2, \frac{3}{2}$, and 1, so the second, third, and fourth terms of the harmonic sequence are $\boxed{\frac{2}{5}, \frac{1}{2}, \text{ and } \frac{2}{3}}$, respectively.

(c) If $x + 4$, $x + 1$, and x form a harmonic progression, then their reciprocals form an arithmetic progression, so

$$\frac{1}{x+1} - \frac{1}{x+4} = \frac{1}{x} - \frac{1}{x+1}.$$

Multiplying both sides by $x(x+1)(x+4)$, we get $x(x+4) - x(x+1) = (x+1)(x+4) - x(x+4)$, which simplifies as $3x = x + 4$, so $x = \boxed{2}$.

(d) If a, b, c is a harmonic sequence, then $1/a, 1/b, 1/c$ is an arithmetic sequence. To show that $a/(b+c)$, $b/(a+c)$, $c/(a+b)$ is a harmonic sequence, we must show that $(b+c)/a$, $(a+c)/b$, $(a+b)/c$ is an arithmetic sequence. Adding 1 to each term, we see that this is equivalent to showing that $(a+b+c)/a$, $(a+b+c)/b$, $(a+b+c)/c$ is an arithmetic sequence. But this follows from the fact that $1/a, 1/b, 1/c$ is an arithmetic sequence, and then multiplying each term by $a + b + c$.

Challenge Problems

10.65 Let the number in the square above the 0 be a. Then the number in the square above a is $2a$. Similarly, let the number in the square to the right of 0 be b. Then the number in the square to the right of b is $2b$. Let c be the number in the square between a and 103, and let d be the number in the square above 103.

		★	
	74		
$2a$		d	186
a	c	103	
0	b	$2b$	

Since the numbers $a, c, 103$ are in arithmetic sequence, we have $c = (a+103)/2$. Then the common difference of the numbers in the second column is $c - b = (a - 2b + 103)/2$. Therefore, the number above c is $c + (c - b) = (a + 103)/2 + (a - 2b + 103)/2 = a - b + 103$, and the number above that is

$$a - b + 103 + (c - b) = a - b + 103 + \frac{a - 2b + 103}{2} = \frac{3a - 4b + 309}{2}.$$

Hence, we have $\dfrac{3a - 4b + 309}{2} = 74$. Multiplying both sides by 2 and simplifying, we get $3a - 4b = -161$.

Similarly, since the numbers $2b, 103, d$ are in arithmetic sequence, we have $103 = (2b + d)/2$, so $d = 2 \cdot 103 - 2b = 206 - 2b$. Furthermore, looking at the third row, the numbers $2a, d, 186$ are in arithmetic sequence, so $d = (2a + 186)/2 = a + 93$. Equating this to our expression for d above, we have $a + 93 = 206 - 2b$, so $a = 113 - 2b$.

52	82	112	142	172
39	74	109	144	179
26	66	106	146	186
13	58	103	148	193
0	50	100	150	200

Substituting this expression for a into the equation $3a - 4b = -161$ from above, we get $3(113 - 2b) - 4b = -161$. Solving, we find $b = 50$, and $a = 113 - 2 \cdot 50 = 13$.

Hence, the common difference in the bottom row is 50, so the fourth number in this row is 150. The common difference in the fourth row down is $(103 - 13)/2 = 45$, so the fourth number in this row is $103 + 45 = 148$. From these numbers, we find that the number in the starred square is $\boxed{142}$. For reference, the complete grid is shown at right.

10.66   The factor $x_1 + 2x_2 + \cdots + kx_k$ can be written as $\displaystyle\sum_{i=1}^{k} ix_i$, so the entire product can be written as $\boxed{\displaystyle\prod_{k=1}^{n}\sum_{i=1}^{k} ix_i}$.

10.67   Let the numbers originally written on the board be $1, 2, \ldots, n$. First, we place bounds on the average of the remaining $n - 1$ numbers, to get bounds on n.

The average is least when the number n is erased, in which case the average is

$$\frac{\frac{n(n+1)}{2} - n}{n-1} = \frac{\frac{n^2+n-2n}{2}}{n-1} = \frac{n(n-1)}{2(n-1)} = \frac{n}{2},$$

so $\frac{n}{2} \le 35\frac{7}{17}$, which means $n \le 70\frac{14}{17}$.

The average is greatest when the number 1 is erased, in which case the average is

$$\frac{\frac{n(n+1)}{2} - 1}{n-1} = \frac{\frac{n^2+n-2}{2}}{n-1} = \frac{(n+2)(n-1)}{2(n-1)} = \frac{n+2}{2},$$

so $35\frac{7}{17} \le \frac{n+2}{2}$, which means $n \ge 68\frac{14}{17}$.

Combining these, we see that $n = 69$ or $n = 70$. Since $35\frac{7}{17}$ is the average of $n - 1$ numbers, $(35\frac{7}{17})(n - 1)$ must be an integer. Therefore, $n = 69$.

Let k be the number that was erased. Then $\dfrac{\frac{69\cdot70}{2} - k}{68} = 35\dfrac{7}{17}$. Solving for k, we find $k = \boxed{7}$.

10.68   In order to use our formula for the sum of an infinite geometric series, we must first confirm that $|a| < 1$ and $|b| < 1$. From the quadratic formula, we find that the roots of the quadratic are $\frac{2\pm\sqrt{26}}{11}$. Since $|2 + \sqrt{26}| < 11$ and $|2 - \sqrt{26}| < 11$, both roots have absolute value less than 1. Therefore, we have

$$(1 + a + a^2 + \cdots)(1 + b + b^2 + \cdots) = \frac{1}{1-a}\cdot\frac{1}{1-b} = \frac{1}{(1-a)(1-b)} = \frac{1}{1-a-b+ab}.$$

By Vieta's Formulas, we have $a + b = 4/11$ and $ab = -2/11$, so $\dfrac{1}{1-a-b+ab} = \dfrac{1}{1-4/11-2/11} = \boxed{\dfrac{11}{5}}$.

10.69   We evaluate the sum from the inside out. First, $\displaystyle\sum_{k=1}^{j} 1 = j$. Therefore,

$$\sum_{j=1}^{i}\sum_{k=1}^{j} 1 = \sum_{j=1}^{i} j = 1 + 2 + \cdots + i = \frac{i(i+1)}{2}.$$

Then $\displaystyle\sum_{i=1}^{5}\sum_{j=1}^{i}\sum_{k=1}^{j} 1 = \sum_{i=1}^{5}\frac{i(i+1)}{2} = 1 + 3 + 6 + 10 + 15 = \boxed{35}$.

10.70 Since a^2, b^2, c^2 is an arithmetic sequence, we have $2b^2 = a^2 + c^2$, so $2b^2 - a^2 - c^2 = 0$. Then to test whether $1/(b+c)$, $1/(a+c)$, $1/(a+b)$ is an arithmetic sequence, we compute

$$\frac{2}{a+c} - \frac{1}{b+c} - \frac{1}{a+b} = \frac{2(a+b)(b+c) - (a+b)(a+c) - (a+c)(b+c)}{(a+b)(a+c)(b+c)}$$

$$= \frac{2ab + 2ac + 2b^2 + 2bc - (a^2 + ab + ac + bc) - (ab + ac + bc + c^2)}{(a+b)(a+c)(b+c)}$$

$$= \frac{2b^2 - a^2 - c^2}{(a+b)(a+c)(b+c)}$$

$$= 0.$$

Therefore, we have $\dfrac{1}{a+c} - \dfrac{1}{b+c} = \dfrac{1}{a+b} - \dfrac{1}{a+c}$, so $\dfrac{1}{b+c}$, $\dfrac{1}{a+c}$, $\dfrac{1}{a+b}$ is an arithmetic sequence.

10.71 The sum of the coefficients of p is equal to $p(1, 1, \ldots, 1) = \displaystyle\prod_{n=1}^{m} \sum_{k=1}^{n} 1 = \prod_{n=1}^{m} n = 1 \cdot 2 \cdots m = \boxed{m!}$.

10.72 Let the roots of the quintic be a, ar, ar^2, ar^3, and ar^4. Then by Vieta's Formulas, we have the equation $80 = a + ar + ar^2 + ar^3 + ar^4 = a(1 + r + r^2 + r^3 + r^4)$, and the sum of the reciprocals of the roots is

$$5 = \frac{1}{a} + \frac{1}{ar} + \frac{1}{ar^2} + \frac{1}{ar^3} + \frac{1}{ar^4} = \frac{1 + r + r^2 + r^3 + r^4}{ar^4}.$$

Dividing $80 = a(1 + r + r^2 + r^3 + r^4)$ by $5 = (1 + r + r^2 + r^3 + r^4)/(ar^4)$, we get $a^2 r^4 = 80/5 = 16$, so $ar^2 = \pm\sqrt{16} = \pm 4$. Then

$$S = a \cdot ar \cdot ar^2 \cdot ar^3 \cdot ar^4 = a^5 r^{10},$$

which means $|S| = |a^5 r^{10}| = |ar^2|^5 = 4^5 = \boxed{1024}$.

10.73 We compute the first few $f(n)$:

$$f(1) = 0,$$
$$f(2) = 1 + 1 = 2,$$
$$f(3) = 2 + 2 + 2 = 6,$$
$$f(4) = 3 + 4 + 4 + 3 = 14,$$
$$f(5) = 4 + 7 + 8 + 7 + 4 = 30,$$
$$f(6) = 5 + 11 + 15 + 15 + 11 + 5 = 62.$$

Each term is approximately double the previous term. In fact, we see that $f(n) = 2f(n-1) + 2$ for each term we calculated above. Now we show that this relation holds in general. To produce the nth row in the triangular array from the $(n-1)^{\text{st}}$ row, each term in the $(n-1)^{\text{st}}$ row is used twice—once in the lower left entry, and once in the lower right entry. The exceptions are the entries at the ends of the row, where we must add 1 to $n-2$ to get $n-1$. Hence, $f(n) = 2f(n-1) + 2$.

Then we can write the terms of the sequence as follows:

$$f(1) = 0,$$
$$f(2) = 2f(1) + 2 = 2,$$
$$f(3) = 2f(2) + 2 = 2^2 + 2,$$
$$f(4) = 2f(3) + 2 = 2^3 + 2^2 + 2,$$

and so on. In general, for $n \geq 2$, we have

$$f(n) = 2^{n-1} + 2^{n-2} + \cdots + 2^2 + 2 = 2(2^{n-2} + 2^{n-3} + \cdots + 2 + 1) = 2 \cdot \frac{2^{n-1} - 1}{2 - 1} = 2^n - 2.$$

Now the problem is to find the remainder when $f(100) = 2^{100} - 2$ is divided by 100. We use modular arithmetic:

$$2^{100} \equiv (2^{10})^{10} \equiv 1024^{10} \equiv 24^{10} \equiv (24^2)^5 \equiv 576^5 \equiv 76^5 \pmod{100}.$$

Since $76^2 \equiv 5776 \equiv 76 \pmod{100}$, we have $76^5 \equiv 76 \pmod{100}$. Therefore, the remainder of $2^{100} - 2$ when divided by 100 is $\boxed{74}$. (By $\pmod{100}$, we basically mean the remainder when the number is divided by 100.)

10.74 Let the consecutive terms of the arithmetic sequence be $a - d$, a, and $a + d$, so

$$k + 36 = (a - d)^2,$$
$$k + 300 = a^2,$$
$$k + 596 = (a + d)^2.$$

Subtracting the first equation from the third equation, we get $560 = 4ad$, so $ad = 140$. Subtracting the second equation from the third equation, we get $296 = 2ad + d^2 = 2 \cdot 140 + d^2$, so $d^2 = 16$. Then

$$a^2 = \frac{140^2}{d^2} = 1225,$$

so from the second equation, we have $k = a^2 - 300 = \boxed{925}$.

10.75 Since S_n is the sum of the infinite geometric series with first term a^n and common ratio r^n, we have

$$S_n = a^n + a^n r^n + a^n r^{2n} + a^n r^{3n} + \cdots = \frac{a^n}{1 - r^n}.$$

Therefore,

$$\frac{1}{S_1} + \frac{1}{S_2} + \frac{1}{S_3} + \cdots = \frac{1 - r}{a} + \frac{1 - r^2}{a^2} + \frac{1 - r^3}{a^3} + \cdots = \left(\frac{1}{a} + \frac{1}{a^2} + \frac{1}{a^3} + \cdots \right) - \left(\frac{r}{a} + \frac{r^2}{a^2} + \frac{r^3}{a^3} + \cdots \right)$$

$$= \frac{1/a}{1 - 1/a} - \frac{r/a}{1 - r/a} = \frac{1}{a - 1} - \frac{r}{a - r} = \boxed{\frac{a(1 - r)}{(a - 1)(a - r)}}.$$

10.76 We begin by writing out the sum:

$$\sum_{j=1}^{\infty} \sum_{k=1}^{\infty} \frac{1}{(j + k)^3} = \sum_{j=1}^{\infty} \left[\frac{1}{(j + 1)^3} + \frac{1}{(j + 2)^3} + \frac{1}{(j + 3)^3} + \cdots \right]$$

$$= \frac{1}{2^3} + \frac{1}{3^3} + \frac{1}{4^3} + \cdots$$
$$+ \frac{1}{3^3} + \frac{1}{4^3} + \frac{1}{5^3} + \cdots$$
$$+ \frac{1}{4^3} + \frac{1}{5^3} + \frac{1}{6^3} + \cdots.$$

We note that the term $1/2^3$ appears once, $1/3^3$ twice, $1/4^3$ three times, and so on. In general, for $n \geq 2$, the term $1/n^3$ will appear $n - 1$ times, because $j + k = n$ for the $n - 1$ ordered pairs $(j, k) = (1, n - 1), (2, n - 2), \ldots, (n - 1, 1)$. Hence,

$$\sum_{j=1}^{\infty} \sum_{k=1}^{\infty} \frac{1}{(j + k)^3} = \sum_{n=2}^{\infty} \frac{n - 1}{n^3} = \sum_{n=1}^{\infty} \frac{n - 1}{n^3} = \sum_{n=1}^{\infty} \frac{n}{n^3} - \sum_{n=1}^{\infty} \frac{1}{n^3} = \sum_{n=1}^{\infty} \frac{1}{n^2} - \sum_{n=1}^{\infty} \frac{1}{n^3} = \boxed{p - q}.$$

(Note that we can change the lower limit from 2 to 1 in our second step without changing the value of the sum because $(n-1)/n^3 = 0$ when $n = 1$.)

10.77 Let the m elements in set A be $a, a+1, \ldots, a+m-1$. Then their sum must be $2m$, so we have

$$2m = a + (a+1) + \cdots + (a+m-1) = \frac{m[2a + (m-1)]}{2}.$$

Solving for a, we find $a = (5-m)/2$, so the greatest element in A is $a + m - 1 = \frac{m+3}{2}$.

Let the $2m$ elements in set B be $b, b+1, \ldots, b+2m-1$. Their sum must be m, so

$$m = b + (b+1) + \cdots + (b+2m-1) = \frac{2m[2b + (2m-1)]}{2}.$$

Solving for b, we find $b = 1 - m$, so the greatest element in B is $b + 2m - 1 = m$. Therefore,

$$\left| m - \frac{m+3}{2} \right| = \left| \frac{m-3}{2} \right| = 99.$$

Multiplying both sides by 2, we get $|m - 3| = 198$. Since m is a positive integer, $m = \boxed{201}$.

10.78 Let the original infinite geometric series have first term a and common ratio r. Then

$$2005 = a + ar + ar^2 + \cdots = \frac{a}{1-r},$$

so $a = 2005(1 - r)$.

Squaring each term, we get $a^2 + a^2r^2 + a^2r^4 + \cdots = \dfrac{a^2}{1-r^2}$, so we have $\dfrac{a^2}{1-r^2} = 20050$. Substituting $a = 2005(1-r)$, we get $20050 = \dfrac{2005^2(1-r)^2}{(1+r)(1-r)} = \dfrac{2005^2(1-r)}{1+r}$. Solving for r, we find $r = \boxed{399/403}$.

10.79 Given that $\binom{n}{k-1}, \binom{n}{k}, \binom{n}{k+1}$ form an arithmetic sequence, we have

$$\binom{n}{k} - \binom{n}{k-1} = \binom{n}{k+1} - \binom{n}{k}$$

$$\Rightarrow \quad \frac{n!}{k!(n-k)!} - \frac{n!}{(k-1)!(n-k+1)!} = \frac{n!}{(k+1)!(n-k-1)!} - \frac{n!}{k!(n-k)!}.$$

All the factors of $n!$ cancel. Furthermore, we can multiply both sides by $(k+1)!(n-k+1)!$, to get

$$(k+1)(n-k+1) - (k+1)k = (n-k+1)(n-k) - (k+1)(n-k+1).$$

Moving all the terms to one side, this simplifies as $n^2 - 4nk + 4k^2 - n - 2 = 0$. Noting that $n^2 - 4nk + 4k^2$ is a perfect square, we can rewrite this equation as $(n - 2k)^2 = n + 2$.

Since the binomial coefficients $\binom{n}{k-1}, \binom{n}{k}, \binom{n}{k+1}$ form an *increasing* arithmetic sequence, we have $2k \leq n$, so $n - 2k \geq 0$. Let $a = n - 2k$. Then $n + 2 = (n - 2k)^2 = a^2$, so $n = a^2 - 2$, and

$$k = \frac{n-a}{2} = \frac{a^2 - a - 2}{2}.$$

Finally, $n > 2$, so $a \geq 3$. Since all of our steps are reversible, we can say that all solutions are given by $(n, k) = \boxed{\left(a^2 - 2, \dfrac{a^2 - a - 2}{2} \right), \text{ where } a \geq 3}$.

10.80 Since a, b, c form an arithmetic progression, $2b - a - c = 0$. But

$$2b - a - c = 2(x^2 + xz + z^2) - (x^2 + xy + y^2) - (y^2 + yz + z^2) = x^2 - 2y^2 + z^2 + 2xz - xy - yz,$$

so we must have $x^2 - 2y^2 + z^2 + 2xz - xy - yz = 0$. To factor the expression on the left, we can try grouping the terms. For example, we see that the terms $x^2 + 2xz + z^2$ form a perfect square, so we have

$$(x + z)^2 - xy - 2y^2 - yz = 0.$$

Then the terms $-xy - yz$ factor as $-y(x + z)$, and we have $(x + z)^2 - y(x + z) - 2y^2 = 0$. Let $s = x + z$, so the equation becomes $s^2 - sy - 2y^2 = 0$. This factors as $(s - 2y)(s + y) = 0$. Hence, we get

$$(x - 2y + z)(x + y + z) = 0.$$

We are given that $x + y + z \neq 0$, so $x - 2y + z = 0$. Therefore, x, y, and z also form an arithmetic progression.

10.81 Let $n = a_2$. The terms $a_1 = 1, a_2 = n, a_3$ are in geometric progression, so $a_3 = a_2^2/a_1 = n^2$. Then the terms $a_2 = n, a_3 = n^2, a_4$ are in arithmetic progression, so $a_4 = 2a_3 - a_2 = 2n^2 - n = n(2n - 1)$.

Continuing, the terms $a_3 = n^2, a_4 = n(2n - 1), a_5$ are in geometric progression, so

$$a_5 = \frac{a_4^2}{a_3} = \frac{n^2(2n - 1)^2}{n^2} = (2n - 1)^2.$$

Then the terms $a_4 = n(2n - 1), a_5 = (2n - 1)^2, a_6$ are in arithmetic progression, so

$$a_6 = 2a_5 - a_4 = 2(2n - 1)^2 - n(2n - 1) = (2n - 1)[2(2n - 1) - n] = (2n - 1)(3n - 2).$$

At this point, we observe that the sequence is determined by n. Furthermore, it appears that

$$a_{2m-1} = [(m - 1)n - (m - 2)]^2,$$
$$a_{2m} = [(m - 1)n - (m - 2)][mn - (m - 1)]$$

for all positive integers m. We prove this formally using induction.

The formulas are easily verified for the base case $m = 1$, so assume that they hold for some positive integer $m = k$. Then

$$a_{2k-1} = [(k - 1)n - (k - 2)]^2,$$
$$a_{2k} = [(k - 1)n - (k - 2)][kn - (k - 1)].$$

The terms $a_{2k-1}, a_{2k}, a_{2k+1}$ are in geometric progression, so

$$a_{2k+1} = \frac{a_{2k}^2}{a_{2k-1}} = \frac{[(k - 1)n - (k - 2)]^2[kn - (k - 1)]^2}{[(k - 1)n - (k - 2)]^2} = [kn - (k - 1)]^2.$$

Then the terms $a_{2k}, a_{2k+1}, a_{2k+2}$ are in arithmetic progression, so

$$a_{2k+2} = 2a_{2k+1} - a_{2k} = 2[kn - (k - 1)]^2 - [(k - 1)n - (k - 2)][kn - (k - 1)]$$
$$= [kn - (k - 1)][2kn - 2(k - 1) - (k - 1)n + (k - 2)] = [kn - (k - 1)][(k + 1)n - k].$$

Hence, the formulas hold for $m = k + 1$, so by induction, they hold for all positive integers m.

In particular, taking $m = 5$, we get

$$a_9 = (4n - 3)^2 = 16n^2 - 24n + 9,$$
$$a_{10} = (4n - 3)(5n - 4) = 20n^2 - 31n + 12,$$

so $a_9 + a_{10} = 36n^2 - 55n + 21$. We are given that $a_9 + a_{10} = 646$, so $36n^2 - 55n + 21 = 646$, which simplifies as $(n - 5)(36n + 125) = 0$. Therefore, $n = 5$.

Checking the terms of the sequence (which are increasing), we find $a_{16} = 957$ and $a_{17} = 1089$. Therefore, the answer is $\boxed{957}$.

10.82 Since the first three terms form an arithmetic progression, we can write them as a, $a + d$, $a + 2d$. Let the fourth term be b. Since the last three terms form a geometric progression, we have $(a + 2d)/(a + d) = b/(a + 2d)$, which means the fourth term is $(a + 2d)^2/(a + d)$. We therefore have $(a + 2d)^2/(a + d) - a = 30$, which means $(a + 2d)^2 = (a + 30)(a + d)$. Rearranging this equation gives $3a(10 - d) = 2d(2d - 15)$. Since a and d are positive, the signs of $10 - d$ and $2d - 15$ must be the same. They are both positive when d is 8 or 9, and they are never both negative. When $d = 8$, we find $a = 8/3$, which is not an integer. When $d = 9$, we have $a = 18$, which gives us the sequence $18, 27, 36, 48$, which has sum $\boxed{129}$.

10.83 To compute the infinite product $(1 + x)(1 + x^2)(1 + x^4)(1 + x^8)\cdots$, we compute the product of the first n factors, for the first few values of n:

$$1 + x = 1 + x,$$
$$(1 + x)(1 + x^2) = 1 + x + x^2 + x^3,$$
$$(1 + x)(1 + x^2)(1 + x^4) = 1 + x + x^2 + x^3 + x^4 + x^5 + x^6 + x^7,$$

and so on. In general, it appears that $(1 + x)(1 + x^2)\cdots(1 + x^{2^n}) = 1 + x + x^2 + \cdots + x^{2^{n+1}-1}$. We can sum the right side as a geometric series:

$$1 + x + x^2 + \cdots + x^{2^{n+1}-1} = \frac{1 - x^{2^{n+1}}}{1 - x}.$$

Thus, we must show that $(1 + x)(1 + x^2)\cdots(1 + x^{2^n}) = \frac{1 - x^{2^{n+1}}}{1 - x}$. The simplest way is to multiply both sides by $1 - x$. Then the left side collapses:

$$(1 - x)(1 + x)(1 + x^2)\cdots(1 + x^{2^n}) = (1 - x^2)(1 + x^2)\cdots(1 + x^{2^n})$$
$$= (1 - x^4)\cdots(1 + x^{2^n})$$
$$= \cdots$$
$$= (1 - x^{2^n})(1 + x^{2^n})$$
$$= 1 - x^{2^{n+1}}.$$

Hence, $(1 + x)(1 + x^2)\cdots(1 + x^{2^n}) = \frac{1 - x^{2^{n+1}}}{1 - x}$ for all nonnegative integers n. Letting n go to infinity, $x^{2^{n+1}}$ goes to 0 (since $|x| < 1$), so the infinite product is equal to $(1 + x)(1 + x^2)(1 + x^4)(1 + x^8)\cdots = \boxed{\frac{1}{1 - x}}$.

10.84 The function $f(x)$ looks like the infinite product

$$(1 + x)(1 + x^2)(1 + x^4)\cdots = 1 + x + x^2 + x^3 + \cdots = \frac{1}{1 - x},$$

except $f(x)$ does not include the terms $1 + x^2$, $1 + x^8$, and so on. However, these missing terms are present in

$$f(x^2) = (1 + x^2)(1 + x^8)(1 + x^{32})\cdots.$$

Thus, $f(x)f(x^2) = (1 + x)(1 + x^2)(1 + x^4)(1 + x^8)\cdots = \frac{1}{1 - x}$ for all $0 < x < 1$.

Since $f(3/8)$ appears in the expression we wish to evaluate, we take $x = 3/8$ to get

$$f\left(\frac{3}{8}\right)f\left(\frac{9}{64}\right) = \frac{1}{1 - 3/8} = \frac{8}{5}.$$

Hence, we have $f^{-1}\left(\dfrac{8}{5f\left(\frac{3}{8}\right)}\right) = f^{-1}\left(f\left(\dfrac{9}{64}\right)\right) = \boxed{\dfrac{9}{64}}$.

10.85 We look at the mth partial sum (which is the sum of the first m terms) for the first few values of m, hoping to find a pattern

$$\sum_{n=0}^{0} \frac{x^{2^n}}{1 - x^{2^{n+1}}} = \frac{x}{1 - x^2},$$

$$\sum_{n=0}^{1} \frac{x^{2^n}}{1 - x^{2^{n+1}}} = \frac{x}{1 - x^2} + \frac{x^2}{1 - x^4} = \frac{x + x^2 + x^3}{1 - x^4},$$

$$\sum_{n=0}^{2} \frac{x^{2^n}}{1 - x^{2^{n+1}}} = \frac{x + x^2 + x^3}{1 - x^4} + \frac{x^4}{1 - x^8} = \frac{x + x^2 + x^3 + x^4 + x^5 + x^6 + x^7}{1 - x^8},$$

and so on. In general, it appears that

$$\sum_{n=0}^{m} \frac{x^{2^n}}{1 - x^{2^{n+1}}} = \frac{x + x^2 + \cdots + x^{2^{m+1}-1}}{1 - x^{2^{m+1}}} = \frac{x}{1 - x^{2^{m+1}}}(1 + x + \cdots + x^{2^{m+1}-2})$$

$$= \frac{x}{1 - x^{2^{m+1}}} \cdot \frac{1 - x^{2^{m+1}-1}}{1 - x} = \frac{x(1 - x^{2^{m+1}-1})}{(1 - x)(1 - x^{2^{m+1}})}.$$

We prove this formally using induction.

For the base case $m = 0$, our formula becomes

$$\frac{x(1 - x^{2^{m+1}-1})}{(1 - x)(1 - x^{2^{m+1}})} = \frac{x(1 - x)}{(1 - x)(1 - x^2)} = \frac{x}{1 - x^2},$$

so the formula holds for $m = 0$.

Now assume that the formula holds for some nonnegative integer $m = k$, so

$$\sum_{n=0}^{k} \frac{x^{2^n}}{1 - x^{2^{n+1}}} = \frac{x(1 - x^{2^{k+1}-1})}{(1 - x)(1 - x^{2^{k+1}})}.$$

Then

$$\sum_{n=0}^{k+1} \frac{x^{2^n}}{1 - x^{2^{n+1}}} = \sum_{n=0}^{k}\left(\frac{x^{2^n}}{1 - x^{2^{n+1}}}\right) + \frac{x^{2^{k+1}}}{1 - x^{2^{k+2}}} = \frac{x(1 - x^{2^{k+1}-1})}{(1 - x)(1 - x^{2^{k+1}})} + \frac{x^{2^{k+1}}}{1 - x^{2^{k+2}}}$$

$$= \frac{x(1 - x^{2^{k+1}-1})(1 + x^{2^{k+1}})}{(1 - x)(1 - x^{2^{k+1}})(1 + x^{2^{k+1}})} + \frac{(1 - x)x^{2^{k+1}}}{(1 - x)(1 - x^{2^{k+2}})} = \frac{x(1 + x^{2^{k+1}} - x^{2^{k+1}-1} - x^{2^{k+2}-1}) + x(x^{2^{k+1}-1} - x^{2^{k+1}})}{(1 - x)(1 - x^{2^{k+2}})}$$

$$= \frac{x(1 - x^{2^{k+2}-1})}{(1 - x)(1 - x^{2^{k+2}})}.$$

Hence, the formula holds for $m = k + 1$, and by induction, it holds for all nonnegative integers m.

Therefore,

$$\sum_{n=0}^{m} \frac{x^{2^n}}{1-x^{2^{n+1}}} = \frac{x(1-x^{2^{m+1}-1})}{(1-x)(1-x^{2^{m+1}})}.$$

Letting n go to infinity, since $0 < x < 1$, both $x^{2^{m+1}-1}$ and $x^{2^{m+1}}$ go to 0, so

$$\sum_{n=0}^{\infty} \frac{x^{2^n}}{1-x^{2^{n+1}}} = \boxed{\frac{x}{1-x}}.$$

Another way to approach this problem is to write each term $\dfrac{x^{2^n}}{1-x^{2^{n+1}}}$ as a geometric series:

$$\frac{x}{1-x^2} = x(1+x^2+x^4+x^6+\cdots),$$

$$\frac{x^2}{1-x^4} = x^2(1+x^4+x^8+x^{12}+\cdots),$$

$$\frac{x^4}{1-x^8} = x^4(1+x^8+x^{16}+x^{24}+\cdots),$$

$$\vdots$$

$$\frac{x^{2^n}}{1-x^{2^{n+1}}} = x^{2^n}(1+x^{2^{n+1}}+x^{2\cdot 2^{n+1}}+x^{3\cdot 2^{n+1}}+\cdots).$$

Our desired sum is the sum of all the right sides of this form. Note that the first right side, when expanded, gives all the odd powers of x. The second right side gives x^k for all k with a prime factorization that has exactly one 2. The third gives all the x^k such that k has exactly two 2s in its prime factorization, and so on. Therefore, adding the right sides gives a sum of x^k for all $k \geq 1$. This means the sum is $x(1+x+x^2+x^3+\cdots) = \frac{x}{1-x}$. (This second solution is not completely rigorous—see if you can perform the induction necessary to make it a rigorous proof.)

CHAPTER 11

Identities, Manipulations, and Induction

Exercises for Section 11.1

11.1.1 Expanding the left side, we get $a^2(b-c) + b^2(c-a) + c^2(a-b) = a^2b - a^2c + b^2c - ab^2 + ac^2 - bc^2$. Expanding the right side gives $bc(b-c) + ca(c-a) + ab(a-b) = b^2c - bc^2 + ac^2 - a^2c + a^2b - ab^2$. Thus, the two sides are equal.

11.1.2 Expanding the left side, we get $2(x-y)^2 - 3x^2 + 3xy = -x^2 - xy + 2y^2$. Expanding the right side, we get $(y-x)(x+2y) = -x^2 - xy + 2y^2$. Thus, the two sides are equal.

11.1.3 Each equation is of the form $a^2 + b^2 + c^2 = d^2$. Furthermore, in each equation, we see that $b = a + 1$, $c = ab = a(a+1) = a^2 + a$, and $d = c + 1 = a^2 + a + 1$. Hence, the identity that is suggested by these examples is

$$a^2 + (a+1)^2 + (a^2+a)^2 = (a^2+a+1)^2.$$

Expanding the left side, we get $a^2 + a^2 + 2a + 1 + a^4 + 2a^3 + a^2 = a^4 + 2a^3 + 3a^2 + 2a + 1$. Expanding the right side, we get $(a^2 + a + 1)^2 = a^4 + 2a^3 + 3a^2 + 2a + 1$. Thus, the two sides are equal, and the identity holds.

11.1.4 Expanding the given expression, we get $\left(x - \dfrac{1}{y}\right)\left(y - \dfrac{1}{z}\right)\left(z - \dfrac{1}{x}\right) = xyz - x - y - z + \dfrac{1}{x} + \dfrac{1}{y} + \dfrac{1}{z} - \dfrac{1}{xyz} =$
$8 - 8 + 8 - \dfrac{1}{8} = \boxed{\dfrac{63}{8}}$.

Exercises for Section 11.2

11.2.1 Let $r = a/b = c/d$, so $a = rb$ and $c = rd$.

(a) Then $\dfrac{a + kc}{b + kd} = \dfrac{rb + krd}{b + kd} = \dfrac{r(b + kd)}{b + kd} = r = \dfrac{a}{b} = \dfrac{c}{d}$.

(b) We have $\dfrac{a + kb}{a - kb} = \dfrac{rb + kb}{rb - kb} = \dfrac{b(r + k)}{b(r - k)} = \dfrac{r + k}{r - k}$, and $\dfrac{c + kd}{c - kd} = \dfrac{rd + kd}{rd - kd} = \dfrac{d(r + k)}{d(r - k)} = \dfrac{r + k}{r - k}$. Hence, we have
$\dfrac{a + kb}{a - kb} = \dfrac{r + k}{r - k} = \dfrac{c + kd}{c - kd}$.

11.2.2 Let $\frac{x_i}{y_i} = k$ for all i with $1 \le i \le n$. Then, we have $x_i = ky_i$, and we have

$$\frac{x_1 + x_2 + \cdots + x_n}{y_1 + y_2 + \cdots + y_n} = \frac{k(y_1 + y_2 + \cdots + y_n)}{y_1 + y_2 + \cdots + y_n} = k,$$

so $\frac{x_1+x_2+\cdots+x_n}{y_1+y_2+\cdots+y_n} = \frac{x_i}{y_i}$ for all i with $1 \le i \le n$, as desired.

11.2.3 $\boxed{\text{Yes}}$. If $\dfrac{a}{x+y-z} = \dfrac{b}{x-y+z} = \dfrac{c}{-x+y+z}$, then each ratio is equal to

$$\frac{a+b+c}{(x+y-z)+(x-y+z)+(-x+y+z)} = \frac{a+b+c}{x+y+z}.$$

11.2.4 We showed in the text that if $\dfrac{a}{b} = \dfrac{p}{q}$, then $\dfrac{a+b}{a-b} = \dfrac{p+q}{p-q}$. Therefore, if $\dfrac{x+y}{x-y} = \dfrac{c^2+d^2}{c^2-d^2}$, then $\dfrac{(x+y)+(x-y)}{(x+y)-(x-y)} = \dfrac{(c^2+d^2)+(c^2-d^2)}{(c^2+d^2)-(c^2-d^2)}$, which simplifies as $\dfrac{2x}{2y} = \dfrac{2c^2}{2d^2}$, so $\dfrac{x}{y} = \dfrac{c^2}{d^2}$.

11.2.5

(a) Since $x/y = y/z$, we have $x^2/y^2 = y^2/z^2$. Therefore, we have $\dfrac{x^2}{y^2} = \dfrac{y^2}{z^2} = \dfrac{x^2+y^2}{y^2+z^2}$.

(b) Let $r = x/y = y/z$, so $x = ry$ and $y = rz$. Therefore, we have

$$\frac{x^2+xy+y^2}{y^2+yz+z^2} = \frac{y^2(1+r+r^2)}{z^2(1+r+r^2)} = \frac{y^2}{z^2} = \left(\frac{y}{z}\right)^2 = r^2 = \frac{x}{y}\cdot\frac{y}{z} = \frac{x}{z},$$

as desired.

11.2.6 We are given that $\dfrac{a+b-c}{c} = \dfrac{a-b+c}{b} = \dfrac{-a+b+c}{a}$. Adding 1 to each fraction, we get $\dfrac{a+b}{c} = \dfrac{a+c}{b} = \dfrac{b+c}{a}$. Let r denote this common ratio, so $rc = a+b$, $rb = a+c$, and $ra = b+c$. Adding all these equations, we get

$$r(a+b+c) = 2a + 2b + 2c = 2(a+b+c),$$

so $(r-2)(a+b+c) = 0$. Therefore, $r = 2$ or $a+b+c = 0$. If $r = 2$, then

$$x = \frac{(a+b)(b+c)(c+a)}{abc} = \frac{a+b}{c}\cdot\frac{a+c}{b}\cdot\frac{b+c}{a} = r^3 = 8.$$

But we are given that $x < 0$, so r is not equal to 2. Therefore, $a+b+c = 0$, so

$$x = \frac{(a+b)(b+c)(c+a)}{abc} = \frac{a+b}{c}\cdot\frac{a+c}{b}\cdot\frac{b+c}{a} = \frac{-c}{c}\cdot\frac{-b}{b}\cdot\frac{-a}{a} = \boxed{-1}.$$

11.2.7 Let $m = x_1/x_2$. Then $m = x_1/x_2 < x_3/x_4 < x_5/x_6$, so $x_1 = mx_2$, $x_3 > mx_4$ and $x_5 > mx_6$. Therefore,

$$\frac{x_1+x_3+x_5}{x_2+x_4+x_6} > \frac{mx_2+mx_4+mx_6}{x_2+x_4+x_6} = \frac{m(x_2+x_4+x_6)}{x_2+x_4+x_6} = m = \frac{x_1}{x_2}.$$

Similarly, let $M = x_5/x_6$. Then $M = x_5/x_6 > x_3/x_4 > x_1/x_2$, so $x_5 = Mx_6$, $x_3 < Mx_4$, and $x_1 < Mx_2$. Therefore,

$$\frac{x_1+x_3+x_5}{x_2+x_4+x_6} < \frac{Mx_2+Mx_4+Mx_6}{x_2+x_4+x_6} = \frac{M(x_2+x_4+x_6)}{x_2+x_4+x_6} = M = \frac{x_5}{x_6}.$$

Exercises for Section 11.3

11.3.1

(a) First, we prove the base case. When $n = 1$, the left side is 1, and the right side is $\frac{1\cdot2}{2} = 1$, so the formula holds for $n = 1$. Now, assume that the formula holds for $n = k$ for some positive integer k, so

$$1 + 4 + 7 + 10 + \cdots + (3k-2) = \frac{k(3k-1)}{2}.$$

Adding $3k + 1$ to both sides, we get

$$1 + 4 + 7 + 10 + \cdots + (3k - 2) + (3k + 1) = \frac{k(3k - 1)}{2} + 3k + 1 = \frac{3k^2 - k + 6k + 2}{2}$$
$$= \frac{3k^2 + 5k + 2}{2} = \frac{(k + 1)(3k + 2)}{2},$$

which is equal to $\frac{n(3n-1)}{2}$ for $n = k + 1$. Hence, the formula holds for $n = k + 1$, and by induction, it holds for all positive integers n.

(b) First, we prove the base case. When $n = 1$, the left side is $1^2 = 1$, and the right side is $\frac{1 \cdot 2}{2} = 1$, so the formula holds for $n = 1$. Now, assume that the formula holds for $n = k$ for some positive integer k, so

$$1^2 + 4^2 + 7^2 + \cdots + (3k - 2)^2 = \frac{k(6k^2 - 3k - 1)}{2}.$$

Adding $(3k + 1)^2$ to both sides, we get

$$1^2 + 4^2 + 7^2 + \cdots + (3k - 2)^2 + (3k + 1)^2 = \frac{k(6k^2 - 3k - 1)}{2} + (3k + 1)^2 = \frac{6k^3 - 3k^2 - k}{2} + 9k^2 + 6k + 1$$
$$= \frac{6k^3 - 3k^2 - k}{2} + \frac{18k^2 + 12k + 2}{2} = \frac{6k^3 + 15k^2 + 11k + 2}{2}$$
$$= \frac{(k + 1)(6k^2 + 9k + 2)}{2} = \frac{(k + 1)[6(k + 1)^2 - 3(k + 1) - 1]}{2},$$

which is equal to $\frac{n(6n^2-3n-1)}{2}$ for $n = k + 1$. Hence, the formula holds for $n = k + 1$, and by induction, it holds for all positive integers n.

(c) First, we prove the base case. When $n = 1$, the left side is $2^2 = 4$, and the right side is $\frac{1 \cdot 8}{2} = 4$, so the formula holds for $n = 1$. Now, assume that the formula holds for $n = k$, where k is a positive integer, so

$$2^2 + 5^2 + 8^2 + \cdots + (3k - 1)^2 = \frac{k(6k^2 + 3k - 1)}{2}.$$

Adding $(3k + 2)^2$ to both sides, we get

$$2^2 + 5^2 + 8^2 + \cdots + (3k - 1)^2 + (3k + 2)^2 = \frac{k(6k^2 + 3k - 1)}{2} + (3k + 2)^2 = \frac{6k^3 + 3k^2 - k}{2} + 9k^2 + 12k + 4$$
$$= \frac{6k^3 + 3k^2 - k}{2} + \frac{18k^2 + 24k + 8}{2} = \frac{6k^3 + 21k^2 + 23k + 8}{2}$$
$$= \frac{(k + 1)(6k^2 + 15k + 8)}{2} = \frac{(k + 1)[6(k + 1)^2 + 3(k + 1) - 1]}{2},$$

which is equal to $\frac{n(6n^2+3n-1)}{2}$ for $n = k + 1$. Hence, the formula holds for $n = k + 1$, and by induction, it holds for all positive integers n.

11.3.2 We wish to prove that $\dfrac{1 \cdot 2}{2} + \dfrac{2 \cdot 3}{2} + \cdots + \dfrac{n(n + 1)}{2} = \dfrac{n(n + 1)(n + 2)}{6}$ for all positive integers n.

First, we prove the base case. For $n = 1$, the left side is $\frac{1 \cdot 2}{2} = 1$, and the right side is $\frac{1 \cdot 2 \cdot 3}{6} = 1$, so the formula holds for $n = 1$.

Now, assume that the formula holds for $n = k$ for some positive integer k, so

$$\frac{1 \cdot 2}{2} + \frac{2 \cdot 3}{2} + \cdots + \frac{k(k + 1)}{2} = \frac{k(k + 1)(k + 2)}{6}.$$

Adding $\frac{(k+1)(k+2)}{2}$ to both sides, we get

$$\frac{1 \cdot 2}{2} + \frac{2 \cdot 3}{2} + \cdots + \frac{k(k+1)}{2} + \frac{(k+1)(k+2)}{2} = \frac{k(k+1)(k+2)}{6} + \frac{(k+1)(k+2)}{2} = \frac{k(k+1)(k+2) + 3(k+1)(k+2)}{6}$$

$$= \frac{(k+1)(k+2)(k+3)}{6} = \frac{(k+1)[(k+1)+1][(k+1)+2]}{6},$$

which is equal to $\frac{n(n+1)(n+2)}{6}$ for $n = k + 1$. Hence, the formula holds for $n = k + 1$, and by induction, it holds for all positive integers n.

11.3.3 The base case is trivial, because for $n = 1$, the given equation becomes $|z^1| = |z|^1$, which is clearly true. Assume that the equation holds for $n = k$ for some positive integer k, so

$$|z^k| = |z|^k.$$

Now, we must prove that $|z^{k+1}| = |z|^{k+1}$. We know that $|ab| = |a||b|$ for any complex numbers a and b. Taking $a = z^k$ and $b = z$, we get

$$|z^{k+1}| = |z^k||z|.$$

By the inductive assumption, we have $|z^k| = |z|^k$, so $|z^k||z| = |z|^k \cdot |z| = |z|^{k+1}$. Therefore, $|z^{k+1}| = |z|^{k+1}$. Hence, the equation holds for $n = k + 1$, and by induction, it holds for all positive integers n.

11.3.4 First, we prove the base case. When $n = 1$, the left side is $\frac{1}{1 \cdot 3} = \frac{1}{3}$, and the right side is $\frac{1}{2 \cdot 1 + 1} = \frac{1}{3}$, so the formula holds for $n = 1$. Now, assume that the formula holds for $n = k$ for some positive integer k, so

$$\frac{1}{1 \cdot 3} + \frac{1}{3 \cdot 5} + \cdots + \frac{1}{(2k-1)(2k+1)} = \frac{k}{2k+1}.$$

Adding $\frac{1}{(2k+1)(2k+3)}$ to both sides, we get

$$\frac{1}{1 \cdot 3} + \frac{1}{3 \cdot 5} + \cdots + \frac{1}{(2k-1)(2k+1)} + \frac{1}{(2k+1)(2k+3)} = \frac{k}{2k+1} + \frac{1}{(2k+1)(2k+3)} = \frac{k(2k+3)}{(2k+1)(2k+3)} + \frac{1}{(2k+1)(2k+3)}$$

$$= \frac{k(2k+3) + 1}{(2k+1)(2k+3)} = \frac{2k^2 + 3k + 1}{(2k+1)(2k+3)} = \frac{(k+1)(2k+1)}{(2k+1)(2k+3)}$$

$$= \frac{k+1}{2k+3} = \frac{k+1}{2(k+1)+1},$$

which is equal to $\frac{n}{2n+1}$ for $n = k + 1$. Hence, the formula holds for $n = k + 1$, and by induction, it holds for all positive integers n.

11.3.5 First, we prove the base case. For $n = 1$, the left side is 1, and the right side is $2\sqrt{1} = 2$, so the inequality holds for $n = 1$. Now, assume that the inequality holds for $n = k$ for some positive integer k, so

$$1 + \frac{1}{\sqrt{2}} + \cdots + \frac{1}{\sqrt{k}} < 2\sqrt{k}.$$

Adding $\frac{1}{\sqrt{k+1}}$ to both sides, we get $1 + \frac{1}{\sqrt{2}} + \cdots + \frac{1}{\sqrt{k}} + \frac{1}{\sqrt{k+1}} < 2\sqrt{k} + \frac{1}{\sqrt{k+1}}$. Hence, to prove the given inequality for $n = k + 1$, it suffices to prove that

$$2\sqrt{k} + \frac{1}{\sqrt{k+1}} < 2\sqrt{k+1}.$$

Multiplying both sides by $\sqrt{k+1}$, we get $2\sqrt{k(k+1)}+1 < 2(k+1) = 2k+2$, which simplifies as $2\sqrt{k(k+1)} < 2k+1$. Squaring both sides, we get $4k^2 + 4k < 4k^2 + 4k + 1$. This inequality is true. Furthermore, all our steps are reversible; thus, the inequality

$$2\sqrt{k} + \frac{1}{\sqrt{k+1}} < 2\sqrt{k+1}$$

is proven. Hence, the given inequality holds for $n = k + 1$, and by induction, it holds for all positive integers n.

11.3.6 The first few terms of the sequence are

$$x_1 = 2,$$
$$x_2 = 2^2 - 2 + 1 = 3,$$
$$x_3 = 3^2 - 3 + 1 = 7,$$
$$x_4 = 7^2 - 7 + 1 = 43,$$

and so on.

To find the infinite sum, we look at the partial sums. Let $S_n = \sum_{k=1}^{n} \frac{1}{x_k}$. Then

$$S_1 = \frac{1}{2},$$
$$S_2 = \frac{1}{2} + \frac{1}{3} = \frac{5}{6},$$
$$S_3 = \frac{1}{2} + \frac{1}{3} + \frac{1}{7} = \frac{41}{42},$$

and so on. In these partial sums, it appears that

$$S_n = \frac{x_{n+1} - 2}{x_{n+1} - 1} = \frac{(x_{n+1} - 1) - 1}{x_{n+1} - 1} = 1 - \frac{1}{x_{n+1} - 1}.$$

We prove this formula using induction.

First, we prove the base case. When $n = 1$, the left side is $S_1 = \frac{1}{2}$, and the right side is $1 - \frac{1}{3-1} = 1 - \frac{1}{2} = \frac{1}{2}$, so the formula holds for $n = 1$. Now, assume that the formula holds for $n = j$ for some positive integer j, so

$$\frac{1}{x_1} + \frac{1}{x_2} + \cdots + \frac{1}{x_j} = 1 - \frac{1}{x_{j+1} - 1}.$$

Adding $\frac{1}{x_{j+1}}$ to both sides, we get

$$\frac{1}{x_1} + \frac{1}{x_2} + \cdots + \frac{1}{x_j} + \frac{1}{x_{j+1}} = 1 - \frac{1}{x_{j+1} - 1} + \frac{1}{x_{j+1}} = 1 - \frac{x_{j+1}}{x_{j+1}(x_{j+1} - 1)} + \frac{x_{j+1} - 1}{x_{j+1}(x_{j+1} - 1)}$$
$$= 1 - \frac{1}{x_{j+1}^2 - x_{j+1}}.$$

From our definition of the sequence $\{x_i\}$, we have $x_{j+2} = x_{j+1}^2 - x_{j+1} + 1$, so $x_{j+1}^2 - x_{j+1} = x_{j+2} - 1$. Substituting this into our expression above gives us

$$\frac{1}{x_1} + \frac{1}{x_2} + \cdots + \frac{1}{x_j} + \frac{1}{x_{j+1}} = 1 - \frac{1}{x_{j+1}^2 - x_{j+1}} = 1 - \frac{1}{x_{j+2} - 1}.$$

Hence, the formula holds for $n = j + 1$, and by induction, it holds for all positive integers n.

Now to compute the infinite sum, we must analyze the behavior of S_n as n goes to infinity. It is clear that as n goes to infinity, x_n goes to infinity, so $1/(x_{n+1}-1)$ goes to 0. Therefore, letting n go to infinity in the formula

$$\sum_{k=1}^{n}\frac{1}{x_k}=1-\frac{1}{x_{n+1}-1},$$

we get $\sum_{k=1}^{\infty}\frac{1}{x_k}=\boxed{1}$.

Exercises for Section 11.4

11.4.1

(a) $(1-x)^5=\binom{5}{5}-\binom{5}{4}x+\binom{5}{3}x^2-\binom{5}{2}x^3+\binom{5}{1}x^4-\binom{5}{0}x^5=\boxed{1-5x+10x^2-10x^3+5x^4-x^5}$.

(b)

$$(2x^2+3y)^4=\binom{4}{4}(2x^2)^4+\binom{4}{3}(2x^2)^3(3y)+\binom{4}{2}(2x^2)^2(3y)^2+\binom{4}{1}(2x^2)(3y)^3+\binom{4}{0}(3y)^4$$
$$=\boxed{16x^8+96x^6y+216x^4y^2+216x^2y^3+81y^4}.$$

11.4.2 Each term in the expansion of $(x+\frac{2}{x^2})^9$ is of the form

$$\binom{9}{k}x^k\left(\frac{2}{x^2}\right)^{9-k}=\binom{9}{k}2^{9-k}\frac{x^k}{x^{18-2k}}=\binom{9}{k}2^{9-k}x^{3k-18}.$$

This term is constant if and only if $3k-18=0$, so $k=6$. Hence, the constant term is $\binom{9}{6}2^{9-6}=\boxed{672}$.

11.4.3 Since $64=2^6$ and 31 and 33 are both 1 away from $32=2^5$, we write $31^{19}+33^{99}$ as $(32-1)^{19}+(32+1)^{99}$. Expanding both of these powers of binomials with the Binomial Theorem gives us

$$(32-1)^{19}=\binom{19}{19}32^{19}-\binom{19}{18}32^{18}+\binom{19}{17}32^{17}-\cdots-\binom{19}{2}32^2+\binom{19}{1}32^1-1,$$
$$(32+1)^{99}=\binom{99}{99}32^{99}+\binom{99}{98}32^{98}+\binom{99}{97}32^{97}+\cdots+\binom{99}{2}32^2+\binom{99}{1}32^1+1.$$

When we add these two, the -1 and $+1$ cancel. Except for $19\cdot 32$ and $99\cdot 32$, all the other terms are divisible by 32^2, so every term except $19\cdot 32$ and $99\cdot 32$ is divisible by 128. Therefore, we can write $31^{19}+33^{99}=128k+19\cdot32+99\cdot32$ for some integer k. Simplifying the right side gives us $128k+118\cdot32$, or $128k+59\cdot64$. Dividing this expression by 64 leaves an integer, namely $2k+59$, but dividing it by 128 does not give an integer. Therefore, $31^{19}+33^{99}$ is divisible by 64, but not divisible by 128.

11.4.4

(a) By the Binomial Theorem, we have $(1+x)^{10}=\binom{10}{0}+\binom{10}{1}x+\binom{10}{2}x^2+\cdots+\binom{10}{10}x^{10}$. Taking $x=1$, we get

$$\binom{10}{0}+\binom{10}{1}+\binom{10}{2}+\cdots+\binom{10}{10}=(1+1)^{10}=2^{10}=\boxed{1024}.$$

(b) We want to find the sum

$$\binom{99}{0} + \binom{99}{2} + \binom{99}{4} + \cdots + \binom{99}{98}.$$

This is like the sum in part (a), except only every other term is present in the binomial expansion. We can begin with the formula

$$(1 + x)^{99} = \binom{99}{0} + \binom{99}{1}x + \binom{99}{2}x^2 + \cdots + \binom{99}{99}x^{99}.$$

Taking $x = 1$, as in part (a), we get

$$\binom{99}{0} + \binom{99}{1} + \binom{99}{2} + \cdots + \binom{99}{99} = (1 + 1)^{99} = 2^{99}.$$

To cancel every other term, we can take $x = -1$, to get

$$\binom{99}{0} - \binom{99}{1} + \binom{99}{2} - \cdots - \binom{99}{99} = (1 - 1)^{99} = 0.$$

Adding these two equations, we get $2\binom{99}{0} + 2\binom{99}{2} + 2\binom{99}{4} + \cdots + 2\binom{99}{98} = 2^{99}$, so $\binom{99}{0} + \binom{99}{2} + \binom{99}{4} + \cdots + \binom{99}{98} =$ $\boxed{2^{98}}$.

11.4.5 By the Binomial Theorem, we have

$$(1 + x)^n = \binom{n}{0} + \binom{n}{1}x + \binom{n}{2}x^2 + \cdots + \binom{n}{n}x^n = 1 + nx + \binom{n}{2}x^2 + \cdots + x^n.$$

Since every term in the expansion of the Binomial Theorem is nonnegative, we have

$$1 + nx + \binom{n}{2}x^2 + \cdots + x^n \geq 1 + nx.$$

Therefore, we have $(1 + x)^n \geq 1 + nx$.

11.4.6 Adding 1 to both sides of the equation, we get $x^4 - 4x^3 + 6x^2 - 4x + 1 = 2006$, so $(x - 1)^4 = 2006$. Hence, using difference of squares, we have

$$[(x - 1)^2 - \sqrt{2006}][(x - 1)^2 + \sqrt{2006}] = 0.$$

The first factor has two real roots, namely $1 \pm \sqrt[4]{2006}$. The second factor has nonreal roots (since $(x - 1)^2 \geq 0$ for all real numbers x), and expands as $x^2 - 2x + 1 + \sqrt{2006} = 0$. Then by Vieta's Formulas, the product of these nonreal roots is $P = 1 + \sqrt{2006}$, so $\lfloor P \rfloor = \lfloor 1 + \sqrt{2006} \rfloor = 1 + \lfloor \sqrt{2006} \rfloor$. Since $44^2 = 1936$ and $45^2 = 2025$, we have $44 < \sqrt{2006} < 45$. Therefore, $\lfloor P \rfloor = 1 + \lfloor \sqrt{2006} \rfloor = 1 + 44 = \boxed{45}$.

Review Problems

11.23

(a) The left side expands as $ab - ac + bc - ab + ac - bc = 0$. Hence, the given equation is $\boxed{\text{an identity}}$.

(b) The left side expands as

$$(x - y)^4 + (x + y)^4 = (x^4 - 4x^3y + 6x^2y^2 - 4xy^3 + y^4) + (x^4 + 4x^3y + 6x^2y^2 + 4xy^3 + y^4) = 2x^4 + 12x^2y^2 + 2y^4,$$

and the right side expands as $2(x^2 + y^2)^2 + 8x^2y^2 = 2(x^4 + 2x^2y^2 + y^4) + 8x^2y^2 = 2x^4 + 12x^2y^2 + 2y^4$. Hence, the given equation is $\boxed{\text{an identity}}$.

(c)   The given equation is $\boxed{\text{not an identity}}$. For example, take $a = 1$ and $b = 4$. Then the left side is $a/b = 1/4$, and the right side is $(a - \sqrt{a})/(b - \sqrt{b}) = (1 - 1)/(4 - 2) = 0/2 = 0$.

(d)   The given equation is $\boxed{\text{not an identity}}$. For example, take $x = y = 1$. Then the left side is $1^2 \cdot (1-1) + 1^2 \cdot (1-1) = 0$, and the right side is $1^3 + 1^3 = 2$.

11.24   We have $\dfrac{1}{xy} + \dfrac{1}{xz} + \dfrac{1}{yz} = \dfrac{x + y + z}{xyz} = \dfrac{x + y + z}{1/18}$, so we have $\dfrac{x + y + z}{1/18} = 12$, which means $x + y + z = \dfrac{12}{18} = \boxed{\dfrac{2}{3}}$.

11.25   Let $s = x + y + z$. We have $\dfrac{x}{3 - x} = \dfrac{y}{5 - y} = \dfrac{z}{16 - z} = \dfrac{x + y + z}{(3 - x) + (5 - y) + (16 - z)} = \dfrac{s}{24 - s}$, so $\dfrac{s}{24 - s} = 2$, which means $s = 48 - 2s$. Then $3s = 48$, so $s = x + y + z = \boxed{16}$.

11.26   Let $r = \frac{a}{b} = \frac{c}{d} = \frac{e}{f}$, so r is positive and $a = rb$, $c = rd$, and $e = rf$. Then

$$\frac{\sqrt[n]{a^n + c^n + e^n}}{\sqrt[n]{b^n + d^n + f^n}} = \frac{\sqrt[n]{r^n b^n + r^n d^n + r^n f^n}}{\sqrt[n]{b^n + d^n + f^n}} = \frac{\sqrt[n]{r^n(b^n + d^n + f^n)}}{\sqrt[n]{b^n + d^n + f^n}} = \frac{r\sqrt[n]{b^n + d^n + f^n}}{\sqrt[n]{b^n + d^n + f^n}} = r.$$

But $r = a/b$, so $\dfrac{a}{b} = \dfrac{\sqrt[n]{a^n + c^n + e^n}}{\sqrt[n]{b^n + d^n + f^n}}$.

11.27   In the text, we showed that if $\dfrac{a}{b} = \dfrac{c}{d}$, then $\dfrac{a + b}{a - b} = \dfrac{c + d}{c - d}$. Therefore, since $\dfrac{p + q - r}{p - q + r} = \dfrac{u + v}{u - v}$, we have $\dfrac{(p + q - r) + (p - q + r)}{(p + q - r) - (p - q + r)} = \dfrac{(u + v) + (u - v)}{(u + v) - (u - v)}$, which simplifies as $\dfrac{2p}{2q - 2r} = \dfrac{2u}{2v}$, so $\dfrac{p}{q - r} = \dfrac{u}{v}$. Cross-multiplying, we get $pv = u(q - r)$.

11.28   Since $\dfrac{a}{b} = \dfrac{x}{y}$, we have $\dfrac{a}{x} = \dfrac{b}{y}$. Let $r = \dfrac{a}{x} = \dfrac{b}{y}$, so $a = rx$ and $b = ry$. Then

$$\frac{a - x}{\sqrt{a^2 + x^2}} = \frac{rx - x}{\sqrt{r^2 x^2 + x^2}} = \frac{x(r - 1)}{\sqrt{x^2(r^2 + 1)}} = \frac{x(r - 1)}{x\sqrt{r^2 + 1}} = \frac{r - 1}{\sqrt{r^2 + 1}},$$

and

$$\frac{b - y}{\sqrt{b^2 + y^2}} = \frac{ry - y}{\sqrt{r^2 y^2 + y^2}} = \frac{y(r - 1)}{\sqrt{y^2(r^2 + 1)}} = \frac{y(r - 1)}{y\sqrt{r^2 + 1}} = \frac{r - 1}{\sqrt{r^2 + 1}}.$$

Hence, we have $\dfrac{a - x}{\sqrt{a^2 + x^2}} = \dfrac{b - y}{\sqrt{b^2 + y^2}}$.

11.29

(a)   First, we prove the base case. When $n = 1$, the left side is 1, and the right side is $1^2 = 1$, so the formula holds for $n = 1$. Now, assume that the formula holds for $n = k$ for some positive integer k, so

$$1 + 3 + 5 + \cdots + (2k - 1) = k^2.$$

Adding $2k + 1$ to both sides, we get

$$1 + 3 + 5 + \cdots + (2k - 1) + (2k + 1) = k^2 + (2k + 1) = k^2 + 2k + 1 = (k + 1)^2,$$

which is equal to n^2 for $n = k + 1$. Hence, the formula holds for $n = k + 1$, and by induction it holds for all positive integers n.

(b) First, we prove the base case. When $n = 1$, the left side is $1^2 = 1$, and the right side is $\frac{1 \cdot 2 \cdot 3}{6} = 1$, so the formula holds for $n = 1$. Now, assume that the formula holds for $n = k$ for some positive integer k, so

$$1^2 + 2^2 + \cdots + k^2 = \frac{k(k+1)(2k+1)}{6}.$$

Adding $(k+1)^2$ to both sides, we get

$$1 + 4 + 9 + \cdots + k^2 + (k+1)^2 = \frac{k(k+1)(2k+1)}{6} + (k+1)^2 = \frac{k(k+1)(2k+1) + 6(k+1)^2}{6}$$
$$= \frac{(k+1)[k(2k+1) + 6(k+1)]}{6} = \frac{(k+1)(2k^2 + k + 6k + 6)}{6} = \frac{(k+1)(2k^2 + 7k + 6)}{6}$$
$$= \frac{(k+1)(k+2)(2k+3)}{6} = \frac{(k+1)[(k+1)+1][2(k+1)+1]}{6},$$

which is equal to $\frac{n(n+1)(2n+1)}{6}$ for $n = k+1$. Hence, the formula holds for $n = k+1$, and by induction it holds for all positive integers n.

11.30 We use induction. First, we prove the base case. When $n = 1$, the left side is $0/1! = 0$, and the right side is $1 - 1/1! = 0$, so the formula holds for $n = 1$. Now, assume that the formula holds for $n = k$ for some positive integer k, so

$$\frac{0}{1!} + \frac{1}{2!} + \cdots + \frac{k-1}{k!} = 1 - \frac{1}{k!}.$$

Adding $k/(k+1)!$ to both sides, we get

$$\frac{0}{1!} + \frac{1}{2!} + \frac{2}{3!} + \cdots + \frac{k-1}{k!} + \frac{k}{(k+1)!} = 1 - \frac{1}{k!} + \frac{k}{(k+1)!} = 1 - \frac{k+1}{(k+1)!} + \frac{k}{(k+1)!} = 1 - \frac{1}{(k+1)!},$$

which is equal to $1 - \frac{1}{n!}$ for $n = k+1$. Hence, the formula holds for $n = k+1$, and by induction it holds for all positive integers n.

11.31 *Solution 1: Brute force.* We "brute force" this problem by examining what happens when we multiply out the right side.

First, we examine terms of the form $a_i b_i$. For all k with $1 \le k \le n-1$, there is a product of the form $(a_{k-1} - a_k)(b_1 + \cdots + b_{k-1})$ on the right side. The only term of the form $a_i b_i$ in the expansion of such a product is $a_{k-1} b_{k-1}$. So, the $(a_{k-1} - a_k)(b_1 + \cdots + b_{k-1})$ produce all $a_i b_i$ terms with $1 \le i \le n-1$, where each term has coefficient 1. The final $a_n b_n$ is produced by the last term on the right, $a_n(b_1 + \cdots + b_n)$, and this term doesn't produce any other $a_i b_i$.

Next, we turn to $a_i b_j$ with $i \ne j$. In each of the products on the right, the subscripts of the a_i equal or exceed those of the b_i. Therefore, there are no terms $a_i b_j$ on the right with $i < j$. If $n > i > j$, there are two terms $a_i b_j$, one for $(a_i - a_{i+1})(b_1 + \cdots b_i)$ and another for $(a_{i-1} - a_i)(b_1 + \cdots b_{i-1})$, and these have opposite signs, so they cancel. There are no other a_i terms with this subscript in any other product, so there are no terms $a_i b_j$ with $n > i > j$ on the right. Finally, if $i = n$ and $j < n$, the $a_i b_j$ term in $a_n(b_1 + \cdots + b_n)$ cancels with the $a_i b_j$ term in $(a_{n-1} - a_n)(b_1 + \cdots + b_{n-1})$.

Therefore, when we expand the right side, the result is $a_1 b_1 + a_2 b_2 + \cdots + a_n b_n$, as desired.

Solution 2: Induction. Let $S_n = a_1 b_1 + a_2 b_2 + \cdots + a_n b_n$, and let

$$T_n = (a_1 - a_2)b_1 + (a_2 - a_3)(b_1 + b_2) + \cdots + (a_3 - a_4)(b_1 + b_2 + b_3) + \cdots$$
$$+ (a_{n-1} - a_n)(b_1 + b_2 + \cdots + b_{n-1}) + a_n(b_1 + b_2 + \cdots + b_n).$$

To show that $S_n = T_n$ for all $n \ge 1$, we use induction. The base case is trivial, because for $n = 1$, we have $S_1 = T_1 = a_1 b_1$. Now, assume that $S_k = T_k$ for some positive integer k. To prove that $S_{k+1} = T_{k+1}$, it suffices to prove that $S_{k+1} - S_k = T_{k+1} - T_k$.

We see that $S_{k+1} - S_k = a_{k+1}b_{k+1}$. The difference between

$$T_k = (a_1 - a_2)b_1 + (a_2 - a_3)(b_1 + b_2) + \cdots + (a_3 - a_4)(b_1 + b_2 + b_3) + \cdots$$
$$+ (a_{k-1} - a_k)(b_1 + b_2 + \cdots + b_{k-1}) + a_k(b_1 + b_2 + \cdots + b_k)$$

and

$$T_{k+1} = (a_1 - a_2)b_1 + (a_2 - a_3)(b_1 + b_2) + \cdots + (a_3 - a_4)(b_1 + b_2 + b_3) + \cdots$$
$$+ (a_{k-1} - a_k)(b_1 + b_2 + \cdots + b_{k-1}) + (a_k - a_{k+1})(b_1 + b_2 + \cdots + b_{k-1} + b_k)$$
$$+ a_{k+1}(b_1 + b_2 + \cdots + b_k + b_{k+1})$$

is

$$T_{k+1} - T_k = (a_k - a_{k+1})(b_1 + b_2 + \cdots + b_{k-1} + b_k) + a_{k+1}(b_1 + b_2 + \cdots + b_k + b_{k+1}) - a_k(b_1 + b_2 + \cdots + b_k)$$
$$= a_k(b_1 + b_2 + \cdots + b_k) - a_{k+1}(b_1 + b_2 + \cdots + b_k) + a_{k+1}(b_1 + b_2 + \cdots + b_{k+1})$$
$$- a_k(b_1 + b_2 + \cdots + b_k)$$
$$= a_{k+1}b_{k+1}.$$

Hence, $S_{k+1} - S_k = T_{k+1} - T_k$, so $S_k = T_k$. By induction, $S_n = T_n$ for all $n \geq 1$.

11.32 Let $x = 2000$ and $y = 3$. Then we can write

$$2003^4 - 1997^4 = (2000 + 3)^4 - (2000 - 3)^4 = (x + y)^4 - (x - y)^4$$
$$= (x^4 + 4x^3y + 6xy + 4xy^3 + y^4) - (x^4 - 4x^3y + 6xy - 4xy^3 + y^4)$$
$$= 8x^3y + 8xy^3 = 8xy(x^2 + y^2) = 8 \cdot 2000 \cdot 3 \cdot (2000^2 + 3^2)$$
$$= 48000 \cdot 4000009 = 192000432000.$$

Therefore, the sum of the digits is $1 + 9 + 2 + 4 + 3 + 2 = \boxed{21}$.

11.33 Note that the first term on the left side of each equation is a square. In fact, the first term in the nth equation is n^2. Furthermore, there are $n + 1$ terms on the left side, so the left side is

$$n^2 + (n^2 + 1) + (n^2 + 2) + \cdots + (n^2 + n).$$

The first term on the right side is equal to $(n^2 + n) + 1 = n^2 + n + 1$, and there are n terms, so the right side is

$$(n^2 + n + 1) + (n^2 + n + 2) + \cdots + (n^2 + 2n).$$

Therefore, the identity that is suggested by these examples is

$$n^2 + (n^2 + 1) + (n^2 + 2) + \cdots + (n^2 + n) = (n^2 + n + 1) + (n^2 + n + 2) + \cdots + (n^2 + 2n).$$

These are both arithmetic series with common difference 1. The left side has $n + 1$ terms and the right side has n terms. Summing the left side gives

$$n^2 + (n^2 + 1) + (n^2 + 2) + \cdots + (n^2 + n) = (n + 1) \cdot \frac{n^2 + (n^2 + n)}{2} = \frac{(n+1)(2n^2 + n)}{2} = \frac{2n^3 + 3n^2 + n}{2}.$$

Summing the right side gives

$$(n^2 + n + 1) + (n^2 + n + 2) + \cdots + (n^2 + 2n) = n \cdot \frac{(n^2 + n + 1) + (n^2 + 2n)}{2} = \frac{2n^3 + 3n^2 + n}{2}.$$

Therefore, the two sides are equal, and we have proved the identity suggested by the equations in the problem.

11.34   The left side expands as

$$2x^4 - x^2y^2 - x^2z^2 + 2y^4 - x^2y^2 - y^2z^2 + 2z^4 - x^2z^2 - y^2z^2 = 2x^4 + 2y^4 + 2z^4 - 2x^2y^2 - 2x^2z^2 - 2y^2z^2,$$

and the right side expands as

$$(x^2 - y^2)^2 + (y^2 - z^2)^2 + (z^2 - x^2)^2 = x^4 - 2x^2y^2 + y^4 + y^4 - 2y^2z^2 + z^4 + z^4 - 2x^2z^2 + x^4 = 2x^4 + 2y^4 + 2z^4 - 2x^2y^2 - 2x^2z^2 - 2y^2z^2.$$

Thus, the two sides are equal, so the identity holds.

11.35   The given expression factors as $[x(2x^2 - 1)]^8 = x^8(2x^2 - 1)^8$, so the coefficient of x^{14} in $(2x^3 - x)^8$ is equal to the coefficient of x^6 in $(2x^2 - 1)^8$. We get an x^6 term in this expansion when $2x^2$ is cubed. This term in the expansion is $\binom{8}{3}(2x^2)^3(-1)^5 = -448x^6$, so the desired coefficient is $\boxed{-448}$.

11.36   We use induction. First, we prove the base case. When $n = 2$, the left side is $\frac{2^3-1}{2^3+1} = \frac{7}{9}$, and the right side is $\frac{2 \cdot 7}{3 \cdot 2 \cdot 3} = \frac{7}{9}$, so the formula holds for $n = 2$. Now, assume that the formula holds for $n = k$ for some positive integer $k \geq 2$, so

$$\frac{2^3 - 1}{2^3 + 1} \cdot \frac{3^3 - 1}{3^3 + 1} \cdots \frac{k^3 - 1}{k^3 + 1} = \frac{2(k^2 + k + 1)}{3k(k + 1)}.$$

Multiplying both sides by $\frac{(k+1)^3-1}{(k+1)^3+1}$, we get

$$\frac{2^3 - 1}{2^3 + 1} \cdot \frac{3^3 - 1}{3^3 + 1} \cdots \frac{k^3 - 1}{k^3 + 1} \cdot \frac{(k + 1)^3 - 1}{(k + 1)^3 + 1} = \frac{2(k^2 + k + 1)}{3k(k + 1)} \cdot \frac{(k + 1)^3 - 1}{(k + 1)^3 + 1}$$

$$= \frac{2(k^2 + k + 1)}{3k(k + 1)} \cdot \frac{k^3 + 3k^2 + 3k + 1 - 1}{[(k + 1) + 1][(k + 1)^2 - (k + 1) + 1]}$$

$$= \frac{2(k^2 + k + 1)}{3k(k + 1)} \cdot \frac{k(k^2 + 3k + 3)}{(k + 2)(k^2 + k + 1)} = \frac{2(k^2 + 3k + 3)}{3(k + 1)(k + 2)}$$

$$= \frac{2[(k + 1)^2 + (k + 1) + 1]}{3(k + 1)[(k + 1) + 1]},$$

which is equal to $\frac{2(n^2+n+1)}{3n(n+1)}$ for $n = k + 1$. Hence, the formula holds for $n = k + 1$, and by induction, it holds for all positive integers n.

11.37

(a)   Expanding the right side, we get

$$\left(x + \frac{1}{x}\right)\left(x^n + \frac{1}{x^n}\right) - \left(x^{n-1} + \frac{1}{x^{n-1}}\right) = x^{n+1} + \frac{1}{x^{n-1}} + x^{n-1} + \frac{1}{x^{n+1}} - \left(x^{n-1} + \frac{1}{x^{n-1}}\right) = x^{n+1} + \frac{1}{x^{n+1}},$$

which is the left side.

(b)   Let $a_n = x^n + \frac{1}{x^n}$ for all $n \geq 0$. Then $a_0 = x^0 + \frac{1}{x^0} = 1 + \frac{1}{1} = 2$ and $a_1 = a$, which are both integers. These values give us the base cases of an induction argument.

Assume that a_n is an integer for $n = k - 1$ and $n = k$, for some positive integer $k \geq 1$. Then by part (a),

$$a_{k+1} = x^{k+1} + \frac{1}{x^{k+1}} = \left(x + \frac{1}{x}\right)\left(x^k + \frac{1}{x^k}\right) - \left(x^{k-1} + \frac{1}{x^{k-1}}\right) = aa_k - a_{k-1}.$$

Since a, a_k, and a_{k-1} are all integers, a_{k+1} is also an integer. Hence, $x^{k+1} + \frac{1}{x^{k+1}}$ is an integer, which completes the induction. Therefore, $x^n + \frac{1}{x^n}$ is an integer for all $n \geq 0$.

11.38 We can write make the expression easier to work with by letting $n = 2004$, and writing each number in terms of n:

$$2005 \cdot 2007^3 - 2006 \cdot 2004^3 = (n+1)(n+3)^3 - (n+2)n^3 = (n+1)(n^3 + 9n^2 + 27n + 27) - (n^4 + 2n^3)$$
$$= n^4 + 10n^3 + 36n^2 + 54n + 27 - (n^4 + 2n^3) = 8n^3 + 36n^2 + 54n + 27 = (2n+3)^3.$$

Thus, the given number is a cube.

11.39

(a) In the solution of Problem 11.4.5, we used the inequality $1 + nx + \binom{n}{2}x^2 + \cdots + x^n \geq 1 + nx$. We knew this was true because $x \geq 0$. However, if x is allowed to be negative, then we can no longer use this argument.

(b) We prove the inequality using induction on n. The base case is trivial, because for $n = 1$, both sides become $1 + x$. Now, assume that the inequality holds for some positive integer $n = k$, so $(1 + x)^k \geq 1 + kx$ for all $x \geq -1$.

Since $x \geq -1$, we have $1 + x \geq 0$. Hence, we can multiply both sides of $(1 + x)^k \geq 1 + kx$ by $1 + x$ (which preserves the direction of the inequality), to get

$$(1 + x)^{k+1} \geq (1 + kx)(1 + x) = 1 + kx + x + kx^2 = 1 + (k+1)x + x^2 \geq 1 + (k+1)x.$$

Thus, the inequality holds for $n = k+1$, and by induction, it holds for all positive integers n. This inequality is called Bernoulli's Inequality.

11.40 *Solution 1.* We would like to simplify the ratios. In the text, we showed that if $\frac{a}{b} = \frac{c}{d}$, then $\frac{a}{b} = \frac{a-c}{b-d}$. Applying this to the ratios in the problem, we have

$$\frac{x^2 - 3x + 4}{x^2 - 5x + 5} = \frac{x^2 - 3x + 9}{x^2 - 5x + 10} = \frac{(x^2 - 3x + 4) - (x^2 - 3x + 9)}{(x^2 - 5x + 5) - (x^2 - 5x + 10)}.$$

The last ratio simplifies to 1. Therefore, we must have $x^2 - 3x + 4 = x^2 - 5x + 5$, which gives us $x = \boxed{1/2}$.

Solution 2. We can rewrite the given equation as $\dfrac{x^2 - 3x + 9}{x^2 - 3x + 4} = \dfrac{x^2 - 5x + 10}{x^2 - 5x + 5}$. Then, we have

$$\frac{(x^2 - 3x + 9) + (x^2 - 3x + 4)}{(x^2 - 3x + 9) - (x^2 - 3x + 4)} = \frac{(x^2 - 5x + 10) + (x^2 - 5x + 5)}{(x^2 - 5x + 10) - (x^2 - 5x + 5)},$$

which simplifies as $\dfrac{2x^2 - 6x + 13}{5} = \dfrac{2x^2 - 10x + 15}{5}$. Hence, $2x^2 - 6x + 13 = 2x^2 - 10x + 15$, which simplifies as $4x = 2$, so $x = \boxed{1/2}$.

11.41 Let $r = \dfrac{a}{b} = \dfrac{b}{c} = \dfrac{c}{d} = \dfrac{d}{e} = \dfrac{e}{a}$. Then $r^5 = \dfrac{a}{b} \cdot \dfrac{b}{c} \cdot \dfrac{c}{d} \cdot \dfrac{d}{e} \cdot \dfrac{e}{a} = 1$, so $r = 1$. In other words, $a = b = c = d = e$. Therefore, $\dfrac{ab^2}{c^2 d} = \boxed{1}$.

Challenge Problems

11.42 Adding x^2 to both sides, we get $x^4 - 4x^3 + 6x^2 - 4x + 1 = x^2$, so we have $(x-1)^4 = x^2$, which means $(x-1)^4 - x^2 = 0$. By difference of squares, we have $[(x-1)^2 + x][(x-1)^2 - x] = 0$, so $(x^2 - x + 1)(x^2 - 3x + 1) = 0$.

The first quadratic equation has no real roots. The second quadratic equation has real roots $\boxed{\dfrac{3 \pm \sqrt{5}}{2}}$.

11.43 By the Binomial Theorem, we have $(n+1)^{n-1} = n^{n-1} + \binom{n-1}{1}n^{n-2} + \binom{n-1}{2}n^{n-3} + \cdots + 1 = \sum_{k=0}^{n-1}\binom{n-1}{k}n^{n-1-k}$.

The binomial coefficient $\binom{n-1}{k}$ expands as $\binom{n-1}{k} = \dfrac{(n-1)(n-2)\cdots(n-k)}{k!}$. Each factor in the numerator is less than n, so we have $\binom{n-1}{k} < n^k$. Therefore, we have

$$(n+1)^{n-1} = \sum_{k=0}^{n-1}\binom{n-1}{k}n^{n-1-k} < \sum_{k=0}^{n-1}n^k \cdot n^{n-1-k} = \sum_{k=0}^{n-1}n^{n-1} = n^n.$$

11.44 We can rewrite the condition $x/(a+b) = y/(a-b)$ as $x/y = (a+b)/(a-b)$. Then

$$\frac{x+y}{x-y} = \frac{(a+b)+(a-b)}{(a+b)-(a-b)} = \frac{2a}{2b} = \frac{a}{b}.$$

To prove the second desired equation holds, let $k = \dfrac{x}{a+b} = \dfrac{y}{a-b}$, so $x = k(a+b)$ and $y = k(a-b)$. Then

$$\frac{x^3+y^3}{x^3-xy^2} = \frac{(x+y)(x^2-xy+y^2)}{x(x^2-y^2)} = \frac{(x+y)(x^2-xy+y^2)}{x(x+y)(x-y)} = \frac{x^2-xy+y^2}{x(x-y)}.$$

Substituting $x = k(a+b)$ and $y = k(a-b)$, we get

$$\frac{x^2-xy+y^2}{x(x-y)} = \frac{k^2(a+b)^2 - k^2(a+b)(a-b) + k^2(a-b)^2}{k(a+b)[k(a+b)-k(a-b)]} = \frac{k^2a^2+2k^2ab+k^2b^2-k^2a^2+k^2b^2+k^2a^2-2k^2ab+k^2b^2}{k(a+b)\cdot 2kb}$$

$$= \frac{k^2a^2+3k^2b^2}{2k^2b(a+b)} = \frac{a^2+3b^2}{2b(a+b)}.$$

Therefore, we have $\dfrac{x^3+y^3}{x^3-xy^2} = \dfrac{x^2-xy+y^2}{x(x-y)} = \dfrac{a^2+3b^2}{2b(a+b)}$, so $\dfrac{x^3+y^3}{a^2+3b^2} = \dfrac{x^3-xy^2}{2b(a+b)}$.

11.45 Expanding each of the powers with the Binomial Theorem will produce a bunch of terms with radicals. We'd like to get rid of these terms. Each such term will appear when the radical is raised to an odd power. We know how to get rid of these terms in a Binomial Expansion! We'll add $(15-\sqrt{220})^{19}$ to $(15+\sqrt{220})^{19}$ and $(15-\sqrt{220})^{82}$ to $(15+\sqrt{220})^{82}$. But what will this do to the units digit we must find? To answer this, we let $a = 15+\sqrt{220}$ and $b = 15-\sqrt{220}$, so we want to find the units digit of $a^{19} + a^{82}$. Since $\sqrt{220}$ is very close to $\sqrt{225}$, the number $15-\sqrt{220}$ is close to 0, so we expect that adding $b^{19}+b^{82}$ will have little effect on the number. To see that this is the case, we note that $ab = 15^2 - 220 = 5$ and $a > 15$, so $b = 5/a < 1/3$. We also have $15 > \sqrt{220}$, which means $b > 0$. Since $0 < b < 1/3$, we have $0 < b^{19}+b^{82} < 1$.

Now, by the Binomial Theorem, we have

$$a^{19}+b^{19} = (15+\sqrt{220})^{19} + (15-\sqrt{220})^{19}$$

$$= 15^{19} + \binom{19}{1}15^{18}\cdot\sqrt{220} + \binom{19}{2}15^{17}(\sqrt{220})^2 + \binom{19}{3}15^{16}(\sqrt{220})^3 + \cdots + (\sqrt{220})^{19}$$

$$+ 15^{19} - \binom{19}{1}15^{18}\cdot\sqrt{220} + \binom{19}{2}15^{17}(\sqrt{220})^2 - \binom{19}{3}15^{16}(\sqrt{220})^3 + \cdots - (\sqrt{220})^{19}$$

$$= 2\cdot15^{19} + 2\binom{19}{2}15^{17}(\sqrt{220})^2 + 2\binom{19}{4}15^{15}(\sqrt{220})^4 + \cdots$$

$$= 2\cdot15^{19} + 2\binom{19}{2}15^{17}\cdot220 + 2\binom{19}{4}15^{15}\cdot220^2 + \cdots.$$

Thus, $a^{19} + b^{19}$ is an integer. Furthermore, this integer is a multiple of 10. (The first term $2 \cdot 15^{19}$ is divisible by 10, and every term thereafter has a factor of 220.) Similarly, $a^{82} + b^{82}$ is also an integer that is a multiple of 10. Hence, $a^{19} + a^{82} + b^{19} + b^{82}$ is an integer that is a multiple of 10.

Since $0 < b^{19} + b^{82} < 1$, the units digit of $a^{19} + a^{82}$ is $\boxed{9}$.

11.46 In each equation, the left side is a sum of four squares. In the nth equation, the first three squares are $(2n)^2$, $(2n + 1)^2$, and $(2n + 2)^2$.

To get a handle on the fourth square on the left side, we look at the numbers $14, 38, 74$, and 122. Note that each is 2 greater than a multiple of 12:

$$14 = 2 + 1 \cdot 12,$$
$$38 = 2 + 3 \cdot 12,$$
$$74 = 2 + 6 \cdot 12,$$
$$122 = 2 + 10 \cdot 12.$$

The numbers $1, 3, 6$, and 10 are the triangular numbers, which are the numbers of the form $n(n + 1)/2$. Hence, the fourth square in the left side is equal to

$$\left(2 + 12 \cdot \frac{n(n + 1)}{2}\right)^2 = (6n^2 + 6n + 2)^2.$$

The square on the right side is then $(6n^2 + 6n + 3)^2$. Therefore, the identity that is suggested by these examples is

$$(2n)^2 + (2n + 1)^2 + (2n + 2)^2 + (6n^2 + 6n + 2)^2 = (6n^2 + 6n + 3)^2.$$

We can rewrite this identity as

$$(2n)^2 + (2n + 1)^2 + (2n + 2)^2 = (6n^2 + 6n + 3)^2 - (6n^2 + 6n + 2)^2.$$

The left side expands as $4n^2 + 4n^2 + 4n + 1 + 4n^2 + 8n + 4 = 12n^2 + 12n + 5$, and by difference of squares, the right side is
$$[(6n^2 + 6n + 3) + (6n^2 + 6n + 2)][(6n^2 + 6n + 3) - (6n^2 + 6n + 2)] = 12n^2 + 12n + 5.$$
Thus, the identity holds.

11.47 To expand $(x + y + z)^9$, we can use the Binomial Theorem on $x + y$ and z:

$$(x + y + z)^9 = [(x + y) + z]^9 = \sum_{k=0}^{9} \binom{9}{k}(x + y)^{9-k}z^k.$$

We then expand $(x + y)^{9-k}$ using the Binomial Theorem. We'll use the form of the Binomial Theorem introduced in the problem statement:

$$(x + y + z)^9 = \sum_{k=0}^{9} \binom{9}{k}(x + y)^{9-k}z^k = \sum_{k=0}^{9} \binom{9}{k}\left(\sum_{\substack{0 \le i, j \le 9-k \\ i+j=9-k}} \frac{(9 - k)!}{i!\,j!} x^i y^j\right) z^k = \sum_{k=0}^{9} \sum_{\substack{0 \le i, j \le 9-k \\ i+j=9-k}} \binom{9}{k}\frac{(9 - k)!}{i!\,j!} x^i y^j z^k$$

$$= \sum_{k=0}^{9} \sum_{\substack{0 \le i, j \le 9-k \\ i+j=9-k}} \frac{9!}{(9 - k)!\,k!} \cdot \frac{(9 - k)!}{i!\,j!} x^i y^j z^k = \sum_{k=0}^{9} \sum_{\substack{0 \le i, j \le 9-k \\ i+j=9-k}} \frac{9!}{i!\,j!\,k!} x^i y^j z^k.$$

The two sums can be combined into one sum that runs over all ordered triples (i, j, k) of nonnegative integers such that $i + j + k = 9$:

$$(x + y + z)^9 = \boxed{\sum_{\substack{0 \le i, j, k \le 9 \\ i + j + k = 9}} \frac{9!}{i! \, j! \, k!} x^i y^j z^k}.$$

There's a slick counting explanation for this formula. To form each term in the expansion of $(x + y + z)^9$, we must choose one term from each of the 9 factors of $(x + y + z)$. When we choose i instances of x, then j instances of y, then take z from the remaining k factors, we form the term $x^i y^j z^k$. We must have $i + j + k = 9$, because there are 9 factors of $(x + y + z)$ to choose from. Counting the number of ways we can choose i instances of x, then j instances of y, then k instances of z, is the same problem as counting the number of words we can form with i copies of x, j copies of y, and k copies of z, where $i + j + k = 9$. Since there are $9!/(i! \, j! \, k!)$ different such words we can form, there are $9!/(i! \, j! \, k!)$ instances of $x^i y^j z^k$ formed in the expansion of $(x + y + z)^9$. Taking all triplets of nonnegative integers (i, j, k) such that $i + j + k = 9$ gives us the summation above.

11.48   We prove the inequality using induction. First, we prove the base case. When $n = 4$, the given inequality becomes $4^4 \le 4^4$, which is true. Now, assume that the inequality holds for $n = k$ for some positive integer $k \ge 4$, so $4^k \ge k^4$.

Multiplying both sides by 4, we get $4^{k+1} \ge 4k^4$. We must show that $4^{k+1} \ge (k+1)^4$, so it suffices to show that $4k^4 \ge (k+1)^4$. Since both sides are positive, this inequality is equivalent to $\sqrt{2}k \ge k + 1$. This inequality holds if $(\sqrt{2} - 1)k \ge 1$, which holds if $k \ge \frac{1}{\sqrt{2}-1} = \frac{1}{\sqrt{2}-1} \cdot \frac{\sqrt{2}+1}{\sqrt{2}+1} = \sqrt{2} + 1$. Since $k \ge 4$, we know that $k > \sqrt{2} + 1$, so we can walk backwards through our steps above (which are all reversible) to conclude that $4k^4 \ge (k+1)^4$.

This completes the induction.

11.49   Taking $x = 1$, we get $a_0 + a_1 + a_2 + \cdots + a_{2n} = (1 + 1 + 1)^n = 3^n$. To cancel every other term, we can take $x = -1$, to get $a_0 - a_1 + a_2 - \cdots + a_{2n} = (1 - 1 + 1)^n = 1$.

Adding both equations, we get $2a_0 + 2a_2 + \cdots + 2a_{2n} = 3^n + 1$, so $a_0 + a_2 + \cdots + a_{2n} = \boxed{\dfrac{3^n + 1}{2}}$.

11.50   To simplify the equations, we can isolate z in each equation:

$$a_1 x + b_1 y = -c_1 z,$$
$$a_2 x + b_2 y = -c_2 z.$$

Multiplying the first equation by c_2 and the second equation by c_1, we get

$$a_1 c_2 x + b_1 c_2 y = -c_1 c_2 z,$$
$$a_2 c_1 x + b_2 c_1 y = -c_1 c_2 z.$$

Therefore, $a_1 c_2 x + b_1 c_2 y = a_2 c_1 x + b_1 c_2 y$, so $(c_1 a_2 - c_2 a_1)x = (b_1 c_2 - b_2 c_1)y$, which gives

$$\frac{x}{b_1 c_2 - b_2 c_1} = \frac{y}{c_1 a_2 - c_2 a_1}.$$

Similarly, starting with the given equations, we can isolate x:

$$b_1 y + c_1 z = -a_1 x,$$
$$b_2 y + c_2 z = -a_2 x.$$

Multiplying the first equation by a_2 and the second equation by a_1, we get

$$a_2 b_1 y + a_2 c_1 z = -a_1 a_2 x,$$
$$a_1 b_2 y + a_1 c_2 z = -a_1 a_2 x.$$

Therefore, $a_2b_1y + a_2c_1z = a_1b_2y + a_1c_2z$, so $(c_1a_2 - c_2a_1)z = (a_1b_2 - a_2b_1)y$, which gives

$$\frac{y}{c_1a_2 - c_2a_1} = \frac{z}{a_1b_2 - a_2b_1}.$$

Hence, $\dfrac{x}{b_1c_2 - b_2c_1} = \dfrac{y}{c_1a_2 - c_2a_1} = \dfrac{z}{a_1b_2 - a_2b_1}.$

11.51   We can establish this identity either through algebraic manipulation or by induction.

Solution 1: We express the left side as the difference of two sums:

$$1 - \frac{1}{2} + \frac{1}{3} - \frac{1}{4} + \cdots + \frac{1}{2n-1} - \frac{1}{2n} = \left(1 + \frac{1}{2} + \frac{1}{3} + \cdots + \frac{1}{2n}\right) - 2\left(\frac{1}{2} + \frac{1}{4} + \frac{1}{6} + \cdots + \frac{1}{2n}\right)$$

$$= \left(1 + \frac{1}{2} + \frac{1}{3} + \cdots + \frac{1}{2n}\right) - \left(1 + \frac{1}{2} + \frac{1}{3} + \cdots + \frac{1}{n}\right)$$

$$= \frac{1}{n+1} + \frac{1}{n+2} + \frac{1}{n+3} + \cdots + \frac{1}{2n-1} + \frac{1}{2n},$$

which is the right side of the identity.

Solution 2: We prove the identity using induction. First, we prove the base case. When $n = 1$, the left side is $1 - \frac{1}{2} = \frac{1}{2}$, and the right side is $\frac{1}{2}$, so the identity holds for $n = 1$.

Now, assume that the identity holds for some $n = k$ where k is a positive integer, so

$$1 - \frac{1}{2} + \frac{1}{3} - \frac{1}{4} + \cdots + \frac{1}{2k-1} - \frac{1}{2k} = \frac{1}{k+1} + \frac{1}{k+2} + \frac{1}{k+3} + \cdots + \frac{1}{2k-1} + \frac{1}{2k}.$$

We must prove that

$$1 - \frac{1}{2} + \frac{1}{3} - \frac{1}{4} + \cdots + \frac{1}{2k-1} - \frac{1}{2k} + \frac{1}{2k+1} - \frac{1}{2k+2} = \frac{1}{k+2} + \frac{1}{k+3} + \cdots + \frac{1}{2k-1} + \frac{1}{2k} + \frac{1}{2k+1} + \frac{1}{2k+2}.$$

The difference in the left sides of the last two equations is

$$\frac{1}{2k+1} - \frac{1}{2k+2},$$

and the difference in the right sides is

$$\frac{1}{2k+1} + \frac{1}{2k+2} - \frac{1}{k+1} = \frac{1}{2k+1} + \frac{1}{2k+2} - \frac{2}{2k+2} = \frac{1}{2k+1} - \frac{1}{2k+2}.$$

Hence, the formula holds for $n = k + 1$, and by induction it holds for all positive integers n.

11.52   By difference of cubes, we have $(x - y)(x^2 + xy + y^2) = 19$. From the second equation, we have $xy(x - y) = 6$. Hence,

$$\frac{19}{x^2 + xy + y^2} = x - y = \frac{6}{xy},$$

so $6(x^2 + xy + y^2) = 19xy$, which simplifies as $6x^2 - 13xy + 6y^2 = 0$, so $(2x - 3y)(3x - 2y) = 0$, which means $2x = 3y$ or $3x = 2y$.

If $2x = 3y$, then $y = 2x/3$, so $x^3 - y^3 = x^3 - \dfrac{8x^3}{27} = \dfrac{19x^3}{27}$, so we have $\dfrac{19x^3}{27} = 19$. Then $x^3 = 27$, so $x = 3$ and $y = 2$.

If $3x = 2y$, then $y = 3x/2$, so $x^3 - y^3 = x^3 - \dfrac{27x^3}{8} = \dfrac{-19x^3}{8}$, so we have $\dfrac{-19x^3}{8} = 19$. Then $x^3 = -8$, so $x = -2$ and $y = -3$.

Therefore, the solutions are $(x, y) = \boxed{(3,2) \text{ and } (-2,-3)}$.

11.53 Note that $1 - x + x^2 - x^3 + \cdots + x^{16} - x^{17}$ is a geometric series, so we may use the formula for a geometric series to get

$$1 - x + x^2 - x^3 + \cdots + x^{16} - x^{17} = \frac{1 - x^{18}}{1 + x}.$$

Since $y = x + 1$, we have $x = y - 1$, so

$$\frac{1 - x^{18}}{1 + x} = \frac{1 - (y - 1)^{18}}{y} = \frac{1 - (1 - y)^{18}}{y}.$$

By the Binomial Theorem,

$$(1 - y)^{18} = 1 - \binom{18}{1}y + \binom{18}{2}y^2 - \binom{18}{3}y^3 + \cdots = 1 - 18y + 153y^2 - 816y^3 + \cdots,$$

so

$$\frac{1 - (1 - y)^{18}}{y} = \frac{1 - (1 - 18y + 153y^2 - 816y^3 + \cdots)}{y} = 18 - 153y + 816y^2 + \cdots.$$

Hence, $a_2 = \boxed{816}$.

11.54 As a sum of fifth powers, we can write $x^5 + y^5 = (x + y)(x^4 - x^3y + x^2y^2 - xy^3 + y^4)$. Therefore,

$$\begin{aligned}
(x + y)^5 - (x^5 + y^5) &= (x + y)[(x + y)^4 - (x^4 - x^3y + x^2y^2 - xy^3 + y^4)] \\
&= (x + y)[(x^4 + 4x^3y + 6x^2y^2 + 4xy^3 + y^4) - (x^4 - x^3y + x^2y^2 - xy^3 + y^4)] \\
&= (x + y)(5x^3y + 5x^2y^2 + 5xy^3) \\
&= \boxed{5xy(x + y)(x^2 + xy + y^2)}.
\end{aligned}$$

11.55 Note that $xy - 3x + 7y - 21 = (x + 7)(y - 3)$, so $(xy - 3x + 7y - 21)^n = (x + 7)^n(y - 3)^n$. When expanded, $(x + 7)^n$ has terms of the form cx^i, where $0 \le i \le n$, and $(y - 3)^n$ has terms of the form cy^j, where $0 \le j \le n$, so $(x + 7)^n(y - 3)^n$ will have terms of the form cx^iy^j, where $0 \le i, j \le n$. Therefore, there will be $(n + 1)^2$ terms in total.

Since $44^2 = 1936$ and $45^2 = 2025$, the smallest positive integer n such that $(n + 1)^2 \ge 1996$ is $n = \boxed{44}$.

11.56 Multiplying both sides by $(a - b)(a - c)(b - c)$, we get

$$bc(b + c)(b - c) + ca(c + a)(c - a) + ab(a + b)(a - b) = (a - b)(a - c)(b - c)(a + b + c).$$

The left side expands as

$$bc(b^2 - c^2) + ac(c^2 - a^2) + ab(a^2 - b^2) = a^3b - ab^3 - a^3c + ac^3 + b^3c - bc^3,$$

and the right side expands as

$$\begin{aligned}
(a - b)(a - c)(b - c)(a + b + c) &= (a^2 - ab - ac + bc)(b - c)(a + b + c) \\
&= (a^2b - a^2c - ab^2 + abc - abc + ac^2 + b^2c - bc^2)(a + b + c) \\
&= (a^2b - ab^2 - a^2c + ac^2 + b^2c - bc^2)(a + b + c) \\
&= a^3b - a^2b^2 - a^3c + a^2c^2 + ab^2c - abc^2 \\
&\quad + a^2b^2 - ab^3 - a^2bc + abc^2 + b^3c - b^2c^2 \\
&\quad + a^2bc - ab^2c - a^2c^2 + ac^3 + b^2c^2 - bc^3 \\
&= a^3b - ab^3 - a^3c + ac^3 + b^3c - bc^3.
\end{aligned}$$

Thus, the two sides are equal.

11.57 We have a binomial raised to a fractional power, so we apply the Binomial Theorem for fractional exponents:

$$(10^{2002} + 1)^{10/7} = 10^{2002(10/7)} + \frac{10}{7} \cdot 10^{2002(3/7)} + \frac{\frac{10}{7} \cdot \frac{3}{7}}{2} \cdot 10^{2002(-4/7)} + \cdots$$

$$= 10^{2860} + \frac{10}{7} \cdot 10^{858} + \frac{15}{49} \cdot 10^{-1144} + \cdots.$$

It's pretty clear that only the second term will affect the first few digits after the decimal point. Expressing this term as a decimal gives us

$$\frac{10}{7} \cdot 10^{858} = 10^{859} \cdot \frac{1}{7} = 10^{859}(0.\overline{142857}).$$

The decimal representation of $\frac{1}{7}$ repeats every 6 digits. Multiplying this fraction by 10^{859} means moving the decimal point 859 digits to the right. Since 859 leaves a remainder of 1 when divided by 6, we get the same repeating decimal by moving the decimal over 1 digit as we do when moving it over 859 digits:

$$.\overline{428571}.$$

We therefore see that 428 are the first three digits after the decimal point of $(10^{2002} + 1)^{10/7}$.

11.58 Since our goal is to prove an inequality for all positive integers n, we give induction a try. The base case where $n = 1$ is simple: $\frac{1}{1+1} = \frac{1}{2}$ and $\frac{2 \cdot 1}{3 \cdot 1 + 1} = \frac{2}{4} = \frac{1}{2}$, so the nonstrict inequality holds when $n = 1$.

Now, we assume that the inequality is true for $n = k$, so

$$\frac{1}{k+1} + \frac{1}{k+2} + \cdots + \frac{1}{2k} \geq \frac{2k}{3k+1}.$$

In order to get the left side of the proposed inequality for $n = k + 1$, we add

$$-\frac{1}{k+1} + \frac{1}{2k+1} + \frac{1}{2k+2} = \frac{1}{2k+1} - \frac{1}{2k+2} = \frac{1}{4k^2 + 6k + 2}$$

to both sides of the assumed inequality to get

$$\frac{1}{k+2} + \frac{1}{k+3} + \cdots + \frac{1}{2k+1} + \frac{1}{2k+2} \geq \frac{2k}{3k+1} + \frac{1}{4k^2 + 6k + 2}.$$

Now, we can finish the inductive step by proving that

$$\frac{2k}{3k+1} + \frac{1}{4k^2 + 6k + 2} \geq \frac{2(k+1)}{3(k+1) + 1}.$$

Simplifying the lesser side, we must show that $\frac{2k}{3k+1} + \frac{1}{4k^2 + 6k + 2} \geq \frac{2k+2}{3k+4}$. Rearranging terms, we need to show that

$$\frac{1}{4k^2 + 6k + 2} \geq \frac{2k+2}{3k+4} - \frac{2k}{3k+1} = \frac{(3k+1)(2k+2) - 2k(3k+4)}{(3k+4)(3k+1)} = \frac{2}{9k^2 + 15k + 4}.$$

Multiplying through by the denominators on the far left side and far right side, we have the equivalent inequality $9k^2 + 15k + 4 \geq 8k^2 + 12k + 4$. This inequality is equivalent to $k^2 + 3k \geq 0$, which is clearly true for all positive k. All our steps are reversible, so we can start with $k^2 + 3k \geq 0$ and deduce that our inequality holds for $n = k + 1$. Therefore, our induction is complete.

11.59 We have

$$1 = A + B + C = \frac{b^2 + c^2 - a^2}{2bc} + \frac{a^2 + c^2 - b^2}{2ac} + \frac{a^2 + b^2 - c^2}{2ab} = \frac{a(b^2 + c^2 - a^2)}{2abc} + \frac{b(a^2 + c^2 - b^2)}{2abc} + \frac{c(a^2 + b^2 - c^2)}{2abc}$$

$$= \frac{-a^3 - b^3 - c^3 + a^2b + ab^2 + a^2c + ac^2 + b^2c + bc^2}{2abc},$$

so $a^3 + b^3 + c^3 - a^2b - ab^2 - a^2c - ac^2 - b^2c - bc^2 + 2abc = 0$.

In the absence of any obvious factors of the left side, we can try substituting specific values. For example, taking $b = 1$ and $c = 4$, the cubic in a above becomes $a^3 - 5a^2 - 9a + 45$. This factors as $(a - 5)(a - 3)(a + 3)$. As another example, taking $b = 2$ and $c = 5$, we get $a^3 - 7a^2 - 9a + 63$. This factors as $(a - 7)(a - 3)(a + 3)$. At this point, we may guess that the roots are always going to be $b + c$, $b - c$, and $-b + c$.

To confirm this, we expand $(a - b - c)(a - b + c)(a + b - c)$. First, by difference of squares, we have

$$(a - b + c)(a + b - c) = a^2 - (b - c)^2 = a^2 - b^2 + 2bc - c^2.$$

Therefore,

$$\begin{aligned}(a - b - c)(a - b + c)(a + b - c) &= (a - b - c)(a^2 - b^2 + 2bc - c^2) \\ &= a^3 - ab^2 + 2abc - ac^2 - a^2b + b^3 - 2b^2c + bc^2 - a^2c + b^2c - 2bc^2 + c^3 \\ &= a^3 + b^3 + c^3 - a^2b - ab^2 - a^2c - ac^2 - b^2c - bc^2 + 2abc.\end{aligned}$$

Thus, the condition $A + B + C = 1$ is equivalent to $(a - b - c)(a - b + c)(a + b - c) = 0$, so $a = b + c$, $b = a + c$, or $c = a + b$. If $a = b + c$, then

$$A = \frac{b^2 + c^2 - a^2}{2bc} = \frac{b^2 + c^2 - (b + c)^2}{2bc} = \frac{-2bc}{2bc} = -1,$$

$$B = \frac{a^2 + c^2 - b^2}{2ac} = \frac{(b + c)^2 + c^2 - b^2}{2(b + c)c} = \frac{2bc + 2c^2}{2bc + 2c^2} = 1,$$

$$C = \frac{a^2 + b^2 - c^2}{2ab} = \frac{(b + c)^2 + b^2 - c^2}{2(b + c)b} = \frac{2b^2 + 2bc}{2b^2 + 2bc} = 1.$$

Therefore, $ABC = (-1) \cdot 1 \cdot 1 = -1$. The cases $b = a + c$ and $c = a + b$ similarly give $ABC = -1$.

11.60 Let $a = x^2 - yz$, $b = y^2 - zx$, and $c = z^2 - xy$. Then we can rewrite the given expression as

$$a^3 + b^3 + c^3 - 3abc,$$

which factors as $(a + b + c)(a^2 + b^2 + c^2 - ab - ac - bc)$. (See Problem 9.54.)

The first factor is $a + b + c = x^2 + y^2 + z^2 - xy - xz - yz$. The second factor can be rewritten as

$$a^2 + b^2 + c^2 - ab - ac - bc = \frac{(a - b)^2 + (a - c)^2 + (b - c)^2}{2},$$

so we look at $a - b$, $a - c$, and $b - c$.

We have $a - b = x^2 - yz - (y^2 - xz) = x^2 - y^2 + xz - yz = (x - y)(x + y) + z(x - y) = (x - y)(x + y + z)$.

Similarly, we have $a - c = (x - z)(x + y + z)$ and $b - c = (y - z)(x + y + z)$. Therefore,

$$\begin{aligned}(a - b)^2 + (a - c)^2 + (b - c)^2 &= (x - y)^2(x + y + z)^2 + (x - z)^2(x + y + z)^2 + (y - z)^2(x + y + z)^2 \\ &= [(x - y)^2 + (x - z)^2 + (y - z)^2](x + y + z)^2 \\ &= 2(x^2 + y^2 + z^2 - xy - xz - yz)(x + y + z)^2.\end{aligned}$$

Hence,

$$\begin{aligned}(a + b + c)(a^2 + b^2 + c^2 - ab - ac - bc) &= (x^2 + y^2 + z^2 - xy - xz - yz) \cdot \frac{(a - b)^2 + (a - c)^2 + (b - c)^2}{2} \\ &= (x^2 + y^2 + z^2 - xy - xz - yz)(x^2 + y^2 + z^2 - xy - xz - yz)(x + y + z)^2 \\ &= [(x^2 + y^2 + z^2 - xy - xz - yz)(x + y + z)]^2 \\ &= \boxed{(x^3 + y^3 + z^3 - 3xyz)^2}.\end{aligned}$$

11.61 Let $S_n = \sum\limits_{k=1}^{n} \dfrac{1}{k^2}$. To analyze why $S_n < 2$ for all $n \geq 1$, we try looking at the first few values of S_n:

$$S_1 = \frac{1}{1^2} = 1,$$

$$S_2 = \frac{1}{1^2} + \frac{1}{2^2} = \frac{5}{4},$$

$$S_3 = \frac{1}{1^2} + \frac{1}{2^2} + \frac{1}{3^2} = \frac{49}{36},$$

$$S_4 = \frac{1}{1^2} + \frac{1}{2^2} + \frac{1}{3^2} + \frac{1}{4^2} = \frac{205}{144},$$

$$S_5 = \frac{1}{1^2} + \frac{1}{2^2} + \frac{1}{3^2} + \frac{1}{4^2} + \frac{1}{5^2} = \frac{5269}{3600}.$$

We confirm that each of these values is less than 2, but otherwise, there are no useful patterns. In particular, there does not seem to be any simple formula for S_n.

In such a case, it may help to compare the sum S_n to another sum T_n, such that T_n has a simple formula, and $S_n < T_n < 2$.

To create such a sum T_n, we adjust each term of S_n so that each term slightly increases. For example, starting with the second term, we replace $1/2^2$ with $1/(1 \cdot 2)$, $1/3^2$ with $1/(2 \cdot 3)$, and so on. Thus, we get the sum

$$T_n = 1 + \frac{1}{1 \cdot 2} + \frac{1}{2 \cdot 3} + \cdots + \frac{1}{(n-1)n}.$$

Then $S_n \leq T_n$ for $n \geq 2$, but does T_n have the other properties we desire?

To check, we look at the first few values of T_n:

$$T_1 = 1,$$

$$T_2 = 1 + \frac{1}{1 \cdot 2} = \frac{3}{2},$$

$$T_3 = 1 + \frac{1}{1 \cdot 2} + \frac{1}{2 \cdot 3} = \frac{5}{3},$$

$$T_4 = 1 + \frac{1}{1 \cdot 2} + \frac{1}{2 \cdot 3} + \frac{1}{3 \cdot 4} = \frac{7}{4},$$

$$T_5 = 1 + \frac{1}{1 \cdot 2} + \frac{1}{2 \cdot 3} + \frac{1}{3 \cdot 4} + \frac{1}{4 \cdot 5} = \frac{9}{5}.$$

Here, the pattern is quite evident. It seems that $T_n = \dfrac{2n-1}{n}$ for all n. If this is true, then $S_n \leq T_n = \dfrac{2n-1}{n} = 2 - \dfrac{1}{n} < 2$, and the problem is solved. It only remains to prove our formula for T_n, which we do by induction.

First, we prove the base case. When $n = 1$, we have $T_1 = 1$ and $\frac{2n-1}{n} = 1$, so the formula holds for $n = 1$. Now, assume that the formula holds for $n = k$ for some positive integer $n = k$, so

$$1 + \frac{1}{1 \cdot 2} + \frac{1}{2 \cdot 3} + \cdots + \frac{1}{(k-1)k} = \frac{2k-1}{k}.$$

Adding $\frac{1}{k(k+1)}$ to both sides, we get

$$1 + \frac{1}{1 \cdot 2} + \frac{1}{2 \cdot 3} + \cdots + \frac{1}{(k-1)k} + \frac{1}{k(k+1)} = \frac{2k-1}{k} + \frac{1}{k(k+1)} = \frac{(2k-1)(k+1)}{k(k+1)} + \frac{1}{k(k+1)} = \frac{2k^2+k-1+1}{k(k+1)}$$

$$= \frac{k(2k+1)}{k(k+1)} = \frac{2k+1}{k+1} = \frac{2(k+1)-1}{k+1},$$

which is equal to $\frac{2n-1}{n}$ for $n = k + 1$. Hence, the formula holds for $n = k + 1$, and by induction, it holds for all positive integers n. In particular, $T_n < 2$, so $S_n < 2$.

11.62

(a) To narrow down the possible choices of f, we can substitute specific values of y_1, y_2, y_3, and y_4. For example, we can take $y_1 = 1$ and $y_2 = y_3 = y_4 = 0$ (taking $y_1 = y_2 = y_3 = y_4 = 0$ would give us 0 on the right side, so these values are not useful), which gives us

$$f(x_1, x_2, x_3, x_4)f(1, 0, 0, 0) = x_1^2 + x_2^2 + x_3^2 + x_4^2.$$

This suggests that we can take $f(a, b, c, d) = a^2 + b^2 + c^2 + d^2$, or at least a constant multiple of $a^2 + b^2 + c^2 + d^2$. To check, we expand the right side. (We could have expanded the right side without substituting values, but now we know what to look for.) The right side becomes

$$(x_1y_1 + x_2y_2 + x_3y_3 + x_4y_4)^2 + (x_1y_2 - x_2y_1 + x_3y_4 - x_4y_3)^2$$
$$+ (x_1y_3 - x_3y_1 + x_4y_2 - x_2y_4)^2 + (x_1y_4 - x_4y_1 + x_2y_3 - x_3y_2)^2$$
$$= x_1^2y_1^2 + x_2^2y_2^2 + x_3^2y_3^2 + x_4^2y_4^2 + 2x_1x_2y_1y_2 + 2x_1x_3y_1y_3 + 2x_1x_4y_1y_4 + 2x_2x_3y_2y_3 + 2x_2x_4y_2y_4 + 2x_3x_4y_3y_4$$
$$+ x_1^2y_2^2 + x_2^2y_1^2 + x_3^2y_4^2 + x_4^2y_3^2 - 2x_1x_2y_1y_2 + 2x_1x_3y_2y_4 - 2x_1x_4y_2y_3 - 2x_2x_3y_1y_4 + 2x_2x_4y_1y_3 - 2x_3x_4y_3y_4$$
$$+ x_1^2y_3^2 + x_3^2y_1^2 + x_4^2y_2^2 + x_2^2y_4^2 - 2x_1x_3y_1y_3 + 2x_1x_4y_2y_3 - 2x_1x_2y_3y_4 - 2x_3x_4y_1y_2 + 2x_2x_3y_1y_4 - 2x_2x_4y_2y_4$$
$$+ x_1^2y_4^2 + x_4^2y_1^2 + x_2^2y_3^2 + x_3^2y_2^2 - 2x_1x_4y_1y_4 + 2x_1x_2y_3y_4 - 2x_1x_3y_2y_4 - 2x_2x_4y_1y_3 + 2x_3x_4y_1y_2 - 2x_2x_3y_2y_3$$
$$= x_1^2y_1^2 + x_1^2y_2^2 + x_1^2y_3^2 + x_1^2y_4^2 + x_2^2y_1^2 + x_2^2y_2^2 + x_2^2y_3^2 + x_2^2y_4^2$$
$$+ x_3^2y_1^2 + x_3^2y_2^2 + x_3^2y_3^2 + x_3^2y_4^2 + x_4^2y_1^2 + x_4^2y_2^2 + x_4^2y_3^2 + x_4^2y_4^2$$
$$= (x_1^2 + x_2^2 + x_3^2 + x_4^2)(y_1^2 + y_2^2 + y_3^2 + y_4^2).$$

This confirms that we can take $f(a, b, c, d) = a^2 + b^2 + c^2 + d^2$.

(b) The identity in part (a) shows that if two integers can be written as the sum of four squares, then so can their product. Then by a straightforward induction argument, we can extend this result to any (finite) set of integers.

Let N be an arbitrary positive integer. If $N = 1$, then we can write $N = 1^2 + 0^2 + 0^2 + 0^2$. Otherwise, $N \geq 2$, so we can write N in the form

$$N = p_1^{e_1} p_2^{e_2} \cdots p_k^{e_k},$$

where the p_i are distinct primes and the e_i are nonnegative integers. Since every prime can be expressed as the sum of four squares, and N is the product of primes, N can also be expressed as the sum of four squares.

The result that every positive integer can be written as the sum of four squares is called Lagrange's Theorem. As we have just shown, its proof hinges on the result that every prime can be written as the sum of four squares. This is a difficult proof, and requires advanced number theory.

CHAPTER 12

Inequalities

Exercises for Section 12.1

12.1.1

(a) Since $x > y$, we can add z to both sides to get $x + z > z + y$. Hence, the given inequality is always $\boxed{\text{true}}$.

(b) Let $x = 2$, $y = 1$, and $z = -1$. Then $(x + y)z = 3 \cdot (-1) = -3 < 0$, so the given inequality is $\boxed{\text{not always true}}$.

(c) Since $x > y$, we can subtract z from both sides to get $x - z > y - z$. Hence, the given inequality is always $\boxed{\text{true}}$.

(d) Let $x = 2$, $y = 1$, and $z = -1$. Then $xz = -2$ and $yz = -1$, and $-2 < -1$, so the given inequality is $\boxed{\text{not always true}}$.

12.1.2 First, since $x < 0$, both x and x^3 are negative while x^2 is positive. Thus, we only need to compare x and x^3. We see that

$$x^3 - x = x(x^2 - 1) = x(x - 1)(x + 1).$$

Since $-1 < x < 0$, the factor x is negative, the factor $x - 1$ is negative, and the factor $x + 1$ is positive, so $x^3 - x = x(x - 1)(x + 1) > 0$, which means $x^3 > x$. Therefore, $\boxed{x < x^3 < x^2}$.

12.1.3 We work backwards. First, we can move all the terms to one side, to obtain the equivalent inequality

$$c\left(1 - \frac{x}{y}\right) - (y - x) \geq 0.$$

Then, we put everything over a common denominator, to get $\dfrac{c(y - x) - y(y - x)}{y} \geq 0$, so we have

$$\frac{(c - y)(y - x)}{y} \geq 0.$$

Since $0 < x \leq y \leq c$, we have $c - y \geq 0$, $y - x \geq 0$, and $y > 0$, so this inequality holds. Since all of our steps are reversible, the original inequality also holds.

12.1.4 Taking the reciprocal of both sides of the inequality $\frac{a}{b} \leq \frac{c}{d}$, we get $\frac{b}{a} \geq \frac{d}{c}$. Adding 1 to both sides, we get $\frac{a+b}{a} \geq \frac{c+d}{c}$. Then, again taking the reciprocal of both sides, we get $\frac{a}{a+b} \leq \frac{c}{c+d}$, as desired.

12.1.5 We work backwards. Because all the variables are positive, we can multiply both sides by $(b + x)(b + y)$ to clear out the denominators and have $(a + y)(b + x) > (a + x)(b + y)$. Expanding both sides gives $ab + ax + by + xy > ab + ay + bx + xy$. Eliminating common terms leaves $ax + by > ay + bx$.

We proved in the text that if $a > b$ and $x > y$, then $ax + by > ay + bx$, so we can now reverse our steps above to complete our proof. Adding $ab + xy$ to both sides of $ax + by > ay + bx$, then factoring the result, gives us $(a + y)(b + x) > (a + x)(b + y)$. Dividing both sides by $(b + x)(b + y)$ then gives the desired inequality.

12.1.6 We wish to prove that $a^2 + 3b^2 + 5c^2 \geq 1$. Since the left side contains quadratic terms, we square the equation $a + b + c = 1$ to get
$$a^2 + b^2 + c^2 + 2ab + 2ac + 2bc = 1.$$

We want $3b^2$ to be in the final expression, so we need to somehow generate a further term of $2b^2$. Since $a \leq b$, we have $2ab \leq 2b^2$. Therefore,
$$1 = a^2 + b^2 + c^2 + 2ab + 2ac + 2bc \leq a^2 + 3b^2 + c^2 + 2ac + 2bc.$$

Similarly, we want $5c^2$ to be in the final expression, so we need to somehow generate a further term of $4c^2$. Since $a \leq b \leq c$, we have $2ac \leq 2bc \leq 2c^2$. Therefore,
$$a^2 + 3b^2 + c^2 + 2ac + 2bc \leq a^2 + 3b^2 + c^2 + 2c^2 + 2c^2 = a^2 + 3b^2 + 5c^2.$$

We conclude that $a^2 + 3b^2 + 5c^2 \geq 1$.

Exercises for Section 12.2

12.2.1 We work backwards. Moving all the terms to one side, we have $x^2 - 2x + 1 \geq 0$. Factoring the left side gives $(x - 1)^2 \geq 0$, which is true by the Trivial Inequality. We can then reverse our steps. Expanding the left side of $(x - 1)^2 \geq 0$, then adding $3x$ to both sides, gives the desired $x^2 + x + 1 \geq 3x$.

12.2.2 We work backwards. Expanding both sides gives $a^2b^2 + a^2 + b^2 + 1 \geq a^2b^2 + 2ab + 1$. Moving all terms to the greater side gives $a^2 + b^2 - 2ab \geq 0$. Factoring gives $(a - b)^2 \geq 0$, which is true by the Trivial Inequality. All our steps are reversible, so we have $(a^2 + 1)(b^2 + 1) \geq (ab + 1)^2$.

12.2.3

(a) We aren't given that a and b are positive. So, we cannot conclude $a^2 > b^2$. For example, we might have $a = 1$ and $b = -2$, for which we have $a > b$, but $a^2 < b^2$. Moreover, we might have $a + b < 0$, so multiplying an inequality by $a + b$ might reverse the inequality sign. We'll have to find another way.

(b) We work backwards. Factoring both sides gives $a^2(a + b) \geq b^2(a + b)$. We can't divide by $a + b$ (it might be negative or 0), but we can move all the terms to the left side, which gives $a^2(a + b) - b^2(a + b) \geq 0$. Factoring gives $(a^2 - b^2)(a + b) \geq 0$, from which we have $(a - b)(a + b)(a + b) \geq 0$, so $(a - b)(a + b)^2 \geq 0$. Now our path to the solution is clear.

We are given $a > b$, and $(a + b)^2 \geq 0$ by the Trivial Inequality, so $(a - b)(a + b)^2 \geq 0$. Expanding the left side gives $a^3 + a^2b - ab^2 - b^3 \geq 0$, and rearranging gives the desired $a^3 + a^2b \geq ab^2 + b^3$. Equality holds when equality holds in $(a + b)^2 \geq 0$, which occurs when $a = -b$.

12.2.4 By the Trivial Inequality, we have $\left(x - \frac{1}{y}\right)^2 \geq 0$, so $x^2 + \frac{1}{y^2} \geq 2 \cdot \frac{x}{y}$. Similarly, we have $y^2 + \frac{1}{z^2} \geq 2 \cdot \frac{y}{z}$ and $z^2 + \frac{1}{x^2} \geq 2 \cdot \frac{z}{x}$. Adding these three together gives the desired inequality.

12.2.5 Moving all the terms to the greater side and expanding, the desired greater side becomes
$$(a^7 + b^7)(a^2 + b^2) - (a^5 + b^5)(a^4 + b^4) = a^9 + a^7b^2 + a^2b^7 + b^9 - (a^9 + a^5b^4 + a^4b^5 + b^9)$$
$$= a^7b^2 + a^2b^7 - a^5b^4 - a^4b^5 = a^2b^2(a^5 + b^5 - a^3b^2 - a^2b^3).$$

The expression $a^5 + b^5 - a^3b^2 - a^2b^3$ can be factored as
$$a^5 + b^5 - a^3b^2 - a^2b^3 = a^5 - a^3b^2 - a^2b^3 + b^5 = a^3(a^2 - b^2) - b^3(a^2 - b^2) = (a^3 - b^3)(a^2 - b^2)$$
$$= (a - b)(a^2 + ab + b^2)(a - b)(a + b) = (a - b)^2(a^2 + ab + b^2)(a + b),$$

so we now must prove that

$$a^2 b^2 (a-b)^2 (a^2 + ab + b^2)(a+b) \geq 0.$$

By the Trivial Inequality, we have $a^2 \geq 0$, $b^2 \geq 0$, and $(a-b)^2 \geq 0$, and it is clear that $a^2 + ab + b^2 \geq 0$ and $a+b \geq 0$ because $a, b \geq 0$. Hence, this inequality holds, and since all of our steps are reversible, the original inequality holds.

12.2.6 Moving all the terms to one side, we obtain the equivalent inequality $\frac{x^2}{4} + y^2 + z^2 - xy + xz - 2yz \geq 0$.

Looking for ways to manipulate the left side, we see that the terms y^2, $2yz$, and z^2 can be combined to form the square of a binomial: $y^2 - 2yz + z^2 = (y-z)^2$. Therefore, we wish to show

$$\frac{x^2}{4} - xy + xz + (y-z)^2 \geq 0.$$

Next, we note that the expression $-xy + xz$ can be factored as $-x(y-z)$. Therefore, we must now show

$$\frac{x^2}{4} - x(y-z) + (y-z)^2 \geq 0.$$

In this form, we recognize that the left side is perfect square, and we have $\left[\frac{x}{2} - (y-z)\right]^2 \geq 0$. This inequality follows from the Trivial Inequality. Since all of our steps are reversible, the original inequality holds.

Exercises for Section 12.3

12.3.1 Suppose $x > 0$. By AM-GM, we have $\frac{x + \frac{1}{x}}{2} \geq \sqrt{x \cdot \frac{1}{x}} = 1$, so $x + \frac{1}{x} \geq 2$, as desired.

12.3.2 By the AM-GM inequality on the numbers a^6 and b^4, we have $\frac{a^6 + b^4}{2} \geq \sqrt{a^6 b^4}$, so $a^3 + b^4 \geq 2a^3 b^2$.

12.3.3 By AM-GM, we have $\frac{ab+cd}{2} \geq \sqrt{abcd}$ and $\frac{ac+bd}{2} \geq \sqrt{abcd}$. Multiplying both inequalities by 2, then taking the product of the two resulting inequalities, we get $(ab+cd)(ac+bd) \geq 4abcd$.

12.3.4 We have sums on the greater side, which is an indication to try AM-GM. Applying AM-GM to the sums inside the radicals, we have $\frac{x + \frac{1}{y}}{2} \geq \sqrt{\frac{x}{y}}$ and $\frac{y + \frac{1}{x}}{2} \geq \sqrt{\frac{y}{x}}$. Therefore, we have $x + \frac{1}{y} \geq 2\sqrt{\frac{x}{y}}$ and $y + \frac{1}{x} \geq 2\sqrt{\frac{y}{x}}$, so

$$\sqrt{x + \frac{1}{y}} + \sqrt{y + \frac{1}{x}} \geq \sqrt{2\sqrt{\frac{x}{y}}} + \sqrt{2\sqrt{\frac{y}{x}}}.$$

Applying AM-GM to the sum on the right, we have

$$\frac{\sqrt{2\sqrt{\frac{x}{y}}} + \sqrt{2\sqrt{\frac{y}{x}}}}{2} \geq \sqrt{\sqrt{2\sqrt{\frac{x}{y}}} \cdot \sqrt{2\sqrt{\frac{y}{x}}}} = \sqrt{\sqrt{2 \cdot 2\sqrt{\frac{x}{y} \cdot \frac{y}{x}}}} = \sqrt{2}.$$

Therefore, we have $\sqrt{2\sqrt{\frac{x}{y}}} + \sqrt{2\sqrt{\frac{y}{x}}} \geq 2\sqrt{2}$, so $\sqrt{x + \frac{1}{y}} + \sqrt{y + \frac{1}{x}} \geq 2\sqrt{2}$, as desired.

We also could have started off by applying AM-GM to $\sqrt{x + \frac{1}{y}}$ and $\sqrt{y + \frac{1}{x}}$, which gives

$$\frac{\sqrt{x + \frac{1}{y}} + \sqrt{y + \frac{1}{x}}}{2} \geq \sqrt{\sqrt{x + \frac{1}{y}} \cdot \sqrt{y + \frac{1}{x}}} = \sqrt{\sqrt{xy + 1 + 1 + \frac{1}{xy}}} = \sqrt[4]{2 + xy + \frac{1}{xy}}.$$

By AM-GM, we have $\frac{xy+\frac{1}{xy}}{2} \geq \sqrt{xy \cdot \frac{1}{xy}} = 1$, so $xy + \frac{1}{xy} \geq 2$. Therefore, we have

$$\frac{\sqrt{x + \frac{1}{y}} + \sqrt{y + \frac{1}{x}}}{2} \geq \sqrt[4]{2 + xy + \frac{1}{xy}} \geq \sqrt[4]{4} = \sqrt{2},$$

and multiplying by 2 gives the desired inequality.

12.3.5   We work backwards. Expanding the left side gives us $1 + \frac{1}{x} + \frac{1}{y} + \frac{1}{xy} \geq 9$. Simplifying this and writing the greater side with a common denominator gives $\frac{x+y+1}{xy} \geq 8$. Since $x + y = 1$, we have $\frac{2}{xy} \geq 8$, or $\frac{1}{xy} \geq 4$. Finally, since xy is positive, we take the reciprocal of both sides to get $xy \leq \frac{1}{4}$. A product on the lesser side—this is a job for AM-GM. We wish to show that $xy \leq \frac{1}{4}$. AM-GM gives us $\frac{x+y}{2} \geq \sqrt{xy}$. Since $x + y = 1$, we have $\sqrt{xy} \leq \frac{1}{2}$. Squaring both sides gives the desired $xy \leq \frac{1}{4}$. Starting from here and walking backwards through our steps above (which are all reversible), we have $\left(1 + \frac{1}{x}\right)\left(1 + \frac{1}{y}\right) \geq 9$, as desired.

12.3.6   Note that $\dfrac{x^2 + 2}{\sqrt{x^2 + 1}} = \dfrac{x^2 + 1 + 1}{\sqrt{x^2 + 1}} = \dfrac{x^2 + 1}{\sqrt{x^2 + 1}} + \dfrac{1}{\sqrt{x^2 + 1}} = \sqrt{x^2 + 1} + \dfrac{1}{\sqrt{x^2 + 1}}$. By AM-GM, we have

$$\frac{\sqrt{x^2 + 1} + \frac{1}{\sqrt{x^2+1}}}{2} \geq \sqrt{\sqrt{x^2 + 1} \cdot \frac{1}{\sqrt{x^2 + 1}}} = 1,$$

so we have the desired $\dfrac{x^2 + 2}{\sqrt{x^2 + 1}} \geq 2$.

12.3.7   By AM-GM, we have $a + b \geq 2\sqrt{ab}$ and $c + d \geq 2\sqrt{cd}$, so

$$\frac{a + b + c + d}{4} \geq \frac{2\sqrt{ab} + 2\sqrt{cd}}{4} = \frac{\sqrt{ab} + \sqrt{cd}}{2}.$$

Then applying the AM-GM Inequality to the numbers $\sqrt{ab}$ and $\sqrt{cd}$, we have

$$\frac{\sqrt{ab} + \sqrt{cd}}{2} \geq \sqrt{\sqrt{ab}\sqrt{cd}} = \sqrt[4]{abcd},$$

so $\dfrac{a + b + c + d}{4} \geq \dfrac{\sqrt{ab} + \sqrt{cd}}{2} \geq \sqrt[4]{abcd}$.

Exercises for Section 12.4

12.4.1   By AM-GM, we have

$$a^2b + ab^2 + a^2c + ac^2 + b^2c + bc^2 \geq 6\sqrt[6]{a^2b \cdot ab^2 \cdot a^2c \cdot ac^2 \cdot b^2c \cdot bc^2} = 6\sqrt[6]{a^6b^6c^6} = 6abc.$$

12.4.2   By AM-GM, we have $\dfrac{a}{b} + \dfrac{b}{c} + \dfrac{c}{a} \geq 3\sqrt[3]{\dfrac{a}{b} \cdot \dfrac{b}{c} \cdot \dfrac{c}{a}} = 3$.

12.4.3   By AM-GM, we have $xy + \dfrac{2}{x} + \dfrac{4}{y} \geq 3\sqrt[3]{xy \cdot \dfrac{2}{x} \cdot \dfrac{4}{y}} = 3\sqrt[3]{8} = 6$. However, we must check whether the expression can achieve the value of 6.

In AM-GM, equality occurs if and only if all the terms are equal. In this case, we must have $xy = 2/x = 4/y = 2$. Solving, we find $x = 1$ and $y = 2$, so the minimum value is $\boxed{6}$.

12.4.4 By AM-GM, we have $p^2 + p + 1 \geq 3\sqrt[3]{p^2 \cdot p \cdot 1} = 3p$. Similarly, $q^2 + q + 1 \geq 3q$, $r^2 + r + 1 \geq 3r$, and $s^2 + s + 1 \geq 3s$. Multiplying all four inequalities, we obtain the desired inequality

$$(p^2 + p + 1)(q^2 + q + 1)(r^2 + r + 1)(s^2 + s + 1) \geq 81pqrs.$$

12.4.5 By AM-GM, we have

$$\frac{a_2}{a_1} + \frac{a_3}{a_2} + \cdots + \frac{a_n}{a_{n-1}} + \frac{a_1}{a_n} \geq n\sqrt[n]{\frac{a_2}{a_1} \cdot \frac{a_3}{a_2} \cdots \frac{a_n}{a_{n-1}} \cdot \frac{a_1}{a_n}} = n,$$

and

$$\frac{a_3}{a_1} + \frac{a_4}{a_2} + \cdots + \frac{a_1}{a_{n-1}} + \frac{a_2}{a_n} \geq n\sqrt[n]{\frac{a_3}{a_1} \cdot \frac{a_4}{a_2} \cdots \frac{a_1}{a_{n-1}} \cdot \frac{a_2}{a_n}} = n.$$

Adding these inequalities, we obtain the desired inequality $\dfrac{a_2 + a_3}{a_1} + \dfrac{a_3 + a_4}{a_2} + \cdots + \dfrac{a_n + a_1}{a_{n-1}} + \dfrac{a_1 + a_2}{a_n} \geq 2n$.

12.4.6 The denominator of the fraction is particularly difficult to deal with. If we had an $a - b$ term in a numerator, then we could use AM-GM to eliminate the denominator in the GM. This suggests writing a as $b + (a - b)$. By AM-GM, we then have:

$$a + \frac{1}{(a-b)b} = (a-b) + b + \frac{1}{(a-b)b} \geq 3\sqrt[3]{(a-b) \cdot b \cdot \frac{1}{(a-b)b}} = 3.$$

However, we must check whether the expression can achieve the value of 3.

In AM-GM, equality occurs if and only if all the terms are equal. In this case, we must have

$$a - b = b = \frac{1}{(a-b)b} = 1.$$

Solving, we find $a = 2$ and $b = 1$. Therefore, the minimum value is $\boxed{3}$.

Exercises for Section 12.5

12.5.1 We have a squared sum on the lesser side and the product of sums on the greater side, so we apply the Cauchy-Schwarz Inequality. Cauchy tells us that

$$(x_1^2 + x_2^2 + \cdots + x_n^2)(y_1^2 + y_2^2 + \cdots + y_n^2) \geq (x_1 y_1 + x_2 y_2 + \cdots + x_n y_n)^2.$$

Comparing this to the greater side of our desired inequality, we let $x_1 = \sqrt{a_1 b_1}$, $x_2 = \sqrt{a_2 b_2}$, $\ldots$, $x_n = \sqrt{a_n b_n}$, $y_1 = \sqrt{\frac{a_1}{b_1}}$, $y_2 = \sqrt{\frac{a_2}{b_2}}$, $\ldots$, $y_n = \sqrt{\frac{a_n}{b_n}}$, and this gives us

$$(a_1 b_1 + a_2 b_2 + \cdots + a_n b_n)\left(\frac{a_1}{b_1} + \frac{a_2}{b_2} + \cdots + \frac{a_n}{b_n}\right) \geq (a_1 + a_2 + \cdots + a_n)^2,$$

as desired.

12.5.2 The sum of squares on the greater side, and the sum of products on the lesser side makes us think of Cauchy-Schwarz. But the sum of products isn't squared. We therefore try squaring the desired inequality, which gives

$$(a^2 + b^2 + c^2)^2 \geq (ab + bc + ca)^2.$$

This looks a lot more like Cauchy; we have a product of sums of squares on the greater side and the square of a sum of products on the lesser. We can see that this inequality is a direct result of Cauchy if we rearrange one of the factors of $a^2 + b^2 + c^2$. Applying Cauchy to $(a^2 + b^2 + c^2)(b^2 + c^2 + a^2)$ gives us

$$(a^2 + b^2 + c^2)(b^2 + c^2 + a^2) \geq (ab + bc + ca)^2.$$

Since a, b, and c are positive, taking the square root of both sides gives the desired inequality.

12.5.3 Let the lengths and the widths of the rectangles be $l_1, l_2, \ldots, l_6$ and $w_1, w_2, \ldots, w_6$, so $l_1^2 + l_2^2 + \cdots + l_6^2 = 40$ and $w_1^2 + w_2^2 + \cdots + w_6^2 = 20$. By Cauchy-Schwarz, we have

$$(l_1 w_1 + l_2 w_2 + \cdots + l_6 w_6)^2 \leq (l_1^2 + l_2^2 + \cdots + l_6^2)(w_1^2 + w_2^2 + \cdots + w_6^2) = 20 \cdot 40 = 800.$$

Therefore, the sum of the areas of the rectangles $l_1 w_1 + l_2 w_2 + \cdots + l_6 w_6$ is at most $\sqrt{800} = \boxed{20\sqrt{2}}$.

To show that the value $20\sqrt{2}$ is achievable, take $w_1, w_2, \ldots, w_6$ to be distinct, positive real numbers such that $w_1^2 + w_2^2 + \cdots + w_6^2 = 20$, and let $l_i = w_i \sqrt{2}$ for $1 \leq i \leq 6$. Then $l_1^2 + l_2^2 + \cdots + l_6^2 = 2(w_1^2 + w_2^2 + \cdots + w_6^2) = 40$, and

$$l_1 w_1 + l_2 w_2 + \cdots + l_6 w_6 = (w_1^2 + w_2^2 + \cdots + w_6^2)\sqrt{2} = 20\sqrt{2}.$$

12.5.4 By Cauchy-Schwarz, we have

$$(x_1^2 + x_2^2 + x_3^2 + x_4^2)(y_1^2 + y_2^2 + y_3^2 + y_4^2) \geq (x_1 y_1 + x_2 y_2 + x_3 x_3 + x_4 y_4)^2$$

for all real numbers x_i, y_i.

Taking $x_1 = \sqrt{a}$, $x_2 = \sqrt{b}$, $x_3 = \sqrt{c}$, $x_4 = \sqrt{d}$, and $y_1 = y_2 = y_3 = y_4 = 1$, we get

$$4(a + b + c + d) \geq (\sqrt{a} + \sqrt{b} + \sqrt{c} + \sqrt{d})^2.$$

Taking the square root of both sides, we obtain the desired inequality $2\sqrt{a + b + c + d} \geq \sqrt{a} + \sqrt{b} + \sqrt{c} + \sqrt{d}$.

12.5.5 By Cauchy-Schwarz, we have

$$(x_1^2 + x_2^2 + x_3^2 + x_4^2)(y_1^2 + y_2^2 + y_3^2 + y_4^2) \geq (x_1 y_1 + x_2 y_2 + x_3 x_3 + x_4 y_4)^2$$

for all real numbers x_i, y_i.

Taking $x_1 = \sqrt{a}$, $x_2 = \sqrt{b}$, $x_3 = \sqrt{c}$, $x_4 = \sqrt{d}$, and $y_1 = 1/\sqrt{a}$, $y_2 = 1/\sqrt{b}$, $y_3 = 2/\sqrt{c}$, $y_4 = 4/\sqrt{d}$, we get

$$(a + b + c + d)\left(\frac{1}{a} + \frac{1}{b} + \frac{4}{c} + \frac{16}{d}\right) \geq (1 + 1 + 2 + 4)^2 = 64.$$

Dividing both sides by $a + b + c + d$, we obtain the desired inequality $\dfrac{1}{a} + \dfrac{1}{b} + \dfrac{4}{c} + \dfrac{16}{d} \geq \dfrac{64}{a + b + c + d}$.

12.5.6 By Cauchy-Schwarz, we have $(a_1^2 + a_2^2)(b_1^2 + b_2^2) \geq (a_1 b_1 + a_2 b_2)^2$ for all real numbers a_i, b_i.

Taking $a_1 = 3$, $a_2 = 4$, and $b_1 = x$, $b_2 = y$, we get $25(x^2 + y^2) \geq (3x + 4y)^2$, so $\dfrac{(3x + 4y)^2}{x^2 + y^2} \leq 25$. Since $(3x + 4y)^2/(x^2 + y^2) = 25$ when $x = 3$ and $y = 4$, we conclude that the maximum value is $\boxed{25}$.

Exercises for Section 12.6

12.6.1 Let r, w, and b be the number of chips that are red, white, and blue, respectively. Then from the given information, we have $w/2 \leq b \leq r/3$ and $b + w \geq 55$. Then $w \leq 2b$, so $55 \leq b + w \leq 3b$. Therefore, $b \geq \frac{55}{3} = 18\frac{1}{3}$. Since b is an integer, we have $b \geq 19$. Then $r \geq 3b \geq 3 \cdot 19 = 57$.

Furthermore, if $r = 57$, $w = 38$, and $b = 19$, then the conditions in the problem are satisfied. Therefore, the minimum number of red chips is $\boxed{57}$.

12.6.2 Adding 9 to everything, we get $a + 2 = b + 12 = c = d + 31$. Therefore, $\boxed{c}$ is largest.

12.6.3 Let the dimensions of the box be a, b, and c. Then the diagonal length of the box is $\sqrt{a^2 + b^2 + c^2} = 12$, so $a^2 + b^2 + c^2 = 12^2 = 144$. By AM-GM, we have

$$\sqrt[3]{a^2b^2c^2} \le \frac{a^2 + b^2 + c^2}{3} = 48,$$

so $abc \le 48^{3/2} = 192\sqrt{3}$.

Equality occurs when $a = b = c$. In such a case, we have $3a^2 = 144$. Then $a^2 = 48$, so $a = b = c = \sqrt{48} = 4\sqrt{3}$. Therefore, the maximum volume is $\boxed{192\sqrt{3}}$.

12.6.4 Dividing the denominator into the numerator, we find

$$\frac{x^2 - 2x + 2}{2x - 2} = \frac{(x-1)^2 + 1}{2(x-1)} = \frac{x-1}{2} + \frac{1}{2(x-1)}.$$

Since $x > 1$, both $(x-1)/2$ and $1/[2(x-1)]$ are positive, so by AM-GM, we have

$$\frac{x-1}{2} + \frac{1}{2(x-1)} \ge 2\sqrt{\frac{x-1}{2} \cdot \frac{1}{2(x-1)}} = 2\sqrt{\frac{1}{4}} = 2 \cdot \frac{1}{2} = 1.$$

Furthermore, when $x = 2$, we have $(x^2 - 2x + 2)/(2x - 2) = 1$, so the minimum value is $\boxed{1}$.

12.6.5 By Cauchy-Schwarz, we have $(a_1^2 + a_2^2)(b_1^2 + b_2^2) \ge (a_1b_1 + a_2b_2)^2$ for all real numbers a_i, b_i.

Taking $a_1 = 5$, $a_2 = 12$, and $b_1 = x$, $b_2 = y$, we get $169(x^2 + y^2) \ge (5x + 12y)^2 = 60^2$, so $x^2 + y^2 \ge \frac{60^2}{169}$. Taking the square root of both sides, we get $\sqrt{x^2 + y^2} \ge \frac{60}{13}$.

To show that equality can be achieved, we turn to the equality condition of Cauchy-Schwarz, which tells us that $5/x = 12/y$, so $x = 5y/12$. Combining this with $5x + 12y = 60$ gives us $x = 5 \cdot 60/169$ and $y = 12 \cdot 60/169$. Substituting these into $\sqrt{x^2 + y^2}$ gives

$$\sqrt{x^2 + y^2} = \sqrt{\frac{5^2 \cdot 60^2}{169^2} + \frac{12^2 \cdot 60^2}{169^2}} = \sqrt{\frac{(25 + 144) \cdot 60^2}{169^2}} = \sqrt{\frac{60^2}{169}} = \frac{60}{13}.$$

Therefore, the minimum value of $\sqrt{x^2 + y^2}$ is $\boxed{60/13}$.

See if you can find another solution using analytic geometry.

Review Problems

12.32

(a) Since $z \ne 0$, $z^2 > 0$. Hence, we can multiply both sides of the inequality $x > y$ by z^2 to get $xz^2 > yz^2$. Therefore, the given inequality is always $\boxed{\text{true}}$.

(b) Since $z \ne 0$, $z^2 > 0$. Hence, we can divide both sides of the inequality $x > y$ by z^2 to get $x/z^2 > y/z^2$. Therefore, the given inequality is always $\boxed{\text{true}}$.

12.33   We work backwards. Because $a > b$, we have $a - b > 0$, so multiplying by $a - b$ doesn't change the direction of the inequality, and gives $1 - ab > a - b$. Moving all terms to the lesser side gives $ab + a - b - 1 < 0$, and factoring gives us $(a - 1)(b + 1) < 0$. Because $a < 1$ and $b > 0$, we have $a - 1 < 0$ and $b + 1 > 0$, so $(a - 1)(b + 1) < 0$. We can now walk backwards through our steps above. Expanding the left side and rearranging gives $1 - ab > a - b$. Because $a - b > 0$, dividing both sides of $1 - ab > a - b$ gives the desired $(1 - ab)/(a - b) > 1$.

12.34   First, we show that $x^2 > ax$. We have $x^2 - ax = x(x - a)$. Because $x < a$, we have $x - a < 0$. Since $x < 0$ also, we have $x(x - a) > 0$. Expanding the left and rearranging gives $x^2 > ax$. Next, we show that $ax > a^2$. We have $ax - a^2 = a(x - a)$. As before, because $a < 0$ and $x - a < 0$, we have $a(x - a) > 0$, so $ax > a^2$. Therefore, we have $x^2 > ax > a^2$.

12.35   Subtracting x from both sides of $x - y > x$, we have $-y > 0$, so $y < 0$. Subtracting y from both sides of $x + y < y$, we have $x < 0$. Therefore, x and y are both negative.

12.36   Moving all the terms to one side, we obtain the equivalent inequality $a^3 - a^2b - ab^2 + b^3 \geq 0$. The left side factors as

$$a^2(a - b) - b^2(a - b) = (a^2 - b^2)(a - b) = (a + b)(a - b)^2.$$

By the Trivial Inequality, we have $(a - b)^2 \geq 0$, and we are given that $a + b \geq 0$, so $(a + b)(a - b)^2 \geq 0$. All our steps are reversible, so the desired inequality holds.

12.37   Taking the reciprocal of everything, we get $\frac{2001}{5} < \frac{a+b}{a} < \frac{2001}{4}$. Subtracting 1 from everything, this inequality chain becomes $\frac{1996}{5} < \frac{b}{a} < \frac{1997}{4}$, so $399 + \frac{1}{5} < \frac{b}{a} < 499 + \frac{1}{4}$. Therefore, $\frac{b}{a}$ must be one of the values $400, 401, \ldots, 499$. There are $\boxed{100}$ possible values.

12.38   Substituting, we find $s^2 - 4p = (x + y)^2 - 4xy = x^2 - 2xy + y^2 = (x - y)^2 \geq 0$. Therefore, $s^2 \geq 4p$.

12.39   By AM-GM, we have $a^2b + b^2c + c^2a \geq 3\sqrt[3]{a^2b \cdot b^2c \cdot c^2a} = 3\sqrt[3]{a^3b^3c^3} = 3abc$.

12.40   By AM-GM on the $n + 1$ numbers $a^{n+1}, b^{n+1}, b^{n+1}, \ldots, b^{n+1}$ (where b^{n+1} appears n times), we have

$$\frac{a^{n+1} + nb^{n+1}}{n + 1} \geq \sqrt[n+1]{a^{n+1}b^{n+1}b^{n+1} \cdots b^{n+1}} = \sqrt[n+1]{a^{n+1}b^{n(n+1)}} = ab^n.$$

12.41   If $d < 100$, then $d \leq 99$. Then $c < 4d \leq 4 \cdot 99 = 396$, so $c \leq 395$. Then $b < 3c \leq 3 \cdot 395 = 1185$, so $b \leq 1184$. Finally, $a < 2b \leq 2 \cdot 1184 = 2368$, so $a \leq 2367$. Furthermore, the values $a = 2367$, $b = 1184$, $c = 395$, and $d = 99$ satisfy the conditions in the problem, so the maximum value of a is $\boxed{2367}$.

12.42   Since $\frac{a}{b} < \frac{c}{d}$ and b and d are positive, cross-multiplying gives $ad < bc$. First, we prove $\frac{a}{b} < \frac{a+c}{b+d}$. Cross-multiplying gives the equivalent inequality $a(b + d) < b(a + c)$, or $ab + ad < ab + bc$. This simplifies as $ad < bc$, which is true.

Next, we prove $\frac{a+c}{b+d} < \frac{c}{d}$. Cross-multiplying gives $d(a + c) < c(b + d)$, or $ad + cd < bc + cd$. Again, this simplifies as $ad < bc$.

All our steps are reversible, so we have $\frac{a}{b} < \frac{a+c}{b+d} < \frac{c}{d}$, as desired.

The inequality $\frac{a}{b} < \frac{a+c}{b+d} < \frac{c}{d}$ has a nice geometric interpretation. If we graph the points (b, a), (d, c) and $(b + d, a + c)$ in the coordinate plane, then along with the origin, these points are the vertices of a parallelogram that lies within the first quadrant (which is the region in the Cartesian plane for which $x, y > 0$).

Then a/b, c/d, are $(a + c)/(b + d)$ the slopes of the lines joining the points to the origin, and we see that the slope $(a + c)/(b + d)$ must always lie between the slopes a/b and c/d.

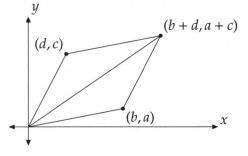

12.43 We have $A = (x + y + z)/3$ and $B = \dfrac{\frac{x+y}{2} + z}{2} = \dfrac{x + y + 2z}{4}$. Therefore, we have

$$B - A = \frac{x + y + 2z}{4} - \frac{x + y + z}{3} = \frac{2z - x - y}{12}.$$

Since $z > y > x$, we have $2z - x - y > 0$. Therefore, $\boxed{B > A}$.

12.44 By AM-GM, we have

$$\frac{a}{2b} + \frac{b}{4c} + \frac{c}{8a} \geq 3\sqrt[3]{\frac{a}{2b} \cdot \frac{b}{4c} \cdot \frac{c}{8a}} = 3\sqrt[3]{\frac{1}{64}} = \frac{3}{4}.$$

Equality occurs if and only if $a/(2b) = b/(4c) = c/(8a) = 1/4$. For example, we may take $(a, b, c) = (1, 2, 2)$. Therefore, the minimum value is $\boxed{3/4}$.

12.45 Let the dimensions of the box be a, b, and c, in centimeters. Then the surface area of the box is $2ab + 2ac + 2bc = 96$, so $ab + ac + bc = 48$. By AM-GM, we have

$$16 = \frac{ab + ac + bc}{3} \geq \sqrt[3]{ab \cdot ac \cdot bc} = \sqrt[3]{a^2 b^2 c^2},$$

so $abc \leq 16^{3/2} = 64$. Equality occurs when $a = b = c = 4$. Therefore, the maximum volume is $\boxed{64 \text{ cm}^3}$.

12.46 Dividing the denominator into the numerator, we find

$$\frac{4x^2 + 8x + 13}{6(1 + x)} = \frac{4(x + 1)^2 + 9}{6(x + 1)} = \frac{2}{3}(x + 1) + \frac{3}{2(x + 1)}.$$

By AM-GM, we have

$$\frac{2}{3}(x + 1) + \frac{3}{2(x + 1)} \geq 2\sqrt{\frac{2}{3}(x + 1) \cdot \frac{3}{2(x + 1)}} = 2.$$

Furthermore, equality occurs if and only if $\frac{2}{3}(x + 1) = \frac{3}{2(x+1)} = 1$. Solving, we find $x = \frac{1}{2}$. Therefore, the minimum value is $\boxed{2}$.

12.47 Let $n = 50 + 8k$, where k is a nonnegative integer. Then the number of red balls is $49 + 7k$, so

$$\frac{49 + 7k}{50 + 8k} \geq \frac{9}{10}.$$

Cross-multiplying, we get $490 + 70k \geq 450 + 72k$, so $40 \geq 2k$, or $k \leq 20$. Then $n = 50 + 8k \leq 50 + 8 \cdot 20 = \boxed{210}$.

12.48 Since the dealer bought n radios for d dollars, each radio cost $\frac{d}{n}$ dollars. The dealer sold the two radios to the bazaar for a total of $2 \cdot \frac{1}{2} \cdot \frac{d}{n} = \frac{d}{n}$ dollars. He sold the remaining $n - 2$ radios for $\frac{d}{n} + 8$ dollars each. The dealer initially paid d dollars for the radios, so his overall profit was

$$\frac{d}{n} + (n - 2)\left(\frac{d}{n} + 8\right) - d = 72$$

dollars.

Expanding the left side, we get $\frac{d}{n} + d + 8n - \frac{2d}{n} - 16 - d = 72$, which simplifies as $8n - \frac{d}{n} = 88$, so $\frac{d}{n} = 8n - 88 = 8(n - 11)$.

Since n is a positive integer and d/n is a positive quantity, n must be at least 12. Furthermore, if $n = 12$, then we get $d/12 = 8$, so $d = 96$, a positive integer. Therefore, n may equal 12, which means the minimum value of n is $\boxed{12}$.

12.49   By Cauchy-Schwarz, we have

$$(1 + 1 + 1 + 1)(a^2 + b^2 + c^2 + d^2) \geq (a + b + c + d)^2,$$

so $4(a^2 + b^2 + c^2 + d^2) \geq 1$. Therefore, $a^2 + b^2 + c^2 + d^2 \geq 1/4$.

12.50   Let x be the amount of money the gambler starts with. Each win multiplies his money by a factor of 3/2, and each loss multiplies it by a factor of 1/2. Hence, if he wins n times and loses n times, the amount of money he has at the end is

$$\left(\frac{3}{2}\right)^n \left(\frac{1}{2}\right)^n x = \left(\frac{3}{4}\right)^n x.$$

Since $3/4 < 1$, we have $(\frac{3}{4})^n < 1$, so the gambler $\boxed{\text{lost}}$ money.

12.51   It doesn't matter which order the three numbers are in, so we can assign k to the smallest number and n to the largest, so that $k \leq m \leq n$.

If $k = 4$, then the smallest possible value of $k + m + n$ is 12, which occurs when $k = m = n = 4$. We clearly can't do better than this if $k > 4$, so we must only check $k = 2$ and $k = 3$.

If $k = 2$, we have $\frac{1}{m} + \frac{1}{n} < \frac{1}{2}$. If $m = 3$, then we must have $n > 6$, so the best we can do if $k = 2$ and $m = 3$ is $k + m + n = 2 + 3 + 7 = 12$. If $m = 4$, then we must have $n > 4$, so the best we can do if $k = 2$ and $m = 4$ is $k + m + n = 2 + 4 + 5 = 11$. If $m \geq 5$, then we have $k + m + n \geq 2 + 2m \geq 12$, so we can't beat $2 + 4 + 5$.

Finally, we check $k = 3$, for which we must have $\frac{1}{m} + \frac{1}{n} < \frac{2}{3}$. If $m = 3$, then we must have $n > 3$, so the best we can do if $k = m = 3$ is $k + m + n = 3 + 3 + 4 = 10$. For all greater m, we have $k + m + n \leq 3 + 2m \leq 3 + 2 \cdot 4 = 11$.

Therefore, the smallest possible value of $k + m + n$ is $\boxed{10}$.

12.52   We work backwards. Expanding both sides gives the equivalent inequality $a^2c^2 - a^2d^2 - b^2c^2 + b^2d^2 \leq a^2c^2 - 2abcd + b^2d^2$. Simplifying this inequality gives $a^2d^2 + b^2c^2 - 2abcd \geq 0$. Factoring then gives $(ad - bc)^2 \geq 0$, which is true by the Trivial Inequality. All our steps are reversible, so we have $(a^2 - b^2)(c^2 - d^2) \leq (ac - bd)^2$.

12.53   We work backwards. We multiply both sides by the denominators of the fractions and we have $(a^2 + b^2 + c^2)(a^4 + b^4 + c^4) \geq (a^3 + b^3 + c^3)^2$ to prove. A sum of squares on the lesser side and product of sums on the greater side—this is a job for Cauchy. Applying Cauchy to the sequences a, b, c and a^2, b^2, c^2 gives us $((a)^2 + (b)^2 + (c)^2)((a^2)^2 + (b^2)^2 + (c^2)^2) \geq (a \cdot a^2 + b \cdot b^2 + c \cdot c^2)^2$, which gives us $(a^2 + b^2 + c^2)(a^4 + b^4 + c^4) \geq (a^3 + b^3 + c^3)^2$. Since all the variables are positive, we can divide both sides by $(a^3 + b^3 + c^3)(a^4 + b^4 + c^4)$ to get the original desired inequality.

12.54

(a)   We have $PQ = PG + GQ = a + b$. Since $\overline{PQ}$ is a diameter, the radius of the semicircle is $PQ/2 = (a + b)/2$. Since $\overline{AM}$ is a radius, we have $AM = (a + b)/2$.

Since $\overline{PQ}$ is a diameter, $\angle PMQ = 90°$. Therefore, $\angle PMG = \angle PMQ - \angle GMQ = 90° - \angle GMQ = \angle MQG$. It follows that right triangles PMG and MQG are similar. Therefore, $\frac{PG}{GM} = \frac{GM}{GQ}$, so $GM^2 = PG \cdot GQ = ab$, which means $GM = \sqrt{ab}$.

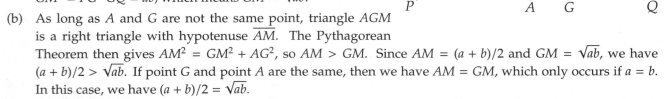

(b)   As long as A and G are not the same point, triangle AGM is a right triangle with hypotenuse $\overline{AM}$. The Pythagorean Theorem then gives $AM^2 = GM^2 + AG^2$, so $AM > GM$. Since $AM = (a + b)/2$ and $GM = \sqrt{ab}$, we have $(a + b)/2 > \sqrt{ab}$. If point G and point A are the same, then we have $AM = GM$, which only occurs if $a = b$. In this case, we have $(a + b)/2 = \sqrt{ab}$.

For any positive values a and b, we can perform our construction described in part (a) and establish that $(a + b)/2 \geq \sqrt{ab}$. Therefore, we have a geometric proof of the AM-GM Inequality. (The cases where $a = 0$ or

$b = 0$ correspond to point M coinciding with P or Q, respectively. In each case, AM equals the radius of the circle and $GM = 0$, so we clearly have $AM > GM$ and $(a + b)/2 \geq \sqrt{ab}$.)

12.55 By Cauchy-Schwarz, we have

$$(3^2 + 4^2 + 12^2)(x^2 + y^2 + z^2) \geq (3x + 4y + 12z)^2.$$

Since $x^2 + y^2 + z^2 = 1$, this simplifies as $169 \geq (3x + 4y + 12z)^2$. Therefore, $3x + 4y + 12z \leq 13$.

From the equality condition of the Cauchy-Schwarz Inequality, we have equality when $x/3 = y/4 = z/12$. Combining this with $x^2 + y^2 + z^2 = 1$ gives $x = 3/13$, $y = 4/13$, and $z = 12/13$. Substituting these values in $3x + 4y + 12z$ gives $3x + 4y + 12z = \frac{3^2 + 4^2 + 12^2}{13} = \frac{169}{13} = 13$. Therefore, the maximum value of $3x + 4y + 12z$ is $\boxed{13}$.

Challenge Problems

12.56 Consider $3a^4 - 4a^3b + b^4$ as a polynomial in a. When $a = b$, then $3a^4 - 4a^3b + b^4 = 3b^4 - 4b^4 + b^4 = 0$, so by the Factor Theorem, $a - b$ is a factor. Dividing by $a - b$, we find

$$3a^4 - 4a^3b + b^4 = (a - b)(3a^3 - a^2b - ab^2 - b^3).$$

Similarly, consider $3a^3 - a^2b - ab^2 - b^3$ as a polynomial in a. When $a = b$, then $3a^3 - a^2b - ab^2 - b^3 = 3b^3 - b^3 - b^3 - b^3 = 0$, so by the Factor Theorem, $a - b$ is again a factor. Dividing by $a - b$, we find

$$(a - b)(3a^3 - a^2b - ab^2 - b^3) = (a - b)^2(3a^2 + 2ab + b^2).$$

We can write $3a^2 + 2ab + b^2 = (a + b)^2 + 2a^2$. By the Trivial Inequality, $(a + b)^2 + 2a^2 \geq 0$, and $(a - b)^2 \geq 0$. Therefore, $3a^4 - 4a^2b + b^4 = (a - b)^2(3a^2 + 2ab + b^2) \geq 0$.

12.57 We are given a product, $xy^2z^3 = 108$, and we want to minimize a sum, $x + y + z$, so we think of using AM-GM. However, we cannot use AM-GM directly. For example, by AM-GM, we have

$$\sqrt[6]{xy^2z^3} = \sqrt[6]{x \cdot y \cdot y \cdot z \cdot z \cdot z} \leq \frac{x + y + y + z + z + z}{6} = \frac{x + 2y + 3z}{6},$$

but the quantity $x + 2y + 3z$ does not directly relate to the quantity $x + y + z$.

To compare a sum to the product xy^2z^3 using AM-GM, we need one factor of x, two factors of y, and three factors of z. We can create these factors by breaking up the sum $x + y + z$ as $x + y/2 + y/2 + z/3 + z/3 + z/3$. Then by AM-GM, we have

$$\frac{x + y + z}{6} = \frac{x + y/2 + y/2 + z/3 + z/3 + z/3}{6} \geq \sqrt[6]{x \cdot \frac{y}{2} \cdot \frac{y}{2} \cdot \frac{z}{3} \cdot \frac{z}{3} \cdot \frac{z}{3}} = \sqrt[6]{\frac{xy^2z^3}{108}} = 1.$$

Therefore, $x + y + z \geq 6$.

Equality occurs if and only if $x = y/2 = z/3 = 1$. Solving, we find that $(x, y, z) = (1, 2, 3)$. Hence, the minimum value of $x + y + z$ is $\boxed{6}$.

12.58 By AM-GM, we have $1 + a_1 \geq 2\sqrt{a_1}$. Similarly, $1 + a_2 \geq 2\sqrt{a_2}, \ldots, 1 + a_n \geq 2\sqrt{a_n}$. Multiplying all n inequalities, we get $(1 + a_1)(1 + a_2) \cdots (1 + a_n) \geq 2^n \sqrt{a_1 a_2 \cdots a_n} = 2^n$.

12.59 Moving all the terms to one side and expanding, we obtain $3x^2 + y^2 + z^2 - 2xy - 2xz = 0$. We can rewrite the left side as

$$x^2 + (x^2 - 2xy + y^2) + (x^2 - 2xz + z^2) = x^2 + (x - y)^2 + (x - z)^2,$$

so we have $x^2 + (x - y)^2 + (x - z)^2 = 0$. If a sum of squares is equal to 0, then each square must be equal to 0. Therefore, $x = x - y = x - z = 0$, so the only solution is $(x, y, z) = \boxed{(0,0,0)}$.

12.60 There are two different variables in each equation, so there is not much we can do with each individual equation. However, adding all three equations gives us the chance to combine all three variables:

$$a^2 + 8a + b^2 + 10b + c^2 + 4c = -45.$$

Completing the square in a, b, and c, we get $(a + 4)^2 + (b + 5)^2 + (c + 2)^2 = 0$.

If a sum of squares is 0, then each square must be 0, so the only possible solution is $(a, b, c) = (-4, -5, -2)$. We must check that this solution satisfies each of the given equations. It does, so the only solution to the system of equations is $(a, b, c) = \boxed{(-4, -5, -2)}$.

12.61 Completing the square in a, we get $a^2 + ab + b^2 = \left(a + \frac{b}{2}\right)^2 + \frac{3b^2}{4} \geq 0$. Equality occurs if and only if $a + \frac{b}{2} = b = 0$, or $(a, b) = \boxed{(0,0)}$.

12.62 Rearranging the given equation gives $9\sqrt{x} + \frac{36}{\sqrt{x}} + \sqrt{y} + \frac{9}{\sqrt{y}} = 42$. By AM-GM, we have

$$9\sqrt{x} + \frac{36}{\sqrt{x}} \geq 2\sqrt{9\sqrt{x} \cdot \frac{36}{\sqrt{x}}} = 2\sqrt{324} = 36,$$

and

$$\sqrt{y} + \frac{9}{\sqrt{y}} \geq 2\sqrt{\sqrt{y} \cdot \frac{9}{\sqrt{y}}} = 2\sqrt{9} = 6,$$

so adding these inequalities gives $9\sqrt{x} + \frac{36}{\sqrt{x}} + \sqrt{y} + \frac{9}{\sqrt{y}} \geq 36 + 6 = 42$. Hence, we must have equality in both of our instances of AM-GM in order for the equation in the problem to hold.

In the first application of AM-GM, equality occurs if and only if $9\sqrt{x} = \frac{36}{\sqrt{x}}$, so $x = 4$. In the second inequality, equality occurs if and only if $\sqrt{y} = \frac{9}{\sqrt{y}}$, so $y = 9$. Therefore, the only solution is $(x, y) = \boxed{(4,9)}$.

12.63 We focus on the complicated fractions on the greater side. In $\frac{x^2 + y^2}{x + y}$, we have a sum of squares in the numerator and a sum in the denominator. If we can relate the sum of squares $x^2 + y^2$ to the sum $x + y$ in some way, perhaps we can simplify the fraction. The sum of squares suggests Cauchy. The two-variable version of Cauchy tells us that $(a_1^2 + a_2^2)(b_1^2 + b_2^2) \geq (a_1 b_1 + a_2 b_2)^2$. Letting $a_1 = x$ and $a_2 = y$ to get a sum of squares on the greater side, we have $(x^2 + y^2)(b_1^2 + b_2^2) \geq (b_1 x + b_2 y)^2$. Now, we see that letting $b_1 = b_2 = 1$ gives us the sum $x + y$ on the lesser side, and we have $2(x^2 + y^2) \geq (x + y)^2$. Aha! Since x and y are positive, dividing by $2(x + y)$ gives us $\frac{x^2 + y^2}{x + y} \geq \frac{x + y}{2}$. Similarly, we have $\frac{y^2 + z^2}{y + z} \geq \frac{y + z}{2}$ and $\frac{z^2 + x^2}{z + x} \geq \frac{z + x}{2}$, and adding these three inequalities gives us the desired result.

12.64 Consider the seven integers $A_1, A_2, \ldots, A_7$. Since each number a_i is between 2 and 3, each integer A_i should be equal to 2 or 3. Let x be the number of 2s, and let y be the number of 3s, so $x + y = 7$. Since the sum of the A_is is equal to 19, we have $2x + 3y = 19$. Solving this system of equations, we find $x = 2$ and $y = 5$.

Two of the A_i must equal 2, so the maximum of the $|A_i - a_i|$ must be at least 0.61, which occurs when we let $A_1 = A_2 = 2$. If we let any of the other A_i equal 2, then we have $|A_i - a_i| > 0.61$ for that value of i.

No matter how we assign the five A_i that equal 3, we cannot have $|A_i - a_i|$ greater than $|3 - 2.56| = 0.44$ if $A_i = 3$. Therefore, $M = 0.61$, so $100M = \boxed{61}$.

12.65 First, we factor the numerator and denominator of the fraction:

$$\frac{x^{2n+1} - 1}{x^{n+1} - x^n} = \frac{(x - 1)(x^{2n} + x^{2n-1} + \cdots + 1)}{x^n(x - 1)} = \frac{x^{2n} + x^{2n-1} + \cdots + 1}{x^n}.$$

Then, we have $\dfrac{x^{2n} + x^{2n-1} + \cdots + 1}{x^n} = x^n + x^{n-1} + \cdots + x + 1 + \dfrac{1}{x} + \cdots + \dfrac{1}{x^{n-1}} + \dfrac{1}{x^n}.$

By AM-GM, we have

$$x^n + \frac{1}{x^n} \geq 2\sqrt{x^n \cdot \frac{1}{x^n}} = 2,$$

$$x^{n-1} + \frac{1}{x^{n-1}} \geq 2\sqrt{x^{n-1} \cdot \frac{1}{x^{n-1}}} = 2,$$

$$\vdots$$

$$x + \frac{1}{x} \geq 2\sqrt{x \cdot \frac{1}{x}} = 2.$$

Adding all n inequalities, and adding 1 to both sides, we obtain the desired inequality

$$x^n + x^{n-1} + \cdots + x + 1 + \frac{1}{x} + \cdots + \frac{1}{x^{n-1}} + \frac{1}{x^n} \geq 2n + 1.$$

12.66 We wish to maximize a product, $x^3(4 - x)$, so we think of using AM-GM. However, we cannot apply AM-GM directly. For example, if we apply AM-GM to the numbers x, x, x, and $4 - x$, then we get

$$\sqrt[4]{x^3(4 - x)} \leq \frac{x + x + x + (4 - x)}{4} = \frac{2x + 4}{4} = \frac{x + 2}{2}.$$

Since $x \leq 4$, we have $(x + 2)/2 \leq (4 + 2)/2 = 3$, so $x^3(4 - x) = 3^4 = 81$. However, equality occurs if and only if $x = 4 - x$, which means $x = 2$, and this $x = 2$ gives us $x^3(4 - x) = 16$, not 81, so this approach does not give us the maximum.

What we really want to do is apply AM-GM to a set of numbers whose sum is a constant. We can achieve this by multiplying the factor $4 - x$ by 3, to get $12 - 3x$. Then by the AM-GM Inequality on the numbers x, x, x, and $12 - 3x$, we get

$$\sqrt[4]{x^3(12 - 3x)} \leq \frac{x + x + x + (12 - 3x)}{4} = 3,$$

so $x^3(12 - 3x) \leq 3^4 = 81$, which means $x^3(4 - x) \leq 27$. In this case, equality occurs if and only if $x = 12 - 3x$, or $x = 3$, which does give $x^3(4 - x) = 27$. Therefore, the maximum value of $x^3(4 - x)$ is $\boxed{27}$.

12.67 To generate the expression $ab + ac + bc$, we can square the equation $a + b + c = 3$ to get

$$a^2 + b^2 + c^2 + 2ab + 2ac + 2bc = 9.$$

In the text, we showed that $a^2 + b^2 + c^2 \geq ab + ac + bc$, so we have

$$9 = a^2 + b^2 + c^2 + 2ab + 2ac + 2bc \geq 3ab + 3ac + 3bc.$$

Therefore, $ab + ac + bc \leq 9/3 = 3$.

12.68 We claim that $\boxed{400! \cdot 400! \cdot 200! < 600! \cdot 300! \cdot 100! < 500! \cdot 500! < 700! \cdot 300! < 1000!}$.

To show that $400! \cdot 400! \cdot 200! < 600! \cdot 300! \cdot 100!$, we consider the quotient of the two numbers:

$$\frac{600! \cdot 300! \cdot 100!}{400! \cdot 400! \cdot 200!} = \frac{600!}{400!} \cdot \frac{300!}{400!} \cdot \frac{100!}{200!} = 600 \cdot 599 \cdots 401 \cdot \frac{1}{400 \cdot 399 \cdots 301} \cdot \frac{1}{200 \cdot 199 \cdots 101}$$

$$= \frac{600 \cdot 599 \cdots 401}{400 \cdot 399 \cdots 301 \cdot 200 \cdot 199 \cdots 101} = \frac{600}{400} \cdot \frac{599}{399} \cdots \frac{501}{301} \cdot \frac{500}{200} \cdot \frac{499}{199} \cdots \frac{401}{101}.$$

Each fraction is greater than 1, so $400! \cdot 400! \cdot 200! < 600! \cdot 300! \cdot 100!$

To show that $600! \cdot 300! \cdot 100! < 500! \cdot 500!$, we consider the quotient of the two numbers:

$$\begin{aligned}
\frac{500! \cdot 500!}{600! \cdot 300! \cdot 100!} &= \frac{500!}{600!} \cdot \frac{500!}{300!} \cdot \frac{1}{100!} \\
&= \frac{1}{600 \cdot 599 \cdots 501} \cdot 500 \cdot 499 \cdots 301 \cdot \frac{1}{100!} \\
&= \frac{500 \cdot 499 \cdots 301}{600 \cdot 599 \cdots 501 \cdot 100 \cdot 99 \cdots 1} \\
&= \frac{500}{600} \cdot \frac{499}{599} \cdots \frac{401}{501} \cdot \frac{400}{100} \cdot \frac{399}{99} \cdots \frac{301}{1}.
\end{aligned}$$

Each of the fractions $500/600$, $499/599$, ..., $401/501$ is greater than $4/5$, and each of the fractions $400/100$, $399/99$, $398/98$, ..., $301/1$ is at least 4, so

$$\frac{500}{600} \cdot \frac{499}{599} \cdots \frac{401}{501} \cdot \frac{400}{100} \cdot \frac{399}{99} \cdots \frac{301}{1} > \left(\frac{4}{5}\right)^{100} \cdot 4^{100} = \left(\frac{16}{5}\right)^{100} > 1.$$

Therefore, $600! \cdot 300! \cdot 100! < 500! \cdot 500!$.

To show that $500! \cdot 500! < 300! \cdot 700!$, we take the quotient of the two numbers:

$$\frac{700! \cdot 300!}{500! \cdot 500!} = \frac{700!/500!}{500!/300!} = \frac{700}{500} \cdot \frac{699}{499} \cdots \frac{501}{301}.$$

Each of the fractions $700/500$, $699/499$, ..., $501/301$ is greater than 1, so $300! \cdot 700! > 500! \cdot 500!$.

Finally, $\frac{1000!}{300! \cdot 700!} = \binom{1000}{300}$ is a positive integer greater than 1, so $300! \cdot 700! < 1000!$.

12.69 We have a sum of n terms on the left side, and an nth root on the right side, so we think of using AM-GM. We can isolate the nth root on the right side by adding n to both sides:

$$\frac{1}{n} + \frac{1}{n+1} + \cdots + \frac{1}{2n-1} + n > n2^{1/n}.$$

To get AM-GM to produce an nth root, we need n terms. We could take the product of the n fractions $1/n$, $1/(n+1)$, ..., $1/(2n-1)$, but this does not produce anything like $2^{1/n}$.

To alter the n fractions and use the summand of n, we write the left side as

$$\frac{1}{n} + \frac{1}{n+1} + \cdots + \frac{1}{2n-1} + n = \left(\frac{1}{n}+1\right) + \left(\frac{1}{n+1}+1\right) + \cdots + \left(\frac{1}{2n-1}+1\right) = \frac{n+1}{n} + \frac{n+2}{n+1} + \cdots + \frac{2n}{2n-1}.$$

Now, we apply the AM-GM Inequality, and we have

$$\frac{n+1}{n} + \frac{n+2}{n+1} + \cdots + \frac{2n}{2n-1} \ge n\sqrt[n]{\frac{n+1}{n} \cdot \frac{n+2}{n+1} \cdots \frac{2n}{2n-1}} = n\sqrt[n]{\frac{2n}{n}} = n2^{1/n}.$$

Furthermore, we have strict inequality, because the n fractions $(n+1)/n$, $(n+2)/(n+1)$, ..., $2n/(2n-1)$ are all different. Thus, the desired inequality holds.

12.70 First, since $0 \le x \le 1$, we have $0 \le x^n \le x^{n-1} \le \cdots \le x^2 \le x \le 1$.

Since $1 - x^{n+1} = (1-x)(1+x+\cdots+x^n)$ and $1-x^n = (1-x)(1+x+\cdots+x^{n-1})$, we may rewrite the given inequality as

$$\frac{(1-x)(1+x+\cdots+x^n)}{n+1} \le \frac{(1-x)(1+x+\cdots+x^{n-1})}{n}.$$

If $x = 1$, then both sides are equal to 0, and so the inequality holds. Otherwise, $x < 1$, so $1 - x$ is positive. Hence, we may divide both sides by $1 - x$, to obtain the equivalent inequality

$$\frac{1 + x + \cdots + x^n}{n + 1} \leq \frac{1 + x + \cdots + x^{n-1}}{n}.$$

Cross-multiplying, we get $n(1 + x + \cdots + x^n) \leq (n + 1)(1 + x + \cdots + x^{n-1})$. Expanding both sides, we get

$$n + nx + \cdots + nx^n \leq (n + 1) + (n + 1)x + \cdots + (n + 1)x^{n-1},$$

which simplifies as $nx^n \leq 1 + x + \cdots + x^{n-1}$. We know this inequality is true because $x^n \leq x^{n-1} \leq \cdots \leq x^2 \leq x \leq 1$, so all n terms in the series $1 + x + \cdots + x^{n-1}$ are at least x^n. Since all of our steps are reversible, the original inequality also holds.

12.71 Let n be the number of elements in S, and let m be their sum. Then the average of the elements in S is $\frac{m}{n} = 56$, and the average of the elements in S without the element 68 is $\frac{m-68}{n-1} = 55$. Solving this system of equations, we find $m = 728$ and $n = 13$.

We know that S contains the element 68, so this leaves 12 elements whose sum is $728 - 68 = 660$. Each element is a positive integer, and hence at least 1. Therefore, the maximum element that can appear in S is $660 - 11 = \boxed{649}$.

12.72 Expanding both sides of the inequality, we get

$$2 + \frac{a}{b} + \frac{b}{c} + \frac{c}{a} + \frac{a}{c} + \frac{c}{b} + \frac{b}{a} \geq 2 + \frac{2(a + b + c)}{\sqrt[3]{abc}}.$$

We can produce terms like $\frac{a}{b}$ by multiplying $a + b + c$ and $\frac{1}{a} + \frac{1}{b} + \frac{1}{c}$. Since $(a + b + c)\left(\frac{1}{a} + \frac{1}{b} + \frac{1}{c}\right) = 3 + \frac{a}{b} + \frac{b}{c} + \frac{c}{a} + \frac{a}{c} + \frac{c}{b} + \frac{b}{a}$, the desired inequality may be rewritten as

$$(a + b + c)\left(\frac{1}{a} + \frac{1}{b} + \frac{1}{c}\right) \geq 3 + \frac{2(a + b + c)}{\sqrt[3]{abc}}.$$

By AM-GM, we have $\frac{1}{a} + \frac{1}{b} + \frac{1}{c} \geq \frac{3}{\sqrt[3]{abc}}$, so

$$(a + b + c)\left(\frac{1}{a} + \frac{1}{b} + \frac{1}{c}\right) \geq \frac{3(a + b + c)}{\sqrt[3]{abc}} = \frac{a + b + c}{\sqrt[3]{abc}} + \frac{2(a + b + c)}{\sqrt[3]{abc}}.$$

Again by AM-GM, we have $a + b + c \geq 3\sqrt[3]{abc}$, so

$$\frac{a + b + c}{\sqrt[3]{abc}} + \frac{2(a + b + c)}{\sqrt[3]{abc}} \geq 3 + \frac{2(a + b + c)}{\sqrt[3]{abc}},$$

and we are done.

12.73 Let the number of people in the m groups be $x_1, x_2, \ldots, x_m$, respectively, and let the side lengths of the square cakes be $s_1, s_2, \ldots, s_m$, in centimeters, respectively. There are a total of n guests, so $x_1 + x_2 + \cdots + x_m = n$. Since each guest may have at most 25 cm² of cake, $s_i^2 \leq 25x_i$, so $s_i \leq 5\sqrt{x_i}$ for all i.

The host will require $4s_1 + 4s_2 + \cdots + 4s_m = 4(s_1 + s_2 + \cdots + s_m)$ of ribbon, in centimeters. Since $s_i \leq 5\sqrt{x_i}$, we have

$$4(s_1 + s_2 + \cdots + s_m) \leq 20(\sqrt{x_1} + \sqrt{x_2} + \cdots + \sqrt{x_m}).$$

We know that $x_1 + x_2 + \cdots + x_m = n$, so we'd like to get rid of those square root signs. Cauchy lets us do so. Applying the Cauchy-Schwarz Inequality to the sequence $\sqrt{x_1}, \sqrt{x_2}, \ldots, \sqrt{x_m}$ and the sequence consisting of m instances of 1 gives us

$$\left(\sqrt{x_1} + \sqrt{x_2} + \cdots + \sqrt{x_m}\right)^2 \leq \left(\left(\sqrt{x_1}\right)^2 + \left(\sqrt{x_2}\right)^2 + \cdots + \left(\sqrt{x_m}\right)^2\right)(1^2 + 1^2 + \cdots 1^2).$$

Simplifying the right side gives

$$\left(\sqrt{x_1} + \sqrt{x_2} + \cdots + \sqrt{x_m}\right)^2 \le (x_1 + x_2 + \cdots + x_m)(m),$$

so we have $\left(\sqrt{x_1} + \sqrt{x_2} + \cdots + \sqrt{x_m}\right)^2 \le mn$. Both sides are positive, so we can take the square root of both sides, then multiply by 20, to give the desired

$$4(s_1 + s_2 + \cdots + s_m) \le 20(\sqrt{x_1} + \sqrt{x_2} + \cdots + \sqrt{x_m}) \le 20\sqrt{mn},$$

which tells us that the host needs no more than $20\sqrt{mn}$ cm of ribbon.

12.74 If $x = 0$ or $y = 0$, then $kxy = 0$, and by the Trivial Inequality, $36x^2 + 25y^2 \ge 0$, so the inequality is satisfied for any value of k.

Now, take the case where x and y have the same sign. By AM-GM, we have

$$36x^2 + 25y^2 \ge 2\sqrt{36x^2 \cdot 25y^2} = 2\sqrt{900x^2y^2} = 60xy.$$

Equality occurs if and only if $36x^2 = 25y^2$, so for example, we may take $x = 5$ and $y = 6$. For these values of x and y, the given inequality becomes $1800 \ge 30k$, so $k \le 60$. Conversely, if $k \le 60$, then

$$36x^2 + 25y^2 \ge 60xy \ge kxy,$$

so the given inequality is satisfied in this case.

Next, we take the case where x is positive and y is negative. Let $a = x$ and $b = -y$, so a and b are positive. By AM-GM, we have

$$36x^2 + 25y^2 = 36a^2 + 25b^2 \ge 2\sqrt{36a^2 \cdot 25b^2} = 2\sqrt{900a^2b^2} = 60ab = -60xy.$$

Equality occurs if and only if $36a^2 = 25b^2$, so as above, we may take $a = 5$ and $b = 6$. Then $x = 5$ and $y = -6$, and for these values of x and y, the given inequality becomes $1800 \ge -30k$, so $k \ge -60$. Conversely, if $k \ge -60$, then

$$36x^2 + 25y^2 \ge -60xy \ge kxy,$$

so the given inequality is satisfied in this case.

Therefore, $36x^2 + 25y^2 \ge kxy$ for all real numbers x and y if and only if $\boxed{-60 \le k \le 60}$.

12.75 We work backwards. Both sides of the given inequality are nonnegative, so we may square both sides to obtain the equivalent inequality

$$a^2 + b^2 + 2\sqrt{a^2 + b^2}\sqrt{x^2 + y^2} + x^2 + y^2 \ge (a - x)^2 + (b - y)^2$$
$$= a^2 - 2ax + x^2 + b^2 - 2by + y^2.$$

This simplifies as $2\sqrt{a^2 + b^2}\sqrt{x^2 + y^2} \ge -2ax - 2by$, or $\sqrt{a^2 + b^2}\sqrt{x^2 + y^2} \ge -(ax + by)$.

If $ax + by$ is positive or 0, then the right side is nonpositive, but the left side is nonnegative, so the inequality clearly holds. Otherwise, assume that $ax + by$ is negative. Then we may square both sides, to obtain the equivalent inequality

$$(a^2 + b^2)(x^2 + y^2) \ge (ax + by)^2.$$

This follows from the Cauchy-Schwarz Inequality. Since all of our steps are reversible, the original inequality holds.

12.76   This inequality doesn't look like any of our inequality tools. However, we can make it look a lot more like Cauchy by multiplying both sides by $a + b + c$. This gives us the equivalent inequality

$$abc(a + b + c)^2 \leq (a + b + c)(bc^3 + ca^3 + ab^3).$$

On the lesser side, we have a square of a sum, and on the greater side, we have a product of sums. This looks a lot like Cauchy. But what's that abc term there for?

The Cauchy-Schwarz Inequality gives us

$$(x_1y_1 + x_2y_2 + x_3y_3)^2 \leq (x_1^2 + x_2^2 + x_3^2)(y_1^2 + y_2^2 + y_3^2).$$

We don't have an exact match with our lesser side above, but pairing up the greater sides, we let $x_1 = \sqrt{a}$, $x_2 = \sqrt{b}$, $x_3 = \sqrt{c}$, $y_1 = \sqrt{bc^3}$, $y_2 = \sqrt{ca^3}$, and $y_3 = \sqrt{ab^3}$. This gives us

$$(\sqrt{abc^3} + \sqrt{a^3bc} + \sqrt{ab^3c})^2 \leq (a + b + c)(bc^3 + ca^3 + ab^3).$$

Aha! Factoring out $\sqrt{abc}$ from the left side gives

$$(\sqrt{abc}(\sqrt{a^2} + \sqrt{b^2} + \sqrt{c^2}))^2 = abc(a + b + c)^2,$$

so the inequality above becomes $abc(a + b + c)^2 \leq (a + b + c)(bc^3 + ca^3 + ab^3)$. Dividing both sides by $a + b + c$ gives the desired $bc^3 + ca^3 + ab^3 \geq abc(a + b + c)$.

12.77   From the given equations, we have $a + b + c + d = 8 - e$ and $a^2 + b^2 + c^2 + d^2 = 16 - e^2$. By Cauchy-Schwarz, we have

$$(1 + 1 + 1 + 1)(a^2 + b^2 + c^2 + d^2) \geq (a + b + c + d)^2,$$

so $4(16 - e^2) \geq (8 - e)^2$. Rearranging gives $e(5e - 16) \leq 0$. Therefore, we have $0 \leq e \leq 16/5$. Furthermore, when $a = b = c = d = 6/5$ and $e = 16/5$, then $a + b + c + d + e = 8$ and $a^2 + b^2 + c^2 + d^2 + e^2 = 16$. Therefore, the maximum value of e is $\boxed{16/5}$.

12.78   If we expand the left side, then we get the terms x_1^{112}, x_2^{112}, ..., x_n^{112}, as well as all the terms of the form $x_i^{19}x_j^{93}$, where $1 \leq i, j \leq n$, and $i \neq j$. In other words, we get

$$x_1^{112} + x_2^{112} + \cdots + x_n^{112} + \sum_{1 \leq i < j \leq n} (x_i^{19}x_j^{93} + x_j^{19}x_i^{93}).$$

Similarly, expanding the right side, we get

$$x_1^{112} + x_2^{112} + \cdots + x_n^{112} + \sum_{1 \leq i < j \leq n} (x_i^{20}x_j^{92} + x_j^{20}x_i^{92}).$$

Hence, the inequality becomes $\sum_{1 \leq i < j \leq n} (x_i^{19}x_j^{93} + x_j^{19}x_i^{93}) \geq \sum_{1 \leq i < j \leq n} (x_i^{20}x_j^{92} + x_j^{20}x_i^{92})$. It suffices to prove that

$$x_i^{19}x_j^{93} + x_j^{19}x_i^{93} \geq x_i^{20}x_j^{92} + x_j^{20}x_i^{92},$$

where $1 \leq i < j \leq n$.

To begin, since $x_i > 0$ for all i, we can divide both sides by $x_i^{19}x_j^{19}$, to get $x_i^{74} + x_j^{74} \geq x_i^{73}x_j + x_ix_j^{73}$. Moving all the terms to one side, the inequality becomes $x_i^{74} + x_j^{74} - x_i^{73}x_j - x_ix_j^{73} \geq 0$. The left side factors as

$$x_i^{74} - x_i^{73}x_j - x_ix_j^{73} + x_j^{74} = x_i^{73}(x_i - x_j) - x_j^{73}(x_i - x_j) = (x_i^{73} - x_j^{73})(x_i - x_j)$$
$$= (x_i - x_j)(x_i^{72} + x_i^{71}x_j + x_i^{70}x_j^2 + \cdots + x_j^{72})(x_i - x_j)$$
$$= (x_i^{72} + x_i^{71}x_j + x_i^{70}x_j^2 + \cdots + x_j^{72})(x_i - x_j)^2.$$

Clearly $x_i^{72} + x_i^{71}x_j + x_i^{70}x_j^2 + \cdots + x_j^{72} \geq 0$, and by the Trivial Inequality, we have $(x_i - x_j)^2 \geq 0$. Hence, we have $x_i^{74} - x_i^{73}x_j - x_ix_j^{73} + x_j^{74} \geq 0$, and we can reverse our earlier steps to conclude that the original inequality holds.

12.79 By Cauchy-Schwarz, we have $(x_1^2 + x_2^2 + x_3^2)(y_1^2 + y_2^2 + y_3^2) \geq (x_1y_1 + x_2y_2 + x_3y_3)^2$ for all real numbers x_i, y_i.

Let

$$x_1^2 = a(a^2 - bc + 1), \qquad\qquad y_1^2 = \frac{a}{a^2 - bc + 1},$$

$$x_2^2 = b(b^2 - ca + 1), \qquad\qquad y_2^2 = \frac{b}{b^2 - ca + 1},$$

$$x_3^2 = c(c^2 - ab + 1), \qquad\qquad y_3^2 = \frac{c}{c^2 - ab + 1}.$$

Then

$$x_1^2 y_1^2 = a(a^2 - bc + 1) \cdot \frac{a}{a^2 - bc + 1} = a^2,$$

so $x_1y_1 = a$. Similarly, $x_2y_2 = b$ and $x_3y_3 = c$, so Cauchy-Schwarz tells us that

$$[a(a^2 - bc + 1) + b(b^2 - ca + 1) + c(c^2 - ab + 1)]\left(\frac{a}{a^2 - bc + 1} + \frac{b}{b^2 - ca + 1} + \frac{c}{c^2 - ab + 1}\right) \geq (a + b + c)^2.$$

Rearranging, we get

$$\frac{a}{a^2 - bc + 1} + \frac{b}{b^2 - ca + 1} + \frac{c}{c^2 - ab + 1} \geq \frac{(a + b + c)^2}{a(a^2 - bc + 1) + b(b^2 - ca + 1) + c(c^2 - ab + 1)}$$

$$= \frac{(a + b + c)^2}{a^3 + b^3 + c^3 - 3abc + a + b + c}.$$

Here are a couple ways to finish from this point:

Solution 1: We have the identity $a^3 + b^3 + c^3 - 3abc = (a + b + c)(a^2 + b^2 + c^2 - ab - ac - bc)$, so the above inequality becomes

$$\frac{a}{a^2 - bc + 1} + \frac{b}{b^2 - ca + 1} + \frac{c}{c^2 - ab + 1} \geq \frac{(a + b + c)^2}{(a + b + c)(a^2 + b^2 + c^2 - ab - ac - bc) + a + b + c}$$

$$= \frac{a + b + c}{a^2 + b^2 + c^2 - ab - ac - bc + 1}$$

$$= \frac{a + b + c}{a^2 + b^2 + c^2 + 2/3}.$$

We would be done if we could show that $\dfrac{a + b + c}{a^2 + b^2 + c^2 + 2/3} \geq \dfrac{1}{a + b + c}$. Cross-multiplying, this inequality becomes $(a + b + c)^2 \geq a^2 + b^2 + c^2 + \frac{2}{3}$. Expanding the left side, we get

$$a^2 + b^2 + c^2 + 2ab + 2ac + 2bc \geq a^2 + b^2 + c^2 + \frac{2}{3}.$$

However, we have $ab + ac + bc = 1/3$, so this inequality is in fact always an equality. In other words,

$$\frac{a + b + c}{a^2 + b^2 + c^2 + 2/3} = \frac{1}{a + b + c}.$$

Hence, we have

$$\frac{a}{a^2 - bc + 1} + \frac{b}{b^2 - ca + 1} + \frac{c}{c^2 - ab + 1} \geq \frac{a + b + c}{a^2 + b^2 + c^2 + 2/3} = \frac{1}{a + b + c}.$$

Solution 2: We must show that

$$\frac{(a+b+c)^2}{a^3+b^3+c^3-3abc+a+b+c} \geq \frac{1}{a+b+c}.$$

Because a, b, and c are positive, AM-GM tells us that $a^3+b^3+c^3 \geq 3abc$, so the denominator of the left side is positive. Therefore, we can multiply both sides of the inequality by the denominators. This gives us

$$(a+b+c)^3 \geq a^3+b^3+c^3-3abc+a+b+c$$

to prove. Expanding the left side and eliminating common terms leaves

$$3(a^2b+a^2c+b^2a+b^2c+c^2a+c^2b+3abc) \geq a+b+c$$

to prove. We appear stuck here, but then we remember that we haven't yet used the fact that $ab+bc+ca = 1/3$. We see that multiplying this by $a+b+c$ will produce a lot of the terms on the greater side of our desired inequality. In fact, we find that $(a+b+c)(ab+bc+ca)$ equals the expression in parentheses above. Therefore, we have

$$3(a^2b+a^2c+b^2a+b^2c+c^2a+c^2b+3abc) = 3(a+b+c)(ab+bc+ca) = 3(a+b+c)(1/3) = a+b+c.$$

Therefore, we can walk backwards through our steps to show that

$$\frac{(a+b+c)^2}{a^3+b^3+c^3-3abc+a+b+c} = \frac{1}{a+b+c},$$

so we have

$$\frac{a}{a^2-bc+1} + \frac{b}{b^2-ca+1} + \frac{c}{c^2-ab+1} \geq \frac{(a+b+c)^2}{a^3+b^3+c^3-3abc+a+b+c} = \frac{1}{a+b+c},$$

as desired.

12.80 To deal with the fractions, we utilize a transformation. Let $x = 1/a$, $y = 1/b$, and $z = 1/c$. Then $xyz = 1/(abc) = 1$, and the left side becomes

$$\frac{x^3}{1/y+1/z} + \frac{y^3}{1/x+1/z} + \frac{z^3}{1/x+1/y} = \frac{x^3yz}{y+z} + \frac{xy^3z}{x+z} + \frac{xyz^3}{x+y}$$
$$= \frac{x^2}{y+z} + \frac{y^2}{x+z} + \frac{z^2}{x+y}.$$

We still have fractions, but these are simpler. And because we still have fractions, we think of using Cauchy-Schwarz. By Cauchy-Schwarz, we have

$$[(y+z)+(x+z)+(x+y)]\left(\frac{x^2}{y+z}+\frac{y^2}{x+z}+\frac{z^2}{x+y}\right) \geq (x+y+z)^2.$$

Therefore, $\dfrac{x^2}{y+z} + \dfrac{y^2}{x+z} + \dfrac{z^2}{x+y} \geq \dfrac{(x+y+z)^2}{2(x+y+z)} = \dfrac{x+y+z}{2}.$

By AM-GM, we have $x+y+z \geq 3\sqrt[3]{xyz} = 3$, so $\dfrac{x^2}{y+z} + \dfrac{y^2}{x+z} + \dfrac{z^2}{x+y} \geq \dfrac{x+y+z}{2} \geq \dfrac{3}{2}$, as desired.

12.81 We don't have any tools to deal with four variables, and we surely don't want to raise both sides of that equation to the fourth power. So, we start with what we know. We have sums on the left and products on the right, so we try AM-GM. We'll split the variables into pairs, and we have

$$\frac{w+x}{2} \geq \sqrt{wx},$$
$$\frac{y+z}{2} \geq \sqrt{yz}.$$

So far, so good. We can form the left side of our desired inequality by adding these two inequalities, then dividing by 2. This gives us

$$\frac{w + x + y + z}{4} \geq \frac{\sqrt{wx} + \sqrt{yz}}{2}.$$

Equality holds when equality holds in both of our original AM-GM inequalities; that is, when $w = x$ and $y = z$. Now, in this inequality, we have the desired greater side, but not the desired lesser side. However, that expression on the lesser side looks familiar—it's an arithmetic mean itself! Applying AM-GM to $\sqrt{wx}$ and $\sqrt{yz}$ gives us

$$\frac{\sqrt{wx} + \sqrt{yz}}{2} \geq \sqrt{(\sqrt{wx})(\sqrt{yz})}.$$

Equality holds here if $\sqrt{wx} = \sqrt{yz}$. Simplifying the right side of the inequality gives us $\sqrt[4]{wxyz}$, so we have

$$\frac{w + x + y + z}{4} \geq \frac{\sqrt{wx} + \sqrt{yz}}{2} \geq \sqrt[4]{wxyz},$$

as desired.

We have equality only if equality holds in each step in our proof. Therefore, all the equality conditions we found earlier must hold. These are $w = x$, $y = z$, and $\sqrt{wx} = \sqrt{yz}$. Combining the first two with the last gives us $\sqrt{x^2} = \sqrt{z^2}$, so $x = z$, which means equality holds in our desired inequality if and only if $w = x = y = z$.

12.82 We've already proved that AM-GM holds for 2^1 and 2^2 variables. We're asked to show it for any 2^n nonnegative variables, where n is a positive integer. Having already seen how to step from $n = 1$ to $n = 2$, this very much feels like a job for induction. We already have our base case, so we move on to the inductive assumption.

Inductive Assumption: If $a_1, a_2, a_3, \ldots, a_{2^k}$ are all nonnegative, then

$$\frac{a_1 + a_2 + \cdots + a_{2^k}}{2^k} \geq \sqrt[2^k]{a_1 a_2 \cdots a_{2^k}},$$

where equality holds if and only if $a_1 = a_2 = a_3 = \cdots = a_{2^k}$.

Inductive Step: We wish to show that in any group of 2^{k+1} nonnegative numbers, $a_1, a_2, a_3, \ldots, a_{2^{k+1}}$, we have

$$\frac{a_1 + a_2 + \cdots + a_{2^{k+1}}}{2^{k+1}} \geq \sqrt[2^{k+1}]{a_1 a_2 \cdots a_{2^{k+1}}}.$$

We use our step from the 2^1-variable case to the 2^2-variable case as our guide. In going from 2^1 variables to 2^2 variables, we broke the 2^2 variables into two groups of 2^1 variables. This allowed us to apply the 2^1-variable version of AM-GM.

We try the same approach here. We divide our 2^{k+1} variables into two groups of 2^k variables. We then apply our inductive assumption to each of these groups. The first group gives us

$$\frac{a_1 + a_2 + \cdots + a_{2^k}}{2^k} \geq \sqrt[2^k]{a_1 a_2 \cdots a_{2^k}}.$$

The second group consists of the 2^k numbers $a_{2^k+1}, a_{2^k+2}, \ldots a_{2^{k+1}}$, so applying the inductive assumption to these 2^k numbers gives us

$$\frac{a_{2^k+1} + a_{2^k+2} + \cdots + a_{2^{k+1}}}{2^k} \geq \sqrt[2^k]{a_{2^k+1} a_{2^k+2} \cdots a_{2^{k+1}}}.$$

Still following our step from 2^1 variables to 2^2 variables as a guide, we add these two inequalities and divide by 2 to get

$$\frac{a_1 + a_2 + \cdots + a_{2^k} + a_{2^k+1} + a_{2^k+2} + \cdots + a_{2^{k+1}}}{2 \cdot 2^k} \geq \frac{\sqrt[2^k]{a_1 a_2 \cdots a_{2^k}} + \sqrt[2^k]{a_{2^k+1} a_{2^k+2} \cdots a_{2^{k+1}}}}{2}. \qquad (12.1)$$

Now, we have an arithmetic mean on the right side. Applying 2-variable AM-GM to the right side gives us

$$\frac{\sqrt[2^k]{a_1 a_2 \cdots a_{2^k}} + \sqrt[2^k]{a_{2^k+1} a_{2^k+2} \cdots a_{2^{k+1}}}}{2} \geq \sqrt{\sqrt[2^k]{a_1 a_2 \cdots a_{2^k}} \cdot \sqrt[2^k]{a_{2^k+1} a_{2^k+2} \cdots a_{2^{k+1}}}}$$

$$= \sqrt{\left(a_1 a_2 a_3 \cdots a_{2^k} a_{2^k+1} a_{2^k+2} \cdots a_{2^{k+1}}\right)^{1/2^k}}$$

$$= \left(a_1 a_2 a_3 \cdots a_{2^k} a_{2^k+1} a_{2^k+2} \cdots a_{2^{k+1}}\right)^{(1/2^k)\cdot(1/2)}$$

$$= \sqrt[2^{k+1}]{a_1 a_2 a_3 \cdots a_{2^k} a_{2^k+1} a_{2^k+2} \cdots a_{2^{k+1}}}.$$

Equality only holds if the two square roots on the left are equal. Combining this inequality with the inequality on line (12.1) gives the desired

$$\frac{a_1 + a_2 + \cdots + a_{2^k} + a_{2^k+1} + a_{2^k+2} + \cdots + a_{2^{k+1}}}{2^{k+1}} \geq \sqrt[2^{k+1}]{a_1 a_2 a_3 \cdots a_{2^k} a_{2^k+1} a_{2^k+2} \cdots a_{2^{k+1}}}.$$

Equality only holds if equality holds in each step. Therefore, we must have each of

$$a_1 = a_2 = a_3 = \cdots = a_{2^k}, \qquad a_{2^k+1} = a_{2^k+2} = \cdots = a_{2^{k+1}}, \qquad \text{and} \qquad \sqrt[2^k]{a_1 a_2 \cdots a_{2^k}} = \sqrt[2^k]{a_{2^k+1} a_{2^k+2} \cdots a_{2^{k+1}}}.$$

Combining the first two equality chains with the last equality gives $\sqrt[2^k]{(a_1)^k} = \sqrt[2^k]{(a_{2^k+1})^k}$, from which we have $a_1 = a_{2^k+1}$, so all the a_i must be equal.

12.83 We might try adding a variable to $\frac{a+b}{2} \geq \sqrt{ab}$, but turning the square root into a cube root is quite difficult. Another option might be to start with

$$\frac{a+b+c+d}{4} \geq \sqrt[4]{abcd}, \qquad (12.2)$$

and try to eliminate a variable. Suppose we want to eliminate d. We'll need a way to eliminate it from both sides. We could let it be the reciprocal of one of the others, but that would eliminate two variables from the right, and we only want to eliminate one.

The flash of genius Cauchy offered was to let d equal the arithmetic mean of a, b, and c. Watch what happens on the left if we do so:

$$\frac{a+b+c+\frac{a+b+c}{3}}{4} = \frac{\frac{4a+4b+4c}{3}}{4} = \frac{a+b+c}{3}.$$

At first, this seems shocking, but then we think about what we're doing. If we start with the group of numbers a, b, and c, then include the average of these three numbers in the group, we don't change the average of the whole group.

The beauty of this substitution is that now we can eliminate d. If $d = (a+b+c)/3$ in the inequality (12.2), then, because $(a+b+c+d)/4 = (a+b+c)/3$, we have

$$d \geq \sqrt[4]{abcd}.$$

Dividing both sides by $\sqrt[4]{d}$ gives us $d/\sqrt[4]{d} = d^{3/4}$ on the left, so we have

$$d^{3/4} \geq \sqrt[4]{abc}.$$

Raising both sides to the 4/3 power then gives us $d \geq \sqrt[3]{abc}$, since $(\sqrt[4]{abc})^{4/3} = (abc)^{(1/4)\cdot(4/3)} = \sqrt[3]{abc}$. Since $d = (a+b+c)/3$, we have

$$\frac{a+b+c}{3} \geq \sqrt[3]{abc}.$$

We only have equality if the equality condition for four-variable AM-GM satisfied; that is, if $a = b = c = (a+b+c)/3$. Therefore, equality holds if and only if $a = b = c$.

12.84 First, we note that we have already proved this statement for all values of n such that $n = 2^m$ for some positive integer m. To show that it holds for all n, we first show that we can always "step down" one variable, as we did in the previous problem.

We start by with the inductive assumption that the AM-GM Inequality holds for any $k+1$ nonnegative numbers $a_1, a_2, \ldots, a_k, a_{k+1}$.

In our previous solution, we let one of the four variables be the arithmetic mean of the remaining three. So, we try something similar here. We consider the $k+1$ numbers $a_1, a_2, \ldots, a_k, x$, where

$$x = \frac{a_1 + a_2 + \cdots + a_k}{k},$$

so $kx = a_1 + a_2 + \cdots + a_k$. Then, applying the inductive assumption that AM-GM holds for any $k+1$ numbers, we have

$$\frac{a_1 + a_2 + \cdots + a_k + x}{k+1} \geq \sqrt[k+1]{a_1 a_2 \cdots a_k x}.$$

Because $a_1 + a_2 + \cdots + a_k = kx$, the left side of our inequality equals $(kx + x)/(k+1) = x$, so our inequality is now

$$x \geq \sqrt[k+1]{a_1 a_2 \cdots a_k x}.$$

Raising both sides to the $k+1$ power gives

$$x^{k+1} \geq a_1 a_2 \cdots a_k x.$$

Dividing both sides by x gives $x^k \geq a_1 a_2 \cdots a_k$. Taking the kth root of both sides and substituting our expression for x gives

$$\frac{a_1 + a_2 + \cdots + a_k}{k} \geq \sqrt[k]{a_1 a_2 \cdots a_k},$$

which is the AM-GM Inequality for k numbers. We only have equality if $a_1 = a_2 = \cdots = a_k = x$, so equality holds if and only if all k numbers are equal. Therefore, if AM-GM holds for any $k+1$ numbers, then it holds for any k numbers. So, we can "induct down" from $k+1$ and show that AM-GM holds for any j numbers, where $j < k+1$ (and $j > 1$).

Now we can put our inductions together to show that AM-GM holds for any n nonnegative numbers, where $n > 1$. For any n such that $n > 1$, there is some integer m such that $2^m > n$. We have already shown that AM-GM holds for 2^m variables, and since $n < 2^m$, we can apply the induction we have just performed to show that AM-GM holds for n variables.

CHAPTER 13

Exponents and Logarithms

Exercises for Section 13.1

13.1.1 We solve the problem by writing each side of the equation with a common base. Since $16 = 2^4$ and $8 = 2^3$, we can rewrite the equation as $2^{4t} = 2^{3(2-t)}$. Hence, $4t = 6 - 3t$, so $t = \boxed{6/7}$.

13.1.2 Since $32 = 2^5$, we can rewrite the equation as $2^{x^2} = 2^{5(3x+8)} = 2^{15x+40}$. Hence, $x^2 = 15x+40$, so $x^2 - 15x - 40 = 0$. By Vieta's Formulas, the product of the roots is $-40/1 = \boxed{-40}$.

13.1.3

(a) To show that g is monotonically decreasing, we must show that if $x > y$, then $g(x) < g(y)$. We know how to deal with an exponential function when the base is greater than 1, so we try to write $g(x)$ in terms of such a function. Because $0 < a < 1$, we have $\frac{1}{a} > 1$. If we let $b = \frac{1}{a}$, we have $g(x) = \left(\frac{1}{b}\right)^x = \frac{1}{b^x}$. Let $h(x) = b^x$, so $g(x) = \frac{1}{h(x)}$.

Because $b > 1$, the function h is monotonically increasing. Therefore, if $x > y$, then $h(x) > h(y)$. Since $h(x)$ and $h(y)$ are both positive, we can divide both sides of our inequality by $h(x)$ and by $h(y)$ to get $\frac{1}{h(y)} > \frac{1}{h(x)}$, which means $g(y) > g(x)$, as desired. So, g is monotonically decreasing.

(b) Because g is decreasing, we know that $g(x) < g(y)$ for any x and y such that $x > y$. So, to test if we must have $x = y$ when $g(x) = g(y)$, we consider what happens in each of the three possible cases, $x > y$, $x < y$, and $x = y$.

If $x > y$, then we must have $g(x) < g(y)$ because g is decreasing. Therefore, if $x > y$, then $a^x < a^y$.

If $x < y$, then we must have $g(x) > g(y)$ because g is decreasing. Therefore, if $x < y$, then $a^x > a^y$.

If $x = y$, then $g(x) = g(y)$, so $a^x = a^y$.

Only in the final case do we have $g(x) = g(y)$, so we conclude that if $g(x) = g(y)$, then we must have $x = y$.

13.1.4 Since 3^x is an increasing function in x, maximizing 3^{-x^2+2x+3} is equivalent to maximizing $-x^2 + 2x + 3$. Completing the square, we get $-x^2 + 2x + 3 = -(x-1)^2 + 4$. Hence, the maximum value of $-x^2 + 2x + 3$ is 4, so the maximum value of 3^{-x^2+2x+3} is $3^4 = \boxed{81}$.

13.1.5 We can rewrite the equation as $2^{2y} - 12 \cdot 2^y + 32 = 0$. This is a quadratic in 2^y, and hence factors as $(2^y - 4)(2^y - 8) = 0$, so either $2^y = 4$, in which case $y = 2$, or $2^y = 8$, in which case $y = 3$. Therefore, the solutions are $y = \boxed{2 \text{ and } 3}$.

Exercises for Section 13.2

13.2.1

(a) Since $4^4 = 256$, we have $\log_4 256 = \boxed{4}$.

(b) Since $3^{-2} = 1/9$, we have $\log_3(1/9) = \boxed{-2}$.

(c) Since $125^{1/3} = 5$, we have $\log_{125} 5 = \boxed{1/3}$.

13.2.2

(a) Let $x = \log_{125} 25$, so $125^x = 25$. We can rewrite this equation as $5^{3x} = 5^2$, so $3x = 2$, or $x = \boxed{2/3}$.

(b) Let $x = \log_{2\sqrt{2}} 16$, so $(2\sqrt{2})^x = 16$. Squaring both sides, we get $8^x = 16^2 = 256$. We can rewrite this equation as $2^{3x} = 2^8$, so $3x = 8$, which means $x = \boxed{8/3}$.

(c) Let $x = \log_{1/3} 9$, so $(1/3)^x = 9$. Taking the reciprocal of both sides, we get $3^x = 1/9 = 3^{-2}$. Therefore, $x = \boxed{-2}$.

13.2.3 Solving for 2^x, we get $2^x = (14 + 9)/3 = 23/3$. Therefore, $x = \boxed{\log_2(23/3)}$.

13.2.4 The equation $\log_2(x^2 - 2x - 7) = 3$ is equivalent to $x^2 - 2x - 7 = 2^3 = 8$, so $x^2 - 2x - 15 = 0$. Factoring gives $(x - 5)(x + 3) = 0$. Therefore, the solutions are $x = \boxed{5 \text{ and } -3}$.

13.2.5

(a) The function $\log_{1/5} x$ is defined for all $x > 0$, so the domain is $\boxed{(0, +\infty)}$. The range is $\boxed{\text{all real numbers}}$.

(b) The expression $\log_6(\log_{1/5} x)$ is defined if and only if $\log_{1/5} x > 0$. Since the base $1/5$ is less than 1, the inequality $\log_{1/5} x > 0$ is equivalent to $x < (1/5)^0 = 1$. We must also have $x > 0$; otherwise, $\log_{1/5} x$ is not defined. Combining these restrictions, the domain is $\boxed{(0, 1)}$. Under the restriction $0 < x < 1$, the range of $\log_{1/5} x$ is all positive real numbers. Therefore, the range of $\log_6(\log_{1/5} x)$ is $\boxed{\text{all real numbers}}$.

13.2.6 First, we note that $f(x)$ is only defined for $x > 1$. So, the domain of $g(f(x))$ cannot include any values of x with $x \leq 1$. To check if there are other restrictions on the domain of $g(f(x))$, we examine the domain of g.

In order for $g(x)$ to be defined, we must have $\frac{x}{x-1} \geq 0$. The expression $\frac{x}{x-1}$ is negative if $0 < x < 1$, undefined if $x = 1$, equal to 0 if $x = 0$, and positive otherwise. Therefore, $g(x)$ is defined if and only if $x \in (-\infty, 0] \cup (1, +\infty)$. Hence, $g(f(x))$ is defined if and only if $f(x) \in (-\infty, 0] \cup (1, +\infty)$, which means $\log_3(x - 1) \leq 0$ or $\log_3(x - 1) > 1$.

The inequality $\log_3(x - 1) \leq 0$ is equivalent to $x - 1 \leq 3^0 = 1$, or $x \leq 2$, and the inequality $\log_3(x - 1) > 1$ is equivalent to $x - 1 > 3^1 = 3$, or $x > 4$. We must also remember the restriction $x > 1$ on the domain of $f(x)$. We therefore find that the domain of $g(f(x))$ is $\boxed{(1, 2] \cup (4, +\infty)}$.

Exercises for Section 13.3

13.3.1

(a) Since $27 = 3^3$, we can rewrite the expression as $\log_3(27^{2007}) = \log_3(3^{3 \cdot 2007}) = 3 \cdot 2007 = \boxed{6021}$.

(b) From the identity $a^{\log_a n} = n$, we have $6^{\log_6 418} = \boxed{418}$.

(c) Both logarithms have base 2, so we may combine them as follows:

$$\log_2 \frac{2}{3} + \log_2 6 = \log_2 \left(\frac{2}{3} \cdot 6\right) = \log_2 4 = \boxed{2}.$$

(d) From the identity $(\log_a c)(\log_c b) = \log_a b$, it follows that

$$\left(\log_2 5\right)\left(\log_5 12\right) + \left(\log_2 7\right)\left(\log_7 \frac{8}{3}\right) = \log_2 12 + \log_2 \frac{8}{3} = \log_2\left(12 \cdot \frac{8}{3}\right) = \log_2 32 = \boxed{5}.$$

(e) From the identity $\dfrac{\log_a b}{\log_a c} = \log_c b$, it follows that $\dfrac{\log_2 125}{\log_2 25} = \log_{25} 125$. Let $x = \log_{25} 125$. Then $25^x = 125$, which we may rewrite as $5^{2x} = 5^3$. Therefore, $2x = 3$, or $x = \boxed{3/2}$.

(f) First, we can write $\dfrac{2^{\log_4 108}}{2^{\log_4 3}} = 2^{\log_4 108 - \log_4 3}$. Then $\log_4 108 - \log_4 3 = \log_4 \frac{108}{3} = \log_4 36$, so $2^{\log_4 108 - \log_4 3} = 2^{\log_4 36}$. Here, we are raising a number to a logarithm, so we would like to use the identity $a^{\log_a n} = n$. However, the base is 2 and the base of the logarithm is 4. We may make 4 the base of the exponential by writing $2 = 4^{1/2}$, so

$$2^{\log_4 36} = 4^{\frac{1}{2}\log_4 36}.$$

Then $\frac{1}{2}\log_4 36 = \log_4 36^{1/2} = \log_4 6$. Therefore, we have $4^{\frac{1}{2}\log_4 36} = 4^{\log_4 6} = \boxed{6}$.

We also could have used the identity $\log_{a^n} b^n = \log_a b$ to write $\log_4 36 = \log_{2^2} 6^2 = \log_2 6$, so $2^{\log_4 36} = 2^{\log_2 6} = 6$.

13.3.2 $\log_b \frac{1}{a} = \log_b a^{-1} = -\log_b a$.

13.3.3 Applying the identity $\log_{a^n} b^n = \log_a b$ to $d = 2\log_{y^3}(b^3)$ gives $d = 2\log_y b$, which makes the logarithms the same in our expressions for c and d. Then, dividing $d = 2\log_y b$ by $c = \log_y b$ gives $d/c = \boxed{2}$.

13.3.4 Raising both sides of the equation $a^x = c^q$ to the power of z, we get $a^{xz} = c^{qz}$. Raising both sides of the equation $a^z = c^y$ to the power of x, we get $a^{xz} = c^{xy}$. Therefore, $c^{xy} = c^{qz}$, so $xy = qz$.

13.3.5 To compute $\log_5 10$ when we are only given the values of logarithms with base 10, we use the formula

$$\log_5 10 = \frac{1}{\log_{10} 5}.$$

Since $5 = 10/2$, we have $\log_{10} 5 = \log_{10} 10 - \log_{10} 2 = 1 - 0.301 = 0.699$, to the nearest hundredth. Then

$$\log_5 10 = \frac{1}{0.699},$$

which is approximately $1/0.7 = \boxed{10/7}$.

13.3.6 To prove that both sides are equal, we first convert every logarithm to the same base, say base a. We have

$$\log_c d = \frac{\log_a d}{\log_a c},$$

so the left side becomes $\log_a b \cdot \log_c d = \dfrac{\log_a b \cdot \log_a d}{\log_a c}$.

Similarly, $\log_c b = \dfrac{\log_a b}{\log_a c}$ so the right side becomes $\log_c b \cdot \log_a d = \dfrac{\log_a b \cdot \log_a d}{\log_a c}$. Hence, both sides are equal.

Exercises for Section 13.4

13.4.1 We have $\log_2 r + \log_2 s = \log_2(rs)$. By Vieta's Formulas, $rs = 12/3 = 4$, so $\log_2(rs) = \log_2 4 = \boxed{2}$.

13.4.2   First, we can rewrite the expression as

$$\frac{\log_{10}\frac{1600}{17}}{\log_{10}136} = \frac{\log_{10}1600 - \log_{10}17}{\log_{10}136}.$$

Then, to express each of the logarithms in this equation in terms of r and s, we factor each argument: $1600 = 2^4 \cdot 10^2$ and $136 = 2^3 \cdot 17$. Therefore,

$$\frac{\log_{10}1600 - \log_{10}17}{\log_{10}136} = \frac{\log_{10}(2^4 \cdot 10^2) - \log_{10}17}{\log_{10}(2^3 \cdot 17)} = \frac{\log_{10}2^4 + \log_{10}10^2 - \log_{10}17}{\log_{10}2^3 + \log_{10}17}$$

$$= \frac{4\log_{10}2 + 2 - \log_{10}17}{3\log_{10}2 + \log_{10}17} = \boxed{\frac{4s - r + 2}{r + 3s}}.$$

13.4.3   To solve for t, we rewrite both sides as a logarithm with base 3. The left side becomes

$$2\log_3(1 - 5t) = \log_3(1 - 5t)^2 = \log_3(25t^2 - 10t + 1),$$

and the right side becomes

$$\log_3(2t + 5) + 2 = \log_3(2t + 5) + \log_3 9 = \log_3(18t + 45).$$

Then $25t^2 - 10t + 1 = 18t + 45$, which simplifies as $25t^2 - 28t - 44 = 0$, so $(t - 2)(25t + 22) = 0$, which means $t = 2$ or $t = -22/25$.

We then check these solutions. For $t = 2$, the left side becomes $2\log_3(-9)$, which is not defined. For $t = -22/25$, the left side becomes

$$2\log_3\left(1 + \frac{22}{5}\right) = 2\log_3\frac{27}{5} = 2(\log_3 27 - \log_3 5) = 6 - 2\log_3 5,$$

and the right side becomes

$$\log_3\left(-\frac{44}{25} + 5\right) + 2 = \log_3\frac{81}{25} + 2 = \log_3 81 - \log_3 25 + 2 = 4 - 2\log_3 5 + 2 = 6 - 2\log_3 5.$$

Hence, the only solution is $t = \boxed{-22/25}$.

There are many other ways to do this problem. For example, we could have moved the logarithms to one side of the equation to give $2\log_3(1 - 5t) - \log_3(2t + 5) = 2$. Then, we apply logarithm identities to the left side to find

$$2\log_3(1 - 5t) - \log_3(2t + 5) = \log_3(1 - 5t)^2 - \log_3(2t + 5) = \log_3\frac{(1 - 5t)^2}{2t + 5},$$

so we have $\log_3\frac{(1-5t)^2}{2t+5} = 2$. Writing this in exponential notation gives $\frac{(1-5t)^2}{2t+5} = 3^2 = 9$, which leads to the same quadratic as before.

13.4.4   Let the first term of the geometric sequence be a and let the common ratio be r, so $x_k = ar^{k-1}$ for $1 \le k \le n$. Then

$$\log_b x_k = \log_b(ar^{k-1}) = \log_b a + \log_b r^{k-1} = \log_b a + (k - 1)\log_b r.$$

Hence, $\log_b x_1, \log_b x_2, \ldots, \log_b x_n$ is an arithmetic sequence with first term $\log_b a$ and common difference $\log_b r$.

13.4.5   Multiplying both sides by $2\log_{10}(x/y)$, the equation becomes

$$2\log_{10}(xy) = \log_{10}\frac{x}{y},$$

so $2\log_{10} x + 2\log_{10} y = \log_{10} x - \log_{10} y$. Rearranging this gives $\log_{10} x + 3\log_{10} y = 0$. Applying logarithm identities to the left side gives

$$\log_{10} x + 3\log_{10} y = \log_{10} x + \log_{10} y^3 = \log_{10}(xy^3),$$

so $\log_{10}(xy^3) = 0$, which means $xy^3 = 1$. Hence, if y increases by 50%, meaning that y is multiplied by 3/2, then x must be multiplied by a factor of $(2/3)^3 = \boxed{8/27}$.

13.4.6 The line $x = k$ intersects the graph $y = \log_5 x$ at the point $(k, \log_5 k)$, and the graph $y = \log_5(x + 4)$ at the point $(k, \log_5(k + 4))$. These two points have the same x-coordinate, so the distance between the two points is the difference between the y-coordinates. Therefore, we have

$$\frac{1}{2} = \log_5(k + 4) - \log_5 k = \log_5 \frac{k + 4}{k}.$$

Hence, $(k + 4)/k = 5^{1/2} = \sqrt{5}$. Then $k + 4 = \sqrt{5}k$, so

$$k = \frac{4}{\sqrt{5} - 1} = \frac{4}{\sqrt{5} - 1} \cdot \frac{\sqrt{5} + 1}{\sqrt{5} + 1} = \frac{4(\sqrt{5} + 1)}{4} = \boxed{1 + \sqrt{5}}.$$

13.4.7 First, we write $a_n = \dfrac{1}{\log_n 2002} = \log_{2002} n$. Now we compute $b - c$, where all the logarithms have a common base of 2002:

$$\begin{aligned}
b - c &= (a_2 + a_3 + a_4 + a_5) - (a_{10} + a_{11} + a_{12} + a_{13} + a_{14}) \\
&= (\log_{2002} 2 + \log_{2002} 3 + \log_{2002} 4 + \log_{2002} 5) \\
&\quad - (\log_{2002} 10 + \log_{2002} 11 + \log_{2002} 12 + \log_{2002} 13 + \log_{2002} 14) \\
&= \log_{2002}(2 \cdot 3 \cdot 4 \cdot 5) - \log_{2002}(10 \cdot 11 \cdot 12 \cdot 13 \cdot 14) \\
&= \log_{2002}\left(\frac{2 \cdot 3 \cdot 4 \cdot 5}{10 \cdot 11 \cdot 12 \cdot 13 \cdot 14}\right) \\
&= \log_{2002}\left(\frac{1}{2002}\right) \\
&= \boxed{-1}.
\end{aligned}$$

Note how keeping the products factored until the end made the calculations easier.

13.4.8 To solve for k, we rewrite both sides as a logarithm with base 10. The left side becomes

$$\begin{aligned}
\log_{10}(k - 2)! + \log_{10}(k - 1)! + 2 &= \log_{10}(k - 2)! + \log_{10}(k - 1)! + \log_{10} 100 \\
&= \log_{10}[100(k - 2)!(k - 1)!],
\end{aligned}$$

and the right side becomes $2\log_{10} k! = \log_{10} k!^2$. Therefore, $100(k - 2)!(k - 1)! = k!^2$. Dividing both sides by $(k - 2)!(k - 1)!$, we get

$$100 = \frac{k! \cdot k!}{(k - 2)!(k - 1)!} = \frac{k \cdot (k - 1) \cdot (k - 2)! \cdot k \cdot (k - 1)!}{(k - 2)!(k - 1)!} = k^2(k - 1).$$

Rearranging $k^2(k - 1) = 100$ gives $k^3 - k^2 - 100 = 0$, and factoring gives $(k - 5)(k^2 + 4k + 20) = 0$. The quadratic factor has no integer roots, so the solution is $k = \boxed{5}$.

Exercises for Section 13.5

13.5.1 Writing the given equations in exponential form, we have $xy^3 = 10^1 = 10$ and $x^2y = 10^1 = 10$. Dividing the square of the first equation by the second equation, we find $y^5 = 10$, so $y = \sqrt[5]{10}$. Substituting this into $xy^3 = 10$ gives $x = \sqrt[5]{10^2}$, so $\log_{10} xy = \log_{10} \sqrt[5]{10^3} = \log_{10}(10)^{3/5} = \boxed{3/5}$.

See if you can also find a solution using logarithm identities.

13.5.2 Since $60^a = 3$ and $60^b = 5$, we have $a = \log_{60} 3$ and $b = \log_{60} 5$.

Let $x = 12^{[(1-a-b)/2(1-b)]}$. This is an expression with a complicated exponent, so we take the base 12 logarithm of it, which gives us

$$\log_{12} x = \log_{12} 12^{[(1-a-b)/2(1-b)]} = \frac{1 - a - b}{2(1 - b)} = \frac{1 - \log_{60} 3 - \log_{60} 5}{2(1 - \log_{60} 5)}.$$

Since $\log_{60} 60 = 1$, the numerator simplifies as

$$1 - \log_{60} 3 - \log_{60} 5 = \log_{60} 60 - \log_{60} 3 - \log_{60} 5 = \log_{60} \frac{60}{3 \cdot 5} = \log_{60} 4,$$

and the denominator simplifies as

$$2(1 - \log_{60} 5) = 2(\log_{60} 60 - \log_{60} 5) = 2 \log_{60} \frac{60}{5} = 2 \log_{60} 12.$$

Hence,

$$\log_{12} x = \frac{1 - \log_{60} 3 - \log_{60} 5}{2(1 - \log_{60} 5)} = \frac{\log_{60} 4}{2 \log_{60} 12} = \frac{1}{2} \log_{12} 4 = \log_{12} 4^{1/2} = \log_{12} 2.$$

Therefore, $x = \boxed{2}$.

13.5.3 Let $x = \log_2 p = \log_3 r = \log_{36} 17$. Then we may express these relations exponentially: $p = 2^x$, $r = 3^x$, and $17 = 36^x$. Hence, $pr = 2^x \cdot 3^x = 6^x = \sqrt{36^x} = \boxed{\sqrt{17}}$.

13.5.4 We first recognize the base of the left side as a perfect square. Factoring then gives $((a^2 - b^2)^2)^{y-1} = (a - b)^{2y}(a + b)^{-2}$, so $(a^2 - b^2)^{2y-2} = (a - b)^{2y}(a + b)^{-2}$. Next, we get y out of the exponent by taking the logarithm of both sides. Either $a - b$ or $a + b$ is a good choice for the base, since both are positive, and both are factors of both sides. We'll choose $a + b$ here. Taking the base $a + b$ logarithm of both sides, then applying some of our logarithmic identities, gives

$$(2y - 2) \log_{(a+b)}(a^2 - b^2) = \log_{(a+b)}((a - b)^{2y}(a + b)^{-2}) = \log_{(a+b)}(a - b)^{2y} + \log_{(a+b)}(a + b)^{-2} = 2y \log_{(a+b)}(a - b) - 2.$$

Rearranging $(2y - 2) \log_{(a+b)}(a^2 - b^2) = 2y \log_{(a+b)}(a - b) - 2$ and applying some more identities gives

$$-2 = (2y - 2) \log_{(a+b)}(a^2 - b^2) - 2y \log_{(a+b)}(a - b) = (2y - 2) \log_{(a+b)}((a + b)(a - b)) - 2y \log_{(a+b)}(a - b)$$
$$= (2y - 2)(\log_{a+b}(a + b) + \log_{a+b}(a - b)) - 2y \log_{(a+b)}(a - b) = (2y - 2)(1 + \log_{a+b}(a - b)) - 2y \log_{(a+b)}(a - b)$$
$$= 2y - 2 + 2y \log_{a+b}(a - b) - 2 \log_{a+b}(a - b) - 2y \log_{a+b}(a - b) = 2y - 2 - 2 \log_{a+b}(a - b).$$

Solving for y then gives $y = \boxed{\log_{(a+b)}(a - b)}$. (See if you can also tackle this problem by rearranging the original equation to the form $(a + b)^y = a - b$.)

13.5.5 Let $y = \log_{10} x$. Then $x = 10^y$, so substituting, we get

$$10^{y^2} = \frac{10^{3y}}{100} = 10^{3y-2}.$$

Then $y^2 = 3y - 2$, which simplifies as $(y - 1)(y - 2) = 0$. If $y = 1$, then we have $\log_{10} x = 1$, so $x = 10$. If $y = 2$, we have $\log_{10} x = 2$, so $x = 10^2 = 100$. Therefore, the solutions are $x = \boxed{10 \text{ and } 100}$.

Exercises for Section 13.6

13.6.1

(a) Let $t_{1/2}$ denote the half-life, in minutes. Then $\left(\dfrac{1}{2}\right)^{10/t_{1/2}} = \dfrac{5}{20} = \dfrac{1}{4}$. Therefore, $10/t_{1/2} = 2$, so $t_{1/2} = 10/2 = \boxed{5}$.

(b) After 24 minutes, the fraction of the substance that is left is

$$\left(\frac{1}{2}\right)^{24/5} \approx 0.036.$$

Therefore, the amount that is left is $0.036 \cdot 20 = \boxed{0.72 \text{ grams}}$.

13.6.2 From the given equation, we have $B + Ce^{-kt} = \dfrac{A}{D}$, so $Ce^{-kt} = \dfrac{A}{D} - B = \dfrac{A - BD}{D}$. Dividing both sides by C, we get $e^{-kt} = \dfrac{A - BD}{CD}$, so $e^{kt} = \dfrac{CD}{A - BD}$. Taking the natural logarithm of both sides, we get $kt = \ln\left(\dfrac{CD}{A - BD}\right)$, so $t = \boxed{\dfrac{1}{k}\ln\left(\dfrac{CD}{A - BD}\right)}$.

13.6.3 Let t denote the number of years the material was buried. Then $\left(\dfrac{1}{2}\right)^{t/120.3} = \dfrac{96.2}{400}$. Taking the natural logarithm of both sides, we get $\ln\left(\dfrac{1}{2}\right)^{t/120.3} = \ln\left(\dfrac{96.2}{400}\right)$, so we have

$$\frac{t}{120.3}\ln\frac{1}{2} = \ln\left(\frac{96.2}{400}\right).$$

Solving for t gives $t = \dfrac{120.3 \cdot \ln(96.2/400)}{\ln(1/2)} \approx 247.32$. Hence, the material was buried in the year $2314 - 247 = \boxed{2067}$.

13.6.4 The function $f(t)$ describes exponential decay. After 2 units of time, 1/9th of the original sample remains. Since $1/9 = (1/3)^2$, we may take $f(t) = \boxed{3^{-t}}$. (Note: This is not the only function that satisfies the equation $f(t + 2) = \frac{1}{9}f(t)$. The function $f(t) = a \cdot 3^{-t}$ satisfies the equation for any constant a, as does $f(t) = e^{-t\ln 3}$.)

13.6.5 Since the original amount is 3 grams, the half-life is the time t at which the amount left is 3/2:

$$3e^{-5t} = \frac{3}{2} \quad \Rightarrow \quad e^{-5t} = \frac{1}{2} \quad \Rightarrow \quad e^{5t} = 2.$$

Taking the natural logarithm of both sides, we get $5t = \ln 2$, so $t = \frac{\ln 2}{5} \approx 0.14$. Therefore, the half-life is $\boxed{0.14 \text{ seconds}}$.

13.6.6 We have $y = \ln(2 - e^y) - \ln 3 = \ln\frac{2 - e^y}{3}$. Raising e to the power of both sides, we get $e^y = \frac{2 - e^y}{3}$, so $3e^y = 2 - e^y$. Then $4e^y = 2$, so $e^y = 1/2$. Hence, $y = \ln(1/2) = \boxed{-\ln 2}$.

Review Problems

13.38 The left side is equal to $6^2 \cdot (6^n)^n = 6^2 \cdot 6^{n^2} = 6^{n^2+2}$, and the right side is equal to $6^n \cdot 6^n \cdot 6^n = 6^{3n}$, so $n^2 + 2 = 3n$. This simplifies as $(n - 1)(n - 2) = 0$, so the solutions are $n = \boxed{1 \text{ and } 2}$.

13.39 Let $x = 2^t$. Then the left side is equal to $2^{t+1} + 2^t = 2x + x = 3x$, and the right side is equal to $2^3 + 2^5 + 2^{7-t} = 8 + 32 + 2^7/2^t = 40 + 128/x$. Therefore, $3x = 40 + 128/x$, which simplifies as $3x^2 - 40x - 128 = 0$. Factoring gives

$(x-16)(3x+8) = 0$, so $x = 16$ or $x = -8/3$. Since $x = 2^t$, we must have $x > 0$, so we discard $x = -8/3$ as extraneous. From $x = 16$, we have $2^t = 16$, so $t = \boxed{4}$.

13.40 We can rewrite the equation as $4 \cdot 3^{2a} - 25 \cdot 3^a - 21 = 0$. This is a quadratic in 3^a, and it factors as $(3^a - 7)(4 \cdot 3^a + 3) = 0$, so $3^a = 7$ or $3^a = -3/4$. Since $3^a > 0$ for all a, we have $3^a = 7$. Therefore, $a = \boxed{\log_3 7}$.

13.41

(a) Since $243 = 3^5$, we have $\log_3 243 = \boxed{5}$.

(b) Let $x = \log_4 \frac{1}{32}$. Then $4^x = \frac{1}{32}$, which we can rewrite as $2^{2x} = 2^{-5}$. Therefore, $2x = -5$, so $x = \boxed{-5/2}$.

(c) $\log_{10} \sqrt[3]{0.01} = \frac{1}{3} \log_{10} \frac{1}{100} = \boxed{-2/3}$.

13.42

(a) $\log_4 12 + 2\log_4 3 - 3\log_4 6 = \log_4 12 + \log_4 3^2 - \log_4 6^3 = \log_4 \frac{12 \cdot 9}{216} = \log_4 \frac{1}{2} = \boxed{-\frac{1}{2}}$.

(b) $\log_2 7 \cdot \log_7 2 = \log_7 7 \cdot \log_2 2 = \boxed{1}$.

13.43 We repeatedly apply the fact that $\log_2 a = b$ means that $2^b = a$. From $\log_2(\log_2(\log_2 x)) = 2$, we have $\log_2(\log_2 x)) = 2^2 = 4$. From this equation, we have $\log_2 x = 2^4 = 16$, so $x = 2^{16} = \boxed{65536}$.

13.44 We can rewrite the first equation as $2^{2a} = 2^{3b}$, so $2a = 3b$. Taking the logarithm of the second equation with respect to base 3, we get $b = \log_3(2 \cdot 3^a) = \log_3 2 + \log_3 3^a = \log_3 2 + a$. Substituting $a = 3b/2$, this equation becomes $b = \log_3 2 + 3b/2$. Solving for b, we find $b = \boxed{-2\log_3 2}$.

13.45

(a) The inequality $\log_5(3 - x) \geq 7$ is equivalent to $3 - x \geq 5^7 = 78125$. Therefore, the solution is $\boxed{x \leq -78122}$. (Note that the logarithm is defined for all x such that $x < 3$, so we do not include any values in our answer that are not in the domain of the logarithm.)

(b) Since the base $1/2$ is less than 1, the inequality $\log_{1/2}(2x) > 3$ is equivalent to $2x < (1/2)^3 = 1/8$, or $x < 1/16$. Since $\log_{1/2}(2x)$ is only defined for $2x > 0$, the solution is $\boxed{0 < x < 1/16}$.

13.46 The function $f(x) = \log((1 - x)(x - n))$ is defined if and only if $(1 - x)(x - n) > 0$. In turn, this inequality is satisfied if and only if $1 < x < n$. The integers in this range are $2, 3, \ldots, n - 1$, for a total of $n - 2$ integers. Hence, $n - 2 = 2n - 6$, which means $n = \boxed{4}$.

13.47 We can simplify A as follows:

$$A = \frac{(\log_3 1 - \log_3 4)(\log_3 9 - \log_3 2)}{(\log_3 1 - \log_3 9)(\log_3 8 - \log_3 4)} = \frac{(-\log_3 2^2)(\log_3 \frac{9}{2})}{(-2)(\log_3 \frac{8}{4})} = \frac{-2\log_3 2 \cdot \log_3 \frac{9}{2}}{-2\log_3 2} = \log_3 \frac{9}{2}.$$

Hence, $3^A = \boxed{9/2}$.

13.48 We can convert $\log_{a^n} b$ to a logarithm with base a as follows: $\log_{a^n} b = \dfrac{\log_a b}{\log_a a^n} = \dfrac{\log_a b}{n} = \dfrac{1}{n} \log_a b$.

13.49 From the equation $3\log_a b = -3$, we get $\log_a b = -1$, so $b = a^{-1}$. In other words, $ab = 1$. Then

$$\log_b(9a) = \log_b 9 + \log_b a = \log_b 9 + \log_b b^{-1} = \log_b 9 - 1,$$

so $\log_b(9a) = -3$ gives us $\log_b 9 = -2$. Hence, $b^{-2} = 9$, so $b^2 = 1/9$. Since b must be positive, we have $b = 1/3$, and then $a = 1/b = 3$. Therefore, the solution is $(a, b) = \boxed{(3, 1/3)}$.

13.50 We have $5^{\log_{10}2} \cdot 2^{\log_{10}3} \cdot 2^{\log_{10}6} \cdot 5^{\log_{10}9} = 2^{\log_{10}3+\log_{10}6} \cdot 5^{\log_{10}2+\log_{10}9} = 2^{\log_{10}18} \cdot 5^{\log_{10}18} = 10^{\log_{10}18} = \boxed{18}$.

13.51 To express $\log_2 5$ in terms of x, we convert $x = \log_{10} 5$ to base 2 as follows:

$$x = \log_{10} 5 = \frac{\log_2 5}{\log_2 10} = \frac{\log_2 5}{\log_2(2 \cdot 5)} = \frac{\log_2 5}{\log_2 2 + \log_2 5} = \frac{\log_2 5}{\log_2 5 + 1}.$$

Multiplying both sides by $\log_2 5 + 1$, we get $(\log_2 5 + 1)x = \log_2 5$. Solving for $\log_2 5$ gives $\log_2 5 = \boxed{\dfrac{x}{1-x}}$.

Alternatively, we can take the reciprocal of both sides of $x = \log_{10} 5$ to give $\frac{1}{x} = \log_5 10 = \log_5 5 + \log_5 2 = 1 + \log_5 2 = 1 + \frac{1}{\log_2 5}$. Solving this equation for $\log_2 5$ gives the same result as above.

13.52 To find the intersection points of the graphs $y = \log_2(x + 3)$ and $y = 3 - \log_2(x - 1)$, we solve the equation $\log_2(x + 3) = 3 - \log_2(x - 1)$. Since

$$\log_2(x + 3) = 3 - \log_2(x - 1) = \log_2 8 - \log_2(x - 1) = \log_2 \frac{8}{x-1},$$

it follows that $x + 3 = \frac{8}{x-1}$. Hence, $(x + 3)(x - 1) = 8$, which simplifies as $x^2 + 2x - 11 = 0$. From the quadratic formula, we have

$$x = \frac{-2 \pm \sqrt{48}}{2} = \frac{-2 \pm 4\sqrt{3}}{2} = -1 \pm 2\sqrt{3}.$$

If $x = -1 - 2\sqrt{3}$, then $\log_2(x+3) = \log_2(2 - 2\sqrt{3})$. But $2 - 2\sqrt{3} < 0$, so $\log_2(2 - 2\sqrt{3})$ is not defined. If $x = -1 + 2\sqrt{3}$, then $\log_2(x + 3) = \log_2(2 + 2\sqrt{3})$, and $3 - \log_2(x - 1) = 3 - \log_2(-2 + 2\sqrt{3})$. These values are equal, because

$$\log_2(2 + 2\sqrt{3}) + \log_2(-2 + 2\sqrt{3}) = \log_2[(2 + 2\sqrt{3})(-2 + 2\sqrt{3})] = \log_2(12 - 4) = \log_2 8 = 3.$$

Hence, the two graphs intersect at the point $\boxed{(-1 + 2\sqrt{3}, \log_2(2 + 2\sqrt{3}))}$.

13.53

$$\frac{1}{\log_2 100!} + \frac{1}{\log_3 100!} + \frac{1}{\log_4 100!} + \cdots + \frac{1}{\log_{100} 100!} = \log_{100!} 2 + \log_{100!} 3 + \log_{100!} 4 + \cdots + \log_{100!} 100$$

$$= \log_{100!}(2 \cdot 3 \cdot 4 \cdots 100) = \log_{100!} 100! = \boxed{1}.$$

13.54 Taking the logarithm of both sides with respect to base 3, we get

$$n > \log_3 2^{102} = 102 \log_3 2 \approx 102 \cdot 0.631 = 64.362.$$

Hence, the smallest possible positive integer n is $\boxed{65}$.

13.55 The sum on the right side is

$$\sum_{k=0}^{39} \log_{10} x^{2^k} = \sum_{k=0}^{39} 2^k \log_{10} x = (\log_{10} x) \sum_{k=0}^{39} 2^k = (\log_{10} x)(1 + 2 + \cdots + 2^{39}) = (\log_{10} x)(2^{40} - 1).$$

Hence, we have $\log_{10} x = \dfrac{2^{41} - 2}{2^{40} - 1} = \dfrac{2(2^{40} - 1)}{2^{40} - 1} = 2$, so $x = 10^2 = \boxed{100}$.

13.56 Subtracting a from both sides, we get $b \ln c = d - a$, so $\ln c = \frac{d-a}{b}$. Then taking e to the power of both sides, we find $c = \boxed{e^{(d-a)/b}}$.

13.57 Let x be the initial amount, in ounces. Then we have $\frac{4.2}{x} = \left(\frac{1}{2}\right)^{12/4.3}$, so $x = \frac{4.2}{(1/2)^{12/4.3}} \approx 29.1$. Therefore, there were initially $\boxed{29.1 \text{ ounces}}$.

13.58 The time is initially 1:01 PM, so the first change in height will be a 50% increase. The first time I am less than 1 foot tall will be immediately after a decrease in height, so there will have been an equal number of increases and decreases in height. Let n be this common number.

Each increase multiplies my height by 1.5 and each decrease multiplies it by 0.6. So, after n of each, my height in feet is $6 \cdot 1.5^n \cdot 0.6^n = 6 \cdot (1.5 \cdot 0.6)^n = 6 \cdot 0.9^n$. We seek the smallest n such that $6 \cdot 0.9^n < 1$, which gives us $0.9^n < 1/6$. We could use trial-and-error, together with the fact that $f(x) = 0.9^x$ is monotonically decreasing, to find n. Or, we could use logarithms. We have $a < b$ if and only if $\ln a < \ln b$. (We could use any base greater than 1 for our logarithm here, not just the natural logarithm.) Therefore, from $0.9^n < 1/6$, we have $\ln 0.9^n < \ln(1/6)$, so $n \ln 0.9 < \ln(1/6)$. Since $\ln 0.9 < 0$, dividing both sides of the inequality by $\ln 0.9$ gives $n > (\ln(1/6))/(\ln 0.9) \approx 17.006$. So, the desired smallest n is 18.

The first decrease in height occurs at 1:32 PM, the second at 2:32 PM, the third at 3:32 PM, and so on, until the 18th decrease in height occurs at $\boxed{6:32 \text{ AM}}$.

13.59 Let x denote the initial amount, in grams, and let $t_{1/2}$ denote the half-life, in seconds. Then

$$\frac{12.2}{x} = \left(\frac{1}{2}\right)^{63/t_{1/2}} \qquad \text{and} \qquad \frac{4.1}{x} = \left(\frac{1}{2}\right)^{107/t_{1/2}}.$$

Taking the quotient of these equations, we get

$$\frac{4.1}{12.2} = \left(\frac{1}{2}\right)^{107/t_{1/2}-63/t_{1/2}} = \left(\frac{1}{2}\right)^{44/t_{1/2}}.$$

Taking the logarithm base e of both sides, we get

$$\ln\frac{4.1}{12.2} = \ln\left(\frac{1}{2}\right)^{44/t_{1/2}} = \frac{44}{t_{1/2}}\ln\frac{1}{2},$$

so

$$t_{1/2} = \frac{44\ln(1/2)}{\ln(4.1/12.2)} \approx 27.97.$$

Then from the first equation, we have $x = \frac{12.2}{(1/2)^{63/t_{1/2}}} \approx 58.1$. Hence, there were initially $\boxed{58.1 \text{ grams}}$.

Challenge Problems

13.60 Adding x^2 to both sides, we get $(x+2)^{2y} = x^2 + 3$, so $((x+2)^2)^y = x^2 + 3$. Writing this in logarithmic form gives $y = \boxed{\log_{(x+2)^2}(x^2+3)}$.

From the expression we found for y, we see that we have to investigate what happens in our original equation if x is -1, -2, or -3, since each of these makes the base of the logarithm either 0 or 1. For all other values of x, the base and argument of $\log_{(x+2)^2}(x^2+3)$ are positive, and the base is not equal to 1. If $x = -1$ in the original equation, we have $1^{2y} = 4$, which has no solutions. If $x = -2$, the equation is $0^{2y} = 7$, which has no solutions. And if $x = -3$, the equation is $(-1)^{2y} = 12$, which has no solutions. Therefore, the values of x for which there are no solutions for y are $\boxed{-1, -2, -3}$.

13.61 If the triangle with sides $a \le b \le c$ is log-right, then we must have

$$\log_{10} c^2 = \log_{10}(a^2) + \log_{10}(b^2) = \log_{10}(a^2b^2),$$

so $c^2 = a^2b^2$. By the Pythagorean Theorem, we have $a^2 + b^2 = c^2$, so $a^2b^2 = a^2 + b^2$. Then by applying Simon's Favorite Factoring Trick, we can rewrite $a^2b^2 = a^2 + b^2$ as $(a^2 - 1)(b^2 - 1) = 1$. Since $0 \le a \le b$, we have $a^2 - 1 \le b^2 - 1$, so $(a^2 - 1)^2 \le (a^2 - 1)(b^2 - 1) = 1$. Then $a^2 - 1 \le 1$, so $a^2 \le 2$, and finally $a \le \sqrt{2}$.

The triangle with sides $a = \sqrt{2}$, $b = \sqrt{2}$, and $c = 2$ is both right and log-right, so the maximum value of a is $\boxed{\sqrt{2}}$.

13.62 The point (a, b), under a rotation of $90°$ counterclockwise, goes to the point $(-b, a)$. If (a, b) lies on the graph of $y = \log_{10} x$, then $b = \log_{10} a$, so the point $(-b, a) = (-\log_{10} a, a)$ lies on the graph of G'. Let $x = -\log_{10} a$ and $y = a$, so $x = -\log_{10} y$. Then $\log_{10} y = -x$, so $\boxed{y = 10^{-x}}$. This is the equation corresponding to G'. The graphs are shown at right; G is solid, G' is dashed.

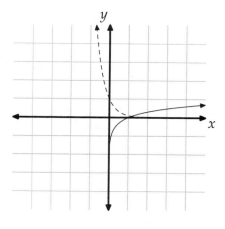

13.63 Let $a = \log_y x$. Then $\log_x y = \frac{1}{a}$, so $a + \frac{1}{a} = \frac{10}{3}$, which simplifies as $3a^2 - 10a + 3 = 0$. Factoring gives $(a - 3)(3a - 1) = 0$, so $a = 3$ or $a = 1/3$. Since the equations are symmetric in x and y, the cases $a = 3$ and $a = 1/3$ are symmetric. We'll address the case $a = 3$ here (the case $a = 1/3$ is essentially the same). If $a = 3$, then $\log_y x = 3$, which means $x = y^3$. Then, $xy = 144$ gives us $y^4 = 144$, which implies $y = \sqrt[4]{144} = 2\sqrt{3}$ (since y must be positive). Then $x = (2\sqrt{3})^3 = 24\sqrt{3}$, and $(x + y)/2 = \boxed{13\sqrt{3}}$. (The case $a = 1/3$ gives $x = 2\sqrt{3}$ and $y = 24\sqrt{3}$.)

13.64 Let $a = \log_x y$, so $\log_y x = \frac{1}{a}$. Then $\log_y x^2 = 2\log_y x = \frac{2}{a}$, so $a + \frac{2}{a} = 3$, which simplifies as $(a - 1)(a - 2) = 0$. Hence, $a = 1$ or $a = 2$.

If $\log_x y = a = 1$, then $y = x$, so the possible ordered pairs (x, y) are $(2, 2)$, $(3, 3)$, $\ldots$, $(100, 100)$, of which there are 99. If $\log_x y = a = 2$, then $y = x^2$, so the possible ordered pairs (x, y) are $(2, 4)$, $(3, 9)$, $\ldots$, $(10, 100)$, of which there are 9. Therefore, there are a total of $9 + 99 = \boxed{108}$ ordered pairs (x, y).

13.65 To deal with the exponent of $\log_{1995} x$, we take the base 1995 logarithm of both sides. Then the left side becomes

$$\log_{1995}(\sqrt{1995}x^{\log_{1995} x}) = \log_{1995}\sqrt{1995} + \log_{1995} x^{\log_{1995} x} = \frac{1}{2} + (\log_{1995} x)^2,$$

and the right side becomes $\log_{1995} x^2 = 2\log_{1995} x$. Let $y = \log_{1995} x$, so we now have $y^2 + \frac{1}{2} = 2y$, which simplifies as $2y^2 - 4y + 1 = 0$. This quadratic has two real roots y_1 and y_2. Then the corresponding values in x are $x_1 = 1995^{y_1}$ and $x_2 = 1995^{y_2}$. Hence, the product of these roots is

$$x_1 x_2 = 1995^{y_1 + y_2}.$$

By Vieta's Formulas, $y_1 + y_2 = 4/2 = 2$, so $x_1 x_2 = 1995^{y_1 + y_2} = 1995^2 = \boxed{3980025}$.

13.66 Since $\log_a b$ is the reciprocal of $\log_b a$, and $\log_a c$ is the reciprocal of $\log_c a$, we have

$$\frac{\log_b a}{\log_c a} = \frac{\log_a c}{\log_a b} = \log_b c = \frac{19}{99},$$

so $c = b^{19/99}$. Then $b = c^{99/19}$, so $\frac{b}{c} = c^{80/19}$. Hence, $k = \boxed{80/19}$.

13.67 We have

$$\log_a \frac{a}{b} + \log_b \frac{b}{a} = \log_a a - \log_a b + \log_b b - \log_b a = 2 - \log_a b - \log_b a.$$

Thus, maximizing $\log_a(a/b) + \log_b(b/a)$ is equivalent to minimizing $\log_a b + \log_b a$. However, $\log_a b$ and $\log_b a$ are reciprocals (and positive, since $a \ge b > 1$), so by AM-GM, we have $\log_a b + \log_b a \ge 2\sqrt{\log_a b \cdot \log_b a} = 2$.

Equality occurs if and only if $\log_a b = \log_b a = 1$; in other words, if $a = b$. Therefore, the maximum value of $\log_a(a/b) + \log_b(b/a)$ is $2 - \log_a a - \log_a a = 2 - 1 - 1 = \boxed{0}$.

13.68 Since points $A(x_1, y_1)$ and $B(x_2, y_2)$ lie on the graph of $y = \log_2 x$, we have $y_1 = \log_2 x_1$ and $y_2 = \log_2 x_2$. Then the y-coordinate of the midpoint of $\overline{AB}$ is

$$\frac{y_1 + y_2}{2} = \frac{\log_2 x_1 + \log_2 x_2}{2} = \frac{\log_2(x_1 x_2)}{2} = \log_2 \sqrt{x_1 x_2}.$$

By definition, the point $C(x_3, y_3)$ and the midpoint of $\overline{AB}$ lie on the same horizontal line, so they have the same y-coordinate. Hence, $y_3 = \log_2 \sqrt{x_1 x_2}$. But $C(x_3, y_3)$ also lies on the graph of $y = \log_2 x$, so $y_3 = \log_2 x_3$. Therefore, $x_3 = \sqrt{x_1 x_2}$, so $x_3^2 = x_1 x_2$.

13.69 Let $b = a^x$. Then $y = (b + b^{-1})/2$, which simplifies as $b^2 - 2yb + 1 = 0$. By the quadratic formula, we have

$$b = \frac{2y \pm \sqrt{4y^2 - 4}}{2} = y \pm \sqrt{y^2 - 1}.$$

Since $b = a^x$, we have $x = \log_a b = \boxed{\log_a(y \pm \sqrt{y^2 - 1})}$.

13.70 To deal with the exponents in the logarithms, we take the logarithms of x and y with respect to base b. This gives $\log_b x = \log_b 2^{\log_b 3} = \log_b 2 \cdot \log_b 3$, and $\log_b y = \log_b 3^{\log_b 2} = \log_b 2 \cdot \log_b 3$. Hence, $x = y$, so $x - y = \boxed{0}$.

13.71 We have $\log_2 ab = \log_2 a + \log_2 b \geq 6$, so $ab \geq 2^6 = 64$. Then by AM-GM,

$$a + b \geq 2\sqrt{ab} \geq 2 \cdot 8 = 16.$$

Equality occurs if and only if $a = b = 8$. Therefore, the minimum value of $a + b$ is $\boxed{16}$.

13.72 We can rewrite the given equation as $2^{333x-2} - 2^{222x+1} + 2^{111x+2} - 1 = 0$. Let $y = 2^{111x}$. Then

$$2^{333x-2} - 2^{222x+1} + 2^{111x+2} - 1 = \frac{y^3}{2^2} - 2y^2 + 4y - 1 = \frac{y^3 - 8y^2 + 16y - 4}{4},$$

so $y^3 - 8y^2 + 16y - 4 = 0$.

Let the three real roots of the original equation be x_1, x_2, and x_3. Then the corresponding values in y are $y_1 = 2^{111x_1}$, $y_2 = 2^{111x_2}$, and $y_3 = 2^{111x_3}$. Therefore, we have

$$y_1 y_2 y_3 = 2^{111(x_1+x_2+x_3)}.$$

But by Vieta's Formulas, we have $y_1 y_2 y_3 = 4$, so

$$2^{111(x_1+x_2+x_3)} = 4 = 2^2,$$

which means $111(x_1 + x_2 + x_3) = 2$. Hence, $x_1 + x_2 + x_3 = \boxed{2/111}$.

13.73 To determine the number of digits of a number N when expressed in decimal, we evaluate $\log_{10} N$. Thus, to determine the number of digits in 4^{18} when expressed in base 3, we determine $\log_3 4^{18}$, which is

$$\log_3 4^{18} = 18 \log_3 4 = 18 \log_3 2^2 = 36 \log_3 2 \approx 36 \cdot 0.631 = 22.716.$$

Therefore, 4^{18} has $\boxed{23}$ digits when expressed in base 3.

13.74 We can rewrite the given equations in exponential form: $\log_x w = 24$ implies $w = x^{24}$, $\log_y w = 40$ implies $w = y^{40}$, and $\log_{xyz} w = 12$ implies $w = (xyz)^{12}$. Then $x = w^{1/24}$, $y = w^{1/40}$, and $xyz = w^{1/12}$.

Hence, we have $z = \dfrac{xyz}{xy} = \dfrac{w^{1/12}}{w^{1/24} \cdot w^{1/40}} = w^{1/60}$, so $w = z^{60}$. Therefore, $\log_z w = \boxed{60}$.

13.75 We can rewrite the given equations in exponential form: $\log_{10}(x + y) = z$ implies $x + y = 10^z$, and $\log_{10}(x^2 + y^2) = z + 1$ implies $x^2 + y^2 = 10^{z+1}$. Squaring $x + y = 10^z$ gives $x^2 + 2xy + y^2 = 10^{2z}$, so $2xy = 10^{2z} - (x^2 + y^2) = 10^{2z} - 10^{z+1}$, or

$$xy = \frac{1}{2} \cdot 10^{2z} - \frac{1}{2} \cdot 10^{z+1}.$$

Then using sum of cubes, we have

$$x^3 + y^3 = (x + y)(x^2 - xy + y^2) = 10^z \left[10^{z+1} - \left(\frac{1}{2} \cdot 10^{2z} - \frac{1}{2} \cdot 10^{z+1} \right) \right]$$
$$= 10^z \left(-\frac{1}{2} \cdot 10^{2z} + \frac{3}{2} \cdot 10^{z+1} \right) = -\frac{1}{2} \cdot 10^{3z} + \frac{3}{2} \cdot 10^{2z+1} = -\frac{1}{2} \cdot 10^{3z} + 15 \cdot 10^{2z}.$$

Thus, $a = -1/2$ and $b = 15$, so $a + b = -1/2 + 15 = \boxed{29/2}$.

13.76 To solve for $(\log_2 x)^2$, we first convert each outer logarithm to a logarithm with base 2. The right side becomes

$$\log_8(\log_2 x) = \frac{\log_2(\log_2 x)}{\log_2 8} = \frac{1}{3}\log_2(\log_2 x),$$

so $\frac{1}{3}\log_2(\log_2 x) = \log_2(\log_8 x)$. Multiplying both sides by 3, we get $\log_2(\log_2 x) = 3\log_2(\log_8 x) = \log_2(\log_8 x)^3$. Therefore, $\log_2 x = (\log_8 x)^3$.

Again, we convert each logarithm to a logarithm with base 2:

$$\log_2 x = (\log_8 x)^3 = \left(\frac{\log_2 x}{\log_2 8} \right)^3 = \left(\frac{\log_2 x}{3} \right)^3 = \frac{(\log_2 x)^3}{27}.$$

Multiplying both sides by 27, we get $(\log_2 x)^3 = 27 \log_2 x$. Since $\log_2 x$ cannot be 0 (otherwise the right side in the original equation would not be defined), we can divide both sides by $\log_2 x$ to get $(\log_2 x)^2 = \boxed{27}$.

13.77 To deal with the exponents in the logarithms, we take the logarithms of both sides of the given equation. Since both 2 and 16 are powers of 2, we use base 2. We start with the first equation, $x^{\log_y x} = 2$. The left side becomes

$$\log_2 x^{\log_y x} = \log_y x \cdot \log_2 x = \frac{\log_2 x}{\log_2 y} \cdot \log_2 x,$$

so we have $(\log_2 x)^2/(\log_2 y) = \log_2 2 = 1$. When we take the base 2 logarithm of both sides of $y^{\log_x y} = 16$, the left side is

$$\log_2 y^{\log_x y} = \log_x y \cdot \log_2 y = \frac{\log_2 y}{\log_2 x} \cdot \log_2 y,$$

so we have $(\log_2 y)^2/(\log_2 x) = \log_2 16 = 4$. Dividing this equation by $(\log_2 x)^2/(\log_2 y) = 1$ gives us the equation $\left(\dfrac{\log_2 y}{\log_2 x} \right)^3 = (\log_x y)^3 = 4$, so $\log_x y = \sqrt[3]{4}$.

Finally, we know that $\log_y x$ and $\log_x y$ are reciprocals, so from the given equation $x^{\log_y x} = 2$, we have

$$x = 2^{1/\log_y x} = 2^{\log_x y} = \boxed{2^{\sqrt[3]{4}}}.$$

13.78 To make the algebra easier, let $a = \log_{10} x$, $b = \log_{10} y$, and $c = \log_{10} z$. Then, the left side of the first equation is

$$\log_{10}(2000xy) - (\log_{10} x)(\log_{10} y) = \log_{10}(2 \cdot 1000 \cdot x \cdot y) - (\log_{10} x)(\log_{10} y)$$
$$= \log_{10} 2 + \log_{10} 1000 + \log_{10} x + \log_{10} y - (\log_{10} x)(\log_{10} y)$$
$$= \log_{10} 2 + 3 + a + b - ab,$$

so the first equation is $\log_{10} 2 + 3 + a + b - ab = 4$. Rearranging gives us $ab - a - b + 1 = \log_{10} 2$, so $(a-1)(b-1) = \log_{10} 2$.

Similarly, in the second equation, the left side is

$$\log_{10}(2yz) - (\log_{10} y)(\log_{10} z) = \log_{10} 2 + \log_{10} y + \log_{10} z - (\log_{10} y)(\log_{10} z)$$
$$= \log_{10} 2 + b + c - bc,$$

so we have $\log_{10} 2 + b + c - bc = 1$. Rearranging this equation gives us $(b-1)(c-1) = \log_{10} 2$.

The left side of the third equation is

$$\log_{10} xz - (\log_{10} x)(\log_{10} z) = \log_{10} x + \log_{10} z - (\log_{10} x)(\log_{10} z) = a + c - ac,$$

so $a + c - ac = 0$. Rearranging gives $ac - a - c = 0$, and applying Simon's Favorite Factoring Trick gives $(a-1)(c-1) = 1$. Thus, we have the system of equations

$$(a-1)(b-1) = \log_{10} 2,$$
$$(a-1)(c-1) = 1,$$
$$(b-1)(c-1) = \log_{10} 2.$$

We therefore have $(a-1)(b-1) = (b-1)(c-1)$, so either $b = 1$ or $a - 1 = c - 1$. If $b = 1$, then $(a-1)(b-1) = 0$, not $\log_{10} 2$, so we cannot have $b = 1$. If $a - 1 = c - 1$, then $a = c$. Letting $a = c$ in $(a-1)(c-1) = 1$ gives $(a-1)^2 = 1$, so $a = 0$ or $a = 2$.

If $a = 0$, then $c = a = 0$ and $b - 1 = -\log_{10} 2$, so $b = 1 - \log_{10} 2 = \log_{10} 10 - \log_{10} 2 = \log_{10} 5$. Therefore, $x = 10^a = 1$, $y = 10^b = 5$ and $z = 10^c = 1$.

If $a = 2$, then $c = a = 2$ and $b - 1 = \log_{10} 2$, so $b = 1 + \log_{10} 2 = \log_{10} 10 + \log_{10} 2 = \log_{10} 20$. Therefore, $x = 10^a = 100$, $y = 10^b = 20$ and $z = 10^c = 100$.

Hence, the solutions are $(x, y, z) = \boxed{(100, 20, 100) \text{ and } (1, 5, 1)}$.

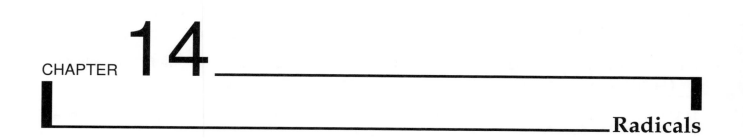

CHAPTER **14**

_____Radicals

Exercises for Section 14.1

14.1.1

(a) It is $\boxed{\text{not true}}$ that $\sqrt{(x-1)^2} = x - 1$ for all real values of x. For example, take $x = 0$. Then $\sqrt{(-1)^2} = \sqrt{1} = 1 \neq -1$. Since $\sqrt{(x-1)^2} = |x-1|$, we have $\sqrt{(x-1)^2} = x - 1$ if and only if $x \geq 1$.

(b) It is $\boxed{\text{true}}$ that $\sqrt[3]{(x-1)^3} = x - 1$ for all real values of x. Suppose $a^3 = b^3$ for some real numbers a and b. Then $a^3 - b^3 = 0$, so by difference of cubes, we have $(a-b)(a^2 + ab + b^2) = 0$, which means $a = b$ or $a^2 + ab + b^2 = 0$. If $a^2 + ab + b^2 = 0$, then completing the square gives

$$\left(a + \frac{b}{2}\right)^2 + \frac{3}{4}b^2 = 0.$$

This is possible if and only if $a + b/2 = b = 0$, so $a = b = 0$. Hence, in either case, $a = b$.

This means that if the cubes of a and b are equal, then a and b must be equal. Therefore, the only number that can be cubed to get $(x-1)^3$ is $x - 1$, so $\sqrt[3]{(x-1)^3} = x - 1$. (Contrast this with the square function, where if $a^2 = b^2$, then it is not necessarily true that $a = b$.)

(c) It is $\boxed{\text{true}}$ that $\sqrt{(x-1)^4} = (x-1)^2$ for all real values of x. Note that $\sqrt{(x-1)^4} = \sqrt{[(x-1)^2]^2} = |(x-1)^2|$. Since $(x-1)^2 \geq 0$ for all x, we have $|(x-1)^2| = (x-1)^2$.

14.1.2 We are given that x is positive, so both $\sqrt{x}$ and $2x$ are positive. Hence, if we square both sides of the inequality $\sqrt{x} < 2x$, we preserve the inequality, to obtain $x < 4x^2$. We can divide both sides by x without reversing the inequality because $x > 0$. This gives us $1 < 4x$, so $\boxed{x > 1/4}$.

14.1.3 Adding $\sqrt{x+1}$ to both sides gives $\sqrt{x-1} + 1 = \sqrt{x+1}$. Then, squaring both sides gives

$$x - 1 + 2\sqrt{x-1} + 1 = x + 1,$$

which simplifies as $2\sqrt{x-1} = 1$. Then squaring both sides again, we get $4(x-1) = 1$, so $x = \boxed{5/4}$. Checking if our answer is extraneous, we note $\sqrt{\frac{1}{4}} - \sqrt{\frac{9}{4}} + 1 = 0$, so the solution is not extraneous.

14.1.4 Let $y = \sqrt{x+8}$. Then the given equation becomes $y - 6/y = 5$, which simplifies as $(y-6)(y+1) = 0$, so $y = 6$ or $y = -1$. Since $y = \sqrt{x+8} \geq 0$, we have $\sqrt{x+8} = y = 6$. Then $x + 8 = 36$, so $x = \boxed{28}$.

14.1.5 Squaring both sides, we get $81(2 - 3x + x^2) = x^6 - 3x^5 + 2x^4$. The left side factors as $81(x-1)(x-2)$, and the right side factors as $x^4(x^2 - 3x + 2) = x^4(x-1)(x-2)$, so the equation becomes $81(x-1)(x-2) = x^4(x-1)(x-2)$. Then $x^4(x-1)(x-2) - 81(x-1)(x-2) = 0$, so $(x^4 - 81)(x-1)(x-2) = 0$. Therefore, $x^4 = 81$, or $x = 1$, or $x = 2$.

If $x = 1$ or $x = 2$, then both sides of the original equation become 0, so these are solutions. If $x^4 = 81$, then $x^2 = 9$, and so $x = \pm 3$. Both of these values satisfy the original equation, so the solutions are $x = \boxed{1, 2, 3, \text{ and } -3}$.

14.1.6 Adding 2 to both sides, we get $\sqrt[3]{5r^2 + 24r + 8} < r + 2$. For any real numbers, if $a < b$, then $a^3 < b^3$, so we can cube both sides to get

$$5r^2 + 24r + 8 < (r + 2)^3 = r^3 + 6r^2 + 12r + 8,$$

which simplifies as $r^3 + r^2 - 12r > 0$, so $r(r - 3)(r + 4) > 0$. Solving the inequality, we find the solution is $\boxed{r \in (-4, 0) \cup (3, +\infty)}$.

14.1.7 Let $a = \sqrt[3]{3x - 5}$, $b = \sqrt[3]{2x - 4}$, and $c = \sqrt[3]{5x - 9}$. Then our equation is $a + b = c$. Also, note that $a^3 + b^3 = (3x - 5) + (2x - 4) = 5x - 9 = c^3$.

Cubing the equation $a + b = c$, we get $a^3 + 3a^2b + 3ab^2 + b^3 = c^3$. Subtracting the equation $a^3 + b^3 = c^3$, we get $3a^2b + 3ab^2 = 0$, so $3ab(a + b) = 0$, which means $a = 0$, $b = 0$, or $a + b = 0$.

If $a = \sqrt[3]{3x - 5} = 0$, then $x = 5/3$. If $b = \sqrt[3]{2x - 4} = 0$, then $x = 4/2 = 2$. Finally, if $a + b = 0$, then $c = \sqrt[3]{5x - 9} = 0$, so $x = 9/5$. Therefore, the solutions are $x = \boxed{5/3, 2, \text{ and } 9/5}$. Checking these solutions, we find that each works.

Exercises for Section 14.2

14.2.1 Let $x = \sqrt{6 + \sqrt{11}} + \sqrt{6 - \sqrt{11}}$. Then

$$x^2 = 6 + \sqrt{11} + 2\sqrt{6 + \sqrt{11}}\sqrt{6 - \sqrt{11}} + 6 - \sqrt{11} = 12 + 2\sqrt{(6 + \sqrt{11})(6 - \sqrt{11})}$$
$$= 12 + 2\sqrt{36 - 11} = 12 + 2\sqrt{25} = 12 + 2 \cdot 5 = 22,$$

so $x = \boxed{\sqrt{22}}$.

14.2.2

(a) Let $\sqrt{49 + 28\sqrt{3}} = \sqrt{a} + \sqrt{b}$. Squaring both sides, we get $a + 2\sqrt{ab} + b = 49 + 28\sqrt{3}$. Therefore, we let $a + b = 49$ and $2\sqrt{ab} = 28\sqrt{3}$, so $\sqrt{ab} = 14\sqrt{3}$, or $ab = 3 \cdot 14^2 = 588$. Then by Vieta's Formulas, a and b are the roots of the quadratic $t^2 - 49t + 588$, which equals $(t - 21)(t - 28)$. Hence, a and b are equal to 21 and 28 in some order, which means $\sqrt{49 + 28\sqrt{3}} = \sqrt{21} + \sqrt{28} = \boxed{2\sqrt{7} + \sqrt{21}}$.

(b) We extract the fourth root by taking the square root twice. Let $\sqrt{49 + 20\sqrt{6}} = \sqrt{a} + \sqrt{b}$. Squaring both sides, we get $a + 2\sqrt{ab} + b = 49 + 20\sqrt{6}$. Therefore, we let $a + b = 49$ and $2\sqrt{ab} = 20\sqrt{6}$, so $\sqrt{ab} = 10\sqrt{6}$, or $ab = 6 \cdot 10^2 = 600$. Then by Vieta's Formulas, a and b are the roots of the quadratic $t^2 - 49t + 600$, which equals $(t - 24)(t - 25)$. Hence, a and b are equal to 24 and 25 in some order, which means $\sqrt{49 + 20\sqrt{6}} = \sqrt{24} + \sqrt{25} = 5 + 2\sqrt{6}$.

Now, let $\sqrt{5 + 2\sqrt{6}} = \sqrt{c} + \sqrt{d}$. Squaring both sides, we get $c + 2\sqrt{cd} + d = 5 + 2\sqrt{6}$. Therefore, we let $c + d = 5$ and $2\sqrt{cd} = 2\sqrt{6}$, so $\sqrt{cd} = \sqrt{6}$, or $cd = 6$. Then by Vieta's Formulas, c and d are the roots of the quadratic $t^2 - 5t + 6$, which equals $(t - 2)(t - 3)$. Hence, c and d are equal to 2 and 3 in some order, which means $\sqrt{5 + 2\sqrt{6}} = \boxed{\sqrt{2} + \sqrt{3}}$.

14.2.3 First, we find the square root of $21 + 12\sqrt{3}$. Let $\sqrt{21 + 12\sqrt{3}} = \sqrt{s} + \sqrt{t}$. Squaring both sides, we get $21 + 12\sqrt{3} = s + 2\sqrt{st} + t$. Therefore, we let $s + t = 21$ and $2\sqrt{st} = 12\sqrt{3}$, so $\sqrt{st} = 6\sqrt{3}$, or $st = 3 \cdot 6^2 = 108$. Then by

Vieta's Formulas, s and t are the roots of the quadratic $x^2 - 21x + 108$, which equals $(x-9)(x-12)$. Hence, s and t are equal to 9 and 12 in some order, which means $\sqrt{21 + 12\sqrt{3}} = \sqrt{9} + \sqrt{12} = 3 + 2\sqrt{3}$, so

$$\sqrt{1 + \sqrt{21 + 12\sqrt{3}}} = \sqrt{1 + 3 + 2\sqrt{3}} = \sqrt{4 + 2\sqrt{3}}.$$

We guess that $\sqrt{4 + 2\sqrt{3}} = c + d\sqrt{3}$ for some integers c and d. Squaring both sides gives us $c^2 + 3d^2 = 4$ and $cd = 1$. A natural guess now is $c = d = 1$, which works. Therefore, the given expression equals $\sqrt{1} + \sqrt{3}$, which means $(a, b) = \boxed{(1, 3)}$.

14.2.4 We must compare $\sqrt{2} + \sqrt{3}$ and $\sqrt{10}$. Since both quantities are positive, this is the same as comparing $(\sqrt{2} + \sqrt{3})^2 = 2 + 2\sqrt{2}\sqrt{3} + 3 = 5 + 2\sqrt{6}$ and $(\sqrt{10})^2 = 10$. Subtracting 5 from both quantities, we get the values $2\sqrt{6}$ and 5.

Squaring both of these values, we get $(2\sqrt{6})^2 = 24$ and 25. Since $24 < 25$, we have $2\sqrt{6} < 5$, so $5 + 2\sqrt{6} < 10$, which tells us that $\sqrt{2} + \sqrt{3} < \sqrt{10}$. In other words, $\boxed{\sqrt{10} \text{ is greater}}$.

14.2.5 We sure don't want to cube those radical expressions, so we first try finding the square roots of each. We guess that $\sqrt{52 + 6\sqrt{43}} = a + b\sqrt{43}$ for some integers a and b. Squaring both sides gives $52 + 6\sqrt{43} = a^2 + 43b^2 + 2ab\sqrt{43}$, so $52 = a^2 + 43b^2$ and $3 = ab$. The first equation strongly suggests $b = 1$, which gives us $a = 3$, and $(a, b) = (3, 1)$ satisfies $ab = 3$. So, we have $\sqrt{52 + 6\sqrt{43}} = 3 + \sqrt{43}$. Similarly, we have $\sqrt{52 - 6\sqrt{43}} = -3 + \sqrt{43}$ (remember, square roots must be nonnegative!), so

$$(52 + 6\sqrt{43})^{3/2} - (52 - 6\sqrt{43})^{3/2} = (3 + \sqrt{43})^3 - (-3 + \sqrt{43})^3.$$

Rather than cubing out the radicals on the right, we use the difference of cubes factorization, and we have

$$(3 + \sqrt{43})^3 - (-3 + \sqrt{43})^3 = [(3 + \sqrt{43}) - (-3 + \sqrt{43})][(3 + \sqrt{43})^2 + (3 + \sqrt{43})(-3 + \sqrt{43}) + (-3 + \sqrt{43})^2]$$
$$= (6)(52 + 6\sqrt{43} + 34 + 52 - 6\sqrt{43}) = (6)(138) = \boxed{828}.$$

Exercises for Section 14.3

14.3.1

(a) $\dfrac{4}{\sqrt{11} + 3} = \dfrac{4}{\sqrt{11} + 3} \cdot \dfrac{\sqrt{11} - 3}{\sqrt{11} - 3} = \dfrac{4(\sqrt{11} - 3)}{11 - 9} = \boxed{2\sqrt{11} - 6}.$

(b) Let $a = \sqrt[3]{5}$ and $b = \sqrt[3]{3}$. To rationalize the denominator $a + b$, we must multiply it by some factor which gives a rational number. We can use sum of cubes to find this factor:

$$(a + b)(a^2 - ab + b^2) = a^3 + b^3 = 8.$$

Hence,

$$\frac{8}{\sqrt[3]{5} + \sqrt[3]{3}} = \frac{8}{a + b} = \frac{8(a^2 - ab + b^2)}{(a + b)(a^2 - ab + b^2)} = \frac{8(a^2 - ab + b^2)}{a^3 + b^3}$$
$$= \frac{8(a^2 - ab + b^2)}{8} = a^2 - ab + b^2 = \boxed{\sqrt[3]{25} - \sqrt[3]{15} + \sqrt[3]{9}}.$$

(c) To rationalize a denominator of the form $a + \sqrt{b}$, we multiply top and bottom by $a - \sqrt{b}$. We try something similar here, by letting $a = \sqrt{2} + \sqrt{3}$, and multiplying top and bottom by $\sqrt{2} + \sqrt{3} - \sqrt{5}$:

$$\frac{2\sqrt{6}}{\sqrt{2} + \sqrt{3} + \sqrt{5}} = \frac{2\sqrt{6}(\sqrt{2} + \sqrt{3} - \sqrt{5})}{(\sqrt{2} + \sqrt{3} + \sqrt{5})(\sqrt{2} + \sqrt{3} - \sqrt{5})} = \frac{2\sqrt{6}(\sqrt{2} + \sqrt{3} - \sqrt{5})}{(\sqrt{2} + \sqrt{3})^2 - 5}$$

$$= \frac{2\sqrt{6}(\sqrt{2} + \sqrt{3} - \sqrt{5})}{2 + 2\sqrt{6} + 3 - 5} = \frac{2\sqrt{6}(\sqrt{2} + \sqrt{3} - \sqrt{5})}{2\sqrt{6}} = \boxed{\sqrt{2} + \sqrt{3} - \sqrt{5}}.$$

We got a little lucky here. We can't always rationalize a denominator of the form $\sqrt{a} + \sqrt{b} + \sqrt{c}$ in one step by multiplying top and bottom by $\sqrt{a} + \sqrt{b} - \sqrt{c}$. Sometimes, we'll need a second step because we won't have a convenient canceling with the numerator as we do in this problem.

(d) We multiply by the "radical conjugate" of the denominator:

$$\frac{\sqrt{4 + 2\sqrt{3}} + \sqrt{4 - 2\sqrt{3}}}{\sqrt{4 + 2\sqrt{3}} - \sqrt{4 - 2\sqrt{3}}} = \frac{\sqrt{4 + 2\sqrt{3}} + \sqrt{4 - 2\sqrt{3}}}{\sqrt{4 + 2\sqrt{3}} - \sqrt{4 - 2\sqrt{3}}} \cdot \frac{\sqrt{4 + 2\sqrt{3}} + \sqrt{4 - 2\sqrt{3}}}{\sqrt{4 + 2\sqrt{3}} + \sqrt{4 - 2\sqrt{3}}}$$

$$= \frac{4 + 2\sqrt{3} + 2\sqrt{4 + 2\sqrt{3}}\sqrt{4 - 2\sqrt{3}} + 4 - 2\sqrt{3}}{4 + 2\sqrt{3} - (4 - 2\sqrt{3})}$$

$$= \frac{8 + 2\sqrt{16 - 12}}{4\sqrt{3}} = \frac{12}{4\sqrt{3}} = \frac{3}{\sqrt{3}} = \boxed{\sqrt{3}}.$$

14.3.2 The product of the first two factors is

$$(\sqrt{5} + \sqrt{6} + \sqrt{7})(\sqrt{5} + \sqrt{6} - \sqrt{7}) = (\sqrt{5} + \sqrt{6})^2 - 7 = 5 + 2\sqrt{30} + 6 - 7 = 4 + 2\sqrt{30}.$$

The product of the last two factors is

$$(\sqrt{5} - \sqrt{6} + \sqrt{7})(-\sqrt{5} + \sqrt{6} + \sqrt{7}) = 7 - (\sqrt{5} - \sqrt{6})^2 = 7 - (5 - 2\sqrt{30} + 6) = 2\sqrt{30} - 4.$$

Therefore, the entire product is

$$(\sqrt{5} + \sqrt{6} + \sqrt{7})(\sqrt{5} + \sqrt{6} - \sqrt{7})(\sqrt{5} - \sqrt{6} + \sqrt{7})(-\sqrt{5} + \sqrt{6} + \sqrt{7}) = (2\sqrt{30} + 4)(2\sqrt{30} - 4) = 120 - 16 = \boxed{104}.$$

14.3.3 We can partially rationalize the denominator by multiplying the top and bottom by $\sqrt{2} + \sqrt[3]{2}$:

$$\frac{1}{\sqrt{2} - \sqrt[3]{2}} = \frac{\sqrt{2} + \sqrt[3]{2}}{(\sqrt{2} - \sqrt[3]{2})(\sqrt{2} + \sqrt[3]{2})} = \frac{\sqrt{2} + \sqrt[3]{2}}{2 - \sqrt[3]{4}}.$$

Since there is a cube root, we can then rationalize the denominator in this fraction by viewing $2 - \sqrt[3]{4}$ as a factor in a difference of cubes:

$$\frac{\sqrt{2} + \sqrt[3]{2}}{2 - \sqrt[3]{4}} = \frac{(\sqrt{2} + \sqrt[3]{2})(2^2 + 2\sqrt[3]{4} + \sqrt[3]{4^2})}{(2 - \sqrt[3]{4})(2^2 + 2\sqrt[3]{4} + \sqrt[3]{4^2})} = \frac{(\sqrt{2} + \sqrt[3]{2})(4 + 2\sqrt[3]{4} + 2\sqrt[3]{2})}{2^3 - 4} = \frac{(2^{1/2} + 2^{1/3})(2^2 + 2^{5/3} + 2^{4/3})}{4}$$

$$= \frac{2^{5/2} + 2^{13/6} + 2^{11/6} + 2^{7/3} + 2^2 + 2^{5/3}}{2^2} = \boxed{2^{1/2} + 2^{1/3} + 2^{1/6} + 2^0 + 2^{-1/6} + 2^{-1/3}}.$$

14.3.4 First, to clear the fraction, we multiply both sides by 10 to get $1 < 10(\sqrt{101} - \sqrt{99})$. Then, using difference of squares, we multiply both sides by $\sqrt{101} + \sqrt{99}$ to get

$$\sqrt{101} + \sqrt{99} < 10(\sqrt{101} - \sqrt{99})(\sqrt{101} + \sqrt{99}) = 10(101 - 99) = 20.$$

To show that $20 > \sqrt{101} + \sqrt{99}$, we square both sides, to get

$$400 > 101 + 2\sqrt{101}\sqrt{99} + 99 = 200 + 2\sqrt{9999},$$

which is equivalent to $200 > 2\sqrt{9999}$. Dividing both sides by 2, we get $100 > \sqrt{9999}$. Finally, we can square both sides again, to obtain the inequality $10000 > 9999$, which is true.

Since all our steps are reversible, the original inequality holds.

Review Problems

14.16

(a) $\dfrac{1}{\sqrt{3}-\sqrt{2}} = \dfrac{\sqrt{3}+\sqrt{2}}{(\sqrt{3}-\sqrt{2})(\sqrt{3}+\sqrt{2})} = \dfrac{\sqrt{3}+\sqrt{2}}{3-2} = \boxed{\sqrt{3}+\sqrt{2}}$.

(b) $\dfrac{4}{3+\sqrt{7}} = \dfrac{4(3-\sqrt{7})}{(3+\sqrt{7})(3-\sqrt{7})} = \dfrac{4(3-\sqrt{7})}{3^2-7} = \boxed{6-2\sqrt{7}}$.

(c) Using difference of cubes, we get

$$\frac{1}{\sqrt[3]{7}-1} = \frac{\sqrt[3]{7^2}+\sqrt[3]{7}+1}{(\sqrt[3]{7}-1)(\sqrt[3]{7^2}+\sqrt[3]{7}+1)} = \frac{\sqrt[3]{7^2}+\sqrt[3]{7}+1}{7-1} = \boxed{\frac{\sqrt[3]{49}+\sqrt[3]{7}+1}{6}}.$$

(d) Using difference of cubes, we get

$$\frac{2}{\sqrt[3]{25}+\sqrt[3]{5}+1} = \frac{2(\sqrt[3]{5}-1)}{(\sqrt[3]{5}-1)(\sqrt[3]{5^2}+\sqrt[3]{5}+1)} = \frac{2(\sqrt[3]{5}-1)}{5-1} = \boxed{\frac{\sqrt[3]{5}-1}{2}}.$$

14.17

(a) We guess that $\sqrt{6+2\sqrt{5}} = a+b\sqrt{5}$ for some integers a and b. Squaring both sides, we get $a^2 + 5b^2 + 2ab\sqrt{5} = 6+2\sqrt{5}$. So, we guess that $a^2 + 5b^2 = 6$ and $ab = 1$. By inspection, both $a = b = 1$ and $a = b = -1$ are solutions. Our square root must be nonnegative, so $\sqrt{6+2\sqrt{5}} = \boxed{1+\sqrt{5}}$.

(b) We extract the fourth root by taking the square root twice. Let $\sqrt{89+28\sqrt{10}} = \sqrt{a} + \sqrt{b}$. Squaring both sides, we get $a + 2\sqrt{ab} + b = 89 + 28\sqrt{10}$. Therefore, we let $a + b = 89$ and $2\sqrt{ab} = 28\sqrt{10}$, so $\sqrt{ab} = 14\sqrt{10}$, or $ab = 10 \cdot 14^2 = 1960$. Then by Vieta's Formulas, a and b are the roots of the quadratic $t^2 - 89t + 1960$, which equals $(t-40)(t-49)$. Hence, a and b are equal to 40 and 49 in some order, which means $\sqrt{89+28\sqrt{10}} = \sqrt{40} + \sqrt{49} = 7 + 2\sqrt{10}$.

Now, let $\sqrt{7+2\sqrt{10}} = \sqrt{c} + \sqrt{d}$. Squaring both sides, we get $c + 2\sqrt{cd} + d = 7 + 2\sqrt{10}$. Therefore, we let $c + d = 7$ and $2\sqrt{cd} = 2\sqrt{10}$, so $\sqrt{cd} = \sqrt{10}$, or $cd = 10$. Then by Vieta's Formulas, c and d are the roots of the quadratic $t^2 - 7t + 10$, which is the same as $(t-2)(t-5)$. Hence, c and d are equal to 2 and 5 in some order, which means $\sqrt{7+2\sqrt{10}} = \boxed{\sqrt{2}+\sqrt{5}}$.

14.18 Let $x = \sqrt{4+\sqrt{7}} - \sqrt{4-\sqrt{7}}$. Then

$$x^2 = 4 + \sqrt{7} - 2\sqrt{4+\sqrt{7}}\sqrt{4-\sqrt{7}} + 4 - \sqrt{7} = 8 - 2\sqrt{(4+\sqrt{7})(4-\sqrt{7})} = 8 - 2\sqrt{4^2-7}$$
$$= 8 - 2\sqrt{9} = 8 - 2\cdot 3 = 2.$$

Since $x > 0$, we have $x = \boxed{\sqrt{2}}$.

14.19 Subtracting $\sqrt{x-1}$ from both sides, we get $\sqrt{5x-1} = 2 - \sqrt{x-1}$. Squaring both sides, we get

$$5x - 1 = 4 - 4\sqrt{x-1} + x - 1,$$

which simplifies as $4\sqrt{x-1} = -4x+4$, so $\sqrt{x-1} = -x+1$. Squaring both sides, we get $x-1 = (-x+1)^2 = x^2 - 2x + 1$, so $x^2 - 3x + 2 = 0$, which means $(x-1)(x-2) = 0$. Hence, $x = 1$ or $x = 2$. Checking these values, we find only $x = \boxed{1}$ satisfies the original equation.

14.20 Squaring both sides, we get $16(x^3 - 2x^2 + x) = x^3 - x^2$. The left side factors as $16x(x^2 - 2x + 1) = 16x(x-1)^2$, and the right side factors as $x^2(x-1)$, so the equation becomes $16x(x-1)^2 = x^2(x-1)$. Subtracting $x^2(x-1)$ from both sides gives $16x(x-1)^2 - x^2(x-1) = 0$, so $x(x-1)[16(x-1) - x] = 0$, which means $x(x-1)(15x-16) = 0$. Therefore, the solutions are $x = \boxed{0, 1, \text{ and } 16/15}$. Checking, we find that all of these values satisfy the original equation.

14.21 Squaring both sides of the given equation, we get

$$2\sqrt{3} - 3 = x\sqrt{3} - 2\sqrt{x\sqrt{3}}\sqrt{y\sqrt{3}} + y\sqrt{3} = (x+y)\sqrt{3} - 2\sqrt{3xy}.$$

Since x and y are rational, the only way we can achieve equality is if $x + y = 2$ and $2\sqrt{3xy} = 3$. Then $\sqrt{3xy} = 3/2$, so $3xy = 9/4$, or $xy = 3/4$. Therefore, by Vieta's Formulas, x and y are the roots of the quadratic $t^2 - 2t + 3/4 = 0$. Multiplying by 4 gives $4t^2 - 8t + 3 = 0$, so $(2t-1)(2t-3) = 0$. Hence, x and y are equal to 1/2 and 3/2 in some order.

Since $\sqrt{2\sqrt{3}-3}$ is positive, $\sqrt{x\sqrt{3}}$ must be greater than $\sqrt{y\sqrt{3}}$. Therefore, the only solution is $(x,y) = \boxed{(3/2, 1/2)}$.

14.22 To simplify the expression, we simplify each part of the expression. So, let

$$x = \frac{\sqrt{\sqrt{5}+2} + \sqrt{\sqrt{5}-2}}{\sqrt{\sqrt{5}+1}} \quad \text{and} \quad y = \sqrt{3 - 2\sqrt{2}}.$$

Then

$$x^2 = \frac{\sqrt{5}+2+2\sqrt{\sqrt{5}+2}\sqrt{\sqrt{5}-2}+\sqrt{5}-2}{\sqrt{5}+1} = \frac{2\sqrt{5}+2\sqrt{(\sqrt{5}+2)(\sqrt{5}-2)}}{\sqrt{5}+1} = \frac{2\sqrt{5}+2\sqrt{5-4}}{\sqrt{5}+1}$$

$$= \frac{2\sqrt{5}+2}{\sqrt{5}+1} = 2.$$

Since $x > 0$, we have $x = \sqrt{2}$.

To simplify y, we guess $\sqrt{3 - 2\sqrt{2}} = a + b\sqrt{2}$. Squaring both sides gives $3 - 2\sqrt{2} = a^2 + 2b^2 + 2ab\sqrt{2}$, from which we have $a^2 + 2b^2 = 3$ and $ab = -1$. We quickly see that the solutions to this system are $(a,b) = (1,-1)$ and $(-1,1)$. Since the square root must be positive, we have $y = \sqrt{3 - 2\sqrt{2}} = -1 + \sqrt{2}$.

Therefore, the expression is equal to $x - y = \sqrt{2} - (-1 + \sqrt{2}) = \boxed{1}$.

14.23

(a) Squaring both sides, we get $x^2 + 7x + 10 = x^2 + 4x + 4 + 2(x+2)\sqrt{x+2} + x + 2$, which simplifies as $2x + 4 = 2(x+2)\sqrt{x+2}$, so $x + 2 = (x+2)\sqrt{x+2}$. Subtracting $x+2$ from both sides gives

$$(x+2)(\sqrt{x+2} - 1) = 0.$$

Hence, $x = -2$ or $\sqrt{x+2} = 1$.

If $x = -2$, then both sides of the original equation become 0, so it is a solution. If $\sqrt{x+2} = 1$, then $x + 2 = 1$, so $x = -1$. This value also satisfies the original equation, so the solutions are $x = \boxed{-1 \text{ and } -2}$.

(b) First, we note that the lesser side is only defined for $x \geq -2$. Checking where the greater side is defined, we note that $\sqrt{x^2 + 7x + 10} = \sqrt{(x+5)(x+2)}$, so the greater side is defined for $x \in (-\infty, -5] \cup [-2, +\infty)$. Combining this with the restrictions on the lesser side, we must have $x \geq -2$. Both sides of the inequality are nonnegative, so we can square both sides to get the equivalent inequality

$$x^2 + 7x + 10 > x^2 + 4x + 4 + 2(x+2)\sqrt{x+2} + x + 2,$$

which simplifies as $2x + 4 > 2(x+2)\sqrt{x+2}$, so $x + 2 > (x+2)\sqrt{x+2}$. Subtracting $x+2$ gives the inequality $(x+2)\sqrt{x+2} - (x+2) < 0$, so $(x+2)(\sqrt{x+2}-1) < 0$. We know that $x+2$ is nonnegative. If $x+2 = 0$, then the inequality is not satisfied, so $x \neq -2$. Therefore, $x + 2$ is positive, so we must have $\sqrt{x+2} - 1 < 0$ in order to have $(x+2)(\sqrt{x+2}-1) < 0$. This gives us $\sqrt{x+2} < 1$. Squaring both sides gives $x + 2 < 1$, so $x < -1$. Therefore, the solution is $\boxed{-2 < x < -1}$.

14.24 Let $\alpha = \sqrt[3]{18 + 5\sqrt{13}}$ and $\beta = \sqrt[3]{18 - 5\sqrt{13}}$, and let $x = \alpha + \beta$. Then

$$x^3 = \alpha^3 + 3\alpha^2\beta + 3\alpha\beta^2 + \beta^3 = 18 + 5\sqrt{13} + 3\alpha\beta(\alpha+\beta) + 18 - 5\sqrt{13}$$

$$= 36 + 3x\sqrt[3]{18 + 5\sqrt{13}}\sqrt[3]{18 - 5\sqrt{13}} = 36 + 3x\sqrt[3]{(18 + 5\sqrt{13})(18 - 5\sqrt{13})}$$

$$= 36 + 3x\sqrt[3]{18^2 - 5^2 \cdot 13} = 36 + 3x\sqrt[3]{-1} = 36 - 3x,$$

which simplifies as $x^3 + 3x - 36 = 0$, so $(x-3)(x^2 + 3x + 12) = 0$, which means $x = 3$ or $x^2 + 2x + 12 = 0$. The quadratic $x^2 + 3x + 12 = 0$ has no real roots, so $x = \boxed{3}$.

14.25 Squaring the left side, we get

$$x + \sqrt{2x-1} - 2\sqrt{x + \sqrt{2x-1}}\sqrt{x - \sqrt{2x-1}} + x - \sqrt{2x-1} = 2x - 2\sqrt{(x+\sqrt{2x-1})(x-\sqrt{2x-1})}$$
$$= 2x - 2\sqrt{x^2 - (2x-1)} = 2x - 2\sqrt{x^2 - 2x + 1}$$
$$= 2x - 2\sqrt{(x-1)^2}.$$

Since $x > 1$, we have $x - 1 > 0$, so $2x - 2\sqrt{(x-1)^2} = 2x - 2(x-1) = 2$. Therefore, $k = \boxed{2}$.

14.26 We want a, b, and c to satisfy

$$\sqrt{104\sqrt{6} + 468\sqrt{10} + 144\sqrt{15} + 2006} = a\sqrt{2} + b\sqrt{3} + c\sqrt{5}.$$

Squaring both sides, we get

$$104\sqrt{6} + 468\sqrt{10} + 144\sqrt{15} + 2006 = 2a^2 + 3b^2 + 5c^2 + 2ab\sqrt{6} + 2ac\sqrt{10} + 2bc\sqrt{15}.$$

Hence, a, b, and c must satisfy the system of equations

$$2a^2 + 3b^2 + 5c^2 = 2006,$$
$$ab = 104/2 = 52,$$
$$ac = 468/2 = 234,$$
$$bc = 144/2 = 72.$$

Multiplying the last three equations, we get $a^2b^2c^2 = 52 \cdot 234 \cdot 72 = 2^2 \cdot 13 \cdot 2 \cdot 3^2 \cdot 13 \cdot 2^3 \cdot 3^2 = 2^6 \cdot 3^4 \cdot 13^2$, so $abc = \sqrt{2^6 \cdot 3^4 \cdot 13^2} = 2^3 \cdot 3^2 \cdot 13 = \boxed{936}$.

14.27 Rationalizing the denominator on the left side of the given equation, we get

$$x + \sqrt{x^2 - 1} + \frac{1}{x - \sqrt{x^2 - 1}} = x + \sqrt{x^2 - 1} + \frac{x + \sqrt{x^2 - 1}}{(x - \sqrt{x^2 - 1})(x + \sqrt{x^2 - 1})} = 2x + 2\sqrt{x^2 - 1},$$

so $2x + 2\sqrt{x^2 - 1} = 20$. Rearranging gives $\sqrt{x^2 - 1} = 10 - x$. Squaring both sides, we get $x^2 - 1 = x^2 - 20x + 100$, so $x = \frac{101}{20}$.

Now we rationalize the denominator of the expression we wish to evaluate:

$$x^2 + \sqrt{x^4 - 1} + \frac{1}{x^2 + \sqrt{x^4 - 1}} = x^2 + \sqrt{x^4 - 1} + \frac{x^2 - \sqrt{x^4 - 1}}{(x^2 + \sqrt{x^4 - 1})(x^2 - \sqrt{x^4 - 1})}$$

$$= x^2 + \sqrt{x^4 - 1} + x^2 - \sqrt{x^4 - 1} = 2x^2 = \boxed{\frac{10201}{200}}.$$

14.28 Let $\alpha = \sqrt[3]{n + \sqrt{n^2 + 8}}$ and $\beta = \sqrt[3]{n - \sqrt{n^2 + 8}}$, so the given equation is $\alpha + \beta = 8$. Cubing both sides of this equation, the left side is

$$\alpha^3 + 3\alpha^2\beta + 3\alpha\beta^2 + \beta^3 = n + \sqrt{n^2 + 8} + 3\alpha\beta(\alpha + \beta) + n - \sqrt{n^2 + 8} = 2n + 24\sqrt[3]{n + \sqrt{n^2 + 8}}\sqrt[3]{n - \sqrt{n^2 + 8}}$$

$$= 2n + 24\sqrt[3]{(n + \sqrt{n^2 + 8})(n - \sqrt{n^2 + 8})} = 2n + 24\sqrt[3]{n^2 - (n^2 + 8)} = 2n + 24\sqrt[3]{-8}$$

$$= 2n - 48.$$

So, we have $2n - 48 = 8^3$, which gives us $n = \boxed{280}$.

Challenge Problems

14.29 Let $\alpha = \sqrt[3]{60 - x}$ and $\beta = \sqrt[3]{x - 11}$, so the given equation is $\alpha + \beta = \sqrt[3]{4}$. Cubing this equation, the left side becomes

$$\alpha^3 + 3\alpha^2\beta + 3\alpha\beta^2 + \beta^3 = 60 - x + 3\alpha\beta(\alpha + \beta) + x - 11 = 49 + 3\alpha\beta(\alpha + \beta) = 49 + 3\sqrt[3]{4}\alpha\beta,$$

so $49 + 3\sqrt[3]{4}\alpha\beta = 4$. Therefore, we have $3\sqrt[3]{4}\alpha\beta = 4 - 49 = -45$, which implies $\alpha\beta = -\frac{45}{3\sqrt[3]{4}} = -\frac{15}{\sqrt[3]{4}}$, so $\alpha^3\beta^3 = \left(-\frac{15}{\sqrt[3]{4}}\right)^3 = -\frac{3375}{4}$.

Since $\alpha^3 + \beta^3 = 60 - x + x - 11 = 49$, we see that α^3 and β^3 are the roots of the quadratic

$$t^2 - 49t - \frac{3375}{4} = \frac{4t^2 - 196t - 3375}{4} = \frac{(2t - 125)(2t + 27)}{4}.$$

Hence, either $60 - x = 125/2$ and $x - 11 = -27/2$, which gives $x = -5/2$, or $60 - x = -27/2$ and $x - 11 = 125/2$, which gives $x = 147/2$. So, our two solutions for x are $\boxed{147/2 \text{ and } -5/2}$.

14.30 First, we simplify $\sqrt{33 + \sqrt{128}} = \sqrt{33 + 8\sqrt{2}}$. We guess that $\sqrt{33 + 8\sqrt{2}} = a + b\sqrt{2}$ for some integers a and b. Squaring both sides, we get $33 + 8\sqrt{2} = a^2 + 2b^2 + 2ab\sqrt{2}$, so $a^2 + 2b^2 = 33$ and $ab = 4$. Since $a^2 + 2b^2$ must be odd, and the only odd factor of 4 is 1, we guess $a = 1$, which gives $b = 4$. Fortunately, $(a, b) = (1, 4)$ satisfies our system, and we have $\sqrt{33 + 8\sqrt{2}} = 1 + 4\sqrt{2}$. (We exclude $(a, b) = (-1, -4)$ because square roots must be positive.)

Hence, we have

$$\left(\sqrt{33 + \sqrt{128}} + \sqrt{2} - 8\right)^{-1} = \frac{1}{1 + 4\sqrt{2} + \sqrt{2} - 8} = \frac{1}{-7 + 5\sqrt{2}} = \frac{-7 - 5\sqrt{2}}{(-7 + 5\sqrt{2})(-7 - 5\sqrt{2})} = \frac{-7 - 5\sqrt{2}}{49 - 50}$$
$$= 7 + 5\sqrt{2}.$$

Thus, the problem has become computing the greatest integer less than $7 + 5\sqrt{2}$.

Since $(5\sqrt{2})^2 = 50$, which is between $7^2 = 49$ and $8^2 = 64$, we know that $5\sqrt{2}$ is between 7 and 8. Then $7 + 5\sqrt{2}$ is between 14 and 15, so the greatest such value of N is $\boxed{14}$.

14.31 From the given equation, we have $3\sqrt{a} + \sqrt{b} - 2\sqrt{c} = \dfrac{12}{\sqrt{5} + \sqrt{2} - \sqrt{3}}$.

Rationalizing the denominator, we get

$$\frac{12}{\sqrt{5} + \sqrt{2} - \sqrt{3}} = \frac{12(\sqrt{5} - \sqrt{2} + \sqrt{3})}{(\sqrt{5} + \sqrt{2} - \sqrt{3})(\sqrt{5} - \sqrt{2} + \sqrt{3})} = \frac{12(\sqrt{5} - \sqrt{2} + \sqrt{3})}{5 - (\sqrt{2} - \sqrt{3})^2} = \frac{12(\sqrt{5} - \sqrt{2} + \sqrt{3})}{5 - (5 - 2\sqrt{6})}$$
$$= \frac{12(\sqrt{5} - \sqrt{2} + \sqrt{3})}{2\sqrt{6}} = \sqrt{6}(\sqrt{5} - \sqrt{2} + \sqrt{3}) = \sqrt{30} - 2\sqrt{3} + 3\sqrt{2} = 3\sqrt{2} + \sqrt{30} - 2\sqrt{3}.$$

Therefore, $(a, b, c) = \boxed{(2, 30, 3)}$.

14.32 To simplify the expression, let $y = \sqrt[16]{5}$. Then

$$x = \frac{4}{(\sqrt{5} + 1)(\sqrt[4]{5} + 1)(\sqrt[8]{5} + 1)(\sqrt[16]{5} + 1)} = \frac{4}{(y^8 + 1)(y^4 + 1)(y^2 + 1)(y + 1)}.$$

Multiplying the top and bottom by $y - 1$, we find the fraction collapses nicely:

$$x = \frac{4(y - 1)}{(y^8 + 1)(y^4 + 1)(y^2 + 1)(y + 1)(y - 1)} = \frac{4(y - 1)}{(y^8 + 1)(y^4 + 1)(y^2 + 1)(y^2 - 1)} = \frac{4(y - 1)}{(y^8 + 1)(y^4 + 1)(y^4 - 1)}$$
$$= \frac{4(y - 1)}{(y^8 + 1)(y^8 - 1)} = \frac{4(y - 1)}{y^{16} - 1} = \frac{4(y - 1)}{4} = y - 1.$$

Therefore, $(x + 1)^{48} = y^{48} = 5^3 = \boxed{125}$.

14.33 Rearranging the desired equation, we must show that $\sqrt{x + 4\sqrt{x - 4}} - \sqrt{x - 4\sqrt{x - 4}} = 4$.

Squaring $\sqrt{x + 4\sqrt{x - 4}} - \sqrt{x - 4\sqrt{x - 4}}$, we get

$$x + 4\sqrt{x - 4} - 2\sqrt{x + 4\sqrt{x - 4}}\sqrt{x - 4\sqrt{x - 4}} + x - 4\sqrt{x - 4}$$
$$= 2x - 2\sqrt{(x + 4\sqrt{x - 4})(x - 4\sqrt{x - 4})} = 2x - 2\sqrt{x^2 - 16(x - 4)}$$
$$= 2x - 2\sqrt{x^2 - 16x + 64} = 2x - 2\sqrt{(x - 8)^2}.$$

Since $x \geq 8$, we have $\sqrt{(x - 8)^2} = x - 8$. Therefore,

$$\left(\sqrt{x + 4\sqrt{x - 4}} - \sqrt{x - 4\sqrt{x - 4}}\right)^2 = 2x - 2\sqrt{(x - 8)^2} = 2x - 2(x - 8) = 16.$$

And since $\sqrt{x+4\sqrt{x-4}} - \sqrt{x-4\sqrt{x-4}} \ge 0$, we have

$$\sqrt{x+4\sqrt{x-4}} - \sqrt{x-4\sqrt{x-4}} = \sqrt{16} = 4,$$

so $\sqrt{x-4\sqrt{x-4}} + 2 = \sqrt{x+4\sqrt{x-4}} - 2$, as desired

14.34   Let $a = (\sqrt{2}+1)^x$ and $b = (\sqrt{2}-1)^x$. Then $a+b=6$, and

$$ab = (\sqrt{2}+1)^x(\sqrt{2}-1)^x = 1^x = 1,$$

so by Vieta's Formulas, a and b are roots of the quadratic $t^2 - 6t + 1 = 0$. The roots of this quadratic are $t = \frac{6\pm\sqrt{32}}{2} = 3 \pm 2\sqrt{2}$.

Thus, we must find all values of x such that $(\sqrt{2}+1)^x = 3 \pm 2\sqrt{2}$. To do so, we check the first few powers of $\sqrt{2}+1$, and find that

$$(\sqrt{2}+1)^2 = 2 + 2\sqrt{2} + 1 = 3 + 2\sqrt{2}.$$

Hence, $x = 2$ is the solution to $(\sqrt{2}+1)^x = 3 + 2\sqrt{2}$. Since $3 - 2\sqrt{2}$ is the reciprocal of $3 + 2\sqrt{2}$, the solution to $(\sqrt{2}+1)^x = 3 - 2\sqrt{2}$ is $x = -2$. Since $(\sqrt{2}+1)^x$ is monotonically increasing, there are no more values of x such that $(\sqrt{2}+1)^x = 3 + 2\sqrt{2}$ or $(\sqrt{2}+1)^x = 3 - 2\sqrt{2}$.

Therefore, the solutions are $x = \boxed{2 \text{ and } -2}$.

14.35   Because of the cube roots, we should look for an expression that contains cubes. Recall the factorization

$$p^3 + q^3 + r^3 - 3pqr = (p+q+r)(p^2+q^2+r^2-pq-pr-qr).$$

(See Problem 9.54 for the derivation of this factorization.) Taking $p = a$, $q = b\sqrt[3]{2}$, and $r = c\sqrt[3]{4}$, we get

$$a^3 + 2b^3 + 4c^3 - 6abc = (a + b\sqrt[3]{2} + c\sqrt[3]{4})(p^2+q^2+r^2-pq-pr-qr).$$

(We leave the expression $p^2+q^2+r^2-pq-pr-qr$ as is, for the moment.) The expression $a^3 + 2b^3 + 4c^3 - 6abc$ has integer coefficients, and if $a + b\sqrt[3]{2} + c\sqrt[3]{4} = 0$, then $a^3 + 2b^3 + 4c^3 - 6abc = 0$, which gives us partially what we want.

We claim that $p^2+q^2+r^2-pq-pr-qr \ge 0$ for all real numbers p, q, and r. This is because

$$p^2+q^2+r^2-pq-pr-qr = \frac{(p-q)^2 + (p-r)^2 + (q-r)^2}{2} \ge 0.$$

This also shows that $p^2+q^2+r^2-pq-pr-qr = 0$ if and only if $p = q = r$.

But if $p = a$, $q = b\sqrt[3]{2}$, and $r = c\sqrt[3]{4}$, then the equation $p = q = r$ becomes $a = b\sqrt[3]{2} = c\sqrt[3]{4}$. Since $\sqrt[3]{2}$ is irrational, this equation can hold if and only if $a = b = c = 0$, in which case both $a^3 + 2b^3 + 4c^3 - 6abc$ and $a + b\sqrt[3]{2} + c\sqrt[3]{4}$ are equal to 0. Otherwise, $p^2+q^2+r^2-pq-pr-qr > 0$, so $a^3 + 2b^3 + 4c^3 - 6abc$ and $a + b\sqrt[3]{2} + c\sqrt[3]{4}$ have the same sign.

Thus, we can take $f(x,y,z) = \boxed{x^3 + 2y^3 + 4z^3 - 6xyz}$.

CHAPTER 15

Special Classes of Functions

Exercises for Section 15.1

15.1.1

(a) The function $f(x)$ is defined for all x, except where the denominator is 0. The denominator is $x - 1$, so the domain of f is $\boxed{\text{all real numbers except 1}}$.

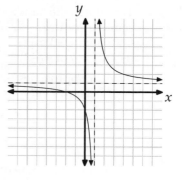

To find the range of f, let $y = f(x) = \frac{x+2}{x-1}$. Multiplying both sides by $x - 1$, we get $yx - y = x + 2$. Solving for x, we get

$$x = \frac{y+2}{y-1}.$$

Hence, the range of f is also $\boxed{\text{all real numbers except 1}}$.

As x approaches 1, the value of $f(x)$ becomes very far from 0, and as x becomes far from 0, then $f(x)$ approaches 1. Therefore, the asymptotes are $\boxed{x = 1 \text{ and } y = 1}$.

(b) The denominator is $2x + 3$, so the domain of f is $\boxed{\text{all real numbers except } -3/2}$.

To find the range of f, let $y = f(x) = \frac{x^2-1}{2x+3}$. Multiplying both sides by $2x + 3$, we get $y(2x + 3) = x^2 - 1$. Writing this as a quadratic in x, we get

$$x^2 - 2yx - (3y + 1) = 0.$$

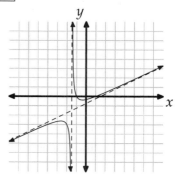

This quadratic has a real root in x if and only if its discriminant is nonnegative. The discriminant is

$$(2y)^2 - 4[-(3y + 1)] = 4y^2 + 12y + 4 = 4(y^2 + 3y + 1).$$

The roots of the quadratic $y^2 + 3y + 1 = 0$ are $y = \frac{-3 \pm \sqrt{5}}{2}$, so the range of $y = f(x)$ is

$$y \in \boxed{\left(-\infty, \frac{-3-\sqrt{5}}{2}\right] \cup \left[\frac{-3+\sqrt{5}}{2}, +\infty\right)}.$$

To find the asymptotes of $f(x)$, we divide the denominator $2x + 3$ into the numerator $x^2 - 1$, and find

$$f(x) = \frac{5/4}{2x + 3} + \frac{1}{2}x - \frac{3}{4}.$$

As x approaches $-3/2$, the value of $f(x)$ becomes very far from 0, so $x = -3/2$ is an asymptote. As x becomes very far from 0, $\dfrac{5/4}{2x+3}$ approaches 0, so $y = x/2 - 3/4$ is an asymptote. Therefore, the asymptotes are

$$\boxed{x = -3/2 \text{ and } y = \frac{1}{2}x - \frac{3}{4}}.$$

(c) The denominator is $(x-2)(x+3)$, so the domain of f is $\boxed{\text{all real numbers except 2 and } -3}$.

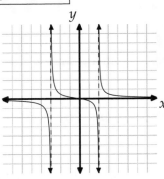

To find the range of f, let $y = f(x) = \dfrac{x}{(x-2)(x+3)}$. Multiplying both sides by $(x-2)(x+3)$, we get $y(x-2)(x+3) = x$. Writing this as a quadratic in x, we get $yx^2 + (y-1)x - 6y = 0$. This quadratic has a real root for x if and only if its discriminant is nonnegative. This discriminant is

$$(y-1)^2 - 4y(-6y) = y^2 - 2y + 1 + 24y^2 = 25y^2 - 2y + 1.$$

The quadratic $25y^2 - 2y + 1 = 0$ has no real roots, so $25y^2 - 2y + 1 > 0$ for all y. This means that for any value of y, we can solve the quadratic $yx^2 + (y-1)x - 6y = 0$ for x to find a value (or values) of x such that $f(x) = y$. Therefore, the range of f is $\boxed{\text{all real numbers}}$. (Note that the quadratic cannot produce $x = 2$ or $x = -3$ as a root. Letting $x = 2$ in the equation gives $-2 = 0$, which has no solutions. Letting $x = -3$ gives $3 = 0$, which has no solutions. So, we never have a case in which our value of y gives us $f(2) = y$ or $f(-3) = y$.)

As x approaches -3 or 2, the value of $f(x)$ becomes very far from 0, so $x = -3$ and $x = 2$ are asymptotes. As x becomes very far from 0, $f(x)$ approaches 0, so $y = 0$ is an asymptote. Therefore, the asymptotes are $\boxed{x = -3, x = 2, \text{ and } y = 0}$.

(d) The denominator is $x^2 - 1 = (x-1)(x+1)$, so the domain of f is $\boxed{\text{all real numbers except 1 and } -1}$.

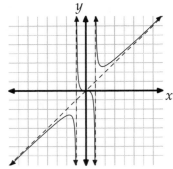

To find the range of f, let

$$y = f(x) = \frac{x^3}{x^2 - 1}.$$

Multiplying both sides by $x^2 - 1$, we get $yx^2 - y = x^3$, so

$$x^3 - yx^2 + y = 0.$$

This is a cubic equation in x, and every cubic equation has at least one real root, which means that for any value of y, there is a real value of x that satisfies our cubic. Therefore, for any value of y, there is a real value of x that satisfies $f(x) = y$. Therefore, the range of f is $\boxed{\text{all real numbers}}$. (Note what happens in the cubic when $x = 1$ or $x = -1$; both give equations with no solution, so there is not a value of y for which the cubic in x has 1 or -1 as a root.)

To find the asymptotes of $f(x)$, we divide the denominator $x^2 - 1$ into the numerator x^3, and find

$$f(x) = x + \frac{x}{x^2 - 1}.$$

As x approaches 1 or -1, the value of $f(x)$ becomes very far from 0, so $x = 1$ and $x = -1$ are asymptotes. As x becomes very far from 0, $\frac{x}{x^2-1}$ approaches 0, so $y = x$ is an asymptote. Therefore, the asymptotes are $\boxed{x = 1, x = -1, \text{ and } y = x}$.

15.1.2 The simplest rational function that has vertical asymptotes $x = 1$ and $x = -2$ is

$$f(x) = \frac{1}{(x-1)(x+2)}.$$

The graph is shown to the right.

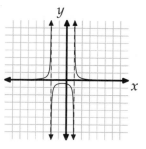

15.1.3 The denominator factors as $(x-3)(x-5)$. Therefore, $f(x)$ has a vertical asymptote at $x = 3$ and $x = 5$, unless there is a factor of $x - 3$ or $x - 5$ in the numerator that cancels the corresponding factor in the denominator.

By the Factor Theorem, if the numerator $x^2 - x + c$ has a factor of $x - 3$, then $3^2 - 3 + c = 0$, so $c = -6$. If the numerator $x^2 - x + c$ has a factor of $x - 5$, then $5^2 - 5 + c = 0$, so $c = -20$. Therefore, the function $f(x)$ has exactly one vertical asymptote for $\boxed{c = -6 \text{ and } c = -20}$. (In both cases, one factor of the numerator of $f(x)$ is also a factor of the denominator. We know the other factors of the numerator and denominator are different because the quadratics in the numerator and denominator are different for any c.)

15.1.4 If f has a vertical asymptote at $x = -1$ and g has a vertical asymptote at $x = 3$, then it is possible that $f(x)g(x)$ has a vertical asymptote at both $x = -1$ and $x = 3$. For example, if $f(x) = \frac{1}{x+1}$ and $g(x) = \frac{1}{x-3}$, then $f(x)g(x) = \frac{1}{(x+1)(x-3)}$.

However, it is also possible that the function $f(x)g(x)$ does not have any vertical asymptotes. For example if $f(x) = \frac{x-3}{x+1}$ and $g(x) = \frac{x+1}{x-3}$, then $f(x)g(x) = \frac{x-3}{x+1} \cdot \frac{x+1}{x-3} = 1$ for $x \neq -1, 3$.

Finally, it is possible that the function $f(x)g(x)$ has exactly one vertical asymptote. For example, if $f(x) = \frac{x-3}{x+1}$ and $g(x) = \frac{1}{x-3}$, then $f(x)g(x) = \frac{x-3}{x+1} \cdot \frac{1}{x-3} = \frac{1}{x+1}$ for $x \neq 3$. Thus, we cannot say anything definitive about the vertical asymptotes of $f(x)g(x)$, except that no lines other than $x = -1$ and $x = 3$ can possibly be vertical asymptotes.

Now we look at the function $f(x) + g(x)$. Since f has a vertical asymptote at $x = -1$, the value of $f(x)$ becomes very far from 0 as x approaches -1. Hence, the function $f(x) + g(x)$ must also be very far from 0 as x approaches -1, so $f(x) + g(x)$ has a vertical asymptote at $x = -1$.

Similarly, since g has a vertical asymptote at $x = 3$, $g(x)$ becomes very far from 0 as x approaches 3. Hence, the function $f(x) + g(x)$ must also become very far from 0 as x approaches 3, so $f(x) + g(x)$ has a vertical asymptote at $x = 3$.

15.1.5

(a) Dividing the denominator into the numerator, we find $f(x) = \frac{2x-1}{x+2} = 2 - \frac{5}{x+2}$. As x becomes very far from zero, $\frac{5}{x+2}$ approaches 0, so $f(x)$ has the horizontal asymptote $\boxed{y = 2}$.

(b) Dividing the denominator into the numerator, we find

$$f(x) = \frac{(x-2)(3x-1)}{x+1} = \frac{3x^2 - 7x + 2}{x+1} = 3x - 10 + \frac{12}{x+1}.$$

As x becomes very far from zero, $\frac{12}{x+1}$ approaches 0, so $f(x)$ has the slant asymptote $y = 3x - 10$, which is not horizontal.

(c) Dividing the denominator into the numerator, we find

$$f(x) = \frac{(x-2)(3x-1)}{(x+1)(x-3)} = \frac{3x^2 - 7x + 2}{x^2 - 2x - 3} = 3 - \frac{x-11}{x^2 - 2x - 3}.$$

As x becomes far from zero, $\frac{x-11}{x^2-2x-3}$ approaches 0, so $f(x)$ has the horizontal asymptote $\boxed{y = 3}$.

(d) Since the degree of the numerator (which is 1) is less than the degree of the denominator (which is 2), the function is already written in the form we want. As x becomes very far from zero, $\frac{x-2}{x(x+1)}$ approaches 0, so $f(x)$ has the horizontal asymptote $\boxed{y = 0}$.

(e) Suppose we divide $h(x)$ into $g(x)$, so that we have polynomials $q(x)$ and $r(x)$ such that

$$g(x) = h(x)q(x) + r(x),$$

and either $\deg r < \deg h$ or $r(x) = 0$ for all x. Hence,

$$f(x) = \frac{g(x)}{h(x)} = \frac{h(x)q(x) + r(x)}{h(x)} = \frac{r(x)}{h(x)} + q(x).$$

Since $\deg r < \deg h$, as x becomes very far from zero, $r(x)/h(x)$ approaches 0, so whether or not $f(x)$ has a horizontal asymptote depends on $q(x)$.

If $\deg q \geq 1$, then $q(x)$ becomes very far from 0 as x becomes very far from 0, so $f(x)$ cannot have a horizontal asymptote. Otherwise, $q(x)$ is a constant, say c, so $f(x)$ will have the horizontal asymptote $y = c$.

Now we will determine when $q(x)$ is a constant. If $\deg g < \deg h$, then $q(x) = 0$ and $r(x) = g(x)$. In this case, $f(x) = g(x)/h(x)$ has the horizontal asymptote $y = 0$. Conversely, if $q(x) = 0$, then $g(x) = r(x)$, and $\deg r < \deg h$, so $\deg g < \deg h$. We conclude that $q(x) = 0$ if and only if $\deg g < \deg h$.

Now, assume that $\deg g \geq \deg h$, and that $q(x)$ is a nonzero polynomial. We begin with

$$g(x) = h(x)q(x) + r(x).$$

Since $\deg r < \deg h$, we can say that $\deg r(x) < \deg(h(x)q(x))$. Hence,

$$\deg g(x) = \deg(h(x)q(x) + r(x)) = \deg(h(x) + q(x)) = \deg h(x) + \deg q(x),$$

so $\deg q(x) = \deg g(x) - \deg h(x)$. The polynomial $q(x)$ is a constant if and only if $\deg q = 0$. Equivalently, we have $\deg g = \deg h$.

To summarize, $f(x) = g(x)/h(x)$ has a horizontal asymptote if and only if $\boxed{\deg g \leq \deg h}$.

Exercises for Section 15.2

15.2.1 Multiplying both sides by $5(2t - 3)(3t - 1)$, the given equation becomes

$$5(3t - 1) + 5t(2t - 3) = 7(2t - 3)(3t - 1),$$

which simplifies as $32t^2 - 77t + 26 = 0$, so $(t - 2)(32t - 13) = 0$, which means the solutions are $t = \boxed{2 \text{ and } 13/32}$.

15.2.2 First, the left side simplifies as $\frac{3r^2 - 12r}{r^2 - 3r - 4} = \frac{3r(r - 4)}{(r - 4)(r + 1)} = \frac{3r}{r + 1}$, so we have $\frac{3r}{r + 1} = \frac{2r - 3}{r + 7} - 1$. Multiplying both sides by $(r + 1)(r + 7)$, the equation becomes

$$3r(r + 7) = (r + 1)(2r - 3) - (r + 1)(r + 7),$$

which simplifies as $2(r^2 + 15r + 5) = 0$. By the quadratic formula, the roots of the quadratic equation $r^2 + 15r + 5 = 0$ are $r = \boxed{\dfrac{-15 \pm \sqrt{205}}{2}}$.

15.2.3 Note that $x^2 + 3x$ factors as $x(x + 3)$. Hence, we multiply both sides by $x(x + 3)$ to get the linear equation $x - 12 - 2(x + 3) = 5x$. This equation gives $x = -3$. However, if we try to substitute $x = -3$ into the original equation, we obtain denominators that are 0, which is not allowed. The solution $x = -3$ is extraneous. Therefore, there are $\boxed{\text{no solutions}}$.

15.2.4 Let $f(x) = \dfrac{x^2 - 9}{x^2 - 25} = \dfrac{(x-3)(x+3)}{(x-5)(x+5)}$. The sign of $f(x)$ depends on the sign of $x-3$, $x+3$, $x-5$, and $x+5$. We construct a table as shown at right. From the table, we see that $f(x) > 0$ for $x \in$ $\boxed{(-\infty, -5) \cup (-3, 3) \cup (5, +\infty)}$.

	$x+5$	$x+3$	$x-3$	$x-5$	$f(x)$
$x < -5$	$-$	$-$	$-$	$-$	$+$
$-5 < x < -3$	$+$	$-$	$-$	$-$	$-$
$-3 < x < 3$	$+$	$+$	$-$	$-$	$+$
$3 < x < 5$	$+$	$+$	$+$	$-$	$-$
$5 < x$	$+$	$+$	$+$	$+$	$+$

15.2.5 Subtracting 1 from both sides, and putting everything over a common denominator, the left side of our inequality is

$$\frac{1}{x+1} + \frac{6}{x+5} - 1 = \frac{x+5+6(x+1)-(x+1)(x+5)}{(x+1)(x+5)} = \frac{-x^2+x+6}{(x+1)(x+5)} = \frac{-(x^2-x-6)}{(x+1)(x+5)}$$
$$= -\frac{(x-3)(x+2)}{(x+1)(x+5)}.$$

So, our inequality is $-\dfrac{(x-3)(x+2)}{(x+1)(x+5)} \geq 0$, or equivalently $\dfrac{(x-3)(x+2)}{(x+1)(x+5)} \leq 0$.

Let $f(x) = \dfrac{(x-3)(x+2)}{(x+1)(x+5)}$. The sign of $f(x)$ depends on the sign of $x-3$, $x+2$, $x+1$, and $x+5$. We construct a table as shown at right.

From the table, we see that $f(x) \leq 0$ for $x \in \boxed{(-5, -2] \cup (-1, 3]}$.

	$x+5$	$x+2$	$x+1$	$x-3$	$f(x)$
$x < -5$	$-$	$-$	$-$	$-$	$+$
$-5 < x < -2$	$+$	$-$	$-$	$-$	$-$
$x = -2$	$+$	0	$-$	$-$	0
$-2 < x < -1$	$+$	$+$	$-$	$-$	$+$
$-1 < x < 3$	$+$	$+$	$+$	$-$	$-$
$x = 3$	$+$	$+$	$+$	0	0
$3 < x$	$+$	$+$	$+$	$+$	$+$

Exercises for Section 15.3

15.3.1

(a) Note that $f(-x) = (-x)^4 + (-x)^6 + (-x)^8 + 2008 - |-x| = x^4 + x^6 + x^8 + 2008 - |x| = f(x)$. Therefore, $f(x)$ is $\boxed{\text{even}}$.

(b) Note that $g(1) = 3^0 = 1$ and $g(-1) = 3^2 = 9$. Since $g(-1)$ is not equal to $g(1)$ or $-g(1)$, $g(x)$ is $\boxed{\text{neither even nor odd}}$.

(c) Note that $h(-x) = |3(-x)^3 - 7(-x)| = |-3x^3 + 7x| = |-(3x^3 - 7x)| = |3x^3 - 7x| = h(x)$. Therefore, $h(x)$ is $\boxed{\text{even}}$.

(d) We have $f(-x) = \sqrt{(-x)^2 - 4} = \sqrt{x^2 - 4} = f(x)$, so $f(x)$ is $\boxed{\text{even}}$.

15.3.2

(a) Let $h(x) = f(x) + g(x)$. Then $h(-x) = f(-x) + g(-x) = -f(x) - g(x) = -[f(x) + g(x)] = -h(x)$. Therefore, $f(x) + g(x)$ is $\boxed{\text{odd}}$.

(b) Let $h(x) = f(x)g(x)$. Then $h(-x) = f(-x)g(-x) = (-f(x)) \cdot (-g(x)) = f(x)g(x) = h(x)$. Therefore, $f(x)g(x)$ is $\boxed{\text{even}}$.

15.3.3 Since $g(x)$ is an odd function, we have $g(-x) = -g(x)$ for all x. In particular, for $x = 0$, we have $g(0) = -g(0)$, so $g(0) = 0$. Therefore, $x = 0$ is a root of $g(x) = 0$, so no matter what the other roots are, the product of the roots is always $\boxed{0}$.

15.3.4 Suppose both $g(x)$ and $h(x)$ are odd functions. Then, we have $g(-x) = -g(x)$ and $h(-x) = -h(x)$, so $f(-x) = g(-x)/h(-x) = (-g(x))/(-h(x)) = g(x)/h(x) = f(x)$, which means that $f(x)$ is even. So, we cannot conclude that $g(x)$ and $h(x)$ are even if $g(x)/h(x)$ is even.

Exercises for Section 15.4

15.4.1 We claim that $f(x)g(x)$ is an increasing function. To prove this, we must show that

$$f(x+c)g(x+c) - f(x)g(x) > 0,$$

where c is a positive constant.

Since f is increasing, we have $f(x + c) > f(x)$. Since $g(x + c)$ is positive, we can multiply both sides by $g(x + c)$ to get

$$f(x + c)g(x + c) > f(x)g(x + c).$$

Similarly, since g is increasing, $g(x + c) > g(x)$. Since $f(x)$ is positive, we can multiply both sides by $f(x)$ to get

$$f(x)g(x + c) > f(x)g(x).$$

Therefore, we have $f(x + c)g(x + c) > f(x)g(x + c) > f(x)g(x)$, as desired.

15.4.2 Let $f(x) = x(15 - x)$. Let $x > 7.5$, and let c be a positive constant. Then

$$f(x + c) - f(x) = (x + c)(15 - x - c) - x(15 - x) = (15x - x^2 - cx + 15c - cx - c^2) - (15x - x^2)$$
$$= -c^2 - 2cx + 15c = -c(c + 2x - 15).$$

Since $2x > 2 \cdot 7.5 = 15$, we have $2x - 15 > 0$, so $c + 2x - 15 > 0$. Therefore, $-c(c + 2x - 15) < 0$, so we have $f(x + c) - f(x) < 0$, which means $f(x + c) < f(x)$. This tells us that f is decreasing for $x > 7.5$. (Alternatively, we could have completed the square on $f(x)$.)

15.4.3 If $8^x(3x + 1) = 4$, then $3x + 1 = 4/8^x$. Since $4/8^x$ is positive, $3x + 1$ is positive, so $x > -1/3$.

The function $3x + 1$ is clearly increasing for $x > -1/3$. The function 8^x is increasing for all x. Since both $3x + 1$ and 8^x are positive and increasing, the function $f(x) = 8^x(3x + 1)$ is increasing for $x > -1/3$. Now, trying different values of x in order to find a nice root, we find that

$$f\left(\frac{1}{3}\right) = 8^{1/3}\left(\frac{3}{3} + 1\right) = 2 \cdot 2 = 4.$$

Since the function $f(x) = 8^x(3x + 1)$ is increasing, the only solution to $8^x(3x + 1) = 4$ is $x = \boxed{1/3}$.

15.4.4 Let c be a positive constant. First, we cover the case $x \geq 0$, so $0 \leq x < x + c$. By the Binomial Theorem,

$$f(x + c) = x^n + \binom{n}{1}x^{n-1}c + \binom{n}{2}x^{n-2}c^2 + \cdots + c^n.$$

Since every term is nonnegative, and $c^n > 0$, we have

$$f(x + c) = x^n + \binom{n}{1}x^{n-1}c + \binom{n}{2}x^{n-2}c^2 + \cdots + c^n \geq x^n + c^n > x^n = f(x).$$

Now assume $x < 0$, so $f(x) = x^n < 0$. If $x + c \geq 0$, then $f(x + c) = (x + c)^n \geq 0$, so it is clear that $f(x) < f(x + c)$. Otherwise, $x < x + c < 0$, so $0 < -(x + c) < -x$. Then from our work above, we know that $f(-(x + c)) < f(-x)$, so

$$[-(x + c)]^n < (-x)^n,$$

which becomes $-(x + c)^n < -x^n$ because n is odd. Therefore, $(x + c)^n > x^n$, or $f(x + c) > f(x)$.

We have shown that $f(x + c) > f(x)$ for all real numbers x, so f is an increasing function.

15.4.5

(a) We claim that $f(x)$ is a decreasing function for $x \geq 3$. To see this, we rewrite $f(x)$ as follows:

$$f(x) = \sqrt{x+3} - \sqrt{x-3} = \frac{(\sqrt{x+3} - \sqrt{x-3})(\sqrt{x+3} + \sqrt{x-3})}{\sqrt{x+3} + \sqrt{x-3}} = \frac{(x+3) - (x-3)}{\sqrt{x+3} + \sqrt{x-3}} = \frac{6}{\sqrt{x+3} + \sqrt{x-3}}.$$

Clearly, both $\sqrt{x+3}$ and $\sqrt{x-3}$ are increasing for $x \geq 3$, so $\sqrt{x+3} + \sqrt{x-3}$ is increasing. We conclude that $f(x)$ is decreasing for $x \geq 3$. Hence, the maximum value of $f(x)$ is $f(3) = \boxed{\sqrt{6}}$.

(b) We claim that $g(x)$ is a decreasing function for $x \geq 0$. To see this, we rewrite $g(x)$ using difference of cubes as follows:

$$g(x) = \sqrt[3]{x+1} - \sqrt[3]{x} = \frac{(\sqrt[3]{x+1} - \sqrt[3]{x})[(\sqrt[3]{x+1})^2 + \sqrt[3]{x+1}\sqrt[3]{x} + (\sqrt[3]{x})^2]}{(\sqrt[3]{x+1})^2 + \sqrt[3]{x+1}\sqrt[3]{x} + (\sqrt[3]{x})^2}$$

$$= \frac{(\sqrt[3]{x+1})^3 - (\sqrt[3]{x})^3}{(\sqrt[3]{x+1})^2 + \sqrt[3]{x+1}\sqrt[3]{x} + (\sqrt[3]{x})^2} = \frac{(x+1) - x}{(\sqrt[3]{x+1})^2 + \sqrt[3]{x+1}\sqrt[3]{x} + (\sqrt[3]{x})^2}$$

$$= \frac{1}{(\sqrt[3]{x+1})^2 + \sqrt[3]{x+1}\sqrt[3]{x} + (\sqrt[3]{x})^2}.$$

Clearly, $(\sqrt[3]{x+1})^2$, $\sqrt[3]{x+1}\sqrt[3]{x}$, and $(\sqrt[3]{x})^2$ are increasing for $x \geq 0$, so $(\sqrt[3]{x+1})^2 + \sqrt[3]{x+1}\sqrt[3]{x} + (\sqrt[3]{x})^2$ is increasing. We conclude that $g(x)$ is decreasing for $x \geq 0$. Hence, the maximum value of $g(x)$ is $g(0) = \boxed{1}$.

Review Problems

15.24

(a) The denominator is $3 - x$, so the domain of f is $\boxed{\text{all real numbers except } 3}$.

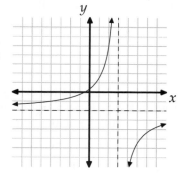

To find the range of f, let $y = f(x) = \frac{2x+1}{3-x}$. Multiplying both sides by $3 - x$, we get $y(3 - x) = 2x + 1$. Solving for x, we get $x = \frac{3y-1}{y+2}$. Hence, the range of f is $\boxed{\text{all real numbers except } -2}$.

To find the asymptotes of $f(x)$, we divide the denominator $3 - x$ into the numerator $2x + 1$, and find

$$f(x) = -\frac{7}{x-3} - 2.$$

As x approaches 3, $f(x)$ becomes very far from 0, so $x = 3$ is an asymptote. As x becomes very far from 0, $7/(x-3)$ approaches 0, so $y = -2$ is an asymptote. Therefore, the asymptotes are $\boxed{x = 3 \text{ and } y = -2}$.

(b) The denominator is $x - 2$, so the domain of f is $\boxed{\text{all real numbers except } 2}$.

To find the range of f, let $y = f(x) = \frac{x^2-4x+3}{x-2}$. Multiplying both sides by $x - 2$, we get $y(x - 2) = x^2 - 4x + 3$. Writing this as a quadratic in x, we get $x^2 - (y + 4)x + 2y + 3 = 0$. This quadratic has a real root in x if and only if its discriminant is nonnegative. This discriminant is

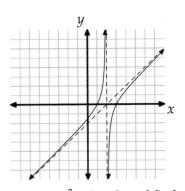

$$(y + 4)^2 - 4(2y + 3) = y^2 + 8y + 16 - 8y - 12 = y^2 + 4.$$

Since $y^2 + 4$ is positive for all y, we know that for every value of y, we can find a real value of x that satisfies the quadratic. Therefore, for every real value of y, we can find a value of x such that $f(x) = y$, which means that the range of f is $\boxed{\text{all real numbers}}$.

To find the asymptotes of $f(x)$, we divide the denominator $x - 2$ into the numerator $x^2 - 4x + 3$, and find

$$f(x) = -\frac{1}{x - 2} + x - 2.$$

As x approaches 2, $f(x)$ becomes very far from 0, so $x = 2$ is an asymptote. As x becomes very far from 0, $1/(x - 2)$ approaches 0, so $y = x - 2$ is an asymptote. Therefore, the asymptotes are $\boxed{x = 2 \text{ and } y = x - 2}$.

15.25 We have

$$f(x) = \frac{p(x)}{q(x)} = \frac{a_n x^n + a_{n-1} x^{n-1} + \cdots + a_0}{b_n x^n + b_{n-1} x^{n-1} + \cdots + b_0}.$$

Dividing both the numerator and denominator by x^n, we get

$$f(x) = \frac{a_n + \frac{a_{n-1}}{x} + \cdots + \frac{a_0}{x^n}}{b_n + \frac{b_{n-1}}{x} + \cdots + \frac{b_0}{x^n}}.$$

Hence, as x becomes very far from 0, $f(x)$ approaches a_n/b_n. Therefore, the horizontal asymptote is $\boxed{y = a_n/b_n}$.

15.26 First, we note that the numerator factors as $x^3 - 2x^2 - 15x = x(x - 5)(x + 3)$, so the graph of $f(x)$ passes through the points $(0,0)$, $(5,0)$, and $(-3,0)$. This observation eliminates (c).

We can also factor the denominator, and we have

$$f(x) = \frac{x(x - 5)(x + 3)}{3(x + 2)^2}.$$

Therefore, the denominator is positive for all x except $x = -2$, at which the graph has a vertical asymptote. All four of the proposed graphs appear to have a vertical asymptote just to the left of the y-axis, so we'll have to look at where $f(x)$ is positive and negative.

Since $x(x-5)(x+3)$ is negative for $0 < x < 5$, we know that $f(x)$ is also negative on this interval. This observation eliminates (a) and (d), so the graph of $f(x)$ can only be $\boxed{\text{(b)}}$. Note that $f(x) < 0$ for $x < -3$, $f(x) > 0$ for $x > 5$, and $f(x) > 0$ for $x \in (-3, -2) \cup (-2, 0)$, as indicated in graph (b).

15.27 We see that $g(x) = \frac{x - 3}{7 - x} = -\frac{x - 3}{x - 7} = -f(x)$, so the graph of $g(x)$ is the the graph of $f(x)$ reflected over the x-axis.

15.28 For each part, to determine whether or not the function has a slant asymptote, we divide the denominator into the numerator.

(a)

$$f(x) = \frac{2x - 1}{x + 2} = 2 - \frac{5}{x + 2}.$$

The quotient is 2, which is a constant. Therefore, the graph of f has the horizontal asymptote $y = 2$, but no slant asymptote.

(b)

$$f(x) = \frac{(x-2)(3x-1)}{x+1} = \frac{3x^2 - 7x + 2}{x+1} = 3x - 10 + \frac{12}{x+1}.$$

The quotient is $3x - 10$, so $\boxed{y = 3x - 10}$ is the slant asymptote of the graph of f.

(c)

$$f(x) = \frac{(x-2)(3x-1)}{(x+1)(x-3)} = \frac{3x^2 - 7x + 2}{x^2 - 2x - 3} = 3 + \frac{11 - x}{x^2 - 2x - 3}.$$

The quotient is 3, which is a constant. Therefore, the graph of f has the horizontal asymptote $y = 3$, but no slant asymptote.

(d)

$$f(x) = \frac{x-2}{x(x+1)} = \frac{x-2}{x^2 + x}.$$

Since the degree of the denominator is greater than the degree of the numerator, the quotient is 0. Therefore, the graph of f has the horizontal asymptote $y = 0$, but no slant asymptote.

(e)

$$f(x) = \frac{2x^3 + 5x^2 - 4x - 3}{x^2 + 2x} = 2x + 1 - \frac{6x + 3}{x^2 + 2x}.$$

The quotient is $2x + 1$, so $\boxed{y = 2x + 1}$ is the equation of the slant asymptote of the graph of f.

(f)

$$f(x) = \frac{2x^3 + 5x^2 - 4x - 3}{x+2} = 2x^2 + x - 6 + \frac{9}{x+2}.$$

The quotient is $2x^2 + x - 6$, which is quadratic, not linear. Therefore, the graph of f has no slant asymptotes.

In general, the graph of $f(x) = g(x)/h(x)$ has a slant asymptote if and only if the quotient of the fraction $g(x)/h(x)$ has degree 1. In other words, the degree of g is exactly one greater than the degree of h.

15.29

(a) $\boxed{\text{No}}$, the graph of a rational function cannot have two different horizontal asymptotes. Recall that we can find the horizontal asymptote of a rational function by dividing the denominator into the numerator. We have a horizontal asymptote if and only if the quotient is a constant. For example, the graph of

$$\frac{3x}{x-2} = \frac{3(x-2)+6}{x-2} = \frac{6}{x-2} + 3$$

has the horizontal asymptote $y = 3$. Since the quotient is uniquely determined, a rational function has at most one horizontal asymptote.

(b) $\boxed{\text{Yes}}$, the graph of a rational function may have two vertical asymptotes. For example, the graph of the function $\frac{1}{x(x-1)}$ has the vertical asymptotes $x = 0$ and $x = 1$.

(c) $\boxed{\text{Yes}}$, the graph of a rational function may have a horizontal asymptote and a vertical asymptote. For example, the graph of the function $\frac{1}{x}$ has the horizontal asymptote $y = 0$ and the vertical asymptote $x = 0$.

(d) $\boxed{\text{No}}$, the graph of a rational function cannot have a horizontal asymptote and a slant asymptote. As in part (a), we find the horizontal or slant asymptote of a rational function by dividing the denominator into the numerator. Then the graph has a horizontal asymptote if the quotient is constant, and a slant asymptote if the quotient is a non-constant linear function. Clearly, both cannot occur simultaneously.

(e) $\boxed{\text{Yes}}$, the graph of a rational function may have a slant asymptote and a vertical asymptote. For example, the graph of the function $x + \frac{1}{x}$ has the slant asymptote $y = x$ and the vertical asymptote $x = 0$.

(f) $\boxed{\text{No}}$, the graph of a rational function cannot have two slant asymptotes. As in part (d), we find the slant asymptote of a rational function by dividing the denominator into the numerator. Then the graph has a slant asymptote if the quotient is a non-constant linear function. Since the quotient is uniquely determined, a rational function has at most one slant asymptote.

15.30   To find the range of $f(x)$, let

$$y = f(x) = \frac{x^2 - 2x + 3}{x^2 + 2x - 3}.$$

Multiplying both sides by $x^2 + 2x - 3$, we get $y(x^2 + 2x - 3) = x^2 - 2x + 3$. Writing this as a quadratic in x, we get

$$(y - 1)x^2 + (2y + 2)x - (3y + 3) = 0.$$

This quadratic has a real root in x if and only if its discriminant is nonnegative. This discriminant is

$$(2y + 2)^2 + 4(y - 1)(3y + 3) = 16y^2 + 8y - 8 = 8(2y^2 + y - 1) = 8(2y - 1)(y + 1).$$

Thus, the discriminant is nonnegative if and only if $y \leq -1$ or $y \geq 1/2$. Therefore, the range of $y = f(x)$ is $\boxed{(-\infty, -1] \cup [1/2, +\infty)}$.

15.31   We have

$$\frac{1}{x + \sqrt{x}} + \frac{1}{x - \sqrt{x}} = \frac{x - \sqrt{x} + x + \sqrt{x}}{(x + \sqrt{x})(x - \sqrt{x})} = \frac{2x}{x^2 - x} = \frac{2}{x - 1}.$$

Hence, the given inequality is equivalent to $\frac{2}{x-1} \leq 1$. Subtracting 1 from both sides gives $\frac{2}{x-1} - 1 \leq 0$. Writing the left side with a common denominator gives us $\frac{3-x}{x-1} \leq 0$, and multiplying by -1 gives $\frac{x-3}{x-1} \geq 0$. We must have either $x - 3 \geq 0$ and $x - 1 > 0$, or $x - 3 \leq 0$ and $x - 1 < 0$. The former occurs for $x \geq 3$, and the latter occurs for $x < 1$. We must also have $x \geq 0$ in order for $\sqrt{x}$ to be defined, and we cannot have $x = 0$, since this makes denominators equal to 0. Therefore, our solution is $\boxed{x \in (0, 1) \cup [3, +\infty)}$.

15.32   We could simply multiply the equation by $(x^2 - 4)(x - 2)$ to get rid of the fractions. However, we notice that we can factor the first denominator as $x^2 - 4 = (x - 2)(x + 2)$. So, our equation is now

$$\frac{15}{(x - 2)(x + 2)} - \frac{2}{x - 2} = 1.$$

Now, we see that we just have to multiply by $(x - 2)(x + 2)$ to get rid of the fractions. Doing so gives us $15 - 2(x + 2) = (x - 2)(x + 2)$. Simplifying this equation gives $x^2 + 2x - 15 = 0$, and the quadratic factors as $(x + 5)(x - 3) = 0$. Therefore, the solutions are $x = \boxed{3 \text{ and } -5}$.

15.33   Multiplying both sides by $(x + 5)(1 - x)$, the given equation becomes

$$(1 - x)(2 - x) - (x + 5)(3 + 2x) = (x + 5)(1 - x)(3x + 1),$$

which simplifies as $3(x^3 + 4x^2 - 9x - 6) = 0$. Using the usual techniques, we find that $x - 2$ is a factor, and we have $(x - 2)(x^2 + 6x + 3) = 0$. We can use the quadratic formula to find the roots of $x^2 + 6x + 3 = 0$. We find that the solutions are $x = \boxed{2, -3 + \sqrt{6}, \text{ and } -3 - \sqrt{6}}$.

15.34   Since $x^2 \geq 0$ for all x, we have $4x^2 + 1 \geq 1$ for all x. Therefore, $f(x) = \frac{1}{4x^2+1} \leq 1$ for all x, so the maximum value of $f(x)$ is $f(0) = \boxed{1}$.

15.35   Subtracting 1 from both sides of the inequality, the lesser side becomes

$$\frac{(2x^2 - 16) - (x^2 + 3x + 2)}{x^2 + 3x + 2} = \frac{x^2 - 3x - 18}{x^2 + 3x + 2} = \frac{(x - 6)(x + 3)}{(x + 2)(x + 1)},$$

so our inequality is equivalent to $\dfrac{(x-6)(x+3)}{(x+2)(x+1)} < 0$. Let $f(x) = \dfrac{(x-6)(x+3)}{(x+2)(x+1)}$, so the sign of $f(x)$ depends on the signs of $x-6$, $x+3$, $x+2$, and $x+1$. We construct a table as follows:

	$x+3$	$x+2$	$x+1$	$x-6$	$f(x)$
$x < -3$	$-$	$-$	$-$	$-$	$+$
$-3 < x < -2$	$+$	$-$	$-$	$-$	$-$
$-2 < x < -1$	$+$	$+$	$-$	$-$	$+$
$-1 < x < 6$	$+$	$+$	$+$	$-$	$-$
$6 < x$	$+$	$+$	$+$	$+$	$+$

Hence, we have $f(x) < 0$ for $x \in \boxed{(-3,-2) \cup (-1,6)}$.

15.36   As seen in Problem 15.13, every polynomial that is even is of the form $f(x) = a_{2n}x^{2n} + a_{2n-2}x^{2n-2} + \cdots + a_2x^2 + a_0$. In other words, only the even powers of x appear. Since the coefficient of x^{2n-1} is 0, by Vieta's Formulas, the sum of the roots must be $\boxed{0}$.

15.37

(a)   Let $h = g \circ f$. Then we have $h(-x) = (g \circ f)(-x) = g(f(-x))$. Since $f(x)$ is even, we have $g(f(-x)) = g(f(x))$. Therefore, we have $h(-x) = g(f(-x)) = g(f(x)) = (g \circ f)(x) = h(x)$. Hence, $h = g \circ f$ is $\boxed{\text{even}}$.

(b)   The function $g \circ f$ need not be even nor odd. For example, take $f(x) = x$, which is odd. Then $(g \circ f)(x) = g(f(x)) = g(x)$, which may be any function with domain $\mathbb{R}$.

15.38

(a)   Since f is even, $f(-x) = f(-(-x)) = f(x)$. In other words, $f(-x)$ is the same function as $f(-(-x))$, so $f(-x)$ is $\boxed{\text{even}}$.

(b)   Substituting $-x$ for x, we get $g((-x)^2) = g(x^2)$, so $g(x^2)$ is $\boxed{\text{even}}$.

(c)   Substituting $-x$ for x, we get $f(-x)g(-x) = f(x) \cdot (-g(x)) = -f(x)g(x)$, so $f(x)g(x)$ is $\boxed{\text{odd}}$.

(d)   The function $f(x) + g(x)$ need not be even or odd. For example, take $f(x) = 1$ (which is even) and $g(x) = x$ (which is odd). Then $f(x) + g(x) = x + 1$, which is $\boxed{\text{neither even nor odd}}$.

(e)   Substituting $-x$ for x, we get $f(g(2(-x))) = f(g(-2x)) = f(-g(2x)) = f(g(2x))$, so $f(g(2x))$ is $\boxed{\text{even}}$.

15.39

(a)   Substituting $-x$ for x, we get $g(-x) = |f(-x)| = |f(x)| = g(x)$, so g must be $\boxed{\text{even}}$.

(b)   Substituting $-x$ for x, we get $g(-x) = |f(-x)| = |-f(x)| = |f(x)| = g(x)$, so g must be $\boxed{\text{even}}$.

(c)   If g is even, then $g(-x) = g(x)$, so $|f(-x)| = |f(x)|$. Hence, $f(-x) = -f(x)$ or $f(-x) = f(x)$, but this must hold for each individual value of x. It is not necessarily the case that $f(-x) = -f(x)$ for all x, or that $f(-x) = f(x)$ for all x. We can have $f(-x) = -f(x)$ for some x, and $f(-x) = f(x)$ for other x.

15.40   It is $\boxed{\text{not necessarily true}}$ that the function $1/f(x)$ is monotonically decreasing. For example, let $f(x) = x$, but only on the domain $[-2, -\frac{1}{2}] \cup [\frac{1}{2}, 2]$. The graphs of $y = f(x)$ and $y = 1/f(x)$ are shown below.

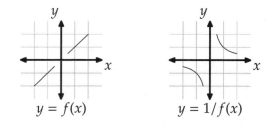

$y = f(x)$ $y = 1/f(x)$

The function $1/f(x)$ is not monotonically decreasing, because $1/f(-1/2) = -2$ is less than $1/f(1/2) = 2$.

15.41 Let c be a positive constant, so $x < x+c$. Then to prove that $\sqrt[3]{x}$ is increasing, we must show that $\sqrt[3]{x+c} > \sqrt[3]{x}$.

For the sake of contradiction, suppose that $\sqrt[3]{x+c} \le \sqrt[3]{x}$. By Problem 15.4.4, the function $f(x) = x^3$ is increasing. Hence, cubing both sides preserves the inequality, so $x + c \le x$, which is a contradiction. Therefore, $\sqrt[3]{x+c} > \sqrt[3]{x}$, so $\sqrt[3]{x}$ is an increasing function.

Alternatively, we could have considered the three possible cases $\sqrt[3]{x+c} > \sqrt[3]{x}$, $\sqrt[3]{x+c} = \sqrt[3]{x}$, and $\sqrt[3]{x+c} < \sqrt[3]{x}$. For any given c, only one of these can be true. Cubing all three give $x + c > x$, $x + c = x$, and $x + c < x$. Only in the first case can we possibly have $x + c > x$, so we conclude that if $x + c > x$, then $\sqrt[3]{x+c} > \sqrt[3]{x}$, as desired.

15.42 Since a is greater than 1, the function $\log_a x$ is increasing. Since $\log_a 1 = 0$, we have $\log_a x < 0$ for $0 < x < 1$ and $\log_a x > 0$ for $x > 1$. Since b is greater than 1, the function $\log_b x$ has the same properties.

Now, let $f(x) = \log_a x \cdot \log_b x$. Since both $\log_a x$ and $\log_b x$ are increasing and positive for $x > 1$, $f(x)$ is also increasing for $x > 1$. Furthermore, as x becomes very far from 0, so does $f(x)$. Therefore, the equation $f(x) = \log_a x \cdot \log_b x = 2007$ has exactly one solution for $x > 1$.

We claim that $f(x)$ is decreasing on the interval $(0, 1)$. To see why, let $0 < x < y < 1$. Then $\log_a x < \log_a y < 0$ and $\log_b x < \log_b y < 0$, so $-(\log_a x) > -(\log_a y)$ and $-(\log_b x) > -(\log_b y)$. Since both sides in both $-(\log_a x) > -(\log_a y)$ and $-(\log_b x) > -(\log_b y)$ are positive, we can multiply these inequalities to give

$$\log_a x \cdot \log_b x > \log_a y \cdot \log_b y > 0,$$

so $f(x) > f(y) > 0$. Hence, $f(x)$ is decreasing on $(0, 1)$.

Furthermore, as x approaches 0, both $\log_a x$ and $\log_b x$ become much less than 0, so $f(x)$ becomes very large. We also have $f(1) = 0$, so as x goes from a positive number very close to 0 up to $x = 1$, the value of $f(x)$ goes from a very large number down to 0. Somewhere between $x = 0$ and $x = 1$, we have $f(x) = 2007$. Because $f(x)$ is decreasing for $0 < x < 1$, we cannot have more than one value of x in this interval for which $f(x) = 2007$. Therefore, the equation $f(x) = \log_a x \cdot \log_b x = 2007$ has exactly one solution for $0 < x < 1$. Hence, there are a total of $\boxed{2}$ solutions.

15.43 Dividing the denominator $x + 10^7$ into the numerator x, we get $f(x) = \frac{x}{x+10^7} = 1 - \frac{10^7}{x+10^7}$. We know that $x + 10^7$ is increasing, so $\frac{10^7}{x+10^7}$ is decreasing. We conclude that $f(x) = 1 - \frac{10^7}{x+10^7}$ is increasing.

15.44 We can rewrite the numbers as

$$10 - 3\sqrt{11} = \sqrt{100} - \sqrt{99},$$
$$7 - 4\sqrt{3} = \sqrt{49} - \sqrt{48},$$
$$5\sqrt{41} - 32 = \sqrt{1025} - \sqrt{1024},$$
$$9 - 4\sqrt{5} = \sqrt{81} - \sqrt{80},$$

so each number is of the form $\sqrt{n+1} - \sqrt{n}$, where n is a positive integer. To make these numbers easier to compare, we can write $\sqrt{n+1} - \sqrt{n}$ as

$$\sqrt{n+1} - \sqrt{n} = (\sqrt{n+1} - \sqrt{n}) \cdot \frac{\sqrt{n+1} + \sqrt{n}}{\sqrt{n+1} + \sqrt{n}} = \frac{1}{\sqrt{n+1} + \sqrt{n}}.$$

Since the function $\sqrt{x}$ is increasing, the function $\sqrt{x+1} + \sqrt{x}$ is also increasing, which means that $1/(\sqrt{x+1} + \sqrt{x})$ is decreasing, so $\sqrt{x+1} - \sqrt{x}$ is decreasing. Therefore, we have

$$\sqrt{49} - \sqrt{48} > \sqrt{81} - \sqrt{80} > \sqrt{100} - \sqrt{99} > \sqrt{1025} - \sqrt{1024}.$$

In terms of the original numbers, we have $\boxed{7 - 4\sqrt{3} > 9 - 4\sqrt{5} > 10 - 3\sqrt{11} > 5\sqrt{41} - 32}$.

Challenge Problems

15.45 Let $f(x) = \sqrt{a + bx} + \sqrt{b + cx} + \sqrt{c + ax}$ and $g(x) = \sqrt{b - ax} + \sqrt{c - bx} + \sqrt{a - cx}$. First, we note that $f(0)$ and $g(0)$ are both equal to $\sqrt{a} + \sqrt{b} + \sqrt{c}$.

Furthermore, the function $f(x)$ is increasing and the function $g(x)$ is decreasing, which means the function $-g(x)$ is increasing. Hence, the function $f(x) + [-g(x)] = f(x) - g(x)$ is increasing. Therefore, the equation $f(x) - g(x) = 0$ has at most one solution, and we found above that $x = 0$ is a solution. We conclude that the solution $x = 0$ is unique.

15.46 Let $y = \sqrt{x} \geq 0$, so $x = y^2 \geq 0$. Then the given inequality becomes

$$\frac{y^2 + y + 4}{y^2 - 1} < 2.$$

Subtracting 2 from both sides, we get $\dfrac{y^2 + y + 4 - 2(y^2 - 1)}{y^2 - 1} < 0$. Simplifying this gives $\dfrac{(y - 3)(y + 2)}{(y - 1)(y + 1)} > 0$.

Let $f(y) = \dfrac{(y - 3)(y + 2)}{(y - 1)(y + 1)}$. Since $y \geq 0$ (remember, $y = \sqrt{x}$), both $y + 1$ and $y + 2$ are always positive. Therefore, the sign of $f(y)$ depends only on the sign of $y - 3$ and $y - 1$. We construct a table at right. From the table, we see that $f(y) > 0$ for $0 \leq y < 1$ and $y > 3$. Since $x = y^2$, the original inequality is satisfied for $x \in \boxed{[0, 1) \cup (9, +\infty)}$.

	$y - 1$	$y - 3$	$f(y)$
$0 \leq y < 1$	$-$	$-$	$+$
$1 < y < 3$	$+$	$-$	$-$
$3 < y$	$+$	$+$	$+$

15.47 By Problem 15.41, the function $\sqrt[3]{x}$ is increasing, so the functions $\sqrt[3]{x + 2}$, $\sqrt[3]{x + 3}$, and $\sqrt[3]{x + 4}$ are all increasing. Therefore, the function

$$f(x) = \sqrt[3]{x + 2} + \sqrt[3]{x + 3} + \sqrt[3]{x + 4}$$

is increasing. Furthermore, $f(-3) = \sqrt[3]{-1} + \sqrt[3]{0} + \sqrt[3]{1} = -1 + 0 + 1 = 0$, so $x = \boxed{-3}$ is a root. Since f is increasing, it is the only root.

15.48 Multiplying both sides by $(x + 2)(Ax + 1)$, the given equation becomes

$$3x(Ax + 1) + 2(x + 2)(Ax + 1) = 2(x + 2),$$

which simplifies as $5Ax^2 + (4A + 3)x = 0$, so $x[5Ax + (4A + 3)] = 0$. Therefore, $x = 0$ or $5Ax + (4A + 3) = 0$.

If $A = 0$, then the equation $5Ax + (4A + 3) = 0$ becomes $3 = 0$, which has no solutions, so the original equation has only the solution $x = 0$ if $A = 0$. If $A \neq 0$, the equation $5Ax + (4A + 3) = 0$ has the solution

$$x = -\frac{4A + 3}{5A}.$$

The original equation always have $x = 0$ as a solution. So, for the equation to have exactly 1 solution, either our above expression for x equals 0, or that expression equals a value for which expressions in the given equation are undefined.

Setting $-(4A + 3)/(5A) = 0$ gives $4A + 3 = 0$, so the equation has only one solution for $A = -3/4$.

The expression $3x/(x + 2)$ in the original equation is undefined for $x = -2$, so the equation has only one solution when $-(4A + 3)/(5A) = -2$, which yields $A = 1/2$. Finally, $2/(Ax + 1)$ is undefined when $x = -1/A$. Solving $-(4A + 3)/(5A) = -1/A$ also gives $A = 1/2$. (Think a bit about why these two cases yield the same result!)

In conclusion, the values of A for which the given equation has exactly one solution are $A = \boxed{0, 1/2 \text{ and } -3/4}$.

15.49 First, we note that $\log_{10}(x - 3)$ is defined only for $x > 3$. To see where $\log_{10}(x^3 - 5x^2 + 11x - 15)$ is defined, we try to factor $x^3 - 5x^2 + 11x - 15$ using the usual techniques, and we find that

$$x^3 - 5x^2 + 11x - 15 = (x - 3)(x^2 - 2x + 5).$$

The quadratic $x^2 - 2x + 5$ has no real roots and the coefficient of x^2 is positive, so $x^2 - 2x + 5 > 0$ for all real numbers x. Therefore, $\log_{10}(x^3 - 5x^2 + 11x - 15) = \log_{10}[(x-3)(x^2 - 2x + 5)]$ is also defined only for $x > 3$.

On this domain, we have

$$f(x) = \log_{10}[(x-3)(x^2 - 2x + 5)] - \log_{10}(x-3) = \log_{10}(x^2 - 2x + 5).$$

The function $\log_{10} x$ is increasing, so $\log_{10}(x^2 - 2x + 5)$ is minimized when $x^2 - 2x + 5$ is minimized. In turn, the function $x^2 - 2x + 5 = (x-1)^2 + 4$ is minimized when $x = 1$. However, this value lies outside the domain of $f(x)$.

Since $x^2 - 2x + 5$ is increasing for $x > 3$ and $\log_{10} x$ is an increasing function, we know that $\log_{10}(x^2 - 2x + 5)$ is increasing for $x > 3$. Therefore, $f(x)$ is increasing for $x > 3$. So, the minimum possible value of $f(x)$ occurs when x is minimized. However, x can only approach 3 without being equal to 3, so the function $f(x)$ has $\boxed{\text{no minimum}}$. (In other words, x has no minimum value—no matter how close x is to 3, we can always find another value of x that is closer.)

15.50 We can rewrite $f(x)$ as

$$f(x) = \sqrt{x(8-x)} - \sqrt{(8-x)(x-6)},$$

which shows that $f(x)$ is defined only for $6 \le x \le 8$. We can then write

$$f(x) = \sqrt{8-x}(\sqrt{x} - \sqrt{x-6}) = \sqrt{8-x}(\sqrt{x} - \sqrt{x-6}) \cdot \frac{\sqrt{x} + \sqrt{x-6}}{\sqrt{x} + \sqrt{x-6}}$$

$$= \sqrt{8-x} \cdot \frac{x - (x-6)}{\sqrt{x} + \sqrt{x-6}} = \sqrt{8-x} \cdot \frac{6}{\sqrt{x} + \sqrt{x-6}}.$$

For $6 \le x \le 8$, the function $\sqrt{8-x}$ is decreasing and nonnegative. Also, the function $\sqrt{x} + \sqrt{x-6}$ is increasing and nonnegative, so the function $\frac{1}{\sqrt{x}+\sqrt{x-6}}$ is decreasing and nonnegative. Therefore,

$$f(x) = \sqrt{8-x} \cdot \frac{6}{\sqrt{x} + \sqrt{x-6}}$$

is decreasing for $6 \le x \le 8$. Hence, the maximum value of $f(x)$ is $f(6) = \sqrt{2} \cdot \frac{6}{\sqrt{6}} = \boxed{2\sqrt{3}}$.

15.51 To simplify the equation, we divide both sides by 789^x, to obtain the equation

$$\left(\frac{567}{789}\right)^x + \left(\frac{678}{789}\right)^x = 1.$$

This step isolates the variable x on one side of the equation.

Let $f(x) = \left(\frac{567}{789}\right)^x + \left(\frac{678}{789}\right)^x$. Since $567/789 < 1$ and $678/789 < 1$, the function $f(x)$ is decreasing.

Furthermore, $f(x)$ becomes large as x becomes much less than 0, and $f(x)$ becomes very close to 0 as x becomes very large. Therefore, $f(x) = 1$ for exactly $\boxed{1}$ value of x.

CHAPTER 16

Piecewise Defined Functions

Exercises for Section 16.1

16.1.1

(a) Since $0 \le 3/2 < 2$, we have $f(3/2) = 3 \cdot 3/2 = \boxed{9/2}$.

(b) Graphing the function, we see that the function is discontinuous at $x = 2$. As x approaches 2 from the left, $f(x)$ approaches 6, but as x approaches 2 from the right, $f(x)$ approaches 7. Therefore, f is $\boxed{\text{not continuous}}$. The graph is shown at right.

(c) As x ranges over the interval $(-\infty, -2)$, $f(x)$ ranges over $(-\infty, -4)$. As x ranges over the interval $[-2, 0)$, $f(x)$ ranges over the interval $[-4, 0)$. As x ranges over the interval $[0, 2)$, $f(x)$ ranges over $[0, 6)$. Finally, as x ranges over the interval $[2, +\infty)$, $f(x)$ ranges over $[7, +\infty)$. Therefore, the range of $f(x)$ is $\boxed{(-\infty, 6) \cup [7, +\infty)}$.

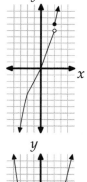

16.1.2 The function f is clearly continuous on the interval $(-\infty, 3]$ and on the interval $(3, +\infty)$, so the only question is whether f is continuous at $x = 3$.

As x approaches 3 from the left, $f(x)$ approaches $2c + 9$. As x approaches 3 from the right, $f(x)$ approaches $6c + 1$. Hence, for f to be continuous at $x = 3$, we require that $2c + 9 = 6c + 1$, so $c = \boxed{2}$. The graph of f is shown at right.

16.1.3 First we check that f has an inverse. As x ranges over the interval $(-\infty, 0)$, $f(x)$ is decreasing and ranges over $(\sqrt{2}, +\infty)$. As x ranges over the interval $[0, \infty)$, $f(x)$ is again decreasing, and ranges over $(-\infty, 1]$. The intervals $(\sqrt{2}, +\infty)$ and $(-\infty, 1]$ don't share any values, so f has an inverse.

If $x < 0$, then $f(x) = \sqrt{2 - x}$. As we noted above, this piece takes numbers in the interval $(-\infty, 0)$ as input and outputs numbers in the interval $(\sqrt{2}, +\infty)$. So, the inverse of this piece must take inputs in the interval $(\sqrt{2}, +\infty)$ and it must output numbers in the interval $(-\infty, 0)$. To find the inverse of this piece, we solve $f(y) = x$ for y. Squaring both sides of $\sqrt{2 - y} = x$ gives $2 - y = x^2$, from which we find $y = 2 - x^2$.

Similarly, if $x \ge 0$, then $f(x) = 1 - x^2$. This pieces takes numbers in the interval $[0, +\infty)$ as input and outputs all numbers in the interval $(-\infty, 1]$, so the domain of the inverse of this piece is $(-\infty, 1]$. Solving $f(y) = x$ for y, we get $y = \sqrt{1 - x}$. (Note that we don't have $y = \pm\sqrt{1 - x}$ because the range of the inverse of this piece is $[0, +\infty)$.)

Therefore, the inverse of f is given by

$$f^{-1}(x) = \boxed{\begin{cases} \sqrt{1 - x} & \text{if } x \le 1, \\ 2 - x^2 & \text{if } x > \sqrt{2}. \end{cases}}$$

Note that $f^{-1}(x)$ is not defined for $1 < x \le \sqrt{2}$, since $f(x)$ never achieves values in the interval $(1, \sqrt{2}]$.

16.1.4 The price per book is the same for up to 24 books, but it suddenly steps down from 12 dollars to 11 if we buy 25. Therefore, it's possible for the total cost of 24 books to be more than the total cost of 25 books. The cost of 25 books is $11(25) = 275$ dollars, but the cost of 24 books is $12(24) = 288$. Since 288 is 13 more than 275, we see that buying 23 books, at a cost of $288 - 12 = 276$ dollars, is also more expensive than buying 25 books. So, if $n = 23$ or $n = 24$, then it is possible to buy more than n books more cheaply than buying exactly n books.

We also must check the next point at which the price per book drops. It costs $10(49) = 490$ dollars to buy 49 books. If we buy n books, and $25 \le n \le 48$, then it costs $11n$ dollars. So, we seek the integer values of n for which $11n > 490$ and $n \le 48$. Dividing both sides of $11n > 490$ by 11 gives $n > 44\frac{6}{11}$. So, we have four more values of n (45, 46, 47, and 48) for which it is cheaper to buy more than n books than to buy exactly n books.

Combining these two cases, we have $\boxed{6}$ values of n for which it is cheaper to buy more than n books than to buy exactly n books.

Exercises for Section 16.2

16.2.1

(a) Since $x \le -4$, the expression $x + 2$ is negative. Therefore, $|x + 2| = \boxed{-x - 2}$.

(b) From part (a), $|x + 2| = -x - 2$, so $|2 - |x + 2|| = |2 - (-x - 2)| = |x + 4|$. Since $x \le -4$, $x + 4$ is nonpositive, so $|x + 4| = \boxed{-x - 4}$.

(c) Since $\pi^2 \approx 9.87$, we find $\pi^2 - 10$ is negative, so $|\pi^2 - 10| = \boxed{10 - \pi^2}$.

(d) We can write $x + \pi^2 - 6$ as $(x + 4) + (\pi^2 - 10)$. We know $x + 4$ is nonpositive, and $\pi^2 - 10$ is negative. Therefore, their sum is negative, and hence $|x + \pi^2 - 6| = \boxed{-x - \pi^2 + 6}$.

16.2.2 The quantity $|x + 2| = |x - (-2)|$ represents the distance between x and -2, and the quantity $|x + 5| = |x - (-5)|$ represents the distance between x and -5. Hence, if x lies in the interval $[-5, -2]$, then $|x + 2| + |x + 5|$ would equal $-2 - (-5) = 3$, so x must lie outside this interval.

Let d be the distance from x to the closer of -2 or -5. Then the distance from x to the farther of -2 or -5 is $d + 3$, so $d + (d + 3) = 8$, which means $d = 5/2$. Therefore, the solutions are $x = -2 + 5/2 = \boxed{1/2}$ and $x = -5 - 5/2 = \boxed{-15/2}$.

16.2.3 If $x \ge 2$, then $|x - 2| = x - 2$, and the inequality $1 \le x - 2 \le 7$ becomes $3 \le x \le 9$. If $x < 2$, then $|x - 2| = -x + 2$, and the inequality $1 \le -x + 2 \le 7$ becomes $-5 \le x \le 1$. Therefore, the set of solutions is $x \in \boxed{[-5, 1] \cup [3, 9]}$.

16.2.4 If $x \ge -1/2$, then $|2x + 1| = 2x + 1$, so $x - |2x + 1| = x - (2x + 1) = -x - 1$. For $x \ge -1/2$, we have $-x - 1 \le -1/2$, so $-x - 1$ is negative. Therefore, we have

$$|x - |2x + 1|| = |-x - 1| = x + 1,$$

so our original equation is $x + 1 = 3$, which means $x = 2$. This value satisfies $x \ge -1/2$.

If $x < -1/2$, then $|2x + 1| = -2x - 1$, so $x - |2x + 1| = x - (-2x - 1) = 3x + 1$. For $x < -1/2$, we have $3x + 1 < -1/2$, so $3x + 1$ is negative. Therefore, we have

$$|x - |2x + 1|| = |3x + 1| = -3x - 1,$$

so our original equation is $-3x - 1 = 3$, which means $x = -4/3$. This value satisfies $x < -1/2$.

Therefore, the solutions are $x = \boxed{2 \text{ and } -4/3}$.

16.2.5 If $x < 0$, then $|x| = -x$, so the first equation becomes $y = 10$. Then the second equation becomes $x = 12$, which does not satisfy $x < 0$, so we must have $x \ge 0$. Hence, the first equation becomes $2x + y = 10$.

If $y \geq 0$, then the second equation becomes $x = 12$. Then the first equation becomes $12 + 12 + y = 10$, so $y = -14$, which does not satisfy $y \geq 0$, so we must have $y < 0$. Hence, the second equation becomes $x - 2y = 12$.

Solving the system of equations $2x + y = 10$ and $x - 2y = 12$, we get $x = 32/5$ and $y = -14/5$. Note that these values satisfy $x \geq 0$ and $y < 0$, so this is a valid solution. Therefore, $x + y = \boxed{18/5}$.

16.2.6 If $y = 0$, then the fraction is not defined. If $y > 0$, then $|y| = y$, so

$$\frac{|y - |y||}{y} = \frac{0}{y} = 0,$$

which is not positive. Finally, if $y < 0$, then $|y| = -y$, so

$$\frac{|y - |y||}{y} = \frac{|2y|}{y} = \frac{2|y|}{y} = -2.$$

Therefore, $|y - |y||/y$ can never be positive, so there are $\boxed{\text{no solutions}}$.

16.2.7 First, since we are dealing with the expression $|f(x)|$, we check where $f(x) \geq 0$. Since $f(x) = (x - 3)^2 - 1$, we have $f(x) \geq 0$ if and only if $x \leq 2$ or $x \geq 4$.

If $x \geq 0$, then $|x| = x$, so the equation $f(|x|) = |f(x)|$ becomes $f(x) = |f(x)|$. This is satisfied if and only if $f(x) \geq 0$, which occurs when $0 \leq x \leq 2$ or $x \geq 4$.

If $x < 0$, then $|x| = -x$. Furthermore, $f(x) \geq 0$ for $x < 0$, so the equation $f(|x|) = |f(x)|$ becomes $f(-x) = f(x)$, or $(-x - 3)^2 - 1 = (x - 3)^2 - 1$. This simplifies as $12x = 0$, so $x = 0$. Hence, there are no solutions for $x < 0$.

Therefore, the set of solutions is $x \in \boxed{[0, 2] \cup [4, +\infty)}$.

See if you can find another solution by considering the graphs of $y = f(|x|)$ and $y = |f(x)|$.

Exercises for Section 16.3

16.3.1 When $2 - 4x \geq 0$, we have both $x \leq 1/2$ and $f(x) = |2 - 4x| - 4 = 2 - 4x - 4 = -4x - 2$. When $2 - 4x < 0$, we have both $x < 1/2$ and $f(x) = |2 - 4x| - 4 = -(2 - 4x) - 4 = 4x - 6$. So, to graph f, we graph $y = -4x - 2$ for $x \leq 1/2$ and $y = 4x - 6$ for $x > 1/2$. The graph is shown at right.

The y-intercept is given by $(0, f(0)) = \boxed{(0, -2)}$. The graph intersects the x-axis when we have $f(x) = |2 - 4x| - 4 = 0$, or $|2 - 4x| = 4$. Dividing both sides by 2, we get $|1 - 2x| = 2$. Then $1 - 2x = 2$ or $1 - 2x = -2$, which leads to $x = -1/2$ or $x = 3/2$. Therefore, the x-intercepts are $\boxed{(-1/2, 0) \text{ and } (3/2, 0)}$.

16.3.2 The graphs of $f(x)$ and $g(x)$ are shown below.

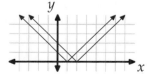

The graphs intersect when $|x - 1| = |x - 2|$. In other words, x is equidistant from the numbers 1 and 2 on the real number line. This is only true for $x = 3/2$, and $f(3/2) = g(3/2) = 1/2$. Therefore, the graphs intersect at $\boxed{(3/2, 1/2)}$.

16.3.3   First we look at the inequality $xy \geq 0$. If $x = 0$ or $y = 0$, then this inequality is satisfied. If $x > 0$, then $xy \geq 0$ if and only if $y \geq 0$. Similarly, if $x < 0$, then $xy \geq 0$ if and only if $y \leq 0$. The Cartesian plane is divided into four quadrants by the coordinate axes. The graph of $xy \geq 0$ consists of the upper right $(x > 0, y > 0)$ and lower left $(x < 0, y < 0)$ quadrants, together with the coordinate axes.

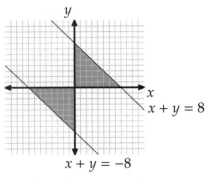

Now, if $|x + y| \leq 8$, then $-8 \leq x + y \leq 8$. Hence, the set of points (x, y) that satisfy $|x + y| \leq 8$ is the set of points bound between the two lines $x + y = 8$ and $x + y = -8$. At right, we've shaded those points that fall between these lines and are in one of the two quadrants for which $xy \geq 0$.

The graph consists of two isosceles right triangles with legs of length 8. Therefore, the area of the region is $2 \cdot \frac{1}{2} \cdot 8 \cdot 8 = \boxed{64}$.

16.3.4   The graph of $x^2 + y^2 = 4 = 2^2$ is the circle with radius 2 centered at the origin. The graphs of $y = |x|$ and $x^2 + y^2 = 4$ are shown below.

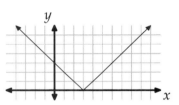

The smallest region bounded by these two graphs is a quarter-circle with radius 2, and its area is $(1/4) \cdot 2^2 \pi = \boxed{\pi}$.

16.3.5

(a)   We first write f as a piecewise defined function. For $x \leq 3$, we have $3 - x \geq 0$, so $f(x) = 3 - x$. For $x > 3$, we have $3 - x < 0$, so $f(x) = -(3 - x) = x - 3$. Therefore, we have

$$f(x) = \begin{cases} -x + 3 & \text{if } x \leq 3, \\ x - 3 & \text{if } 3 < x. \end{cases}$$

Using this definition, we graph f at right.

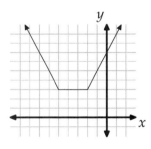

(b)   As with part (a), we first write f as a piecewise defined function. We have $x + 2 \geq 0$ and $x + 5 \geq 0$ for $x \geq -2$, so $f(x) = |x + 2| + |x + 5| = x + 2 + x + 5 = 2x + 7$ for $x \geq -2$. We have $x + 2 < 0$ and $x + 5 > 0$ for $-5 < x < -2$, so $f(x) = |x + 2| + |x + 5| = -(x + 2) + x + 5 = 3$ for $-5 < x < -2$. Finally, we have $x + 2 \leq 0$ and $x + 5 \leq 0$ for $x \leq -5$, so $f(x) = -(x + 2) - (x + 5) = -2x - 7$ for $x \leq -5$. Putting these together gives

$$f(x) = \begin{cases} -2x - 7 & \text{if } x \leq -5, \\ 3 & \text{if } -5 < x < -2, \\ 2x + 7 & \text{if } -2 \leq x. \end{cases}$$

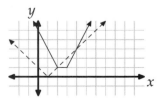

The graph of $f(x)$ is shown at right.

(c)   We rewrite the equation as $|x - 2| + |x - 3| = |x - 1|$. We then graph both $f(x) = |x - 2| + |x - 3|$ and $g(x) = |x - 1|$ and look for the points where the graphs intersect. These graphs are shown at right. The graph of f is solid and the graph of g is dashed.

The graphs appear to intersect at only $(2, 1)$ and $(4, 3)$. Of course, looks can be deceiving, so we must make sure both that the graphs meet at these two points, and that they meet at no other points. First, we confirm that $f(2) = g(2) = 1$ and $f(4) = g(4) = 3$. Next, we note that all three pieces of the graph of f are linear, as are both pieces of the graph of g.

The slope of the left piece of g is less steep than that of the left piece of f, so these two pieces will not intersect. Clearly, the left piece of g cannot intersect either of the other pieces of f, since the domain of the left piece of g is entirely within the domain of the left piece of f. The right piece of g intersects each of the pieces of f in at most one point each, because all the pieces are linear. The right piece of g intersects both the left and the middle pieces of f at $(2, 1)$, and intersects the right piece of g at $(4, 3)$. Therefore, looks have not deceived us; we have found all points where the graphs intersect, which means that the solutions to the original equation are $\boxed{x = 2 \text{ and } 4}$.

16.3.6   First, we write $h(x)$ as a piecewise function:

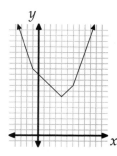

$$h(x) = \begin{cases} -3x + 9 & \text{if } x \le -1, \\ -x + 11 & \text{if } -1 < x \le 4, \\ x + 3 & \text{if } 4 < x \le 6, \\ 3x - 9 & \text{if } 6 \le x. \end{cases}$$

The graph of $h(x)$ is shown at right.

The function $h(x)$ is decreasing on the interval $(-\infty, 4]$ and increasing on the interval $[4, +\infty)$, so the minimum value of $h(x)$ is $h(4) = \boxed{7}$.

16.3.7   If $x \ge 0$ and $y \ge 0$, then the inequality $|5x| + |6y| \le 30$ becomes $5x + 6y \le 30$. Hence, the portion of the region described by $|5x| + |6y| \le 30$ that is contained in region with $x \ge 0$ and $y \ge 0$ is bound by the line $5x + 6y = 30$. Similarly, if $x \le 0$ and $y \ge 0$, we have $|5x| = -5x$ and $|6y| = 6y$, so $|5x| + |6y| \le 30$ becomes $-5x + 6y \le 30$. This tells us that the portion of the graph for which $x \le 0$ and $y \ge 0$ is bounded by the line $-5x + 6y = 30$. Similarly, we find that the region is also bound by the lines $-5x - 6y = 30$ and $5x - 6y = 30$. The graph of the four boundary lines and the resulting region is shown at right.

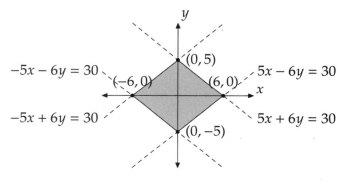

This region consists of four right triangles whose legs have length 5 and 6. Therefore, the area of the region is $4 \cdot \frac{1}{2} \cdot 5 \cdot 6 = \boxed{60}$.

Exercises for Section 16.4

16.4.1   Let $n = \lfloor x \rfloor$, so n is an integer. If $x = n$, then $\lceil x \rceil = n$, so the given equation becomes $2n = 9$, which does not have any solutions in integers n. Hence, x lies strictly between the integers n and $n + 1$. Therefore, $\lceil x \rceil = n + 1$, so the given equation becomes $n + (n + 1) = 9$, or $n = 4$. The set of all x such that $\lfloor x \rfloor = 4$ and x is not an integer is given by $\boxed{4 < x < 5}$.

16.4.2   From $\lfloor n/6 \rfloor = 5$, we get $5 \le n/6 < 6$, or $30 \le n < 36$. The integers n that satisfy this inequality are $n = \boxed{30, 31, 32, 33, 34, \text{ and } 35}$.

16.4.3   We have $\log_3 1000 > \log_3 81 = \log_3 3^4 = 4$, so $\sqrt{\log_3 1000} > 2$. Also, $1000 = 10^3 < 27^3 = 3^9$, so $\log_3 1000 < \log_3 3^9 = 9$, which means $\sqrt{\log_3 1000} < 3$. Therefore, $\left\lfloor \sqrt{\log_3 1000} \right\rfloor = \boxed{2}$.

16.4.4   Let $n = \lfloor x \rfloor$. If $x = n$, then $\lceil x \rceil = n$, so $f(x) = n + n = 2n$. Otherwise, x lies strictly between n and $n + 1$, so $\lceil x \rceil = n + 1$, which means $f(x) = 2n + 1$. The graph of $f(x)$ is shown below to the left.

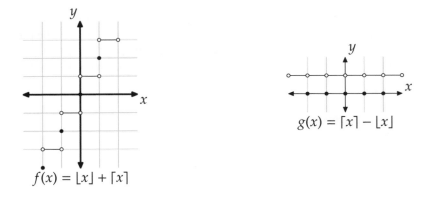

$$f(x) = \lfloor x \rfloor + \lceil x \rceil$$

$$g(x) = \lceil x \rceil - \lfloor x \rfloor$$

If $x = n$, then $g(x) = n - n = 0$. Otherwise, x lies strictly between n and $n + 1$, so $g(x) = (n + 1) - n = 1$. The graph of $g(x)$ is shown above to the right.

16.4.5 If $-2 \le \lfloor 3x + 2 \rfloor \le 2$, then $-2 \le 3x + 2 < 3$. This is equivalent to the inequality $\boxed{-4/3 \le x < 1/3}$.

16.4.6 Since $\lfloor \sqrt{2} \rfloor = 1$, we have $\{\sqrt{2}\} = \sqrt{2} - 1$. Since $\lfloor \sqrt{3} \rfloor = 1$, we have $\{\sqrt{3}\} = \sqrt{3} - 1$. Therefore,

$$\frac{\{\sqrt{3}\}^2 - 2\{\sqrt{2}\}^2}{\{\sqrt{3}\} - 2\{\sqrt{2}\}} = \frac{(\sqrt{3} - 1)^2 - 2(\sqrt{2} - 1)^2}{(\sqrt{3} - 1) - 2(\sqrt{2} - 1)} = \frac{3 - 2\sqrt{3} + 1 - 2(2 - 2\sqrt{2} + 1)}{\sqrt{3} - 1 - 2\sqrt{2} + 2}$$

$$= \frac{4\sqrt{2} - 2\sqrt{3} - 2}{-2\sqrt{2} + \sqrt{3} + 1} = \frac{-2(-2\sqrt{2} + \sqrt{3} + 1)}{-2\sqrt{2} + \sqrt{3} + 1}$$

$$= \boxed{-2}.$$

16.4.7 Let $x = (3^{31} + 2^{31})/(3^{29} + 2^{29})$. Since 3^{31} is much greater than 2^{31} and 3^{29} is much greater than 2^{29}, we expect x to be close to $3^{31}/3^{29} = 9$.

To compare the numerator with the denominator, we write $3^{31} + 2^{31} = 3^2 \cdot 3^{29} + 2^2 \cdot 2^{29} = 9 \cdot 3^{29} + 4 \cdot 2^{29}$. Therefore,

$$x = \frac{3^{31} + 2^{31}}{3^{29} + 2^{29}} = \frac{9 \cdot 3^{29} + 4 \cdot 2^{29}}{3^{29} + 2^{29}} < \frac{9 \cdot 3^{29} + 9 \cdot 2^{29}}{3^{29} + 2^{29}} = 9.$$

Hence, we expect x to be between 8 and 9.

To show that x is greater than 8, we compute $x - 8$:

$$x - 8 = \frac{3^{31} + 2^{31}}{3^{29} + 2^{29}} - 8 = \frac{9 \cdot 3^{29} + 4 \cdot 2^{29} - 8 \cdot 3^{29} - 8 \cdot 2^{29}}{3^{29} + 2^{29}} = \frac{3^{29} - 4 \cdot 2^{29}}{3^{29} + 2^{29}}$$

Thus, to show that x is greater than 8, it suffices to show that $3^{29} > 4 \cdot 2^{29}$. We start with the inequality $3^2 > 2^3$. Raising both sides to the power of 14, we get $3^{28} > 2^{42}$. Therefore, we have $3^{29} > 3 \cdot 2^{42} = 6 \cdot 2^{41} > 4 \cdot 2^{29}$. We conclude that $x > 8$. Therefore, $\lfloor x \rfloor = \boxed{8}$.

Exercises for Section 16.5

16.5.1 Since $x = \lfloor x \rfloor + \{x\}$, we can write the given equation as $2\lfloor x \rfloor + \{x\} = 5.7$. Then $\{x\}$ must be the fractional part of 5.7, so $\{x\} = 0.7$, which means $2\lfloor x \rfloor = 5$. However, 5 is not even, so there are $\boxed{\text{no solutions}}$.

16.5.2 Since $x = \lfloor x \rfloor + \{x\}$ and $y = \lfloor y \rfloor + \{y\}$, the given equations become $\lfloor x \rfloor + \lfloor y \rfloor + \{x\} = 5.3$ and $\lfloor x \rfloor + \lfloor y \rfloor + \{y\} = 5.7$. Then $\{x\}$ must be the fractional part of 5.3, so $\{x\} = 0.3$, and $\{y\}$ must be the fractional part of 5.7, so $\{y\} = 0.7$. Hence, $\lfloor x \rfloor + \lfloor y \rfloor = 5$.

Since $x > 0$ and $y > 0$, we have $\lfloor x \rfloor \geq 0$ and $\lfloor y \rfloor \geq 0$. Therefore, $(\lfloor x \rfloor, \lfloor y \rfloor)$ must be one of the pairs $(0, 5)$, $(1, 4)$, $(2, 3)$, $(3, 2)$, $(4, 1)$, and $(5, 0)$. Then the solutions in (x, y) are $(0.3, 5.7)$, $(1.3, 4.7)$, $(2.3, 3.7)$, $(3.3, 2.7)$, $(4.3, 1.7)$, and $(5.3, 0.7)$, for a total of $\boxed{6}$ solutions.

16.5.3 Let $x = 2t$. Then the equation becomes $7x/2 + \lfloor x \rfloor = 52$. Multiplying both sides by 2, we get $7x + 2\lfloor x \rfloor = 104$. Since $x = \lfloor x \rfloor + \{x\}$, this equation becomes $7(\lfloor x \rfloor + \{x\}) + 2\lfloor x \rfloor = 104$, so $9\lfloor x \rfloor + 7\{x\} = 104$, which gives $\{x\} = \frac{104 - 9\lfloor x \rfloor}{7}$.

We know that $0 \leq \{x\} < 1$, so $0 \leq \dfrac{104 - 9\lfloor x \rfloor}{7} < 1$. This leads to the bounds

$$\frac{97}{9} < \lfloor x \rfloor \leq \frac{104}{9}.$$

The only integer in this interval is 11, so $\lfloor x \rfloor = 11$. Then

$$\{x\} = \frac{104 - 9 \cdot 11}{7} = \frac{5}{7}.$$

Therefore, $x = 11 + 5/7 = 82/7$, so $t = (82/7)/2 = \boxed{41/7}$.

16.5.4 Since $n < 1000$, we have $\log_2 n < \log_2 1000 < \log_2 1024 = 10$. Hence, if $\lfloor \log_2 n \rfloor$ is an even integer, it must be one of 2, 4, 6, or 8.

If $k = \lfloor \log_2 n \rfloor$, then $k \leq \log_2 n < k + 1$, or $2^k \leq n < 2^{k+1}$. The number of positive integers n in this range is $2^{k+1} - 2^k = 2^k$, so there are 2^k integers n such that $\lfloor \log_2 n \rfloor = k$ for any nonnegative integer k. Therefore, the number of positive integers $n < 1000$ such that $\lfloor \log_2 n \rfloor$ is even is $2^2 + 2^4 + 2^6 + 2^8 = \boxed{340}$.

16.5.5 Note that $44^2 = 1936$ and $45^2 = 2025$, so for all k in the interval $1936 \leq k < 1997$, we have $\lfloor \sqrt{k} \rfloor = 44$, which means
$$S_{1997} - S_k = \lfloor \sqrt{k+1} \rfloor + \lfloor \sqrt{k+2} \rfloor + \cdots + \lfloor \sqrt{1997} \rfloor = 44 + 44 + \cdots + 44 = 44(1997 - k).$$
Since 44 factors as $2^2 \cdot 11$, $44(1997 - k)$ is a perfect square if and only if $11(1997 - k)$ is a perfect square. The largest $k < 1997$ such that $11(1997 - k)$ is a perfect square is given by $1997 - k = 11$, so $k = \boxed{1986}$.

16.5.6 Since $x = \lfloor x \rfloor + \{x\}$, the given equation becomes $\dfrac{\lfloor x \rfloor + \{x\}}{\lfloor x \rfloor} = \dfrac{2002}{2003}$, so $\{x\} = -\dfrac{\lfloor x \rfloor}{2003}$.

We know that $0 \leq \{x\} < 1$, so $0 \leq -\frac{\lfloor x \rfloor}{2003} < 1$. This leads to the bounds $-2003 < \lfloor x \rfloor \leq 0$. Then the smallest value of $\lfloor x \rfloor$ (which must lead to the smallest value of x) is -2002, in which case $\{x\} = \frac{2002}{2003}$.

Therefore, the smallest possible value of x is $-2002 + \dfrac{2002}{2003} = -2002\left(1 - \dfrac{1}{2003}\right) = \boxed{-\dfrac{2002^2}{2003}}$.

Review Problems

16.29

(a) Since $-2 \leq 3$, we have $f(-2) = (-2)^3 - (-2) = \boxed{-6}$. Since $3 \leq 3$, we have $f(3) = 3^3 - 3 = \boxed{24}$. Since $3 < 8 \leq 8$, we have $f(8) = 2\sqrt{8} = \boxed{4\sqrt{2}}$.

(b) We solve for $f(c) = 6$ on each individual piece. If $c \leq 3$, then $f(c) = c^3 - c$. The equation $c^3 - c = 6$ simplifies as $c^3 - c - 6 = 0$, and factoring gives $(c - 2)(c^2 + 2c + 3) = 0$. The root $c = 2$ satisfies $c \leq 3$, but the roots of $c^2 + 2c + 3 = 0$ are not real, so the only solution in this interval is $c = 2$.

If $3 < c \leq 8$, then $f(c) = 2\sqrt{c}$. If $2\sqrt{c} = 6$, then $c = (6/2)^2 = 9$. This value does not satisfy $3 < c \leq 8$, so there are no solutions in this interval.

Finally, if $c > 8$, then $f(c) = 2|17 - c|$. If $2|17 - c| = 6$, then $|17 - c| = 3$. The solutions to this equation are $c = 14$ and $c = 20$. Both of these values satisfy $c > 8$.

Therefore, the solutions are $c = \boxed{2, 14, \text{ and } 20}$.

16.30 The graphs of f and g intersect when $f(x) = g(x)$, which we solve on each individual piece.

If $x \le -3$, then $f(x) = x^2 - 12$ and $g(x) = -4x$. The equation $x^2 - 12 = -4x$ simplifies as $x^2 + 4x - 12 = 0$, so $(x - 2)(x + 6) = 0$. The only root that satisfies $x \le -3$ is $x = -6$.

If $-3 < x < 2$, then $f(x) = x + 7$ and $g(x) = -4x$. The solution to $x + 7 = -4x$ is $x = -7/5$, which does satisfy $-3 < x < 2$.

If $2 \le x \le 4$, then $f(x) = 5$ and $g(x) = -4x$. The solution to $5 = -4x$ is $x = -5/4$, which does not satisfy $2 \le x \le 4$, so there are no solutions in this interval.

If $4 < x \le 9$, then $f(x) = 5$ and $g(x) = x - 9$. The solution to $5 = x - 9$ is $x = 14$, which does not satisfy $4 < x \le 9$, so there are no solutions in this interval.

If $x = -6$, then $f(x) = g(x) = 24$, and if $x = -7/5$, then $f(x) = g(x) = 28/5$. Hence, the graphs of f and g intersect at $\boxed{(-6, 24) \text{ and } (-7/5, 28/5)}$.

16.31 If $x < 0$, then $x - 9$ is negative, so $\sqrt{(x - 9)^2} = -x + 9$. Then $|x - \sqrt{(x - 9)^2}| = |x - (-x + 9)| = |2x - 9|$. Since $2x - 9$ is negative for $x < 0$, we have $|2x - 9| = \boxed{-2x + 9}$.

16.32 If $x \ge -1$, then $|x + 1| = x + 1$, so the given equation becomes $x(x + 1) - 1 = x - (x + 1)$, which simplifies as $x^2 + x = 0$. The roots are $x = 0$ and $x = -1$, both of which satisfy $x \ge -1$.

If $x < -1$, then $|x + 1| = -x - 1$, so the given equation becomes $x(-x - 1) - 1 = x - (-x - 1)$, which simplifies as $x^2 + 3x + 2 = 0$. Factoring gives $(x + 2)(x + 1) = 0$. The only root satisfying $x < -1$ is $x = -2$.

Therefore, the solutions are $x = \boxed{0, -1, \text{ and } -2}$.

16.33 Let

$$S = \frac{a}{|a|} + \frac{b}{|b|} + \frac{c}{|c|} + \frac{abc}{|abc|}.$$

First, we note that a, b, and c must all be nonzero. If x is positive, then $x/|x| = x/x = 1$. If x is negative, then $x/|x| = x/(-x) = -1$. Hence, each of the terms $a/|a|$, $b/|b|$, $c/|c|$, and $abc/|abc|$ is either 1 or -1. Furthermore,

$$\frac{a}{|a|} \cdot \frac{b}{|b|} \cdot \frac{c}{|c|} = \frac{abc}{|abc|}.$$

In other words, the fourth term is equal to the product of the first three terms. We use symmetry to divide into the following cases:

If a, b, and c are all positive, then $S = 1 + 1 + 1 + 1 = 4$.

If two of a, b, and c are positive and one is negative, then $S = 1 + 1 - 1 - 1 = 0$.

If one of a, b, and c is positive and the other two are negative, then $S = 1 - 1 - 1 + 1 = 0$.

Finally, if a, b, and c are all negative, then $S = -1 - 1 - 1 - 1 = -4$.

Hence, the possible values of S are $\boxed{-4, 0, \text{ and } 4}$.

16.34 Let $f(x) = |x| + |x + 3|$. We write $f(x)$ as a piecewise function:

$$f(x) = \begin{cases} -2x - 3 & \text{if } x \le -3, \\ 3 & \text{if } -3 < x \le 0, \\ 2x + 3 & \text{if } 0 < x. \end{cases}$$

We see that $f(x)$ is decreasing on the interval $(-\infty, -3]$ and increasing on the interval $(0, +\infty)$. Therefore, the minimum value of $f(x)$ is $\boxed{3}$.

16.35   The equation $|x - a| = |x - b|$ states that x is equidistant from both a and b on the real number line. The only real number that satisfies this condition is $x = \boxed{(a + b)/2}$.

16.36   One way to deal with absolute value signs is to take cases. Another way is to square, because $|x|^2 = x^2$ for all real numbers x. Squaring each of the given equations, we get

$$a^2 = b^2 - 4b + 4,$$
$$b^2 = c^2 - 4c + 4,$$
$$c^2 = a^2 - 4a + 4.$$

Adding all three equations, we find that all the quadratic terms cancel, to get $4a + 4b + 4c = 12$. Therefore, $a + b + c = \boxed{3}$. As an additional challenge, find all triples (a, b, c) of real numbers that satisfy the given equations.

16.37   Let $f(x) = |x - 4| + |x + 5|$. Then for a real number a, there exists a real number x such that $f(x) < a$ as long as a is greater than the minimum value of $f(x)$.

To find the minimum value of x, we write $f(x)$ as a piecewise function:

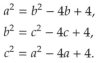

$$f(x) = \begin{cases} -2x - 1 & \text{if } x \le -5, \\ 9 & \text{if } -5 < x \le 4, \\ 2x + 1 & \text{if } 4 < x. \end{cases}$$

We see that $f(x)$ is decreasing on the interval $(-\infty, -5]$ and increasing on the interval $(4, +\infty)$. Therefore, the minimum value of $f(x)$ is 9. Hence, the possible values of a are given by $\boxed{a > 9}$.

16.38   If $x > 0$, then $y = 24/x > 0$, so $|x| + |y| = x + y$, and we have $x + y = 10$. We are also given $xy = 24$, so, by Vieta's Formulas, x and y are the roots of the equation $t^2 - 10t + 24 = 0$. Factoring gives $(t - 4)(t - 6) = 0$, so the solutions in this case are $(x, y) = (4, 6)$ and $(6, 4)$.

If $x < 0$, then $y = 24/x < 0$, so $|x| + |y| = -x - y$, which means we have $x + y = -10$. Combining this with $xy = 24$, Vieta's Formulas tell us that x and y are the roots of the equation $t^2 + 10t + 24 = 0$. Factoring gives $(t + 4)(t + 6) = 0$, so the solutions in this case are $(x, y) = (-4, -6)$ and $(-6, -4)$.

Putting the two cases together, the solutions are

$$(x, y) = \boxed{(4, 6),\ (6, 4),\ (-4, -6),\ \text{and}\ (-6, -4)}.$$

16.39   First, we view $f(x)$ as a piecewise defined function. For $x \ge 0$, we have $f(x) = x|x| + 2 = x^2 + 2$. For $x < 0$, we have $f(x) = x|x| + 2 = x(-x) + 2 = -x^2 + 2$. We find the inverse of each piece separately.

We have $f(x) = x^2 + 2$ for $x \ge 0$, so $f(x)$ returns outputs in the interval $[2, +\infty)$ for inputs in the interval $[0, +\infty)$. Therefore, the inverse of this piece has $[2, +\infty)$ as its domain and $[0, +\infty)$ as its range. We solve $f(y) = x$ for y in terms of x and find $y = \sqrt{x - 2}$. (We don't have $\pm\sqrt{x - 2}$ because the range of this piece of the inverse must be $[0, +\infty)$.)

Next, we have $f(x) = -x^2 + 2$ for $x < 0$, so the other piece of f has domain $(-\infty, 0)$ and range $(-\infty, 2)$. Therefore, the inverse of this piece has $(-\infty, 2)$ as its domain and $(-\infty, 0)$ as its range. We solve $f(y) = x$, for y in terms of x and find $y = -\sqrt{2 - x}$. (We don't have $\pm\sqrt{2 - x}$ because the range of this piece of the inverse must be $(-\infty, 0)$.)

Combining these two pieces, we have

$$f^{-1}(x) = \boxed{\begin{cases} -\sqrt{2 - x} & \text{if } x < 2, \\ \sqrt{x - 2} & \text{if } x \ge 2. \end{cases}}$$

16.40 We get rid of the absolute value signs by considering cases. We ignore the case $x = 0$ because we want nonzero solutions.

- *Case 1: $x > 0$.* When $x > 0$, we have $|x| = x$, so our first equation becomes $x + y = 3$, or $y = 3 - x$. Our second equation becomes $xy + x^3 = 0$. Substitution then gives us

$$x(3 - x) + x^3 = 0 \quad \Rightarrow \quad x^3 - x^2 + 3x = 0 \quad \Rightarrow \quad x(x^2 - x + 3) = 0.$$

Since x must be positive in this case, we disregard the solution $x = 0$. The discriminant of the quadratic is negative, so there are no real solutions in this case.

- *Case 2: $x < 0$.* When $x < 0$, we have $|x| = -x$, so our first equation becomes $-x + y = 3$, or $y = 3 + x$. The second equation becomes $-xy + x^3 = 0$. Substitution then gives us

$$-x(3 + x) + x^3 = 0 \quad \Rightarrow \quad x^3 - x^2 - 3x = 0 \quad \Rightarrow \quad x(x^2 - x - 3) = 0.$$

Again, we disregard the $x = 0$ solution since we are told that x and y are nonzero. The discriminant of the quadratic is positive, and one of the roots of the quadratic is negative, because $-b - \sqrt{b^2 - 4ac}$ is negative. We don't have to find x and y to evaluate $x - y$, however. We can multiply the equation $-x + y = 3$ by -1 to find $x - y = \boxed{-3}$.

16.41

(a) We must have $x^2 + 8x - 3 = 19$ or $x^2 + 8x - 3 = -19$. Rearranging both, we must have $x^2 + 8x - 22 = 0$ or $x^2 + 8x + 16 = 0$. From the quadratic formula, the first equation has solutions $x = -4 \pm \sqrt{38}$. Factoring the second quadratic gives $(x + 4)^2 = 0$, so $x = -4$. All three solutions work, so our solutions are $\boxed{x = -4 - \sqrt{38}, -4, \text{ and }, -4 + \sqrt{38}}$.

(b) Subtracting 3 from both sides gives $|z^2 - z - 18| = z - 3$. So, we must have $z^2 - z - 18 = z - 3$ or $z^2 - z - 18 = -(z - 3)$. Rearranging both, we must have $z^2 - 2z - 15 = 0$ or $z^2 - 21 = 0$. The first gives $z = -3$ or $z = 5$, and the second gives $z = \pm\sqrt{21}$. However, we're not finished. Because we must have $|z^2 - z - 18| = z - 3$, we must have $z - 3 \geq 0$. Therefore, the solutions $z = -3$ and $z = -\sqrt{21}$ are extraneous. Our only solutions are $\boxed{z = 5 \text{ and } \sqrt{21}}$.

(c) Rearranging gives us $|2 - 4r - r^2| \geq -3$. But $|2 - 4r - r^2| \geq 0$ for any value of $2 - 4r - r^2$. Therefore, the inequality is true for $\boxed{\text{all values of } r}$.

(d) We must have $-6 \leq t^2 - 4t + 5 \leq 6$. Completing the square in the middle expression gives us $-6 \leq (t-2)^2 + 1 \leq 6$. Subtracting 1 from all parts of the inequality chain gives $-7 \leq (t - 2)^2 \leq 5$. We have $(t - 2)^2 \geq 0$ for all t, so we only need to solve $(t - 2)^2 \leq 5$. We have $(t - 2)^2 \leq 5$ if and only if $-\sqrt{5} \leq t - 2 \leq \sqrt{5}$, and adding 2 to all parts of this inequality chain gives us $\boxed{2 - \sqrt{5} \leq t \leq 2 + \sqrt{5}}$.

16.42 Let $A = (a, |a|)$ and $B = (b, |b|)$. If both A and B lie on the positive branch of $y = |x|$, then the midpoint of $\overline{AB}$ would also lie on the positive branch, but the point $(3, 4)$ does not. Similarly, A and B cannot both lie on the negative branch. Therefore, one point lies on the positive branch, and the other lies on the negative branch. Suppose $a \geq 0$ and $b \leq 0$. (The case $a \leq 0$ and $b \geq 0$ is essentially the same.) Then $|a| = a$ and $|b| = -b$, so $A = (a, a)$ and $B = (b, -b)$.

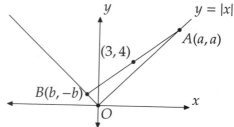

Then the midpoint of $\overline{AB}$ is given by $((a+b)/2, (a-b)/2)$, so $(a+b)/2 = 3$ and $(a - b)/2 = 4$. Solving for a and b, we find $a = 7$ and $b = -1$, so $A = (7, 7)$ and $B = (-1, 1)$. By the distance formula, the length of $\overline{AB}$ is $\sqrt{(7 + 1)^2 + (7 - 1)^2} = \sqrt{64 + 36} = \sqrt{100} = \boxed{10}$.

Another solution is as follows. As above, one point lies on the positive branch of $y = |x|$, and the other lies on the negative branch, so $\angle AOB = 90°$. In such a case, the midpoint of $\overline{AB}$ is also the circumcenter of triangle AOB. The circumradius R of the triangle is then the distance between the origin and $(3, 4)$, which is $R = \sqrt{3^2 + 4^2} = \sqrt{25} = 5$, so $AB = 2R = 10$.

16.43 Since $31^2 = 961$ and $32^2 = 1024$, we have $31 < \sqrt{999} < 32$. Then $\log_2 31 < \log_2 \sqrt{999} < \log_2 32 = 5$. Since $\log_2 31 > \log_2 16 = 4$, we have $4 < \log_2 \sqrt{999} < 5$. Therefore, $\lfloor \log_2 \sqrt{999} \rfloor = \boxed{4}$.

16.44 If $\lfloor \sqrt{2r - 7} \rfloor = 9$, then $9 \le \sqrt{2r - 7} < 10$. Squaring everything, we get $81 \le 2r - 7 < 100$, which becomes $\boxed{44 \le r < 107/2}$.

16.45 Let $n = \lfloor x \rfloor$. If $x = n$, then $\lfloor -x \rfloor = \lfloor -n \rfloor = -n$, since $-n$ is an integer, and the given equation becomes $1990n - 1989n = 1$, so $n = 1$.

Otherwise, x lies strictly between the integers n and $n + 1$. Hence, $-(n + 1) < -x < -n$, so $\lfloor -x \rfloor = -(n + 1)$, and the given equation becomes $1990n + 1989(-n - 1) = 1$, so $n - 1989 = 1$, which gives $n = 1990$. Hence, the solution in this case is all x such that $1990 < x < 1991$.

Therefore, the solutions are given by $\boxed{x = 1 \text{ and } 1990 < x < 1991}$.

16.46 First, we note that if $x < -1$, then $x^3 - \lfloor x \rfloor < 0$, so we don't have to check for solutions less than -1. Next, for $-1 \le x < 0$, we have $x^3 - \lfloor x \rfloor < 0 - (-1) = 1$, so there are no solutions in this case, either. We therefore only have to check positive values of x.

If $0 \le x < 1$, then $x^3 - \lfloor x \rfloor < 1 - 0 = 1$, so there are no solutions in this case.

If $1 \le x < 2$, then $x^3 - \lfloor x \rfloor < 8 - 1 = 7$, so there are no solutions in this case.

If $2 \le x < 3$, then $x^3 - \lfloor x \rfloor < 27 - 2 = 25$, so there are no solutions in this case.

If $3 \le x < 4$, then $x^3 - \lfloor x \rfloor < 64 - 3 = 61$, so there might be solutions in this case. For $3 \le x < 4$, we have $\lfloor x \rfloor = 3$, so our equation is $x^3 = 50$, which gives us $x = \sqrt[3]{50}$.

Intuitively, it seems obvious that there are no solutions for $x \ge 4$. To prove it, we rewrite the equation as $x^3 = 47 + \lfloor x \rfloor$. Since $\lfloor x \rfloor < x + 1$, we have $47 + \lfloor x \rfloor < 48 + x$. For $x \ge 4$, we have $x^3 \ge 16x$. We also have $16x > 48 + x$ for all $x \ge 4$. Therefore, if $x \ge 4$, we have $x^3 \ge 16x \ge 48 + x > 47 + \lfloor x \rfloor$, which means we cannot have $x^3 = 47 + \lfloor x \rfloor$ for $x \ge 4$. So, the only solution is $x = \boxed{\sqrt[3]{50}}$

16.47 Let $n = \lfloor x \rfloor$. If $x = n$, then $\lceil x \rceil = n$, and $-\lfloor -x \rfloor = -\lfloor -n \rfloor = -(-n) = n$.

Otherwise, x lies strictly between n and $n + 1$, so $-x$ lies strictly between $-(n + 1)$ and $-n$. Then $\lceil x \rceil = n + 1$, and $-\lfloor -x \rfloor = -(-(n + 1)) = n + 1$. Hence, $\lceil x \rceil = -\lfloor -x \rfloor$ for all real numbers x.

16.48 First, we graph the functions $y = 4x + 1$, $y = x + 2$, and $y = -2x + 4$. We see that

$$f(x) = \begin{cases} 4x + 1 & \text{if } x \le 1/3, \\ x + 2 & \text{if } 1/3 \le x \le 2/3, \\ -2x + 4 & \text{if } 2/3 \le x. \end{cases}$$

On the interval $(-\infty, 1/3]$, we have $f(x) = 4x + 1$, so $f(x)$ is increasing. On the interval $[1/3, 2/3]$, we have $f(x) = x + 2$, so $f(x)$ is increasing. Finally, on the interval $[2/3, +\infty)$, $f(x) = -2x + 4$, so $f(x)$ is decreasing. Therefore, the maximum value of $f(x)$ is $f(2/3) = \boxed{8/3}$.

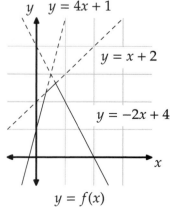

Challenge Problems

16.49 There are $91 - 19 + 1 = 73$ terms on the left side, and each must be an integer. Let $n = \lfloor r + 19/100 \rfloor$. The difference between $r + 19/100$ and $r + 91/100$ is $(91 - 19)/100 = 72/100$, which is less than 1, so the last term $\lfloor r + 91/100 \rfloor$ is equal to either n or $n + 1$.

If the last term is equal to n, then all the terms must be equal to n, and we get $73n = 546$. However, 546 is not divisible by 73, so the last term must be equal to $n + 1$.

Among the 73 terms, let the number of terms that are equal to n be k. Then the remaining $73 - k$ terms are equal to $n + 1$. Hence, we have $kn + (73 - k)(n + 1) = 546$. This equation simplifies as $73n - k = 473$. Since $1 \le k \le 72$, we know that $73n$ must be the smallest multiple of 73 that is greater than 473, which is $7 \cdot 73 = 511$, so $n = 7$, and $k = 511 - 473 = 38$.

Hence, $\left\lfloor r + \dfrac{19}{100} \right\rfloor, \left\lfloor r + \dfrac{20}{100} \right\rfloor, \ldots, \left\lfloor r + \dfrac{56}{100} \right\rfloor$ are all equal to 7, and $\left\lfloor r + \dfrac{57}{100} \right\rfloor, \left\lfloor r + \dfrac{58}{100} \right\rfloor, \ldots, \left\lfloor r + \dfrac{91}{100} \right\rfloor$ are all equal to 8.

Since $\lfloor r + 56/100 \rfloor = 7$ and $\lfloor r + 57/100 \rfloor = 8$, we have $r + \frac{56}{100} < 8 \le r + \frac{57}{100}$. Multiplying everything by 100, we get $100r + 56 < 800 \le 100r + 57$. Hence, $743 \le 100r < 744$, so $\lfloor 100r \rfloor = \boxed{743}$.

16.50 Completing the square, we get $n^2 - 10n + 29 = (n - 5)^2 + 4$. Then $n^2 - 10n + 29 > (n - 5)^2$. The next square after $(n - 5)^2$ is $(n - 4)^2$. We have $(n - 4)^2 - (n - 5)^2 = 2n - 9$. So, when $n = 19941994$, these two squares are much more than 4 apart. This means we have $(n - 5)^2 < n^2 - 10n + 29 < (n - 4)^2$ for $n = 19941994$, which means $n - 5 < \sqrt{n^2 - 10n + 29} < n - 4$. Therefore, we have $\lfloor \sqrt{n^2 - 10n + 29} \rfloor = n - 5 = \boxed{19941989}$.

16.51 The condition $|\sqrt{x} - 2| < 0.1$ is equivalent to $-0.1 < \sqrt{x} - 2 < 0.1$, or $1.9 < \sqrt{x} < 2.1$. Squaring, we get $3.61 < x < 4.41$, so $-0.39 < x - 4 < 0.41$. Therefore, the largest δ we can take is $\boxed{0.39}$.

16.52 We can rewrite the given equation as $|y| = |x/4| - |x - 60|$. Let $f(x) = |x/4| - |x - 60|$, so the given equation is $|y| = f(x)$. We write $f(x)$ as a piecewise function:

$$f(x) = \begin{cases} \dfrac{3x}{4} - 60 & \text{if } x \le 0, \\[2mm] \dfrac{5x}{4} - 60 & \text{if } 0 < x \le 60, \\[2mm] 60 - \dfrac{3x}{4} & \text{if } 60 < x. \end{cases}$$

In order for there to be a solution to $|y| = f(x)$, we must have $f(x) \ge 0$. The first piece of $f(x)$ is negative for all x in its domain. Graphing the portions of $f(x)$ for which $f(x) \ge 0$ gives us the graph at left below.

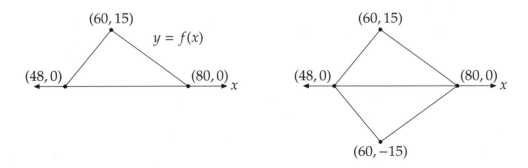

However, for every point $(f(x), y)$ on the graph of $|y| = f(x)$, the point $(f(x), -y)$ is also on the graph, because $|y| = |-y|$. The point $(f(x), -y)$ is the reflection of $(f(x), y)$ over the x-axis, so the graph at right above is the graph of $|y| = |x/4| - |x - 60|$. The region enclosed by the graph is composed of two triangles, with base $80 - 48 = 32$ and height 15. Therefore, its area is

$$2 \cdot \frac{1}{2} \cdot 32 \cdot 15 = \boxed{480}.$$

16.53 First, we note that y cannot be 0, and y cannot be negative, because the left side of the inequality is always at least 0. Hence, y must be positive. And since y is positive, $y + 3$ is positive. Hence, we can write the given inequality as

$$\left|\frac{2 - 3y}{3 + y}\right| = \frac{|3y - 2|}{|y + 3|} = \frac{|3y - 2|}{y + 3} \le \frac{1}{y}.$$

If $0 < y \le 2/3$, then $3y - 2 \le 0$, and $|3y - 2| = 2 - 3y$, and the inequality becomes

$$\frac{2 - 3y}{y + 3} \le \frac{1}{y}.$$

Since both y and $y + 3$ are positive, we can multiply both sides by $y(y + 3)$ to get $y(2 - 3y) \le y + 3$, which simplifies as $3y^2 - y + 3 \ge 0$. The discriminant of $3y^2 - y + 3$ is $(-1)^2 - 4 \cdot 3 \cdot 3 = -35$, which is negative. Hence, $3y^2 - y + 3 \ge 0$ for all y. Therefore, all real numbers in the interval $0 < y \le 2/3$ satisfy the given inequality.

If $y > 2/3$, then $3y - 2 \ge 0$, and $|3y - 2| = 3y - 2$, and the inequality becomes

$$\frac{3y - 2}{y + 3} \le \frac{1}{y}.$$

Both y and $y + 3$ are positive, so we can multiply both sides by $y(y + 3)$ to get $y(3y - 2) \le y + 3$, which simplifies as $3(y^2 - y - 1) \le 0$. By the quadratic formula, the roots of $y^2 - y - 1$ are

$$y = \frac{1 \pm \sqrt{5}}{2}.$$

Hence, $y^2 - y - 1 \le 0$ for $(-1 + \sqrt{5})/2 \le y \le (1 + \sqrt{5})/2$. The real numbers y that satisfy this condition and $y > 2/3$ are given by $2/3 < y \le (1 + \sqrt{5})/2$.

Therefore, the solutions in y are given by $\boxed{0 < y \le (1 + \sqrt{5})/2}$.

16.54 To help visualize the problem, we graph $S(x)$ and $g(x)$ below.

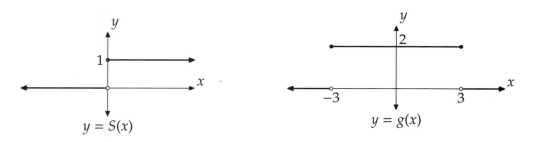

$$y = S(x) \qquad\qquad y = g(x)$$

To express $g(x)$ in terms of $S(x)$, we can begin by translating $S(x)$ to the left three units, and multiplying by 2. The resulting function $y = 2S(x + 3)$ produces the same discontinuity at $x = -3$ as we see in $g(x)$.

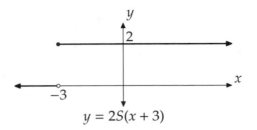

$$y = 2S(x + 3)$$

Similarly, the function $y = 2S(3 - x)$ produces the same discontinuity at $x = 3$ as we see in $g(x)$.

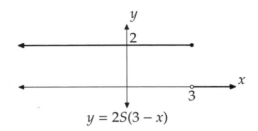

$$y = 2S(3 - x)$$

Hence, including both $S(x + 3)$ and $S(3 - x)$ as factors results in a function that includes both discontinuities:

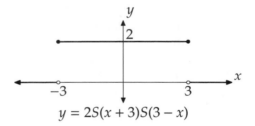

$$y = 2S(x + 3)S(3 - x)$$

Thus, we can take $g(x) = \boxed{2S(x + 3)S(3 - x)}$. An alternative answer is $g(x) = 2S(x + 3) + 2S(3 - x) - 2$.

16.55 The function $f(x)$ is defined for all x such that both $64 - x^2 \geq 0$ and $16 - |2x + 5| > 0$.

The inequality $64 - x^2 \geq 0$ holds if and only if $-8 \leq x \leq 8$. The inequality $16 - |2x + 5| > 0$ holds if and only if $|2x + 5| < 16$. Dividing both sides by 2, we get $|x + 5/2| < 8$. Therefore, $-8 < x + \frac{5}{2} < 8$, or $-\frac{21}{2} < x < \frac{11}{2}$. Hence, $64 - x^2 \geq 0$ and $16 - |2x + 5| > 0$ if and only if $\boxed{-8 \leq x < \frac{11}{2}}$.

16.56 Adding all three equations, we get $x + \lfloor x \rfloor + \{x\} + y + \lfloor y \rfloor + \{y\} + z + \lfloor z \rfloor + \{z\} = 7.38$. Since $x = \lfloor x \rfloor + \{x\}$, $y = \lfloor y \rfloor + \{y\}$, and $z = \lfloor z \rfloor + \{z\}$, this equation becomes $2x + 2y + 2z = 7.38$. Multiplying both sides by 50, we get $100x + 100y + 100z = \boxed{369}$.

16.57 Let $n = \lfloor x \rfloor$. Then $x = \lfloor x \rfloor + \{x\} = n + \{x\}$ and $x - \lfloor x \rfloor = \{x\}$. Hence, $n + \{x\}$, n, $\{x\}$ form a geometric sequence, which means $n^2 = (n + \{x\})\{x\} = \{x\}^2 + n\{x\}$, so $\{x\}^2 + n\{x\} - n^2 = 0$. Solving for $\{x\}$ using the quadratic formula, we get

$$\{x\} = \frac{-n \pm \sqrt{5n^2}}{2} = \frac{-1 \pm \sqrt{5}}{2} \, n.$$

Since $\{x\} \geq 0$, the $\pm$ sign must be a plus sign, so $\{x\} = \frac{-1 + \sqrt{5}}{2} \, n$. Furthermore, $\{x\} < 1$, and $\frac{-1 + \sqrt{5}}{2} \cdot 2 = -1 + \sqrt{5} > 1$.

Hence, the only possible value of n is $n = 1$, in which case $\{x\} = \frac{-1 + \sqrt{5}}{2}$. Therefore, $x = n + \{x\} = 1 + \frac{-1 + \sqrt{5}}{2} = \boxed{\frac{1 + \sqrt{5}}{2}}$.

16.58 Let $x = 19n + r$, where n and r are integers, and $0 \le r \le 18$. (In other words, when x is divided by 19, n is the quotient and r is the remainder.) Then

$$\lfloor x \rfloor - 19 \cdot \left\lfloor \frac{x}{19} \right\rfloor = 19n + r - 19 \cdot \left\lfloor \frac{19n + r}{19} \right\rfloor = 19n + r - 19 \cdot \left\lfloor n + \frac{r}{19} \right\rfloor = 19n + r - 19n = r.$$

Similarly, let $x = 89m + s$, where m and s are integers, and $0 \le s \le 88$. (In other words, when x is divided by 89, m is the quotient and s is the remainder.) Then

$$\lfloor x \rfloor - 89 \cdot \left\lfloor \frac{x}{89} \right\rfloor = 89m + s - 89 \cdot \left\lfloor \frac{89m + s}{89} \right\rfloor = 89m + s - 89 \cdot \left\lfloor m + \frac{s}{89} \right\rfloor = 89m + s - 89m = s.$$

The given equation then gives us $r = s = 9$. Hence, x is of the form $19n + 9$ and $89m + 9$, where m and n are integers. In other words, $x - 9$ is divisible by both 19 and 89. Therefore, $x - 9$ is divisible by $\text{lcm}(19, 89) = 19 \cdot 89 = 1691$. Therefore, the smallest possible value for x is $1691 + 9 = \boxed{1700}$.

16.59 First, we graph at right the piecewise defined function described in the problem.

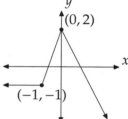

Since this function is piecewise linear, we will look for a solution where $f(x)$, $g(x)$, and $h(x)$ are all linear. The graph of this function exhibits "corners" at $x = -1$ and $x = 0$, so the graph of $|f(x)| - |g(x)| + h(x)$ must also do the same. Since the graphs of polynomials are always smooth, the only way to generate a corner is via the absolute value function.

The graph of $y = |cx + d|$, where c and d are constants, has a "corner" at the value of x such that $cx + d = 0$. Our graph has a corner at $x = 0$ and a corner at $x = -1$, so we want one of $f(x)$ and $g(x)$ to be 0 at $x = -1$, and the other to be 0 at $x = 0$. We look at the case where $f(x) = 0$ at $x = -1$ and $g(x) = 0$ at $x = 0$.

Since $f(x)$ is linear, we have $f(x) = a(x + 1)$ for some constant a. Furthermore, since we are taking the absolute value of $f(x)$, we can assume that $a \ge 0$. Similarly, $g(x) = bx$ for some constant $b \ge 0$. Thus,

$$|f(x)| - |g(x)| + h(x) = a|x + 1| - b|x| + h(x) = \begin{cases} (b - a)x - a + h(x) & \text{if } x < -1, \\ (a + b)x + a + h(x) & \text{if } -1 \le x \le 0, \\ (a - b)x + a + h(x) & \text{if } 0 < x. \end{cases}$$

Hence, we want a, b, and $h(x)$ to satisfy

$$(b - a)x - a + h(x) = -1,$$
$$(a + b)x + a + h(x) = 3x + 2,$$
$$(a - b)x + a + h(x) = -2x + 2.$$

Adding the first equation and third equation, we get $2h(x) = -2x + 1$, so $h(x) = -x + \frac{1}{2}$. Substituting this expression for $h(x)$ into the above equations and simplifying, we get

$$(b - a)x - a = x - \frac{3}{2},$$
$$(a + b)x + a = 4x + \frac{3}{2},$$
$$(a - b)x + a = -x + \frac{3}{2}.$$

Hence, we want a and b to satisfy the equations $b - a = 1$, $a = \frac{3}{2}$, and $a + b = 4$. This system has the solution $a = \frac{3}{2}$ and $b = \frac{5}{2}$. Thus, we can take

$$\boxed{f(x) = \frac{3}{2}(x + 1), \quad g(x) = \frac{5}{2}x, \quad h(x) = -x + \frac{1}{2}}.$$

16.60 If $\lfloor \sqrt{x} \rfloor = 12$, then $12 \le \sqrt{x} < 13$, so $144 \le x < 169$. If $\lfloor \sqrt{100x} \rfloor = 120$, then $120 \le \sqrt{100x} < 121$, so $14400 \le 100x < 14641$, or $144 \le x < 146.41$. Therefore, the probability that $\lfloor \sqrt{100x} \rfloor = 120$ given that $\lfloor \sqrt{x} \rfloor = 12$ is $(146.41 - 144)/(169 - 144) = \boxed{241/2500}$.

16.61 There are n terms on the right side, and each must be an integer. Let $a = \lfloor x \rfloor$. The difference between $x + (n-1)/n$ and x is $(n-1)/n$, which is less than 1, so the last term $\lfloor x + (n-1)/n \rfloor$ is equal to either a or $a+1$.

Let k be the number of terms on the right side that are equal to a, so the remaining $n-k$ terms are equal to $a+1$. Then the right side is equal to

$$ka + (n-k)(a+1) = ka + na + n - ka - k = na + n - k.$$

Furthermore,

$$\lfloor x \rfloor, \quad \left\lfloor x + \frac{1}{n} \right\rfloor, \quad \ldots, \quad \left\lfloor x + \frac{k-1}{n} \right\rfloor$$

are all equal to a, and

$$\left\lfloor x + \frac{k}{n} \right\rfloor, \quad \left\lfloor x + \frac{k+1}{n} \right\rfloor, \quad \ldots, \quad \left\lfloor x + \frac{n-1}{n} \right\rfloor$$

are all equal to $a+1$.

Since $\lfloor x + (k-1)/n \rfloor = a$ and $\lfloor x + k/n \rfloor = a+1$, we have

$$x + \frac{k-1}{n} < a+1 \le x + \frac{k}{n}.$$

Multiplying everything by n, we get $nx + k - 1 < na + n \le nx + k$. Hence, $na + n - k \le nx < na + n - k + 1$, so $\lfloor nx \rfloor = na + n - k$. Therefore, the two sides are equal, as desired.

16.62 By the identity in the previous problem, we have $\lfloor 2x \rfloor = \lfloor x \rfloor + \left\lfloor x + \frac{1}{2} \right\rfloor$ and $\lfloor 2y \rfloor = \lfloor y \rfloor + \left\lfloor y + \frac{1}{2} \right\rfloor$.

Thus, the given inequality is equivalent to

$$\lfloor x \rfloor + \left\lfloor x + \frac{1}{2} \right\rfloor + \lfloor y \rfloor + \left\lfloor y + \frac{1}{2} \right\rfloor \ge \lfloor x \rfloor + \lfloor y \rfloor + \lfloor x + y \rfloor,$$

or $\left\lfloor x + \frac{1}{2} \right\rfloor + \left\lfloor y + \frac{1}{2} \right\rfloor \ge \lfloor x + y \rfloor$.

To make this expression easier to work with, let $a = x + \frac{1}{2}$ and $b = y + \frac{1}{2}$. Then $x = a - \frac{1}{2}$ and $y = b - \frac{1}{2}$, so $x + y = a + b - 1$, and the inequality becomes $\lfloor a \rfloor + \lfloor b \rfloor \ge \lfloor a + b - 1 \rfloor$. Since $\lfloor a + b - 1 \rfloor = \lfloor a + b \rfloor - 1$, we can rewrite this inequality as $\lfloor a + b \rfloor \le \lfloor a \rfloor + \lfloor b \rfloor + 1$. In the text, we showed that $\lfloor a + b \rfloor$ is equal to either $\lfloor a \rfloor + \lfloor b \rfloor$ or $\lfloor a \rfloor + \lfloor b \rfloor + 1$. Hence, $\lfloor a + b \rfloor \le \lfloor a \rfloor + \lfloor b \rfloor + 1$ holds, which means the original inequality holds.

16.63 If the point (x, y) lies in S, then so do the points $(-x, y)$, $(x, -y)$, and $(-x, -y)$. Hence, we can find the portion of S that lies in the upper right quadrant of the Cartesian plane, and then reflect over the x- and y-axes to obtain the rest of S.

If (x, y) lies in the first quadrant, then $x \ge 0$ and $y \ge 0$, so $|x| = x$ and $|y| = y$. Then the given equation becomes $\left| |x - 2| - 1 \right| + \left| |y - 2| - 1 \right| = 1$.

Let $u = x - 2$ and $v = y - 2$. Then the equation becomes $\left| |u| - 1 \right| + \left| |v| - 1 \right| = 1$. Again, we see that in the uv-plane, if the point (u, v) satisfies this equation, then so do the points $(u, -v)$, $(-u, v)$, and $(-u, -v)$. Hence, we may take the points that satisfy $u \ge 0$ and $v \ge 0$, and then reflect over the u- and v-axes. (Note that this is equivalent to reflecting over the lines $x = 2$ and $y = 2$ in the xy-plane.)

If $u \ge 0$ and $v \ge 0$, then $|u| = u$ and $|v| = v$, so the equation becomes $|u - 1| + |v - 1| = 1$.

Let $a = u - 1$ and $b = v - 1$. Then the equation becomes $|a| + |b| = 1$. Again we use symmetry. In the ab-plane, if the point (a, b) satisfies this equation, then so do the points $(a, -b)$, $(-a, b)$, and $(-a, -b)$. Hence, we may take the points that satisfy $a \ge 0$ and $b \ge 0$, and then reflect over the a- and b-axes.

If $a \geq 0$ and $b \geq 0$, then $|a| = a$ and $|b| = b$, so the equation becomes $a + b = 1$. The portion of this line that satisfies $a \geq 0$ and $b \geq 0$ is the portion that joins $(0, 1)$ to $(1, 0)$. Reflecting this portion in the a- and b-axes, we obtain a square, with side length $\sqrt{2}$.

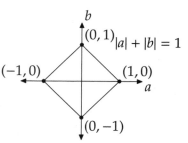

Now we unwind the substitutions that we have made. The square centered at $(0, 0)$ in the ab-plane becomes a square centered at $(1, 1)$ in the uv-plane. Reflecting this square in the u- and v-axes, we obtain four squares. Then translating these four squares up and to the right by 2 units (because $x = u + 2$ and $y = v + 2$), we obtain the portion of the original graph that is in the first quadrant of the xy-plane. Finally, reflecting these four squares in the x- and y-axes, we obtain a graph that can be built with 16 such squares.

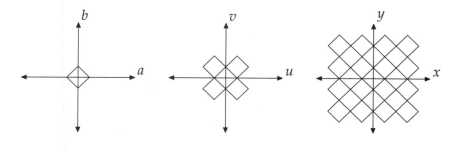

Therefore, the length of wire required is $16 \cdot 4 \cdot \sqrt{2} = \boxed{64\sqrt{2}}$.

16.64

(a) First, we note that $f(x)$ is increasing for $x \geq 1$. Also, if x is a positive integer, then $f(x) = x^4$. Since $6^4 = 1296 < 2001 < 7^4 = 2401$, a solution to $f(x) = 2001$ must satisfy $6 < x < 7$. Then $\lfloor x \rfloor = 6$, so $x\lfloor x \rfloor < 7 \cdot 6 = 42$, which means $\lfloor x\lfloor x\rfloor\rfloor \leq 41$. Since $x < 7$, we have $x\lfloor x\lfloor x\rfloor\rfloor < 7 \cdot 41 = 287$, so $\lfloor x\lfloor x\lfloor x\rfloor\rfloor\rfloor \leq 286$.

Now, since $f(x) = x\lfloor x\lfloor x\lfloor x\rfloor\rfloor\rfloor = 2001$, we have

$$\lfloor x\lfloor x\lfloor x\rfloor\rfloor\rfloor = \frac{2001}{x}.$$

Since $x < 7$, we have

$$\lfloor x\lfloor x\lfloor x\rfloor\rfloor\rfloor = \frac{2001}{x} > \frac{2001}{7} = 285 + \frac{6}{7}.$$

Since $\lfloor x\lfloor x\lfloor x\rfloor\rfloor\rfloor$ is an integer, we have $\lfloor x\lfloor x\lfloor x\rfloor\rfloor\rfloor \geq 286$. We also have $\lfloor x\lfloor x\lfloor x\rfloor\rfloor\rfloor \leq 286$ from above, so $\lfloor x\lfloor x\lfloor x\rfloor\rfloor\rfloor = 286$, which means $x = \boxed{2001/286}$. It is easy to check that this solution works.

(b) If $f(x) = 2002$, then by the same reasoning as in part (a), we have $\lfloor x\lfloor x\lfloor x\rfloor\rfloor\rfloor \leq 286$. However, we must also have

$$\lfloor x\lfloor x\lfloor x\rfloor\rfloor\rfloor = \frac{2002}{x} > \frac{2002}{7} = 286.$$

These inequalities cannot both be satisfied, so there is no x such that $f(x) = 2002$.

16.65 Since $\{x\} = x - \lfloor x \rfloor$, the first equation becomes $\{x\}^2 + y^2 = \{x\}$. Completing the square in $\{x\}$, we get

$$\left(\{x\} - \frac{1}{2}\right)^2 + y^2 = \frac{1}{4}.$$

This looks like the equation for a circle, but we want an equation in x, not $\{x\}$.

Let $n = \lfloor x \rfloor$, which is an integer. Then $\{x\} = x - n$, so the equation becomes

$$\left(x - n - \frac{1}{2}\right)^2 + y^2 = \frac{1}{4}.$$

This equation represents a circle centered at $(n + \frac{1}{2}, 0)$ with radius 1/2. Hence, the first original equation represents a family of circles, one for each integer n. These circles, along with the line $y = x/5$ for the second original equation, are graphed below.

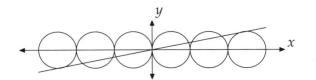

The line clearly intersects the four circles closest to the y-axis (two to the right, two to the left) at a total of seven points. To check whether it intersects the third circle to the right of the y-axis, we solve the system of equations

$$y = \frac{x}{5},$$
$$\left(x - \frac{5}{2}\right)^2 + y^2 = \frac{1}{4}.$$

Substituting the first equation into the second equation, and simplifying, we get $26x^2 - 125x + 150 = 0$, so $(2x - 5)(13x - 30) = 0$, which has solutions $x = 5/2$ and $x = 30/13$. When $x = 5/2$, we have $y = x/5 = 1/2$, and when $x = 30/13$, we have $y = x/5 = 6/13$. Hence, the line $y = x/5$ intersects the rightmost circle in the figure above at the two points $(5/2, 1/2)$ and $(30/13, 6/13)$.

We must check if the line $y = x/5$ intersects the fourth circle to the right of the y-axis, which is the circle with center $(7/2, 0)$ and radius 1/2. For every point (x, y) on this circle, we have $x \geq 3$ and $y \leq 1/2$. However, if (x, y) lies on the line $y = x/5$ and $x \geq 3$, then $y \geq 3/5 > 1/2$. Thus, the line and the fourth circle cannot intersect. For the same reason, the line $y = x/5$ cannot intersect any circles farther to the right of the y-axis.

By symmetry, the line also intersects the third circle to the left of the y-axis at two points, and no others. Therefore, the line intersects the circles at a total of $\boxed{11}$ points.

16.66 Since $2 < a^2 < 3$, we have $\lfloor a^2 \rfloor = 2$. Therefore, $\{a^2\} = a^2 - \lfloor a^2 \rfloor = a^2 - 2$. Since $a^2 > 2 > 1$ and a is positive, we have $0 < a^{-1} < 1$. Therefore, $\{a^{-1}\} = a^{-1} - \lfloor a^{-1} \rfloor = a^{-1}$. Hence, we have $a^{-1} = a^2 - 2$, which simplifies as $a^3 - 2a - 1 = 0$. This factors as $(a + 1)(a^2 - a - 1) = 0$. Since $a^2 > 2$, we have $a \neq -1$, so $a^2 - a - 1 = 0$.

At this point, we can solve for a and calculate $a^{12} - 144a^{-1}$, but we can also use the relations $a^3 = 2a + 1$ and $a^2 = a + 1$. Squaring $a^3 = 2a + 1$, we get

$$a^6 = (2a + 1)^2 = 4a^2 + 4a + 1 = 4(a + 1) + 4a + 1 = 8a + 5.$$

Squaring again, we get

$$a^{12} = (8a + 5)^2 = 64a^2 + 80a + 25 = 64(a + 1) + 80a + 25 = 144a + 89.$$

Hence,

$$a^{12} - 144a^{-1} = 144a + 89 - \frac{144}{a} = \frac{144a^2 + 89a - 144}{a} = \frac{144(a + 1) + 89a - 144}{a} = \frac{233a}{a} = \boxed{233}.$$

The numbers 5, 8, 89, 144, and 233 are all Fibonacci numbers. Is this a coincidence?

16.67 To compute the given sum, we count the number of positive integers k such that $f(k) = n$, for each positive integer n.

If $f(k) = n$, then the closest integer to $\sqrt[4]{k}$ is n. In other words, $n - \frac{1}{2} < \sqrt[4]{k} < n + \frac{1}{2}$. Taking everything to the fourth power, we get

$$n^4 - 2n^3 + \frac{3}{2}n^2 - \frac{1}{2}n + \frac{1}{16} < k < n^4 + 2n^3 + \frac{3}{2}n^2 + \frac{1}{2}n + \frac{1}{16}.$$

Note that

$$n^4 - 2n^3 + \frac{3}{2}n^2 - \frac{1}{2}n = n^4 - 2n^3 + \frac{n(3n-1)}{2}$$

is an integer, since either n or $3n - 1$ must be even. Similarly,

$$n^4 + 2n^3 + \frac{3}{2}n^2 + \frac{1}{2}n = n^4 + 2n^3 + \frac{n(3n+1)}{2}$$

is also an integer. Hence, the condition on k above is equivalent to

$$n^4 - 2n^3 + \frac{3}{2}n^2 - \frac{1}{2}n + 1 \le k \le n^4 + 2n^3 + \frac{3}{2}n^2 + \frac{1}{2}n.$$

The number of integers k in this interval is

$$\left(n^4 + 2n^3 + \frac{3}{2}n^2 + \frac{1}{2}n\right) - \left(n^4 - 2n^3 + \frac{3}{2}n^2 - \frac{1}{2}n\right) = 4n^3 + n.$$

We can then build a table as follows:

n	Integers k such that $f(k) = n$	Number of integers k
1	$1 \le k \le 5$	5
2	$6 \le k \le 39$	34
3	$40 \le k \le 150$	111
4	$151 \le k \le 410$	260
5	$411 \le k \le 915$	505
6	$916 \le k \le 1785$	870
7	$1786 \le k \le 1995$	210

Note that for $n = 7$, we cut off the interval for k at 1995, because this is the upper bound in the given sum.

Hence,

$$\sum_{k=1}^{1995} \frac{1}{f(k)} = \frac{5}{1} + \frac{34}{2} + \frac{111}{3} + \frac{260}{4} + \frac{505}{5} + \frac{870}{6} + \frac{210}{7} = \boxed{400}.$$

16.68 Let $n = 2m + b_0$, where $b_0 = 0$ if n is even, and $b_0 = 1$ if n is odd. (The notation b_0 is meant to denote the last digit of n when written in binary.) So if n is even, then

$$f(n) = f(2m) = \frac{2m}{2} = m.$$

If n is odd, then

$$f(n) = f(2m + 1) = \frac{2m + 1 + 1023}{2} = m + 512.$$

Therefore, because $b_0 = 0$ when n is even and $b_0 = 1$ when n is odd, we can write $f(n)$ as $m + 512b_0$ for all n.

Then to deal with $f(f(f(f(f(n)))))$ (where we are applying f five times), we write n in the form

$$n = 32m + 16b_4 + 8b_3 + 4b_2 + 2b_1 + b_0,$$

where b_4, b_3, b_2, b_1, and b_0 are the last five digits of n when written in binary. Then the last digit of n in binary is b_0, and the remaining "even part" of n is $32m + 16b_4 + 8b_3 + 4b_2 + 2b_1$, so

$$f(n) = \frac{32m + 16b_4 + 8b_3 + 4b_2 + 2b_1}{2} + 512b_0 = 16m + 8b_4 + 4b_3 + 2b_2 + b_1 + 512b_0.$$

We see that the last digit of $f(n)$ is b_1, and the remaining "even part" of $f(n)$ is $16m + 8b_4 + 4b_3 + 2b_2 + 512b_0$. Hence, we have $f^2(n) = 8m + 4b_4 + 2b_3 + b_2 + 512b_1 + 256b_0$.

Then the last digit of $f^2(n)$ in binary is b_2, and so on. We can continue as follows:

$$f^3(n) = 4m + 2b_4 + b_3 + 512b_2 + 256b_1 + 128b_0,$$
$$f^4(n) = 2m + b_4 + 512b_3 + 256b_2 + 128b_1 + 64b_0,$$
$$f^5(n) = m + 512b_4 + 256b_3 + 128b_2 + 64b_1 + 32b_0.$$

Let $b = 16b_4 + 8b_3 + 4b_2 + 2b_1 + b_0$. Then $n = 32m + 16b_4 + 8b_3 + 4b_2 + 2b_1 + b_0 = 32m + b$, and

$$f^5(n) = m + 512b_4 + 256b_3 + 128b_2 + 64b_1 + 32b_0 = m + 32(16b_4 + 8b_3 + 4b_2 + 2b_1 + b_0) = m + 32b.$$

Hence, if $f^5(n) = n$, then $m + 32b = 32m + b$, so $m = b$, which means $n = 32m + b = 33b$. Therefore, the smallest positive integer n for which $f^5(n) = n$ is $n = \boxed{33}$.

16.69 First, we determine an integer that is close to the given expression. Since 10^{20000} is 10^{100} to the power of 200, we can try to simplify the expression by using a difference of 200th powers factorization. Let $x = 10^{100}$ and $y = -3$. Then

$$\frac{10^{20000} - 3^{200}}{10^{100} + 3} = \frac{x^{200} - y^{200}}{x - y} = x^{199} + x^{198}y + x^{197}y^2 + \cdots + y^{199}$$

is an integer.

Furthermore, we have $\dfrac{3^{200}}{10^{100} + 3} = \dfrac{9^{100}}{10^{100} + 3} < 1$, so

$$\left\lfloor \frac{10^{20000}}{10^{100} + 3} \right\rfloor = \frac{10^{20000} - 3^{200}}{10^{100} + 3} = x^{199} + x^{198}y + x^{197}y^2 + \cdots + y^{199}.$$

Since x is divisible by 10, every term in this sum is divisible by 10 except $y^{199} = -3^{199}$. Since we have $3^4 \equiv 81 \equiv 1$ (mod 10), we have

$$-3^{199} \equiv -3^{4\cdot49+3} \equiv -(3^4)^{49} \cdot 3^3 \equiv -81^{49} \cdot 27 \equiv -27 \equiv 3 \quad \text{(mod 10)}.$$

Therefore, the units digit is $\boxed{3}$.

Just in case you aren't familiar with modular arithmetic, here's another way to finish without mods. If a and b are integers, then the last digit of ab is the same as the last digit of the product of the last digits of a and b. Since $3^4 = 81$, the last digit of 3^{4n} is 1 for all positive integers n. Specifically, 3^{196} ends in 1. Therefore, $3^{199} = (27)(3^{196})$ ends in 7. Since all the numbers in our sum above end in 0 except for the term -3^{199}, the units digit in the sum must be the units digit when we subtract a number that ends in 7 from a larger number that ends in 0, which is 3.

CHAPTER 17

More Sequences and Series

Exercises for Section 17.1

17.1.1 Since each term is 2/3 greater than the previous term, the sequence $\{a_n\}$ is an arithmetic sequence. Since $a_0 = 4$, we have $a_n = \boxed{4 + 2n/3}$.

17.1.2 From the given information, we have

$$a_n = \frac{a_1 + a_2 + \cdots + a_{n-1}}{n - 1}$$

for all $n \geq 3$. Multiplying both sides by $n - 1$, we get $(n - 1)a_n = a_1 + a_2 + \cdots + a_{n-1}$. Shifting the index n up by 1, we get

$$na_{n+1} = a_1 + a_2 + \cdots + a_{n-1} + a_n = (a_1 + a_2 + \cdots + a_{n-1}) + a_n = (n - 1)a_n + a_n = na_n,$$

so $a_{n+1} = a_n$ for all $n \geq 3$.

Hence, $a_3 = a_4 = a_5 = \cdots = a_9 = 99$. Since $a_3 = (a_1 + a_2)/2 = (19 + a_2)/2$, we have $(19 + a_2)/2 = 99$. Solving for a_2, we find $a_2 = \boxed{179}$.

It makes sense that the sequence $\{a_n\}$ becomes constant for $n \geq 3$. We know that a_3 is the average of a_1 and a_2. Then a_4 is the average of a_1, a_2, and a_3, which is the same as the average of a_1 and a_2, so $a_4 = a_3$, and so on.

17.1.3 The first few terms of the sequence are

$$t_2 = \frac{2 - 1}{2 + 1} = \frac{1}{3},$$

$$t_3 = \frac{\frac{1}{3} - 1}{\frac{1}{3} + 1} = -\frac{1}{2},$$

$$t_4 = \frac{-\frac{1}{2} - 1}{-\frac{1}{2} + 1} = -3,$$

$$t_5 = \frac{-3 - 1}{-3 + 1} = 2,$$

$$t_6 = \frac{2 - 1}{2 + 1} = \frac{1}{3},$$

and so on. Since $t_5 = t_1$ and each term in the sequence depends only on the previous term, the sequence $\{t_n\}$ is periodic with period 4. Then $t_{999} = t_{995} = t_{991} = \cdots = t_3 = \boxed{-1/2}$.

17.1.4

(a) We seek t_n, so we isolate it in $S_n = t_1 + t_2 + \cdots + t_n$ and we find $t_n = S_n - (t_1 + t_2 + \cdots + t_{n-1})$. If we can simplify $t_1 + t_2 + \cdots + t_{n-1}$, we can find t_n. But this sum is simply S_{n-1}! So, we have

$$t_n = S_n - S_{n-1} = \left(1 - \frac{1}{2^n}\right) - \left(1 - \frac{1}{2^{n-1}}\right) = \frac{1}{2^{n-1}} - \frac{1}{2^n} = \frac{1}{2^n}$$

for $n > 1$. We also have $t_1 = S_1 = \frac{1}{2}$, so $t_n = \boxed{\dfrac{1}{2^n}}$ for all positive integers n.

(b) We try using the same approach, $t_n = S_n - S_{n-1}$, that worked in part (a). We can use $t_n + 2S_n = 3$ to find another expression for $S_n - S_{n-1}$ in terms of t_n and t_{n-1}. This equation gives us

$$t_n + 2S_n = 3,$$
$$t_{n-1} + 2S_{n-1} = 3.$$

We subtract the second from the first to give $t_n - t_{n-1} + 2(S_n - S_{n-1}) = 0$. Because $S_n - S_{n-1} = t_n$, we have $t_n - t_{n-1} + 2t_n = 0$, so $t_{n-1} = 3t_n$, or $t_n = t_{n-1}/3$. So, we see that $\{t_n\}$ is a geometric sequence with common ratio $1/3$. But we don't know the first term! However, we do know that $S_1 = t_1$, and $t_1 + 2S_1 = 3$, so $t_1 = 1$. Therefore, we have

$$\boxed{t_n = \frac{1}{3^{n-1}} \quad \text{and} \quad S_n = \frac{3 - t_n}{2} = \frac{3}{2} - \frac{1}{2 \cdot 3^{n-1}}}.$$

17.1.5 We begin by letting $x = a_1$ and $y = a_2$, and computing the first few terms of the sequence in terms of x and y. As usual with sequence problems, we hope a pattern emerges.

$$a_3 = a_1 + a_2 = x + y,$$
$$a_4 = a_2 + a_3 = x + 2y,$$
$$a_5 = a_3 + a_4 = 2x + 3y,$$
$$a_6 = a_4 + a_5 = 3x + 5y,$$
$$a_7 = a_5 + a_6 = 5x + 8y,$$
$$a_8 = a_6 + a_7 = 8x + 13y.$$

Interestingly, each of the coefficients of x and y are numbers in the Fibonacci sequence. See if you can figure out why! We now have two expressions for a_7 that we can equate: $a_7 = 5x + 8y = 120$.

Our new equation is important because x and y must be positive integers, meaning there is a very limited number of solutions. We could hunt them down in several ways. One way is to solve for one of the variables in terms of the other:

$$x = \frac{120 - 8y}{5} = 24 - \frac{8}{5}y.$$

So, for x to be an integer, y must be a multiple of 5. Furthermore, x is a positive integer, so $24 - \frac{8}{5}y > 0$, which gives us $y < 15$. Therefore, the only possible values of y are 5 and 10.

If $y = 5$, then $x = 16$. But we are given that the sequence is increasing, and in particular, $a_1 < a_2$ (meaning $x < y$), so the pair $(x, y) = (16, 5)$ is not possible. If $y = 10$, then $x = 8$, which satisfies the increasing condition. Then, we have $a_8 = 8x + 13y = 8 \cdot 8 + 13 \cdot 10 = \boxed{194}$.

17.1.6 Let a_n be the answer the nth sheeple gives, so from the given information, we have

$$a_n = \frac{a_1 + a_2 + \cdots + a_{n-1}}{n - 1} + 0.6$$

for all $n \geq 2$. To make this expression easier to work with, we multiply both sides by $n - 1$, we get

$$(n - 1)a_n = a_1 + a_2 + \cdots + a_{n-1} + 0.6(n - 1).$$

Shifting the index n up by 1, we get

$$na_{n+1} = a_1 + a_2 + \cdots + a_{n-1} + a_n + 0.6n = [a_1 + a_2 + \cdots + a_{n-1} + 0.6(n-1)] + a_n + 0.6 = (n-1)a_n + a_n + 0.6 = na_n + 0.6.$$

Hence, we have $a_{n+1} = a_n + \frac{0.6}{n}$ for all $n \geq 1$. (We could also have shifted the index down by 1 to find this relationship.)

Then

$$a_2 = a_1 + 0.6,$$
$$a_3 = a_2 + \frac{0.6}{2} = a_1 + 0.9,$$
$$a_4 = a_3 + \frac{0.6}{3} = a_1 + 1.1,$$
$$a_5 = a_4 + \frac{0.6}{4} = a_1 + 1.25,$$
$$a_6 = a_5 + \frac{0.6}{5} = a_1 + 1.37,$$
$$a_7 = a_6 + \frac{0.6}{6} = a_1 + 1.47.$$

Since $a_7 = 5.47$, we have $a_1 = 5.47 - 1.47 = \boxed{4}$.

17.1.7

(a) For $k = -1$, the first few terms of the sequence are

$$a_2 = -a_1 - a_0,$$
$$a_3 = -a_2 - a_1 = -(-a_1 - a_0) - a_1 = a_0,$$
$$a_4 = -a_3 - a_2 = -a_0 - (-a_1 - a_0) = a_1.$$

Since $a_3 = a_0$ and $a_4 = a_1$, and each term depends only on the previous two terms, the sequence $\{a_n\}$ is periodic.

(b) For $k = \sqrt{2}$, the first few terms of the sequence are

$$
\begin{aligned}
a_2 &= \sqrt{2}a_1 - a_0, \\
a_3 &= \sqrt{2}a_2 - a_1 = \sqrt{2}(\sqrt{2}a_1 - a_0) - a_1 &&= a_1 - \sqrt{2}a_0, \\
a_4 &= \sqrt{2}a_3 - a_2 = \sqrt{2}(a_1 - \sqrt{2}a_0) - (\sqrt{2}a_1 - a_0) &&= -a_0, \\
a_5 &= \sqrt{2}a_4 - a_3 = \sqrt{2}(-a_0) - (a_1 - \sqrt{2}a_0) &&= -a_1, \\
a_6 &= \sqrt{2}a_5 - a_4 = \sqrt{2}(-a_1) - (-a_0) &&= -\sqrt{2}a_1 + a_0, \\
a_7 &= \sqrt{2}a_6 - a_5 = \sqrt{2}(-\sqrt{2}a_1 + a_0) - (a_1) &&= -a_1 + \sqrt{2}a_0, \\
a_8 &= \sqrt{2}a_7 - a_6 = \sqrt{2}(-a_1 + \sqrt{2}a_0) - (-\sqrt{2}a_1 + a_0) &&= a_0, \\
a_9 &= \sqrt{2}a_8 - a_7 = \sqrt{2}a_0 - (-a_1 + \sqrt{2}a_0) &&= a_1.
\end{aligned}
$$

Since $a_8 = a_0$ and $a_9 = a_1$, the sequence $\{a_n\}$ is periodic.

(c) We have $a_2 = ka_1 - a_0$ and $a_3 = ka_2 - a_1 = k(ka_1 - a_0) - a_1 = (k^2 - 1)a_1 - ka_0$. If we keep going like this, it's going to get messy. We go ahead and compute $k^2 - 1$, hoping it will give us something that allows us to simplify our expression for a_3. We get

$$k^2 - 1 = \frac{(1 + \sqrt{5})^2}{2^2} - 1 = \frac{1 + 2\sqrt{5} + 5 - 4}{4} = \frac{2 + 2\sqrt{5}}{4} = \frac{1 + \sqrt{5}}{2} = k.$$

Aha! Because $k^2 - 1 = k$, we can replace $k^2 - 1$ with k as we continue. We can also rearrange $k^2 - 1 = k$ to get both $k + 1 = k^2$ and $k^2 - k = 1$, which will also help us as we continue:

$$
\begin{aligned}
a_2 &= ka_1 - a_0, \\
a_3 &= ka_2 - a_1 &= k(ka_1 - a_0) - a_1 && = (k^2 - 1)a_1 - ka_0 && = ka_1 - ka_0, \\
a_4 &= ka_3 - a_2 &= k(ka_1 - ka_0) - (ka_1 - a_0) && = (k^2 - k)a_1 - (k^2 - 1)a_0 && = a_1 - ka_0, \\
a_5 &= ka_4 - a_3 &= k(a_1 - ka_0) - (ka_1 - ka_0) && = -(k^2 - k)a_0 && = -a_0, \\
a_6 &= ka_5 - a_4 &= k(-a_0) - (a_1 - ka_0) && = -a_1, \\
a_7 &= ka_6 - a_5 &= k(-a_1) - (-a_0) && = -ka_1 + a_0, \\
a_8 &= ka_7 - a_6 &= k(-ka_1 + a_0) - (-a_1) && = -(k^2 - 1)a_1 + ka_0 && = -ka_1 + ka_0, \\
a_9 &= ka_8 - a_7 &= k(-ka_1 + ka_0) - (-ka_1 + a_0) && = -(k^2 - k)a_1 + (k^2 - 1)a_0 && = -a_1 + ka_0, \\
a_{10} &= ka_9 - a_8 &= k(-a_1 + ka_0) - (-ka_1 + ka_0) && = (k^2 - k)a_0 && = a_0, \\
a_{11} &= ka_{10} - a_9 &= ka_0 - (-a_1 + ka_0) && = a_1.
\end{aligned}
$$

Since $a_{10} = a_0$ and $a_{11} = a_1$, the sequence $\{a_n\}$ is periodic.

Exercises for Section 17.2

17.2.1   Writing the product out, we see that most of the terms cancel, and we get

$$
\prod_{n=1}^{13} \frac{n(n+2)}{(n+4)^2} = \frac{1 \cdot 3}{5^2} \cdot \frac{2 \cdot 4}{6^2} \cdot \frac{3 \cdot 5}{7^2} \cdots \frac{12 \cdot 14}{16^2} \cdot \frac{13 \cdot 15}{17^2} = \frac{1 \cdot 2 \cdot 3^2 \cdot 4^2}{14 \cdot 15 \cdot 16^2 \cdot 17^2} = \boxed{\frac{3}{161840}}.
$$

17.2.2   Let k be a positive integer. Then rationalizing the denominator, we get

$$
\frac{1}{\sqrt{k} + \sqrt{k+1}} = \frac{1}{\sqrt{k+1} + \sqrt{k}} \cdot \frac{\sqrt{k+1} - \sqrt{k}}{\sqrt{k+1} - \sqrt{k}} = \frac{\sqrt{k+1} - \sqrt{k}}{(k+1) - k} = \sqrt{k+1} - \sqrt{k}.
$$

Hence, summing over $1 \le k \le 99$, the sum telescopes, and we get

$$
\sum_{k=1}^{99} \frac{1}{\sqrt{k} + \sqrt{k+1}} = (\sqrt{2} - 1) + (\sqrt{3} - \sqrt{2}) + (\sqrt{4} - \sqrt{3}) + \cdots + (\sqrt{100} - \sqrt{99}) = \sqrt{100} - 1 = 10 - 1 = \boxed{9}.
$$

17.2.3   We can average the first 200 terms by finding their sum. These partial sums tell the story. The first two terms add to -1. The next two terms add another -1, as do the next two terms. Similarly, the pattern continues, with each successive pair of terms having sum $n + (-(n+1)) = n + (-n - 1) = -1$. There are 100 such pairs, so the sum of all 200 numbers is -100. Therefore, the average of all 200 numbers is $\boxed{-1/2}$.

17.2.4   The general term of the given series can be written as

$$
\frac{k-1}{k!} = \frac{k}{k!} - \frac{1}{k!} = \frac{1}{(k-1)!} - \frac{1}{k!}.
$$

Hence, summing over $2 \le k \le n$, the sum telescopes, and we get

$$
\sum_{k=2}^{n} \frac{k-1}{k!} = \left(\frac{1}{1!} - \frac{1}{2!} \right) + \left(\frac{1}{2!} - \frac{1}{3!} \right) + \left(\frac{1}{3!} - \frac{1}{4!} \right) + \cdots + \left(\frac{1}{(n-1)!} - \frac{1}{n!} \right) = \boxed{1 - \frac{1}{n!}}.
$$

17.2.5   First, we perform a partial fraction decomposition. There exist constants A and B such that

$$
\frac{1}{(2n-1)(2n+1)} = \frac{A}{2n-1} + \frac{B}{2n+1}.
$$

Multiplying both sides by $(2n-1)(2n+1)$, we get $(2n+1)A + (2n-1)B = 1$. Taking $n = 1/2$ gives $2A = 1$, so $A = 1/2$. Taking $n = -1/2$ gives $-2B = 1$, so $B = -1/2$. Therefore,

$$\frac{1}{(2n-1)(2n+1)} = \frac{1/2}{2n-1} - \frac{1/2}{2n+1}.$$

Then, summing over $1 \le n \le 128$, the sum telescopes, and we get

$$\sum_{n=1}^{128} \frac{1}{(2n-1)(2n+1)} = \left(\frac{1/2}{1} - \frac{1/2}{3}\right) + \left(\frac{1/2}{3} - \frac{1/2}{5}\right) + \left(\frac{1/2}{5} - \frac{1/2}{7}\right) + \cdots + \left(\frac{1/2}{255} - \frac{1/2}{257}\right)$$

$$= \frac{1}{2} - \frac{1}{2 \cdot 257} = \boxed{\frac{128}{257}}.$$

17.2.6 Putting the fraction over a common denominator, we find the product telescopes, and we get

$$\prod_{n=1}^{20} \left(1 + \frac{2n+1}{n^2}\right) = \prod_{n=1}^{20} \frac{n^2 + 2n + 1}{n^2} = \prod_{n=1}^{20} \frac{(n+1)^2}{n^2} = \frac{2^2}{1^2} \cdot \frac{3^2}{2^2} \cdot \frac{4^2}{3^2} \cdots \frac{21^2}{20^2} = 21^2 = \boxed{441}.$$

17.2.7 For a fixed integer k, let $\alpha = \sqrt[3]{k}$ and $\beta = \sqrt[3]{k+1}$. Then

$$\frac{1}{\sqrt[3]{k^2} + \sqrt[3]{k(k+1)} + \sqrt[3]{(k+1)^2}} = \frac{1}{\alpha^2 + \alpha\beta + \beta^2} = \frac{\beta - \alpha}{(\beta - \alpha)(\beta^2 + \alpha\beta + \alpha^2)} = \frac{\beta - \alpha}{\beta^3 - \alpha^3} = \frac{\beta - \alpha}{(k+1) - k} = \beta - \alpha = \sqrt[3]{k+1} - \sqrt[3]{k}.$$

Hence, summing from $k = 1$ to 7, the sum telescopes, and we get

$$\sum_{k=1}^{7} \frac{1}{\sqrt[3]{k^2} + \sqrt[3]{k(k+1)} + \sqrt[3]{(k+1)^2}} = (\sqrt[3]{2} - \sqrt[3]{1}) + (\sqrt[3]{3} - \sqrt[3]{2}) + \cdots + (\sqrt[3]{8} - \sqrt[3]{7}) = \sqrt[3]{8} - 1 = 2 - 1 = \boxed{1}.$$

17.2.8 Note that $H_{n+1} = H_n + \frac{1}{n+1}$, so $H_{n+1} - H_n = \frac{1}{n+1}$. Then

$$T_n = \frac{1}{(n+1)H_n H_{n+1}} = \frac{H_{n+1} - H_n}{H_n H_{n+1}} = \frac{1}{H_n} - \frac{1}{H_{n+1}}.$$

Hence, summing from $k = 1$ to ∞, the sum telescopes, and we get

$$\sum_{k=1}^{\infty} T_k = \left(\frac{1}{H_1} - \frac{1}{H_2}\right) + \left(\frac{1}{H_2} - \frac{1}{H_3}\right) + \left(\frac{1}{H_3} - \frac{1}{H_4}\right) + \cdots = \frac{1}{H_1} = 1.$$

Exercises for Section 17.3

17.3.1 From the formulas in the text, we have

$$\sum_{i=1}^{50} (i^3 - 2i^2) = \sum_{i=1}^{50} i^3 - 2\sum_{i=1}^{50} i^2 = \frac{50^2 \cdot 51^2}{4} - 2 \cdot \frac{50 \cdot 51 \cdot 101}{6} = 1625625 - 85850 = \boxed{1539775}.$$

17.3.2 First, we express the sum using sigma notation:

$$1 \cdot 2 + 3 \cdot 4 + \cdots + (2n-1) \cdot 2n = \sum_{k=1}^{n} (2k-1) \cdot 2k.$$

Then from the formulas in the text, we have

$$\sum_{k=1}^{n} (2k-1) \cdot 2k = \sum_{k=1}^{n} (4k^2 - 2k) = 4\sum_{k=1}^{n} k^2 - 2\sum_{k=1}^{n} k = 4 \cdot \frac{n(n+1)(2n+1)}{6} - 2 \cdot \frac{n(n+1)}{2}$$

$$= \frac{4n^3 + 6n^2 + 2n}{3} - n^2 - n = \boxed{\frac{4n^3 + 3n^2 - n}{3}}.$$

17.3.3 From the formulas in the text, we have

$$\sum_{k=1}^{n} g(k) = \sum_{k=1}^{n} (k^2 - 11k - 28) = \sum_{k=1}^{n} k^2 - 11\sum_{k=1}^{n} k - \sum_{k=1}^{n} 28$$

$$= \frac{n(n+1)(2n+1)}{6} - 11 \cdot \frac{n(n+1)}{2} - 28n = \frac{2n^3 + 3n^2 + n}{6} - \frac{11n^2 + 11n}{2} - 28n$$

$$= \frac{n^3 - 15n^2 - 100n}{3}.$$

The numerator factors as $n^3 - 15n^2 - 100n = n(n^2 - 15n - 100) = n(n-20)(n+5)$, so the sum is 0 for $n = \boxed{20}$.

17.3.4 We find $\sum_{k=1}^{n} k^4$ in the same way we found $\sum_{k=1}^{n} k^3$, $\sum_{k=1}^{n} k^2$, and $\sum_{k=1}^{n} k$. First, we expand $(k+1)^5 - k^5$:

$$(k+1)^5 - k^5 = 5k^4 + 10k^3 + 10k^2 + 5k + 1.$$

Summing over $1 \le k \le n$, the sum telescopes, and we get

$$\sum_{k=1}^{n} (5k^4 + 10k^3 + 10k^2 + 5k + 1) = \sum_{k=1}^{n} [(k+1)^5 - k^5] = (2^5 - 1^5) + (3^5 - 2^5) + \cdots + [(n+1)^5 - n^5]$$

$$= (n+1)^5 - 1 = n^5 + 5n^4 + 10n^3 + 10n^2 + 5n.$$

The sum on the left expands as

$$\sum_{k=1}^{n} (5k^4 + 10k^3 + 10k^2 + 5k + 1) = 5\sum_{k=1}^{n} k^4 + 10\sum_{k=1}^{n} k^3 + 10\sum_{k=1}^{n} k^2 + 5\sum_{k=1}^{n} k + \sum_{k=1}^{n} 1$$

$$= 5\left(\sum_{k=1}^{n} k^4\right) + 10 \cdot \frac{n^2(n+1)^2}{4} + 10 \cdot \frac{n(n+1)(2n+1)}{6} + 5 \cdot \frac{n(n+1)}{2} + n$$

$$= 5\left(\sum_{k=1}^{n} k^4\right) + \frac{5n^4 + 10n^3 + 5n^2}{2} + \frac{10n^3 + 15n^2 + 5n}{3} + \frac{5n^2 + 5n}{2} + n$$

$$= 5\left(\sum_{k=1}^{n} k^4\right) + \frac{15n^4 + 50n^3 + 60n^2 + 31n}{6}.$$

It follows that

$$5\sum_{k=1}^{n} k^4 = n^5 + 5n^4 + 10n^3 + 10n^2 + 5n - \frac{15n^4 + 50n^3 + 60n^2 + 31n}{6} = \frac{6n^5 + 15n^4 + 10n^3 - n}{6},$$

so $\sum_{k=1}^{n} k^4 = \boxed{\frac{6n^5 + 15n^4 + 10n^3 - n}{30}}$, which can also be expressed as $\sum_{k=1}^{n} k^4 = \boxed{\frac{n(n+1)(2n+1)(3n^2 + 3n - 1)}{30}}.$

17.3.5 First, we determine a closed form for the kth term in the sum. The kth term is the product of a fraction and the sum of the first k squares, which is

$$1^2 + 2^2 + \cdots + k^2 = \frac{k(k+1)(2k+1)}{6}.$$

The fractions are $\frac{1}{2} = \frac{1}{1\cdot 2}$, $\frac{1}{6} = \frac{1}{2\cdot 3}$, $\ldots$, $\frac{1}{3660} = \frac{1}{60\cdot 61}$. Hence, the kth term in the sum is

$$\frac{1}{k(k+1)} \cdot \frac{k(k+1)(2k+1)}{6} = \frac{2k+1}{6}.$$

Therefore, the given sum is

$$\sum_{k=1}^{60} \frac{2k+1}{6} = \frac{1}{3}\sum_{k=1}^{60} k + \frac{1}{6}\sum_{k=1}^{60} 1 = \frac{1}{3}\cdot\frac{60\cdot 61}{2} + \frac{1}{6}\cdot 60 = 610 + 10 = \boxed{620}.$$

Exercises for Section 17.4

17.4.1

(a) Let $S = \frac{1}{5} + \frac{2}{5^2} + \frac{3}{5^3} + \frac{4}{5^4} + \cdots$. Then $\frac{S}{5} = \frac{1}{5^2} + \frac{2}{5^3} + \frac{3}{5^4} + \cdots$. Taking the difference of these two equations, we get

$$\frac{4}{5}S = \frac{1}{5} + \frac{1}{5^2} + \frac{1}{5^3} + \cdots = \frac{1/5}{1 - 1/5} = \frac{1}{4},$$

so $S = \boxed{5/16}$.

(b) Let

$$S = \sum_{k=1}^{n}(2k-1)3^{k-1} = 1 + 3\cdot 3 + 5\cdot 3^2 + \cdots + (2n-1)3^{n-1}.$$

Then $3S = 3 + 3\cdot 3^2 + \cdots + (2n-3)3^{n-1} + (2n-1)3^n$. Subtracting our equation for S from our equation for $3S$, we get

$$2S = -1 - 2\cdot 3 - 2\cdot 3^2 - \cdots - 2\cdot 3^{n-1} + (2n-1)3^n = 1 - 2\cdot(1 + 3 + \cdots + 3^{n-1}) + (2n-1)3^n$$

$$= 1 - 2\cdot\frac{3^n - 1}{3 - 1} + (2n-1)3^n = +1 - 3^n + 1 + (2n-1)3^n = (2n-2)3^n + 2,$$

so $S = \boxed{(n-1)3^n + 1}$.

(c) Let $S = \displaystyle\sum_{n=1}^{\infty} \frac{5n-1}{2^n} = \frac{4}{2} + \frac{9}{2^2} + \frac{14}{2^3} + \cdots$. Then $2S = 4 + \frac{9}{2} + \frac{14}{2^2} + \cdots$. Taking the difference of these two equations, we get

$$S = 4 + \frac{5}{2} + \frac{5}{2^2} + \frac{5}{2^3} + \cdots = 4 + \frac{\frac{5}{2}}{1 - \frac{1}{2}} = 4 + 5 = \boxed{9}.$$

17.4.2 Let

$$S = \sum_{n=1}^{\infty}\log_2 \sqrt[2^n]{2^n} = \sum_{n=1}^{\infty}\log_2 2^{n/2^n} = \sum_{n=1}^{\infty}\frac{n}{2^n} = \frac{1}{2} + \frac{2}{4} + \frac{3}{8} + \cdots.$$

Then $2S = 1 + \frac{2}{2} + \frac{3}{4} + \cdots$. Taking the difference of these two equations, we get

$$S = 1 + \frac{1}{2} + \frac{1}{4} + \cdots = \frac{1}{1 - 1/2} = \boxed{2}.$$

17.4.3 Let $S = 1 + 2x + 3x^2 + \cdots + nx^{n-1}$. Multiplying this equation by x gives $xS = x + 2x^2 + 3x^3 + \cdots + nx^n$. Subtracting this from our equation for S, we have $S - Sx = 1 + x + \cdots + x^{n-1} - nx^n$. Excluding the last term from the right, we have a geometric series. Since $x \neq 1$, we therefore have

$$S - Sx = \frac{x^n - 1}{x - 1} - nx^n = \frac{x^n - 1 - nx^n(x-1)}{x-1} = \frac{-nx^{n+1} + (n+1)x^n - 1}{x-1}.$$

Since $S - Sx = (1 - x)S$, we can divide both sides of the equation above by $1 - x$ to find

$$S = \frac{-nx^{n+1} + (n+1)x^n - 1}{(x-1)(1-x)} = \boxed{\frac{nx^{n+1} - (n+1)x^n + 1}{(x-1)^2}}.$$

Exercises for Section 17.5

17.5.1

(a) We construct the difference table as follows:

$$\begin{array}{llll}
p(n): & 1 & 2 & 7 \\
\Delta p(n): & & 1 & 5 \\
\Delta^2 p(n): & & & 4
\end{array}$$

Since $p(n)$ is quadratic, the polynomial $\Delta^2 p(n)$ is constant. Therefore, we have $\Delta^2 p(n) = 4$ for all n, and we can extend the difference table as follows:

$$\begin{array}{lllll}
p(n): & 1 & 2 & 7 & 16 \\
\Delta p(n): & & 1 & 5 & 9 \\
\Delta^2 p(n): & & & 4 & 4
\end{array}$$

Thus, $p(3) = \boxed{16}$.

(b) As observed in part (a), we have $\Delta^2 p(n) = 4$ for all n, so

$$\Delta p(1) - \Delta p(0) = 4,$$
$$\Delta p(2) - \Delta p(1) = 4,$$
$$\vdots$$
$$\Delta p(n) - \Delta p(n-1) = 4.$$

Adding all these equations, we get $\Delta p(n) - \Delta p(0) = 4n$. Therefore, we have $\Delta p(n) = 4n + \Delta p(0) = 4n + 1$. Then

$$p(1) - p(0) = 4 \cdot 0 + 1,$$
$$p(2) - p(1) = 4 \cdot 1 + 1,$$
$$\vdots$$
$$p(n) - p(n-1) = 4 \cdot (n-1) + 1.$$

Adding all these equations, we get

$$p(n) - p(0) = 4 \cdot [0 + 1 + \cdots + (n-1)] + n = 4 \cdot \frac{(n-1)n}{2} + n = 2n^2 - n.$$

Then $p(n) = 2n^2 - n + p(0) = \boxed{2n^2 - n + 1}$.

17.5.2 The completed difference table is shown below:

$$a_n : 1 \quad 3 \quad 9 \quad 27$$
$$\Delta a_n : \quad 2 \quad 6 \quad 18$$
$$\Delta^2 a_n : \quad 4 \quad 12$$
$$\Delta^3 a_n : \quad 8$$

Thus, $a_1 = \boxed{3}$, $a_2 = \boxed{9}$, and $a_3 = \boxed{27}$.

17.5.3 Since $\Delta^3 p(n) = 3n + 5$, we have

$$\Delta^4 p(n) = \Delta^3 p(n+1) - \Delta^3 p(n) = 3(n+1) + 5 - (3n+5) = 3.$$

Thus, $\Delta^4 p(n)$ is a nonzero constant. Therefore, as explained in the text, we have $\deg p = \boxed{4}$.

17.5.4 Since $\Delta^2 a_n = 1$ for all n, a_n is a quadratic in n. Furthermore, $a_{19} = a_{92} = 0$, so $a_n = k(n-19)(n-92)$ for some constant k. Then

$$\Delta a_n = a_{n+1} - a_n = k(n-18)(n-91) - k(n-19)(n-92) = k(n^2 - 109n + 1638) - k(n^2 - 111n + 1748) = 2kn - 110k,$$

and

$$\Delta^2 a_n = \Delta a_{n+1} - \Delta a_n = 2k(n+1) - 110k - 2kn + 110k = 2k.$$

Since $\Delta^2 a_n = 1$ for all n, we have $2k = 1$, so $k = 1/2$. Hence, $a_1 = \frac{1}{2}(1-19)(1-92) = \boxed{819}$.

Review Problems

17.32 We compute the first few terms of the sequence:

$$a_3 = \frac{3}{2}, \quad a_4 = \frac{3/2}{3} = \frac{1}{2}, \quad a_5 = \frac{1/2}{3/2} = \frac{1}{3}, \quad a_6 = \frac{1/3}{1/2} = \frac{2}{3}, \quad a_7 = \frac{2/3}{1/3} = 2, \quad a_8 = \frac{2}{2/3} = 3,$$

and so on. Since $a_7 = a_1 = 2$ and $a_8 = a_2 = 3$, and each term only depends on the previous two terms, the sequence $\{a_n\}$ is periodic with period 6. Since 2006 leaves a remainder of 2 when divided by 6, we have $a_{2006} = a_{2000} = a_{1994} = \cdots = a_2 = \boxed{3}$.

17.33 Each term is r times the previous term, so the sequence can be defined by $t_1 = a$ and $\boxed{t_n = rt_{n-1}}$.

17.34 Let $f^n(x) = c_n x + d_n$. To find formulas for c_n and d_n, we compute the first few $f^n(x)$:

$$f^1(x) = f(x) = ax + b,$$

so $c_1 = a$ and $d_1 = b$. Then

$$f^2(x) = f(f(x)) = f(ax+b) = a(ax+b) + b = a^2 x + ab + b,$$

so $c_2 = a^2$ and $d_2 = ab + b$. Then

$$f^3(x) = f(f^2(x)) = f(a^2 x + ab + b) = a(a^2 x + ab + b) + b = a^3 x + a^2 b + ab + b,$$

so $c_3 = a^3$ and $d_3 = a^2 b + ab + b$.

It appears that $c_n = a^n$ and $d_n = a^{n-1}b + a^{n-2}b + \cdots + b$ Since $a \neq 1$, we can write this expression for d_n as

$$d_n = a^{n-1}b + a^{n-2}b + \cdots + b = (a^{n-1} + a^{n-2} + \cdots + 1)b = \frac{a^n - 1}{a - 1} \cdot b.$$

We prove these formulas for c_n and d_n with induction.

The result is true for $n = 1$, so assume that it is true for some positive integer $n = k$, so $f^k(x) = a^k x + \frac{a^k - 1}{a - 1} \cdot b$. Then

$$f^{k+1}(x) = f(f^k(x)) = f\left(a^k x + \frac{a^k - 1}{a - 1} \cdot b\right) = a\left(a^k x + \frac{a^k - 1}{a - 1} \cdot b\right) + b$$

$$= a^{k+1} x + \frac{a(a^k - 1)}{a - 1} \cdot b + b = a^{k+1} x + \left(\frac{a^{k+1} - a}{a - 1} + 1\right) b = a^{k+1} x + \frac{a^{k+1} - 1}{a - 1} \cdot b.$$

Hence, the result is true for $n = k + 1$, so it is true for all positive integers n. Therefore, our formula for $f^n(x)$ is

$$f^n(x) = \boxed{a^n x + \frac{a^n - 1}{a - 1} \cdot b}.$$

What happens in this problem if $a = 1$?

17.35 Let a_n denote the number of pieces of candy that the nth kid took, so

$$a_n = \left\lceil \frac{a_{n-1} + a_{n-2} + a_{n-3}}{3} \right\rceil$$

for all $n \geq 4$. We compute the first few terms of the sequence:

$$a_4 = \left\lceil \frac{5 + 2 + 3}{3} \right\rceil = \left\lceil \frac{10}{3} \right\rceil = 4,$$

$$a_5 = \left\lceil \frac{4 + 5 + 2}{3} \right\rceil = \left\lceil \frac{11}{3} \right\rceil = 4,$$

$$a_6 = \left\lceil \frac{4 + 4 + 5}{3} \right\rceil = \left\lceil \frac{13}{3} \right\rceil = 5,$$

$$a_7 = \left\lceil \frac{5 + 4 + 4}{3} \right\rceil = \left\lceil \frac{13}{3} \right\rceil = 5,$$

$$a_8 = \left\lceil \frac{5 + 5 + 4}{3} \right\rceil = \left\lceil \frac{14}{3} \right\rceil = 5,$$

$$a_9 = \left\lceil \frac{5 + 5 + 5}{3} \right\rceil = \left\lceil \frac{15}{3} \right\rceil = 5,$$

and so on. We have three consecutive terms, namely a_7, a_8, a_9, that are all equal to 5, so all subsequent terms are also equal to 5. In particular, $a_{2007} = \boxed{5}$.

17.36 Given $n_2 = 5$, we can compute further terms in the sequence in terms of a:

$$n_3 = 2n_2 + a = a + 10,$$
$$n_4 = 2n_3 + a = 3a + 20,$$
$$n_5 = 2n_4 + a = 7a + 40,$$
$$n_6 = 2n_5 + a = 15a + 80,$$
$$n_7 = 2n_6 + a = 31a + 160,$$
$$n_8 = 2n_7 + a = 63a + 320.$$

Since $n_8 = 257$, we have $63a + 320 = 257$, so $a = -1$. Therefore, $n_5 = 7 \cdot (-1) + 40 = \boxed{33}$.

17.37 $\log_{10} \frac{1}{2} + \log_{10} \frac{2}{3} + \log_{10} \frac{3}{4} + \cdots + \log_{10} \frac{99}{100} = \log_{10} \left(\frac{1}{2} \cdot \frac{2}{3} \cdot \frac{3}{4} \cdots \frac{99}{100}\right) = \log_{10} \frac{1}{100} = \boxed{-2}.$

17.38 We can evaluate the expression by grouping the terms as follows:

$$1 \cdot 2 - 2 \cdot 3 + 3 \cdot 4 - 4 \cdot 5 + 5 \cdot 6 - \cdots - 2000 \cdot 2001 + 2001 \cdot 2002$$
$$= 1 \cdot 2 + (4 - 2) \cdot 3 + (6 - 4) \cdot 5 + \cdots + (2002 - 2000) \cdot 2001$$
$$= 2 \cdot 1 + 2 \cdot 3 + 2 \cdot 5 + \cdots + 2 \cdot 2001$$
$$= 2 \cdot (1 + 3 + 5 + \cdots + 2001) = 2 \cdot 1001^2 = \boxed{2004002}.$$

17.39 First, we express each factor as a single fraction:

$$\left(1 - \frac{1}{2^2}\right)\left(1 - \frac{1}{3^2}\right)\cdots\left(1 - \frac{1}{9^2}\right)\left(1 - \frac{1}{10^2}\right) = \frac{2^2 - 1}{2^2} \cdot \frac{3^2 - 1}{3^2} \cdots \frac{9^2 - 1}{9^2} \cdot \frac{10^2 - 1}{10^2}.$$

Then using difference of squares, we find that the product telescopes, and we get

$$\frac{2^2 - 1}{2^2} \cdot \frac{3^2 - 1}{3^2} \cdots \frac{9^2 - 1}{9^2} \cdot \frac{10^2 - 1}{10^2} = \frac{(2-1)(2+1)}{2^2} \cdot \frac{(3-1)(3+1)}{3^2} \cdots \frac{(9-1)(9+1)}{9^2} \cdot \frac{(10-1)(10+1)}{10^2}$$

$$= \frac{1 \cdot 3}{2^2} \cdot \frac{2 \cdot 4}{3^2} \cdots \frac{8 \cdot 10}{9^2} \cdot \frac{9 \cdot 11}{10^2} = \frac{1 \cdot 11}{2 \cdot 10} = \boxed{\frac{11}{20}}.$$

17.40

(a) Let $S = 1 + \frac{2}{7} + \frac{3}{7^2} + \cdots$. Then $7S = 7 + 2 + \frac{3}{7} + \frac{4}{7^2} + \cdots$. Subtracting the first equation from the second, we get

$$6S = 8 + \frac{1}{7} + \frac{1}{7^2} + \cdots = 8 + \frac{1/7}{1 - 1/7} = 8 + \frac{1}{6} = \frac{49}{6}.$$

Therefore, $S = \boxed{49/36}$.

(b) Let $S = \sum_{n=1}^{\infty} \frac{3n - 2}{5^n} = \frac{1}{5} + \frac{4}{5^2} + \frac{7}{5^3} + \frac{10}{5^4} + \cdots$. Then $5S = 1 + \frac{4}{5} + \frac{7}{5^2} + \frac{10}{5^3} + \cdots$. Subtracting the first equation from the second, we get

$$4S = 1 + \frac{3}{5} + \frac{3}{5^2} + \frac{3}{5^3} + \cdots = 1 + \frac{3/5}{1 - 1/5} = 1 + \frac{3}{4} = \frac{7}{4}.$$

Therefore, $S = \boxed{7/16}$.

17.41 Let $S = 2 \cdot 2^2 + 3 \cdot 2^3 + 4 \cdot 2^4 + \cdots + n \cdot 2^n$. Then $2S = 2 \cdot 2^3 + 3 \cdot 2^4 + 4 \cdot 2^5 + \cdots + (n-1) \cdot 2^n + n \cdot 2^{n+1}$. Subtracting the first equation from the second, we get

$$S = -2 \cdot 2^2 - 2^3 - 2^4 - \cdots - 2^{n-1} - 2^n + n \cdot 2^{n+1} = -8 - 2^3 - 2^4 - \cdots - 2^{n-1} - 2^n + n \cdot 2^{n+1}$$
$$= n \cdot 2^{n+1} - 1 - (1 + 2 + 2^2 + 2^3 + 2^4 + \cdots + 2^{n-1} + 2^n) = n \cdot 2^{n+1} - 1 - (2^{n+1} - 1) = (n-1) \cdot 2^{n+1}.$$

Thus, the given equation becomes $(n-1) \cdot 2^{n+1} = 2^{n+10}$. Dividing both sides by 2^{n+1}, we get $n - 1 = 2^9 = 512$, so $n = \boxed{513}$.

17.42 Using difference of squares, we get

$$\frac{(1998^2 - 1996^2)(1998^2 - 1995^2) \cdots (1998^2 - 0^2)}{(1997^2 - 1996^2)(1997^2 - 1995^2) \cdots (1997^2 - 0^2)}$$

$$= \frac{(1998 + 1996)(1998 - 1996)(1998 + 1995)(1998 - 1995) \cdots (1998 + 0)(1998 - 0)}{(1997 + 1996)(1997 - 1996)(1997 + 1995)(1997 - 1995) \cdots (1997 + 0)(1997 - 0)}$$

$$= \frac{3994 \cdot 2 \cdot 3993 \cdot 3 \cdots 1998 \cdot 1998}{3993 \cdot 1 \cdot 3992 \cdot 2 \cdots 1997 \cdot 1997}.$$

In the numerator, each number from 1 to 3994 appears once as a factor, except for 1998, which appears twice. In the denominator, each number from 1 to 3993 appears once as a factor, except for 1997, which appears twice. Hence, the fraction simplifies as

$$\frac{1998 \cdot 3994!}{1997 \cdot 3993!} = \frac{1998 \cdot 3994}{1997} = \frac{1998 \cdot 2 \cdot 1997}{1997} = 2 \cdot 1998 = \boxed{3996}.$$

17.43 $\displaystyle\sum_{k=1}^{20}(2k^2 - k) = 2\sum_{k=1}^{20}k^2 - \sum_{k=1}^{20}k = 2 \cdot \frac{20 \cdot 21 \cdot 41}{6} - \frac{20 \cdot 21}{2} = 5740 - 210 = \boxed{5530}.$

17.44 We can compute the sum by using the formulas we have derived. However, we can instead express a_n in the form $a_n = 3n^2 + 3n + 1 = (n+1)^3 - n^3$. Thus, the sum telescopes, and we get

$$a_1 + a_2 + \cdots + a_{100} = (2^3 - 1^3) + (3^2 - 2^3) + \cdots + (101^3 - 100^3) = 101^3 - 1^3 = \boxed{1030300}.$$

17.45 First, we can express the sum on the left side using sigma notation:

$$1 \cdot 1987 + 2 \cdot 1986 + 3 \cdot 1985 + \cdots + 1987 \cdot 1 = \sum_{k=1}^{1987} k(1988 - k) = \sum_{k=1}^{1987}(1988k - k^2) = 1988\sum_{k=1}^{1987}k - \sum_{k=1}^{1987}k^2$$

$$= 1988 \cdot \frac{1987 \cdot 1988}{2} - \frac{1987 \cdot 1988 \cdot 3975}{6} = 1987 \cdot 1988 \cdot \left(\frac{1988}{2} - \frac{3975}{6}\right)$$

$$= 1987 \cdot 1988 \cdot \frac{1989}{6}.$$

Hence, the given equation becomes $\frac{1987 \cdot 1988 \cdot 1989}{6} = 1987 \cdot 994 \cdot x$, so

$$x = \frac{1987 \cdot 1988 \cdot 1989}{6 \cdot 1987 \cdot 994} = \frac{2 \cdot 994 \cdot 1989}{6 \cdot 994} = \boxed{663}.$$

17.46 To find the polynomial, we construct the difference table, as follows:

$$
\begin{array}{lccccccccc}
p(x): & 3 & & 7 & & 13 & & 21 & & 31 \\
\Delta p(x): & & 4 & & 6 & & 8 & & 10 & \\
\Delta^2 p(x): & & & 2 & & 2 & & 2 & &
\end{array}
$$

Since the second row is a nonzero constant sequence, we know that $p(x)$ is quadratic. We therefore know that $p(x) = ax^2 + bx + c$ for some constants a, b, and c. From the given table, we have the system

$$a + b + c = 3,$$
$$4a + 2b + c = 7,$$
$$9a + 3b + c = 13.$$

Subtracting the first equation from the second gives $3a + b = 4$, and subtracting the first equation from the third gives $8a + 2b = 10$, so $4a + b = 5$. Subtracting $3a + b = 4$ from this equation gives $a = 1$, from which we find $b = 1$ and $c = 1$. Therefore, we have $p(x) = \boxed{x^2 + x + 1}$.

17.47 We construct the difference table of $f(n)$ as follows:

$$
\begin{array}{lcccccc}
f(n): & 5 & & 4 & & 17 & & 56 \\
\Delta f(n): & & -1 & & 13 & & 39 & \\
\Delta^2 f(n): & & & 14 & & 26 & & \\
\Delta^3 f(n): & & & & 12 & & &
\end{array}
$$

Since $f(n)$ is cubic, the polynomial $\Delta^3 f(n)$ is constant. Therefore, $\Delta^3 f(n) = 12$ for all n, and we can extend the difference table as follows:

$$\begin{array}{llllll} f(n): & 5 & 4 & 17 & 56 & 133 \\ \Delta f(n): & & -1 & 13 & 39 & 77 \\ \Delta^2 f(n): & & & 14 & 26 & 38 \\ \Delta^3 f(n): & & & & 12 & 12 \end{array}$$

Thus, $f(4) = \boxed{133}$.

Challenge Problems

17.48 The Cesàro sum of the 99-term sequence $a_1, a_2, \ldots, a_{99}$ is

$$\frac{a_1 + (a_1 + a_2) + (a_1 + a_2 + a_3) + \cdots + (a_1 + a_2 + \cdots + a_{99})}{99} = \frac{99a_1 + 98a_2 + \cdots + a_{99}}{99} = 1000,$$

so $99a_1 + 98a_2 + \cdots + a_{99} = 99000$.

Then the Cesàro sum of the 100-term sequence $1, a_1, a_2, \ldots, a_{99}$ is

$$\frac{1 + (1 + a_1) + (1 + a_1 + a_2) + \cdots + (1 + a_1 + a_2 + \cdots + a_{99})}{100} = \frac{100 + 99a_1 + 98a_2 + 97a_3 + \cdots + a_{99}}{100}$$
$$= \frac{100 + 99000}{100} = \boxed{991}.$$

17.49 We compute the first few polynomials $P_n(x)$:

$$\begin{aligned} P_1(x) &= P_0(x - 1), \\ P_2(x) &= P_1(x - 2) = P_0(x - 1 - 2), \\ P_3(x) &= P_2(x - 3) = P_0(x - 1 - 2 - 3), \\ P_4(x) &= P_3(x - 4) = P_0(x - 1 - 2 - 3 - 4), \end{aligned}$$

and so on. In general, we see that

$$P_n(x) = P_0(x - 1 - 2 - \cdots - n) = P_0\left(x - \frac{n(n+1)}{2}\right),$$

so

$$P_{20}(x) = P_0\left(x - \frac{20 \cdot 21}{2}\right) = P_0(x - 210) = (x - 210)^3 + 313(x - 210)^2 - 77(x - 210) - 8$$
$$= x^3 - 3 \cdot 210 x^2 + 3 \cdot 210^2 x - 210^3 + 313 x^2 - 313 \cdot 2 \cdot 210 x + 313 \cdot 210^2 - 77x + 77 \cdot 210 - 8.$$

Hence, the coefficient of x in $P_{20}(x)$ is $3 \cdot 210^2 - 313 \cdot 2 \cdot 210 - 77 = \boxed{763}$.

17.50 Let $a = a_1$ and $b = a_2$. We compute the first few terms of the sequence $\{a_n\}$ in terms of a and b:

$$\begin{aligned} a_3 &= a_2 - a_1 = b - a, \\ a_4 &= a_3 - a_2 = (b - a) - b = -a, \\ a_5 &= a_4 - a_3 = (-a) - (b - a) = -b, \\ a_6 &= a_5 - a_4 = (-b) - (-a) = a - b, \\ a_7 &= a_6 - a_5 = (a - b) - (-b) = a, \\ a_8 &= a_7 - a_6 = a - (a - b) = b, \end{aligned}$$

and so on. Since $a_7 = a_1 = a$ and $a_8 = a_2 = b$, the sequence $\{a_n\}$ is periodic.

Furthermore, any six consecutive terms are equal to $a, b, b - a, -a, -b, a - b$, in some order, and their sum is 0. Therefore, by grouping the terms in groups of six, we find that the sum of the first 1492 terms is equal to

$$a_1 + a_2 + a_3 + a_4 + a_5 + a_6 + a_7 + a_8 + a_9 + a_{10} + \cdots + a_{1487} + a_{1488} + a_{1489} + a_{1490} + a_{1491} + a_{1492}$$
$$= a_1 + a_2 + a_3 + a_4 + (a_5 + a_6 + a_7 + a_8 + a_9 + a_{10}) + \cdots + (a_{1487} + a_{1488} + a_{1489} + a_{1490} + a_{1491} + a_{1492})$$
$$= a_1 + a_2 + a_3 + a_4 = a + b + (b - a) + (-a) = 2b - a,$$

so $2b - a = 1985$. Similarly, the sum of the first 1985 terms is

$$a_1 + a_2 + a_3 + a_4 + a_5 + a_6 + a_7 + a_8 + a_9 + a_{10} + a_{11} + \cdots + a_{1980} + a_{1981} + a_{1982} + a_{1983} + a_{1984} + a_{1985}$$
$$= a_1 + a_2 + a_3 + a_4 + a_5 + (a_6 + a_7 + a_8 + a_9 + a_{10} + a_{11}) + \cdots + (a_{1980} + a_{1981} + a_{1982} + a_{1983} + a_{1984} + a_{1985})$$
$$= a_1 + a_2 + a_3 + a_4 + a_5 = a + b + (b - a) + (-a) + (-b) = b - a,$$

so $b - a = 1492$.

Solving this system of equations, we find $a = -999$ and $b = 493$. Therefore, the sum of the first 2001 terms is

$$a_1 + a_2 + a_3 + a_4 + a_5 + a_6 + a_7 + a_8 + a_9 + \cdots + a_{1996} + a_{1997} + a_{1998} + a_{1999} + a_{2000} + a_{2001}$$
$$= a_1 + a_2 + a_3 + (a_4 + a_5 + a_6 + a_7 + a_8 + a_9) + \cdots + (a_{1996} + a_{1997} + a_{1998} + a_{1999} + a_{2000} + a_{2001})$$
$$= a_1 + a_2 + a_3 = a + b + (b - a) = 2b = \boxed{986}.$$

17.51 Let S be the sum we seek. Then to find S, we look at another sum that is related. Consider the product

$$(1 + 2 + \cdots + 21)(1 + 2 + \cdots + 21).$$

If we were to expand this product, then every term of the form ij, where i and j are distinct integers from 1 to 21 inclusive, would appear twice. Also, every term of the form i^2 would appear once. Therefore,

$$(1 + 2 + \cdots + 21)(1 + 2 + \cdots + 21) = 2S + (1^2 + 2^2 + \cdots + 21^2).$$

Then

$$2S = (1 + 2 + \cdots + 21)^2 - (1^2 + 2^2 + \cdots + 21^2) = \left(\frac{21 \cdot 22}{2}\right)^2 - \frac{21 \cdot 22 \cdot 43}{6} = 53361 - 3311 = 50050,$$

so $S = 50050/2 = \boxed{25025}$.

17.52 Let $S = \displaystyle\sum_{k=1}^{\infty} \frac{k^3 + k}{2^k} = \frac{1^3 + 1}{2} + \frac{2^3 + 2}{2^2} + \frac{3^3 + 3}{2^3} + \cdots$. Then $2S = 1^3 + 1 + \dfrac{2^3 + 2}{2} + \dfrac{3^3 + 3}{2^2} + \cdots$. Subtracting the first equation from the second, we get

$$S = 2 + \frac{2^3 + 2 - (1^3 + 1)}{2} + \frac{3^3 + 3 - (2^3 + 2)}{2^2} + \cdots = 2 + \sum_{k=1}^{\infty} \frac{(k + 1)^3 + (k + 1) - k^3 - k}{2^k}$$

$$= 2 + \sum_{k=1}^{\infty} \frac{k^3 + 3k^2 + 3k + 1 + k + 1 - k^3 - k}{2^k} = 2 + \sum_{k=1}^{\infty} \frac{3k^2 + 3k + 2}{2^k}$$

$$= 2 + 3\sum_{k=1}^{\infty} \frac{k^2}{2^k} + 3\sum_{k=1}^{\infty} \frac{k}{2^k} + 2\sum_{k=1}^{\infty} \frac{1}{2^k}.$$

From the formulas in the text, we have $\displaystyle\sum_{k=1}^{\infty} kx^k = \frac{x}{(1-x)^2}$ and $\displaystyle\sum_{k=1}^{\infty} k^2 x^k = \frac{x(x+1)}{(1-x)^3}$. Taking $x = 1/2$, we get

$$\sum_{k=1}^{\infty} \frac{k}{2^k} = \frac{1/2}{(1-1/2)^2} = 2,$$

$$\sum_{k=1}^{\infty} \frac{k^2}{2^k} = \frac{1/2 \cdot (1/2 + 1)}{(1-1/2)^3} = 6, \text{ and}$$

$$\sum_{k=1}^{\infty} \frac{1}{2^k} = \frac{1}{2} + \frac{1}{4} + \frac{1}{8} + \cdots = \frac{1/2}{1-1/2} = 1.$$

Therefore, $S = 2 + 3 \cdot 6 + 3 \cdot 2 + 2 \cdot 1 = \boxed{28}$.

17.53 We can rewrite the given equation as $2f(n) + f(n+1) = (-1)^{n+1}n$. Taking $n = 1, 2, \ldots, 1985$, we obtain the system of equations

$$2f(1) + f(2) = 1,$$
$$2f(2) + f(3) = -2,$$
$$2f(3) + f(4) = 3,$$
$$\vdots$$
$$2f(1985) + f(1986) = 1985.$$

Adding all these equations, we get

$$2f(1) + 3f(2) + 3f(3) + \cdots + 3f(1985) + f(1986) = 1 - 2 + 3 - 4 + \cdots + 1985$$
$$= 1 + (-2 + 3) + (-4 + 5) + \cdots + (-1984 + 1985)$$
$$= 1 + \underbrace{1 + 1 + \cdots + 1}_{992 \text{ 1s}}$$
$$= 993.$$

Since $f(1986) = f(1)$, this equation becomes

$$3f(1) + 3f(2) + 3f(3) + \cdots + 3f(1985) = 993,$$

so $f(1) + f(2) + f(3) + \cdots + f(1985) = 993/3 = \boxed{331}$.

17.54 First, we apply partial fraction decomposition. There exist constants A, B and C such that

$$\frac{1}{k(k+1)(k+2)} = \frac{A}{k} + \frac{B}{k+1} + \frac{C}{k+2}.$$

Multiplying both sides by $k(k+1)(k+2)$, we get

$$(k+1)(k+2)A + k(k+2)B + k(k+1)C = 1.$$

Taking $k = 0$ gives $2A = 1$, so $A = 1/2$. Taking $k = -1$ gives $-B = 1$, so $B = -1$. Finally, taking $k = -2$, we get $2C = 1$, so $C = 1/2$. Hence,

$$\frac{1}{k(k+1)(k+2)} = \frac{1/2}{k} - \frac{1}{k+1} + \frac{1/2}{k+2}.$$

Summing over $1 \le k \le n$, the sum telescopes, and we get

$$\sum_{k=1}^{n} \frac{1}{k(k+1)(k+2)} = \sum_{k=1}^{n} \left(\frac{1/2}{k} - \frac{1}{k+1} + \frac{1/2}{k+2} \right)$$

$$= \left(\frac{1/2}{1} - \frac{1}{2} + \frac{1/2}{3} \right) + \left(\frac{1/2}{2} - \frac{1}{3} + \frac{1/2}{4} \right) + \left(\frac{1/2}{3} - \frac{1}{4} + \frac{1/2}{5} \right) + \cdots$$

$$+ \left(\frac{1/2}{n-1} - \frac{1}{n} + \frac{1/2}{n+1} \right) + \left(\frac{1/2}{n} - \frac{1}{n+1} + \frac{1/2}{n+2} \right)$$

$$= \frac{1/2}{1} - \frac{1/2}{2} - \frac{1/2}{n+1} + \frac{1/2}{n+2} = \frac{1}{2} - \frac{1}{4} - \frac{1}{2(n+1)} + \frac{1}{2(n+2)}$$

$$= \frac{1}{4} - \frac{1}{2(n+1)} + \frac{1}{2(n+2)} = \frac{(n+1)(n+2)}{4(n+1)(n+2)} - \frac{2(n+2)}{4(n+1)(n+2)} + \frac{2(n+1)}{4(n+1)(n+2)}$$

$$= \frac{n^2 + 3n + 2 - 2n - 4 + 2n + 2}{4(n+1)(n+2)} = \frac{n^2 + 3n}{4(n+1)(n+2)}$$

$$= \boxed{\frac{n(n+3)}{4(n+1)(n+2)}}.$$

17.55 First, we simplify the expression under the square root in the nth term:

$$1 + \frac{1}{n^2} + \frac{1}{(n+1)^2} = \frac{n^2(n+1)^2 + (n+1)^2 + n^2}{n^2(n+1)^2} = \frac{n^4 + 2n^3 + n^2 + n^2 + 2n + 1 + n^2}{n^2(n+1)^2}$$

$$= \frac{n^4 + 2n^3 + 3n^2 + 2n + 1}{n^2(n+1)^2} = \frac{(n^2 + n + 1)^2}{n^2(n+1)^2}.$$

Hence, $\sqrt{1 + \dfrac{1}{n^2} + \dfrac{1}{(n+1)^2}} = \dfrac{n^2 + n + 1}{n(n+1)}$, which is equal to

$$\frac{n^2 + n + 1}{n(n+1)} = \frac{n(n+1) + 1}{n(n+1)} = 1 + \frac{1}{n(n+1)} = 1 + \frac{1}{n} - \frac{1}{n+1}.$$

Summing over $1 \le n \le 1999$, the sum telescopes, and we get

$$S = \sum_{n=1}^{1999} \sqrt{1 + \frac{1}{n^2} + \frac{1}{(n+1)^2}} = \sum_{n=1}^{1999} \frac{n^2 + n + 1}{n(n+1)} = \sum_{n=1}^{1999} \left(1 + \frac{1}{n} - \frac{1}{n+1} \right)$$

$$= 1999 + \left(1 - \frac{1}{2} \right) + \left(\frac{1}{2} - \frac{1}{3} \right) + \cdots + \left(\frac{1}{1999} - \frac{1}{2000} \right) = 1999 + 1 - \frac{1}{2000} = \boxed{\frac{3999999}{2000}}.$$

17.56 The nth term in the series is given by

$$\frac{1}{(n+2)^2 + n} = \frac{1}{n^2 + 5n + 4} = \frac{1}{(n+1)(n+4)}.$$

Now we apply partial fraction decomposition. There exist constants A and B such that

$$\frac{1}{(n+1)(n+4)} = \frac{A}{n+1} + \frac{B}{n+4}.$$

Multiplying both sides by $(n+1)(n+4)$, we get $A(n+4) + B(n+1) = 1$. Taking $n = -1$ gives $3A = 1$, so $A = 1/3$. Taking $n = -4$ gives $-3B = 1$, so $B = -1/3$. Hence,

$$\frac{1}{(n+1)(n+4)} = \frac{1/3}{n+1} - \frac{1/3}{n+4}.$$

Summing over all positive integers n, the sum telescopes, and we get

$$\sum_{n=1}^{\infty} \frac{1}{(n+1)(n+4)} = \left(\frac{1/3}{2} - \frac{1/3}{5}\right) + \left(\frac{1/3}{3} - \frac{1/3}{6}\right) + \left(\frac{1/3}{4} - \frac{1/3}{7}\right) + \left(\frac{1/3}{5} - \frac{1/3}{8}\right) + \cdots$$

$$= \frac{1}{3\cdot 2} + \frac{1}{3\cdot 3} + \frac{1}{3\cdot 4} = \boxed{\frac{13}{36}}.$$

17.57 The fraction leads us to try to find a way to telescope the series, which in turn leads us to try to factor the denominator. As seen in Problem 9.14 in the text, we can factor $k^4 + k^2 + 1$ as follows:

$$k^4 + k^2 + 1 = (k^4 + 2k^2 + 1) - k^2 = (k^2 + 1)^2 - k^2 = (k^2 + k + 1)(k^2 - k + 1).$$

Now that we have factored the denominator of the general term, we can perform a partial fraction decomposition. Observe that the numerator k is half the difference of the factors $k^2 + k + 1$ and $k^2 - k + 1$. Hence, we may write

$$\frac{k}{k^4 + k^2 + 1} = \frac{(k^2 + k + 1)/2 - (k^2 - k + 1)/2}{(k^2 + k + 1)(k^2 - k + 1)} = \frac{1/2}{k^2 - k + 1} - \frac{1/2}{k^2 + k + 1}.$$

To see how we can take advantage of this decomposition, we write out the first few terms:

$$\sum_{k=1}^{n} \frac{k}{k^4 + k^2 + 1} = \sum_{k=1}^{n} \left(\frac{1/2}{k^2 - k + 1} - \frac{1/2}{k^2 + k + 1}\right)$$

$$= \left(\frac{1/2}{1} - \frac{1/2}{3}\right) + \left(\frac{1/2}{3} - \frac{1/2}{7}\right) + \left(\frac{1/2}{7} - \frac{1/2}{13}\right) + \cdots.$$

The sum appears to telescope. We check this algebraically as follows. The denominator of the second fraction in the kth term is $k^2 + k + 1$. The denominator of the first fraction in the $(k+1)^{\text{st}}$ term is $(k+1)^2 - (k+1) + 1 = k^2 + k + 1$. Thus, the terms cancel, which means the sum telescopes, and we get

$$\sum_{k=1}^{n} \frac{k}{k^4 + k^2 + 1} = \sum_{k=1}^{n} \left(\frac{1/2}{k^2 - k + 1} - \frac{1/2}{k^2 + k + 1}\right) = \left(\frac{1/2}{1} - \frac{1/2}{3}\right) + \left(\frac{1/2}{3} - \frac{1/2}{7}\right) + \cdots + \left(\frac{1/2}{n^2 - n + 1} - \frac{1/2}{n^2 + n + 1}\right)$$

$$= \frac{1}{2} - \frac{1}{2(n^2 + n + 1)} = \frac{n^2 + n}{2(n^2 + n + 1)} = \boxed{\frac{n(n+1)}{2(n^2 + n + 1)}}.$$

17.58 Solving for a_n in the second equation, we get

$$a_n = \frac{a_1 + a_2 + \cdots + a_{n-1}}{n^2 - 1}.$$

This formula allows us to compute the terms of the sequence recursively. To try to find a pattern, we compute the first few terms of the sequence:

$$a_2 = \frac{a_1}{2^2 - 1} = \frac{1/2}{3} = \frac{1}{6},$$

$$a_3 = \frac{a_1 + a_2}{3^2 - 1} = \frac{1/2 + 1/6}{8} = \frac{1}{12},$$

$$a_4 = \frac{a_1 + a_2 + a_3}{4^2 - 1} = \frac{1/2 + 1/6 + 1/12}{15} = \frac{1}{20},$$

and so on. The denominators are $6 = 2\cdot 3$, $12 = 3\cdot 4$, and $20 = 4\cdot 5$, so it appears that $a_n = \boxed{\frac{1}{n(n+1)}}$. We prove this using induction.

For $n = 1$, we have $\frac{1}{n(n+1)} = \frac{1}{2} = a_1$, which establishes the base case. Now assume that the formula holds for $n = 1, 2, \ldots, k$, for some positive integer k. Then

$$a_{k+1} = \frac{a_1 + a_2 + \cdots + a_k}{(k+1)^2 - 1} = \frac{\frac{1}{1\cdot 2} + \frac{1}{2\cdot 3} + \cdots + \frac{1}{k(k+1)}}{k^2 + 2k}.$$

We found in the text that $\frac{1}{1\cdot 2} + \frac{1}{2\cdot 3} + \cdots + \frac{1}{k(k+1)} = 1 - \frac{1}{k+1} = \frac{k}{k+1}$. Therefore, we have

$$a_{k+1} = \frac{k/(k+1)}{k^2 + 2k} = \frac{k}{(k+1)k(k+2)} = \frac{1}{(k+1)(k+2)}.$$

Hence, the formula holds for $n = k + 1$, and by induction, it holds for all positive integers n.

17.59 To get rid of the fraction, we multiply both sides of the given equation by a_k, to get $a_k a_{k+1} = a_{k-1}a_k - 3$. Let $b_k = a_k a_{k+1}$. Then this equation becomes $b_k = b_{k-1} + 3$, which means the sequence $\{b_k\}$ is an arithmetic sequence with first term $b_0 = a_0 a_1 = 37 \cdot 72 = 2664$ and common difference -3. Hence, $b_k = 2664 - 3k$.

We seek the integer m such that $a_m = 0$. From the given recursion, none of the terms $a_0, a_1, \ldots, a_{m-1}$ can be equal to 0. Hence, the integer m such that $a_m = 0$ is the same integer m such that $b_{m-1} = a_{m-1}a_m = 0$. Therefore, $2664 - 3(m-1) = 0$, which means $m - 1 = 2664/3 = 888$, so $m = \boxed{889}$.

17.60 First, using difference and sum of cubes, we can factor the numerator and the denominator of the nth term:

$$\frac{n^3 - 1}{n^3 + 1} = \frac{(n-1)(n^2 + n + 1)}{(n+1)(n^2 - n + 1)}.$$

To see how we can take advantage of this factorization, we write out the first few terms:

$$\prod_{n=2}^{\infty} \frac{n^3 - 1}{n^3 + 1} = \prod_{n=2}^{\infty} \frac{(n-1)(n^2 + n + 1)}{(n+1)(n^2 - n + 1)} = \frac{1 \cdot 7}{3 \cdot 3} \cdot \frac{2 \cdot 13}{4 \cdot 7} \cdot \frac{3 \cdot 21}{5 \cdot 13} \cdot \frac{4 \cdot 31}{6 \cdot 21} \cdots .$$

We see that the product telescopes. It is clear how the factors of $n - 1$ and $n + 1$ cancel, and we saw in the solution to Problem 17.57 how the factors of $n^2 + n + 1$ and $n^2 - n + 1$ cancel. The factor of $n^2 + n + 1$ in the nth term is the same as the factor $n^2 - n + 1$ in the $(n+1)^{\text{st}}$ term. After the factors cancel, all we are left with is

$$\prod_{n=2}^{\infty} \frac{n^3 - 1}{n^3 + 1} = \frac{1 \cdot 7}{3 \cdot 3} \cdot \frac{2 \cdot 13}{4 \cdot 7} \cdot \frac{3 \cdot 21}{5 \cdot 13} \cdot \frac{4 \cdot 31}{6 \cdot 21} \cdots = \boxed{\frac{2}{3}}.$$

17.61 Let $y_n = x_{n-1}x_{n+1} - x_n^2$ for all $n \geq 1$. Since $x_2 = 4x_1 - x_0 = 3$, we have $y_1 = x_0 x_2 - x_1^2 = 1 \cdot 3 - 1^2 = 2$. Then it suffices to show that $y_{n+1} = y_n$ for all $n \geq 1$.

Taking the difference between y_{n+1} and y_n, we get

$$y_{n+1} - y_n = x_n x_{n+2} - x_{n+1}^2 - x_{n-1}x_{n+1} + x_n^2.$$

We can reduce the number of different x_i that appear on the right by writing x_{n+2} and x_{n-1} in terms of x_n and x_{n+1}. Since $x_{n+2} = 4x_{n+1} - x_n$ and $x_{n-1} = 4x_n - x_{n+1}$ (we get this by rearranging $x_{n+1} = 4x_n - x_{n-1}$), this expression becomes

$$y_{n+1} - y_n = x_n(4x_{n+1} - x_n) - x_{n+1}^2 - (4x_n - x_{n+1})x_{n+1} + x_n^2 = 4x_n x_{n+1} - x_n^2 - x_{n+1}^2 - 4x_n x_{n+1} + x_{n+1}^2 + x_n^2 = 0.$$

Hence, we have $y_{n+1} = y_n$ for all $n \geq 1$. Since $y_1 = 2$, we have $y_n = x_{n-1}x_{n+1} - x_n^2 = 2$ for all $n \geq 1$.

17.62 One way to analyze an infinite sum is to look at the nth partial sum, that is, the sum of the first n terms. By difference of squares, $4 - 1$ is a factor of $4^2 - 1$, so the sum of the first two terms is

$$\frac{2}{4 - 1} + \frac{2^2}{4^2 - 1} = \frac{2(4 + 1)}{4^2 - 1} + \frac{4}{4^2 - 1} = \frac{14}{4^2 - 1}.$$

Similarly, $4^2 - 1$ is a factor of $4^4 - 1$, so the sum of the first three terms is

$$\frac{2}{4-1} + \frac{2^2}{4^2-1} + \frac{2^4}{4^4-1} = \frac{14}{4^2-1} + \frac{2^4}{4^4-1} = \frac{14(4^2+1)}{4^4-1} + \frac{16}{4^4-1} = \frac{254}{4^4-1}.$$

Note that $14 = 4^2 - 2$ and $254 = 4^4 - 2$. Hence, it appears that the nth partial sum is

$$\sum_{i=1}^{n} \frac{2^{2^{i-1}}}{4^{2^{i-1}} - 1} = \frac{4^{2^{n-1}} - 2}{4^{2^{n-1}} - 1}.$$

We prove this using induction.

We have already shown the base case, so assume that the formula holds for some positive integer $n = k$. Then

$$\sum_{i=1}^{k+1} \frac{2^{2^{i-1}}}{4^{2^{i-1}} - 1} = \sum_{i=1}^{k} \left(\frac{2^{2^{i-1}}}{4^{2^{i-1}} - 1} \right) + \frac{2^{2^k}}{4^{2^k} - 1} = \frac{4^{2^{k-1}} - 2}{4^{2^{k-1}} - 1} + \frac{2^{2^k}}{4^{2^k} - 1} = \frac{(4^{2^{k-1}} - 2)(4^{2^{k-1}} + 1)}{(4^{2^{k-1}} - 1)(4^{2^{k-1}} + 1)} + \frac{2^{2 \cdot 2^{k-1}}}{4^{2^k} - 1}$$

$$= \frac{4^{2^k} - 4^{2^{k-1}} - 2 + 4^{2^{k-1}}}{4^{2^k} - 1} = \frac{4^{2^k} - 2}{4^{2^k} - 1}.$$

Hence, the formula holds for $n = k + 1$, and by induction, it holds for all positive integers n. We can rewrite this formula as

$$\sum_{i=1}^{n} \frac{2^{2^{i-1}}}{4^{2^{i-1}} - 1} = \frac{4^{2^{n-1}} - 2}{4^{2^{n-1}} - 1} = \frac{4^{2^{n-1}} - 1 - 1}{4^{2^{n-1}} - 1} = 1 - \frac{1}{4^{2^{n-1}} - 1}.$$

As n goes to infinity, $4^{2^{n-1}} - 1$ goes to infinity. Therefore, the infinite sum is equal to $\sum_{i=1}^{\infty} \frac{2^{2^{i-1}}}{4^{2^{i-1}} - 1} = \boxed{1}$.

17.63 Let $S = \frac{1}{3} + \frac{1}{9} + \frac{2}{27} + \cdots + \frac{F_n}{3^n} + \cdots$. Inspired by our approach to arithmetico-geometric series (and geometric series) in the text, we multiply this equation by 3 to get $3S = 1 + \frac{1}{3} + \frac{2}{9} + \cdots + \frac{F_n}{3^{n-1}}$. We write both equations in terms of the F_i:

$$3S = F_1 + \frac{F_2}{3} + \frac{F_3}{9} + \cdots + \frac{F_n}{3^{n-1}} + \cdots,$$

$$S = \frac{F_1}{3} + \frac{F_2}{9} + \cdots + \frac{F_{n-1}}{3^{n-1}} + \cdots.$$

Subtracting the second equation from the first, we have

$$2S = F_1 + \frac{F_2 - F_1}{3} + \frac{F_3 - F_2}{9} + \cdots + \frac{F_n - F_{n-1}}{3^{n-1}} + \cdots.$$

From our recursion, we have $F_2 - F_1 = F_0$, $F_3 - F_2 = F_1$, and so on. In general, we have $F_n - F_{n-1} = F_{n-2}$, so our equation is

$$2S = F_1 + \frac{F_0}{3} + \frac{F_1}{9} + \frac{F_2}{27} + \cdots + \frac{F_{n-2}}{3^{n-1}} + \cdots.$$

Since $F_1 = 1$ and $F_0 = 0$, we have

$$2S = 1 + \frac{F_1}{9} + \frac{F_2}{27} + \cdots + \frac{F_{n-2}}{3^{n-1}} + \cdots.$$

After the leading 1, the remaining series on the right is what we get when we divide S by 3. Therefore, we can replace it with $S/3$ and we have $2S = 1 + \frac{S}{3}$. Solving for S gives $S = \boxed{\frac{3}{5}}$.

CHAPTER 18

_____ More Inequalities

Exercises for Section 18.1

18.1.1 By AM-HM, we have $\dfrac{a+b}{2} \geq \dfrac{2}{1/a + 1/b}$. Taking the reciprocal of both sides, we get $\dfrac{2}{a+b} \leq \dfrac{1}{2}\left(\dfrac{1}{a} + \dfrac{1}{b}\right)$. Similarly,

$$\frac{2}{a+c} \leq \frac{1}{2}\left(\frac{1}{a} + \frac{1}{c}\right) \quad \text{and} \quad \frac{2}{b+c} \leq \frac{1}{2}\left(\frac{1}{b} + \frac{1}{c}\right).$$

Adding all three inequalities, we get $\dfrac{2}{a+b} + \dfrac{2}{a+c} + \dfrac{2}{b+c} \leq \dfrac{1}{a} + \dfrac{1}{b} + \dfrac{1}{c}$. Dividing both sides by 2, we obtain the desired inequality.

18.1.2 By AM-HM on $a+b$, $a+c$, and $b+c$, we have $\dfrac{(a+b) + (a+c) + (b+c)}{3} \geq \dfrac{3}{\frac{1}{a+b} + \frac{1}{a+c} + \frac{1}{b+c}}$, so

$$\frac{2(a+b+c)}{3} \geq \frac{3}{\frac{1}{a+b} + \frac{1}{a+c} + \frac{1}{b+c}}.$$

Taking the reciprocal of both sides, we get

$$\frac{3}{2(a+b+c)} \leq \frac{1}{3}\left(\frac{1}{a+b} + \frac{1}{a+c} + \frac{1}{b+c}\right).$$

Multiplying both sides by 6, we get $\dfrac{9}{a+b+c} \leq 2\left(\dfrac{1}{a+b} + \dfrac{1}{a+c} + \dfrac{1}{b+c}\right)$.

18.1.3 The QM-AM Inequality still holds if some of the variables are negative. This is because if we were to start with all positive variables, and then replace some of the variables with their negatives, then the QM would remain the same, but the AM would decrease.

Neither the AM-GM Inequality nor the GM-HM Inequality necessarily holds if some of the variables are negative. For example, consider the three numbers 1, −2, and −4. Their AM is $(1 - 2 - 4)/3 = -5/3$, their GM is $\sqrt[3]{1 \cdot (-2) \cdot (-4)} = 2$, and their HM is $3/(1 - 1/2 - 1/4) = 12$, and $-5/3 < 2 < 12$.

18.1.4 By QM-AM, we have

$$\sqrt{\frac{(1-x_1)^2 + (x_1 - x_2)^2 + (x_2 - x_3)^2 + x_3^2}{4}} \geq \frac{(1 - x_1) + (x_1 - x_2) + (x_2 - x_3) + x_3}{4} = \frac{1}{4}.$$

Taking the square of both sides, and then multiplying both sides by 4, we get

$$(1 - x_1)^2 + (x_1 - x_2)^2 + (x_2 - x_3)^2 + x_3^2 \geq \frac{1}{4}.$$

Equality occurs if and only if $1 - x_1 = x_1 - x_2 = x_2 - x_3 = x_3$.

In such a case, we have $x_2 - x_3 = x_3$, so $x_2 = 2x_3$. Then $x_1 - x_2 = x_3$, so $x_1 = x_2 + x_3 = 3x_3$. Finally, we have $1 - x_1 = x_3$, so $1 = x_1 + x_3 = 4x_3$, which means $x_3 = 1/4$. Then $x_2 = 2x_3 = 1/2$ and $x_1 = 3x_3 = 3/4$. Therefore, the only solution is $(x_1, x_2, x_3) = \boxed{(3/4, 1/2, 1/4)}$.

18.1.5 Moving all the terms involving square roots to one side, we see that the given inequality is equivalent to

$$\sqrt{\frac{a^2 + b^2}{2}} + \sqrt{ab} \le a + b.$$

Since both sides are nonnegative, we may square both sides to obtain the equivalent inequality

$$\frac{a^2 + b^2}{2} + \sqrt{2ab(a^2 + b^2)} + ab \le a^2 + 2ab + b^2.$$

Isolating the term with the square root, the inequality becomes

$$\sqrt{2ab(a^2 + b^2)} \le \frac{a^2 + 2ab + b^2}{2},$$

which follows from the AM-GM Inequality applied to $2ab$ and $a^2 + b^2$. Since all of our steps are reversible, the original inequality holds.

18.1.6 We cannot apply the AM-GM Inequality directly, except when $m = n$. If we tried to, we would get

$$x^m + \frac{1}{x^n} \ge 2\sqrt{x^m \cdot \frac{1}{x^n}} = 2x^{(m-n)/2}.$$

If $m = n$, then the right side becomes 2. However, if $m \neq n$, then the right side becomes a function of x that does not have any minimum for $x > 0$, so the inequality is useless. Ideally, we would like a constant value on the right side.

To obtain a constant geometric mean, we write

$$x^m + \frac{1}{x^n} = \underbrace{\frac{x^m}{n} + \frac{x^m}{n} + \cdots + \frac{x^m}{n}}_{n \text{ times}} + \underbrace{\frac{1}{mx^n} + \frac{1}{mx^n} + \cdots + \frac{1}{mx^n}}_{m \text{ times}}.$$

When we take the geometric mean of the terms on the right, the x's in the numerators will cancel with those in the denominators. Applying AM-GM to these $m + n$ numbers, we get

$$\underbrace{\frac{x^m}{n} + \frac{x^m}{n} + \cdots + \frac{x^m}{n}}_{n \text{ times}} + \underbrace{\frac{1}{mx^n} + \frac{1}{mx^n} + \cdots + \frac{1}{mx^n}}_{m \text{ times}} \ge (m+n) \sqrt[m+n]{\left(\frac{1}{n}x^m\right)^n \left(\frac{1}{mx^n}\right)^m}$$

$$= (m+n)\sqrt[m+n]{\frac{1}{n^n}x^{mn} \cdot \frac{1}{m^m x^{mn}}}$$

$$= (m+n)\sqrt[m+n]{\frac{1}{n^n m^m}}$$

$$= \frac{m+n}{\sqrt[m+n]{n^n m^m}},$$

which gives us a constant value on the right side.

We must then check whether the expression $x^m + \frac{1}{x^n}$ can achieve this constant value. Equality occurs if and only if all the terms are equal, which in this case means

$$\frac{x^m}{n} = \frac{1}{mx^n} \quad \Leftrightarrow \quad x^{m+n} = \frac{n}{m} \quad \Leftrightarrow \quad x = \sqrt[m+n]{\frac{n}{m}}.$$

We check and find that for this value of x, we have

$$x^m + \frac{1}{x^n} = \frac{m+n}{\sqrt[m+n]{n^n m^m}}.$$

Therefore, the minimum value of $x^m + \frac{1}{x^n}$ is $\boxed{\dfrac{m+n}{\sqrt[m+n]{n^n m^m}}}$.

Exercises for Section 18.2

18.2.1 By the Rearrangement Inequality, we have

$$\frac{1}{a}\cdot\frac{1}{b} + \frac{1}{b}\cdot\frac{1}{c} + \frac{1}{c}\cdot\frac{1}{a} \le \frac{1}{a}\cdot\frac{1}{a} + \frac{1}{b}\cdot\frac{1}{b} + \frac{1}{c}\cdot\frac{1}{c},$$

so $\frac{1}{ab} + \frac{1}{bc} + \frac{1}{ac} \le \frac{1}{a^2} + \frac{1}{b^2} + \frac{1}{c^2}$. Putting the left side over a common denominator, we get $\frac{a+b+c}{abc} \le \frac{1}{a^2} + \frac{1}{b^2} + \frac{1}{c^2}$.

18.2.2 Let $z_1, z_2, \ldots, z_n$ be any permutation (order) of the numbers $y_1, y_2, \ldots, y_n$. Then $-z_1, -z_2, \ldots, -z_n$ is a permutation of the numbers $-y_1, -y_2, \ldots, -y_n$. Furthermore, we have $y_1 \le y_2 \le \cdots \le y_n$, and multiplying this inequality chain by -1 gives $-y_n \le -y_{n-1} \le \cdots \le -y_1$.

Therefore, as proved in the text, we have

$$x_1 \cdot (-y_n) + x_2 \cdot (-y_{n-1}) + \cdots + x_n \cdot (-y_1) \ge x_1 \cdot (-z_1) + x_2 \cdot (-z_2) + \cdots + x_n \cdot (-z_n).$$

Then

$$x_1 y_n + x_2 y_{n-1} + \cdots + x_n y_1 \le x_1 z_1 + x_2 x_2 + \cdots + x_n z_n.$$

Since this holds for any permutation $z_1, z_2, \ldots, z_n$ of the numbers $y_1, y_2, \ldots, y_n$, we conclude that the sum $x_1 y_n + x_2 y_{n-1} + \cdots + x_n y_1$ is the minimum sequence-product of the two sequences.

18.2.3 The given inequality is symmetric in x and y, so without loss of generality, let $0 \le x \le y$. Then $x^4 \le y^4$, so by the Rearrangement Inequality, we have $x^4 \cdot x + y^4 \cdot y \ge x^4 \cdot y + y^4 \cdot x$, or $x^5 + y^5 \ge x^4 y + xy^4$. Also, $x^2 \le y^2$, so again by the Rearrangement Inequality, we have $x^2 \cdot x + y^2 \cdot y \ge x^2 \cdot y + y^2 \cdot x$, or $x^3 + y^3 \ge x^2 y + xy^2$. Multiplying both sides by xy (which is nonnegative), we get $x^4 y + xy^4 \ge x^3 y^2 + x^2 y^3$.

18.2.4 Without loss of generality, let a be greater than or equal to both b and c, so we have two cases to consider, $a \ge b \ge c$ and $a \ge c \ge b$.

Suppose that $a \ge b \ge c$. Then, we have $a^3 \ge b^3 \ge c^3$ and $\frac{1}{c} \ge \frac{1}{b} \ge \frac{1}{a}$. So, the Rearrangement Inequality tells us that the smallest sequence-product of the sequences a^3, b^3, c^3 and $\frac{1}{c}, \frac{1}{b}, \frac{1}{a}$ is $a^3 \cdot \frac{1}{c} + b^3 \cdot \frac{1}{b} + c^3 \cdot \frac{1}{a}$. The greater side of the desired inequality is one of the other sequence-products of these sequence, so we apply the Rearrangement Inequality to find

$$\frac{a^3}{b} + \frac{b^3}{c} + \frac{c^3}{a} \ge a^3 \cdot \frac{1}{a} + b^3 \cdot \frac{1}{b} + c^3 \cdot \frac{1}{c} = a^2 + b^2 + c^2.$$

Since $a \ge b \ge c$, we can apply the Rearrangement Inequality to the sequences a, b, c and a, b, c to find $a^2 + b^2 + c^2 \ge ab + bc + ca$. Combining this with the inequality above gives us $\frac{a^3}{b} + \frac{b^3}{c} + \frac{c^3}{a} \ge ab + bc + ca$.

The case $a \ge c \ge b$ is follows essentially the same steps, giving us the desired inequality.

Exercises for Section 18.3

18.3.1 Each term in the series is less than 5.56, so $S_{100} < 100 \cdot 5.56 = 556$. The first term in the series is 5.5, and each term thereafter is greater than 5.5, so $S_{100} > 100 \cdot 5.5 = 550$.

18.3.2 By the Binomial Theorem, we have

$$1.01^{100} = (1 + 0.01)^{100} = 1 + \binom{100}{1}0.01 + \binom{100}{2}0.01^2 + \binom{100}{3}0.01^3 + \cdots + 0.01^{100}.$$

Since all the terms in this expansion are positive, we have $1.01^{100} > 1 + \binom{100}{1}0.01 = 1 + 100 \cdot 0.01 = 2$.

To prove that $1.01^{100} < e$, we place a bound on each binomial coefficient. For $n \geq 1$, we have

$$\binom{100}{n} = \frac{100!}{n!(100-n)} = \frac{100 \cdot 99 \cdots (101-n)}{n!}.$$

The numerator consists of n factors, each at most 100, so

$$\frac{100 \cdot 99 \cdots (101-n)}{n!} \leq \frac{100^n}{n!}.$$

Furthermore, the inequality is strict for $n \geq 2$. Therefore,

$$
\begin{aligned}
1.01^{100} &= 1 + \binom{100}{1}0.01 + \binom{100}{2}0.01^2 + \binom{100}{3}0.01^3 + \cdots + 0.01^{100} \\
&< 1 + 1 + \frac{100^2}{2!} \cdot 0.01^2 + \frac{100^3}{3!} \cdot 0.01^3 + \cdots + \frac{100^{100}}{100!} \cdot 0.01^{100} \\
&= \frac{1}{0!} + \frac{1}{1!} + \frac{1}{2!} + \cdots + \frac{1}{100!} \\
&< \frac{1}{0!} + \frac{1}{1!} + \frac{1}{2!} + \cdots + \frac{1}{100!} + \cdots \\
&= e.
\end{aligned}
$$

18.3.3 We prove the inequality by grouping the terms appropriately. First, we claim that $x^3 + y^3 \geq x^2 y + xy^2$. Moving all the terms to the greater side and factoring, the greater side becomes

$$x^3 - x^2 y - xy^2 + y^3 = x^2(x-y) - y^2(x-y) = (x^2 - y^2)(x-y) = (x+y)(x-y)^2.$$

Since $x + y \geq 0$ and $(x-y)^2 \geq 0$, we have $(x+y)(x-y)^2 \geq 0$, so the inequality $x^3 + y^3 \geq x^2 y + xy^2$ holds. Similarly, we have $x^3 + z^3 \geq x^2 z + xz^2$ and $y^3 + z^3 \geq y^2 z + yz^2$. Adding all three inequalities, we get

$$2(x^3 + y^3 + z^3) \geq x^2 y + xy^2 + x^2 z + xz^2 + y^2 z + yz^2.$$

18.3.4 We prove the inequality by induction. Taking $n = 7$, the left side becomes $7! = 5040$, and the right side becomes $3^7 = 2187$, so the inequality holds for the base case $n = 7$. Now assume that the inequality holds for some positive integer $n = k$, where $k \geq 7$, so $k! \geq 3^k$. Multiplying both sides by $k + 1$, we get $(k+1)! \geq (k+1)3^k$. Since $k \geq 7$, it is certainly true that $k + 1 \geq 3$, so $(k+1)3^k \geq 3^{k+1}$. Hence, the inequality holds for $n = k + 1$, and by induction, it holds for all positive integers $n \geq 7$.

18.3.5 To get an idea of how S behaves, we evaluate S for various values of x, y, and z. For $x = y = z = 1$, we have $S = 1$. For $x = 2$, $y = 1/2$, and $z = 1$, we have

$$S = \frac{1}{1+2+1} + \frac{1}{1+1/2+1/2} + \frac{1}{1+1+2} = 1.$$

For $x = 2$, $y = 3$, and $z = 1/6$, we have

$$S = \frac{1}{1+2+6} + \frac{1}{1+3+1/2} + \frac{1}{1+1/6+1/3} = 1.$$

These examples suggest that S is always equal to 1. To do so, we must find a way to add the fractions. One way is to make the denominators the same.

Multiplying top and bottom by x and using the fact that $xyz = 1$, the second fraction becomes

$$\frac{1}{1+y+yz} = \frac{x}{x+xy+xyz} = \frac{x}{1+x+xy}.$$

Multiplying top and bottom by xy, the third fraction becomes

$$\frac{1}{1+z+zx} = \frac{xy}{xy+xyz+x^2yz} = \frac{xy}{1+x+xy}.$$

Therefore,

$$S = \frac{1}{1+x+xy} + \frac{1}{1+y+yz} + \frac{1}{1+z+zx} = \frac{1}{1+x+xy} + \frac{x}{1+x+xy} + \frac{xy}{1+x+xy} = \frac{1+x+xy}{1+x+xy} = 1,$$

as claimed. Hence, the minimum and maximum values of S are both equal to $\boxed{1}$.

18.3.6 One way to deal with absolute value signs is to square them. For example, the inequalities $|a-b| \le 2$, and $|b-c| \le 2$, and $|c-a| \le 2$ are equivalent to $(a-b)^2 \le 4$, $(b-c)^2 \le 4$, and $(c-a)^2 \le 4$, respectively. This suggests creating an expression that relates to these squares.

First, we square the inequality $a + b + c \le 4$ to get

$$a^2 + b^2 + c^2 + 2ab + 2ac + 2bc \le 16.$$

Multiplying both sides by 2, we get

$$2a^2 + 2b^2 + 2c^2 + 4ab + 4ac + 4bc \le 32.$$

We are also given that $ab + ac + bc \ge 4$. Multiplying both sides by -6, we get $-6ab - 6ac - 6bc \le -24$. Adding this inequality to the inequality above, we get

$$2a^2 + 2b^2 + 2c^2 - 2ab - 2ac - 2bc \le 8,$$

which we can rewrite as $(a-b)^2 + (c-a)^2 + (b-c)^2 \le 8$.

Now we argue by contradiction. Suppose that at least two of $|a-b|$, $|c-a|$, and $|b-c|$ are greater than 2. Without loss of generality, let $|a-b| > 2$ and $|c-a| > 2$. Then $(a-b)^2 > 4$ and $(c-a)^2 > 4$, so

$$(a-b)^2 + (c-a)^2 + (b-c)^2 > 8,$$

contradicting the inequality above. Therefore, at most one of $|a-b|$, $|c-a|$, and $|b-c|$ is greater than 2, which means the other two are less than or equal to 2.

18.3.7 From the first equation, we have $z = 2 - x - y$. Substituting into the second equation, we get

$$xy + x(2 - x - y) + y(2 - x - y) = 1,$$

which simplifies as $x^2 + xy + y^2 - 2x - 2y + 1 = 0$. We can rewrite this equation as a quadratic in y:

$$y^2 + (x-2)y + x^2 - 2x + 1 = 0.$$

The discriminant of this quadratic is $(x-2)^2 - 4(x^2 - 2x + 1) = -3x^2 + 4x = -x(3x - 4)$, which must be nonnegative. Therefore, $0 \le x \le 4/3$.

Furthermore, for $x = 0$, we have the solution $(x, y, z) = (0, 1, 1)$, and for $x = 4/3$, we have the solution $(x, y, z) = (4/3, 1/3, 1/3)$. Hence, the minimum value of x is $\boxed{0}$, and the maximum value of x is $\boxed{4/3}$.

18.3.8 To build to the inequality with three variables, we first prove the corresponding inequality for two variables. In other words, we will prove that

$$a^a b^b \ge (ab)^{(a+b)/2}.$$

This inequality is symmetric in a and b, so without loss of generality, assume that $a \ge b$. Squaring both sides, we get

$$a^{2a} b^{2b} \ge (ab)^{a+b} = a^{a+b} b^{a+b}.$$

Dividing both sides by $a^{a+b} b^{2b}$, this inequality becomes $a^{a-b} \ge b^{a-b}$. But this inequality is clearly true, because $a \ge b > 0$. Since all of our steps are reversible, the original inequality $a^a b^b \ge (ab)^{(a+b)/2}$ is also true.

Similarly, $a^a c^c \ge (ac)^{(a+c)/2}$ and $b^b c^c \ge (bc)^{(b+c)/2}$. Taking the product of all three inequalities, we get

$$a^{2a} b^{2b} c^{2c} \ge a^{(2a+b+c)/2} b^{(a+2b+c)/2} c^{(a+b+2c)/2}.$$

Squaring both sides to get rid of the fractions in the exponents, we obtain

$$a^{4a} b^{4b} c^{4c} \ge a^{2a+b+c} b^{a+2b+c} c^{a+b+2c}.$$

Dividing both sides by $a^a b^b c^c$ gives

$$a^{3a} b^{3b} c^{3c} \ge a^{a+b+c} b^{a+b+c} c^{a+b+c} = (abc)^{a+b+c}.$$

Finally, taking the cube root of both sides, the inequality becomes $a^a b^b c^c \ge (abc)^{(a+b+c)/3}$, as desired.

Review Problems

18.16 By QM-AM, we have $\sqrt{\dfrac{a^2 + b^2 + c^2}{3}} \ge \dfrac{a + b + c}{3}$. Squaring both sides, we get $\dfrac{a^2 + b^2 + c^2}{3} \ge \dfrac{(a + b + c)^2}{9}$. Then multiplying both sides by 9, we get $3a^2 + 3b^2 + 3c^2 \ge (a + b + c)^2$.

18.17

(a) Let the vertices of the rectangle be A, B, C, and D, and let $x = AB$ and $y = BC$.

Then the perimeter of the rectangle is $2x + 2y$. Furthermore, $\angle ABC = 90°$, so $\overline{AC}$ is a diameter of the circle, which has length 2. Therefore, by the Pythagorean Theorem applied to triangle ABC, we have $x^2 + y^2 = 4$. By QM-AM, we have

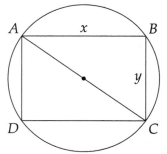

$$\frac{x + y}{2} \le \sqrt{\frac{x^2 + y^2}{2}} = \sqrt{2},$$

so $2x + 2y \le 4\sqrt{2}$. Equality occurs if and only if $x = y = \sqrt{2}$, i.e., if rectangle $ABCD$ is a square. Hence, the maximum perimeter of rectangle $ABCD$ is $\boxed{4\sqrt{2}}$.

(b) Let the vertices of the rectangle be A, B, C, and D, such that A and D lie on the diameter and B and C lie on the arc, and let $x = BC$ and $y = AB$. Then the perimeter of the rectangle is $2x + 2y$, and the center of the circle O is the midpoint of $\overline{AD}$, so $AO = OD = x/2$.

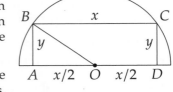

Then $\angle BAO = 90°$, and as a radius of the semicircle, $OB = 1$, so the Pythagorean Theorem applied to triangle OAB gives us $\frac{x^2}{4} + y^2 = 1$. Thus, we want to maximize $2x + 2y$ subject to the condition that $\frac{x^2}{4} + y^2 = 1$.

First, we note that maximizing $2x + 2y$ is the same as maximizing $x + y$. We want to maximize a sum, and we have a sum of squares, so QM-AM is a natural tool to try. However, applying QM-AM to x and y just gives us $\sqrt{\frac{x^2+y^2}{2}} \geq \frac{x+y}{2}$. Unfortunately, we can't do anything with that sum of squares on the greater side. We need $\frac{x^2}{4} + y^2$, since we know this equals 1. This gives us the idea to write x as $\frac{x}{2} + \frac{x}{2}$ and apply QM-AM to $\frac{x}{2}$, $\frac{x}{2}$, and y. This gives us

$$\sqrt{\frac{\frac{x^2}{4} + \frac{x^2}{4} + y^2}{3}} \geq \frac{\frac{x}{2} + \frac{x}{2} + y}{3}.$$

Simplifying the left side gives us $\frac{x^2}{2} + y^2$ inside the radical. Closer, but we're still not there. But we have a good idea what to try next. We went from $x^2 + y^2$ to $\frac{x^2}{2} + y^2$ by writing $x = \frac{x}{2} + \frac{x}{2}$. So, we try $x = \frac{x}{4} + \frac{x}{4} + \frac{x}{4} + \frac{x}{4}$. Now, QM-AM gives

$$\sqrt{\frac{\frac{x^2}{16} + \frac{x^2}{16} + \frac{x^2}{16} + \frac{x^2}{16} + y^2}{5}} \geq \frac{\frac{x}{4} + \frac{x}{4} + \frac{x}{4} + \frac{x}{4} + y}{5}.$$

Simplifying both sides gives

$$\sqrt{\frac{\frac{x^2}{4} + y^2}{5}} \geq \frac{x + y}{5}.$$

Aha! Since $\frac{x^2}{4} + y^2 = 1$, we have $x + y \leq 5\sqrt{\frac{1}{5}} = \sqrt{5}$, which gives us the maximum perimeter of $2(x+y) = 2\sqrt{5}$. Equality holds when $\frac{x}{4} = y$. Substituting this into $\frac{x^2}{4} + y^2 = 1$ gives $x = \frac{4}{\sqrt{5}}$ and $y = \frac{1}{\sqrt{5}}$. Using this, we confirm that the maximum perimeter of $2\sqrt{5}$ is indeed achievable.

18.18 The Rearrangement Inequality applied to a, b, c and a, b, c gives $ab + ac + bc \leq a^2 + b^2 + c^2 = 1$, which proves the right inequality. (We've also seen other methods to prove this in the text, such as using the fact that $(a - b)^2 + (b - c)^2 + (c - a)^2 \geq 0$.)

By the Trivial Inequality, we have $(a + b + c)^2 \geq 0$, so $a^2 + b^2 + c^2 + 2ab + 2ac + 2bc \geq 0$. Therefore, we have $ab + ac + bc \geq -\frac{a^2+b^2+c^2}{2} = -\frac{1}{2}$, which proves the left inequality.

18.19 As shown in the previous problem, we have $a^2 + b^2 + c^2 \geq ab + ac + bc$ for all real numbers a, b, and c. In particular, for $a = xy$, $b = xz$, and $c = yz$, we have $x^2y^2 + x^2z^2 + y^2z^2 \geq x^2yz + xy^2z + xyz^2 = xyz(x + y + z)$.

We also could have directly applied the Rearrangement Inequality to xy, xz, yz and xy, xz, yz, which gives

$$(xy)(xy) + (xz)(xz) + (yz)(yz) \geq (xy)(xz) + (xz)(yz) + (yz)(xy).$$

This turns into the desired inequality $x^2y^2 + x^2z^2 + y^2z^2 \geq xyz(x + y + z)$.

18.20 Let $x = a + \frac{1}{b}$, $y = b + \frac{1}{c}$, and $z = c + \frac{1}{a}$. By QM-AM, we have $\sqrt{\frac{x^2 + y^2 + z^2}{3}} \geq \frac{x + y + z}{3}$. We also have

$$x + y + z = a + b + c + \frac{1}{a} + \frac{1}{b} + \frac{1}{c} = \frac{1}{a} + \frac{1}{b} + \frac{1}{c} + 6.$$

By AM-HM, $\dfrac{a+b+c}{3} \geq \dfrac{3}{1/a+1/b+1/c}$, so $\dfrac{1}{a}+\dfrac{1}{b}+\dfrac{1}{c} \geq \dfrac{9}{a+b+c} = \dfrac{9}{6} = \dfrac{3}{2}$. Therefore, $x+y+z \geq \dfrac{3}{2}+6 = \dfrac{15}{2}$.

Hence, $\sqrt{\dfrac{x^2+y^2+z^2}{3}} \geq \dfrac{x+y+z}{3} \geq \dfrac{5}{2}$. Squaring both sides, we get $\dfrac{x^2+y^2+z^2}{3} \geq \dfrac{25}{4}$, so

$$\left(a+\dfrac{1}{b}\right)^2 + \left(b+\dfrac{1}{c}\right)^2 + \left(c+\dfrac{1}{a}\right)^2 = x^2+y^2+z^2 \geq \dfrac{75}{4}.$$

18.21 We wish to compare the sum $x+y$ with the product x^2y. This makes us think of using AM-GM. However, to get the factor x^2, we need two factors of x. We can obtain two factors of x by writing the sum $x+y$ as $\frac{x}{2}+\frac{x}{2}+y$. Then by AM-GM, we have

$$2 = \dfrac{\frac{x}{2}+\frac{x}{2}+y}{3} \geq \sqrt[3]{\dfrac{x}{2}\cdot\dfrac{x}{2}\cdot y} = \sqrt[3]{\dfrac{x^2y}{4}},$$

so $x^2y \leq 4\cdot 2^3 = 32$. Equality occurs if and only if $\frac{x}{2}=y$, or $x=2y$. Substituting into $x+y=6$, we get $3y=6$, or $y=2$, and $x=4$. Hence, the maximum value of x^2y is $\boxed{32}$.

18.22 Starting with the inequality $ab \leq a+3b$, we move all the terms to one side, to get $ab-a-3b \leq 0$. Then, we can add 3 to both sides and factor the left side (in the spirit of Simon's Favorite Factoring Trick), to get

$$(a-3)(b-1) \leq 3. \tag{$*$}$$

If $a=1$, then the condition $(*)$ becomes $-2(b-1) \leq 3$, or $b-1 \geq -\frac{3}{2}$. The positive integer $b=2$ satisfies both this inequality and $b>a$, so there is a solution for $a=1$.

If $a=2$, then the condition $(*)$ becomes $-(b-1) \leq 3$, or $b-1 \geq -3$. The positive integer $b=3$ satisfies both this inequality and $b>a$, so there is a solution for $a=2$.

If $a=3$, then the condition $(*)$ becomes $0 \leq 3$, so we may take any positive integer $b>a=3$. Thus, there is a solution for $a=3$.

If $a\geq 4$, then $b>a\geq 4$, so $b\geq 5$. Therefore $(a-3)(b-1) \geq 1\cdot 4 = 4$, which means the condition $(*)$ cannot be satisfied. Therefore, the conditions $a<b$ and $ab \leq a+3b$ can only be satisfied for $\boxed{3}$ positive values of a, namely $a=1, 2$, and 3.

18.23 First, we claim that

$$\left(\dfrac{a_1}{b_1}+\dfrac{a_2}{b_2}+\cdots+\dfrac{a_n}{b_n}\right)\left(\dfrac{b_1}{a_1}+\dfrac{b_2}{a_2}+\cdots+\dfrac{b_n}{a_n}\right) \geq n^2.$$

By AM-HM on the n numbers $a_1/b_1, a_2/b_2, \ldots, a_n/b_n$, we have

$$\dfrac{a_1/b_1 + a_2/b_2 + \cdots + a_n/b_n}{n} \geq \dfrac{n}{b_1/a_1 + b_2/a_2 + \cdots + b_n/a_n}.$$

Cross-multiplying, we get $\left(\dfrac{a_1}{b_1}+\dfrac{a_2}{b_2}+\cdots+\dfrac{a_n}{b_n}\right)\left(\dfrac{b_1}{a_1}+\dfrac{b_2}{a_2}+\cdots+\dfrac{b_n}{a_n}\right) \geq n^2$.

This inequality also follows from Cauchy-Schwarz:

$$\left(\dfrac{a_1}{b_1}+\dfrac{a_2}{b_2}+\cdots+\dfrac{a_n}{b_n}\right)\left(\dfrac{b_1}{a_1}+\dfrac{b_2}{a_2}+\cdots+\dfrac{b_n}{a_n}\right) \geq \left(\sqrt{\dfrac{a_1}{b_1}\cdot\dfrac{b_1}{a_1}}+\sqrt{\dfrac{a_2}{b_2}\cdot\dfrac{b_2}{a_2}}+\cdots+\sqrt{\dfrac{a_n}{b_n}\cdot\dfrac{b_n}{a_n}}\right)^2$$
$$= (1+1+\cdots+1)^2$$
$$= n^2.$$

We now argue by contradiction. If $\frac{a_1}{b_1} + \frac{a_2}{b_2} + \cdots + \frac{a_n}{b_n} < n$ and $\frac{b_1}{a_1} + \frac{b_2}{a_2} + \cdots + \frac{b_n}{a_n} < n$, then

$$\left(\frac{a_1}{b_1} + \frac{a_2}{b_2} + \cdots + \frac{a_n}{b_n}\right)\left(\frac{b_1}{a_1} + \frac{b_2}{a_2} + \cdots + \frac{b_n}{a_n}\right) < n^2,$$

which is a contradiction. Therefore, at least one of the inequalities

$$\frac{a_1}{b_1} + \frac{a_2}{b_2} + \cdots + \frac{a_n}{b_n} \geq n \quad \text{and} \quad \frac{b_1}{a_1} + \frac{b_2}{a_2} + \cdots + \frac{b_n}{a_n} \geq n$$

must be satisfied.

18.24 Without loss of generality, let e be the largest positive integer among a, b, c, d, and e, since we wish to maximize e. Then

$$abcde = a + b + c + d + e \leq 5e,$$

so $abcd \leq 5$. We divide into cases, depending on the value of $abcd$.

If $abcd = 5$, then one of a, b, c, and d is equal to 5, and the others are equal to 1. Solving for e in the equation $abcde = a + b + c + d + e$, we get

$$e = \frac{a+b+c+d}{abcd - 1} = \frac{1+1+1+5}{4} = 2.$$

By assumption, e is the greatest positive integer, so we reject this solution.

If $abcd = 4$, then either one of a, b, c, and d is equal to 4, and the others are equal to 1, or two are equal to 2 and two are equal to 1. In the first case, $e = \frac{1+1+1+4}{3} = \frac{7}{3}$, which is not an integer, and in the second case, $e = \frac{1+2+1+2}{3} = 2$.

If $abcd = 3$, then one of a, b, c, and d is equal to 3, and the others are equal to 1, so $e = \frac{1+1+1+3}{2} = 3$.

If $abcd = 2$, then one of a, b, c, and d is equal to 2, and the others are equal to 1, so $e = \frac{1+1+1+2}{1} = 5$.

Finally, if $abcd = 1$, then all of a, b, c, and d are equal to 1. But then the given equation becomes $e = e + 4$, which has no solutions.

Therefore, the only solutions in positive integers are $(a, b, c, d, e) = (1, 1, 2, 2, 2)$, $(1, 1, 1, 3, 3)$, and $(1, 1, 1, 2, 5)$ (and their permutations). We conclude that the maximum value of e is $\boxed{5}$.

18.25 By the AM-HM inequality on the numbers $a + b$, $a + c$, and $b + c$, we have

$$\frac{(a + b) + (a + c) + (b + c)}{3} \geq \frac{3}{\frac{1}{a+b} + \frac{1}{a+c} + \frac{1}{b+c}}.$$

Multiplying both sides by $\frac{1}{a+b} + \frac{1}{a+c} + \frac{1}{b+c}$ and simplifying, we get $\frac{1}{3}(2a + 2b + 2c)\left(\frac{1}{a+b} + \frac{1}{a+c} + \frac{1}{b+c}\right) \geq 3$, so we have $(a + b + c)\left(\frac{1}{a+b} + \frac{1}{a+c} + \frac{1}{b+c}\right) \geq \frac{9}{2}$.

18.26 Let $x = a + \frac{1}{a}$ and $y = b + \frac{1}{b}$. By the QM-AM inequality, we have $\sqrt{\frac{x^2+y^2}{2}} \geq \frac{x+y}{2}$. Squaring both sides gives $\frac{x^2+y^2}{2} \geq \frac{(x+y)^2}{4}$, so $x^2 + y^2 \geq \frac{(x+y)^2}{2}$. In other words,

$$\left(a + \frac{1}{a}\right)^2 + \left(b + \frac{1}{b}\right)^2 \geq \frac{1}{2}\left(a + \frac{1}{a} + b + \frac{1}{b}\right)^2 = \frac{1}{2}\left(a + b + \frac{a+b}{ab}\right)^2 = \frac{1}{2}\left(1 + \frac{1}{ab}\right)^2.$$

By AM-GM, we have $\sqrt{ab} \leq \frac{a+b}{2} = \frac{1}{2}$, so $ab \leq \frac{1}{4}$, which means $\frac{1}{ab} \geq 4$ (because $a, b > 0$). Therefore,

$$\left(a + \frac{1}{a}\right)^2 + \left(b + \frac{1}{b}\right)^2 \geq \frac{1}{2}\left(1 + \frac{1}{ab}\right)^2 \geq \frac{1}{2}(1 + 4)^2 = \frac{25}{2}.$$

18.27 We want to maximize a product, so we think of using AM-GM. The sum of the factors $x + x + (1 - x) = 1 + x$ is not a constant, but the sum $x + x + 2(1 - x) = 2$ is. Therefore, by AM-GM on the three numbers x, x, and $2(1 - x)$, we get

$$\frac{2}{3} = \frac{x + x + 2(1 - x)}{3} \geq \sqrt[3]{2x^2(1 - x)}.$$

Therefore,

$$2x^2(1 - x) \leq \left(\frac{2}{3}\right)^3 = \frac{8}{27},$$

so $x^2(1 - x) \leq 4/27$. Equality occurs if and only if $x = 2(1 - x) = 2/3$, which gives $x = 2/3$. Hence, the maximum value of $x^2(1 - x)$ is $\boxed{4/27}$.

18.28 We move all the terms to the larger side and factor, which produces

$$x^7 - x^4 - x^3 + 1 = x^4(x^3 - 1) - (x^3 - 1) = (x^4 - 1)(x^3 - 1) = (x - 1)(x^3 + x^2 + x + 1)(x - 1)(x^2 + x + 1)$$
$$= (x - 1)^2(x^3 + x^2 + x + 1)(x^2 + x + 1),$$

so we must show $(x - 1)^2(x^3 + x^2 + x + 1)(x^2 + x + 1) \geq 0$. By the Trivial Inequality, we have $(x - 1)^2 \geq 0$, and since $x \geq 0$, we have $x^3 + x^2 + x + 1 \geq 1 > 0$ and $x^2 + x + 1 \geq 1 > 0$. Hence, the inequality holds. (Note that we also could have deduced that $(x^4 - 1)(x^3 - 1) \geq 0$ because the two factors have the same sign, or are 0, for all values of x.)

18.29 Multiplying both sides by a^2b^2 to get rid of the fractions, the given inequality becomes $a^{10} + b^{10} \geq a^8b^2 + a^2b^8$. This inequality is symmetric in both a and b, so without loss of generality, we may assume that $a^2 \leq b^2$. Then $a^8 \leq b^8$. Therefore, by the Rearrangement Inequality, we have $a^8 \cdot a^2 + b^8 \cdot b^2 \geq a^8 \cdot b^2 + b^8 \cdot a^2$, which becomes $a^{10} + b^{10} \geq a^8b^2 + a^2b^8$, as desired.

18.30 First, we take the case where one or more of the factors on the right side is negative. Clearly, any of the factors $a + b - c$, $a - b + c$, and $-a + b + c$ can be negative. However, if two of the factors are negative, say $a + b - c$ and $a - b + c$, then their sum, namely $(a + b - c) + (a - b + c) = 2a$, must also be negative. This is a contradiction, because a is nonnegative. Therefore, at most one of the factors can be negative.

In other words, if one of the factors on the right side is negative, then the other two factors are nonnegative, which means the product is either 0 or negative. But the left side is always nonnegative, so the inequality holds in this case. Henceforth, we assume that each factor $a + b - c$, $a - b + c$, and $-a + b + c$ is nonnegative.

Let $x = a + b - c$, $y = a - b + c$, and $z = -a + b + c$, so $x \geq 0$, $y \geq 0$, and $z \geq 0$. We saw above that $x + y = 2a$. Similarly, $x + z = 2b$, and $y + z = 2c$. Hence, we can rewrite the given inequality as

$$\frac{x + y}{2} \cdot \frac{x + z}{2} \cdot \frac{y + z}{2} \geq xyz.$$

By AM-GM, $\frac{x+y}{2} \geq \sqrt{xy}$, $\frac{x+z}{2} \geq \sqrt{xz}$, and $\frac{y+z}{2} \geq \sqrt{yz}$. Therefore,

$$\frac{x + y}{2} \cdot \frac{x + z}{2} \cdot \frac{y + z}{2} \geq \sqrt{xy} \cdot \sqrt{xz} \cdot \sqrt{yz} = xyz,$$

as desired.

18.31 We argue by contradiction. Suppose that

$$a(1 - b) > \frac{1}{4}, \quad b(1 - c) > \frac{1}{4}, \quad \text{and} \quad c(1 - a) > \frac{1}{4}.$$

If $a > 1$, then $c(1 - a) < 0$, so the inequality $c(1 - a) > 1/4$ cannot hold. Similarly, if $b > 1$, then the inequality $a(1 - b) > 1/4$ cannot hold, and if $c > 1$, then the inequality $b(1 - c) > 1/4$ cannot hold. Therefore, we have $a \leq 1$, $b \leq 1$, and $c \leq 1$.

Then each of the quantities $a(1 - b)$, $b(1 - c)$, and $c(1 - a)$ is positive. Hence, we may take product of all three inequalities, to get $a(1 - a) \cdot b(1 - b) \cdot c(1 - c) > \frac{1}{64}$.

However, by AM-GM, we have

$$a(1-a) \le \left[\frac{a+(1-a)}{2}\right]^2 = \frac{1}{4}.$$

Similarly, $b(1-b) \le \frac{1}{4}$ and $c(1-c) \le \frac{1}{4}$. Taking the product of these three inequalities, we get

$$a(1-a) \cdot b(1-b) \cdot c(1-c) \le \frac{1}{64},$$

so we have a contradiction. Therefore, all three inequalities $a(1-b) > \frac{1}{4}$, $b(1-c) > \frac{1}{4}$, and $c(1-a) > \frac{1}{4}$ cannot hold simultaneously.

18.32 Without loss of generality, we let $x \ge y \ge z$. Because these are all positive, we have $\frac{1}{z} \ge \frac{1}{y} \ge \frac{1}{x}$ and $xy \ge xz \ge yz$. Therefore, the Rearrangement Inequality gives

$$\frac{xy}{z} + \frac{xz}{y} + \frac{yz}{x} \ge \frac{xy}{y} + \frac{yz}{z} + \frac{xz}{x} = x + y + z = 1,$$

as desired.

18.33 By AM-GM, we have $\sqrt[n]{1 \cdot 3 \cdot 5 \cdots (2n-1)} \le \frac{1+3+5+\cdots+(2n-1)}{n} = \frac{n^2}{n} = n$, so $1 \cdot 3 \cdot 5 \cdots (2n-1) \le n^n$.

Challenge Problems

18.34 *Solution 1.* Moving all the terms to one side, we obtain the inequality

$$x^6 - x^3 - x^2 - x + 2 \ge 0.$$

We first find integer roots of the polynomial on the left side, and we find that 1 is a double root. Factoring $x-1$ out twice gives us

$$(x-1)^2(x^4 + 2x^3 + 3x^2 + 3x + 2) \ge 0.$$

By the Trivial Inequality, we have $(x-1)^2 \ge 0$, so it suffices to prove that $x^4 + 2x^3 + 3x^2 + 3x + 2 \ge 0$. Completing the square on the first two terms x^4 and $2x^3$, we can rewrite the inequality as

$$(x^4 + 2x^3 + x^2) + 2x^2 + 3x + 2 = (x^2 + x)^2 + 2x^2 + 3x + 2 \ge 0.$$

Then, completing the square on the next two terms $2x^2$ and $3x$, we get

$$(x^2+x)^2 + 2x^2 + 3x + 2 = (x^2+x)^2 + 2\left(x^2 + \frac{3}{2}x\right) + 2 = (x^2+x)^2 + 2\left[\left(x + \frac{3}{4}\right)^2 - \frac{9}{16}\right] + 2$$

$$= (x^2+x)^2 + 2\left(x + \frac{3}{4}\right)^2 + \frac{7}{8}.$$

Hence, by the Trivial Inequality, we have

$$(x^2+x)^2 + 2x^2 + 3x + 2 = (x^2+x)^2 + 2\left(x + \frac{3}{4}\right)^2 + \frac{7}{8} \ge \frac{7}{8} \ge 0.$$

Solution 2. We divide into cases, whether x is nonnegative or x is negative.

If x is nonnegative, then by AM-GM, we have

$$\frac{x^6 + 1}{2} \geq \sqrt{x^6} = x^3,$$

$$\frac{x^6 + 1 + 1}{3} \geq \sqrt[3]{x^6} = x^2,$$

$$\frac{x^6 + 1 + 1 + 1 + 1 + 1}{6} \geq \sqrt[6]{x^6} = x.$$

Adding all three inequalities, we get $x^6 + 2 \geq x^3 + x^2 + x$.

If x is negative, then let $x = -y$, so $y > 0$. Substituting, the given inequality becomes $y^6 + 2 \geq -y^3 + y^2 - y$, or $y^6 + y^3 + y + 2 \geq y^2$. If $y \geq 1$, then $y^6 \geq y^2$. If $0 < y < 1$, then $2 > 1 > y^2$. In either case, the inequality holds.

18.35   First, note that if the only steps in a walk are to the north or to the east, then it is self-avoiding. Hence, the number of such walks provides a lower bound for $f(n)$. There are two choices for each step, to the north or to the east. Furthermore, any combination of steps is possible. Therefore, the total of number of walks with n steps, where each step is to the north or to the east, is 2^n. This gives us the bound $f(n) > 2^n$.

Next, note that in a self-avoiding walk, the walk never immediately doubles back on itself. For example, if one step is to the north, then the next step cannot be to the south. There are four choices for the first step, in each of the four cardinal directions. For each step thereafter, there are at most three possible choices: the four cardinal directions, except for the direction that takes the walk back to the previous point. Therefore, the total number of walks that do not double back is at most $4 \cdot 3^{n-1}$. This gives us the bound $f(n) \leq 4 \cdot 3^{n-1}$.

18.36   First, we apply the Rearrangement Inequality to two copies of $\frac{1}{\sqrt{bc}}, \frac{1}{\sqrt{ac}}, \frac{1}{\sqrt{ab}}$. We want the smaller side of the desired inequality to show up on the smaller side of our application of the Rearrangement Inequality, so we pair up the terms appropriately to find

$$\frac{1}{\sqrt{bc}} \cdot \frac{1}{\sqrt{bc}} + \frac{1}{\sqrt{ac}} \cdot \frac{1}{\sqrt{ac}} + \frac{1}{\sqrt{ab}} \cdot \frac{1}{\sqrt{ab}} \geq \frac{1}{\sqrt{ab}} \cdot \frac{1}{\sqrt{ac}} + \frac{1}{\sqrt{bc}} \cdot \frac{1}{\sqrt{ab}} + \frac{1}{\sqrt{ac}} \cdot \frac{1}{\sqrt{bc}},$$

so

$$\frac{1}{bc} + \frac{1}{ac} + \frac{1}{ab} \geq \frac{1}{a\sqrt{bc}} + \frac{1}{b\sqrt{ac}} + \frac{1}{c\sqrt{ab}}. \tag{18.1}$$

We then apply Rearrangement Inequality to two copies of $\frac{1}{c}, \frac{1}{b}, \frac{1}{a}$ to give

$$\frac{1}{a^2} + \frac{1}{b^2} + \frac{1}{c^2} \geq \frac{1}{bc} + \frac{1}{ac} + \frac{1}{bc}.$$

Combining this with the inequality on line (18.1) gives the desired inequality.

18.37   By the AM-HM inequality on the n numbers $S - x_1, S - x_2, \ldots, S - x_n$, we have

$$\frac{(S - x_1) + (S - x_2) + \cdots + (S - x_n)}{n} \geq \frac{n}{\frac{1}{S - x_1} + \frac{1}{S - x_2} + \cdots + \frac{1}{S - x_n}}.$$

Since $(S - x_1) + (S - x_2) + \cdots + (S - x_n) = nS - (x_1 + x_2 + \cdots + x_n) = nS - S = (n - 1)S$, this simplifies as

$$\frac{(n - 1)S}{n} \geq \frac{n}{\frac{1}{S - x_1} + \frac{1}{S - x_2} + \cdots + \frac{1}{S - x_n}}.$$

Rearranging, we get $\dfrac{S}{S - x_1} + \dfrac{S}{S - x_2} + \cdots + \dfrac{S}{S - x_n} \geq \dfrac{n^2}{n - 1}$.

Equality occurs if and only if $S - x_1 = S - x_2 = \cdots = S - x_n$, which is equivalent to $x_1 = x_2 = \cdots = x_n$.

18.38 Let $S = \dfrac{xy}{z} + \dfrac{xz}{y} + \dfrac{yz}{x}$. Then $S^2 = \dfrac{x^2y^2}{z^2} + \dfrac{x^2z^2}{y^2} + \dfrac{y^2z^2}{x^2} + 2x^2 + 2y^2 + 2z^2 = \dfrac{x^2y^2}{z^2} + \dfrac{x^2z^2}{y^2} + \dfrac{y^2z^2}{x^2} + 2.$

We have shown both earlier in these solutions and in the text that $a^2 + b^2 + c^2 \geq ab + ac + bc$ for any real numbers a, b, and c. In particular, taking $a = xy/z$, $b = xz/y$, and $c = yz/x$, we get

$$\frac{x^2y^2}{z^2} + \frac{x^2z^2}{y^2} + \frac{y^2z^2}{x^2} \geq x^2 + y^2 + z^2 = 1.$$

Therefore, $S^2 \geq 2 + 1 = 3$, so $S \geq \sqrt{3}$.

Furthermore, if $x = y = z = 1/\sqrt{3}$, then $x^2 + y^2 + z^2 = 1/3 + 1/3 + 1/3 = 1$, and

$$\frac{xy}{z} + \frac{xz}{y} + \frac{yz}{x} = \frac{1}{\sqrt{3}} + \frac{1}{\sqrt{3}} + \frac{1}{\sqrt{3}} = \frac{3}{\sqrt{3}} = \sqrt{3}.$$

Hence, the minimum value of S is $\boxed{\sqrt{3}}$.

18.39 First, by AM-GM, we have $\dfrac{1}{3} = \dfrac{x + y + z}{3} \geq \sqrt[3]{xyz}$, so $xyz \leq (1/3)^3 = 1/27$.

Expanding the left side of the given inequality, we get

$$\left(1 + \frac{1}{x}\right)\left(1 + \frac{1}{y}\right)\left(1 + \frac{1}{z}\right) = 1 + \frac{1}{x} + \frac{1}{y} + \frac{1}{z} + \frac{1}{xy} + \frac{1}{xz} + \frac{1}{yz} + \frac{1}{xyz}.$$

By AM-GM, we have

$$\frac{1}{x} + \frac{1}{y} + \frac{1}{z} \geq \frac{3}{\sqrt[3]{xyz}} \geq 3 \cdot \sqrt[3]{27} = 3 \cdot 3 = 9,$$

$$\frac{1}{xy} + \frac{1}{xz} + \frac{1}{yz} \geq \frac{3}{\sqrt[3]{x^2y^2z^2}} \geq 3 \cdot \sqrt[3]{27^2} = 3 \cdot 3^2 = 27,$$

and $\frac{1}{xyz} \geq 27$. Therefore,

$$\left(1 + \frac{1}{x}\right)\left(1 + \frac{1}{y}\right)\left(1 + \frac{1}{z}\right) = 1 + \frac{1}{x} + \frac{1}{y} + \frac{1}{z} + \frac{1}{xy} + \frac{1}{xz} + \frac{1}{yz} + \frac{1}{xyz} \geq 1 + 9 + 27 + 27 = 64.$$

Equality occurs if and only if $x = y = z = 1/3$.

18.40 If A_k is the largest of the A_i, then we must have $A_k/A_{k-1} \geq 1$. So, we'll start by investigating this inequality, hoping to get some restriction on the value of k. We have

$$\frac{A_k}{A_{k-1}} = \frac{\binom{1000}{k}0.2^k}{\binom{1000}{k-1}0.2^{k-1}} = \frac{0.2 \cdot \frac{1000!}{k!(1000-k)!}}{\frac{1000!}{(k-1)!(1001-k)!}} = \frac{0.2 \cdot (k-1)!(1001-k)!}{k!(1000-k)!} = \frac{0.2 \cdot (1001-k)}{k}.$$

Hence, the inequality $A_k \geq A_{k-1}$ is equivalent to $\frac{0.2 \cdot (1001-k)}{k} \geq 1$. Multiplying both sides by $5k$, we get $1001 - k \geq 5k$, or $6k \leq 1001$. Then $k \leq \frac{1001}{6} = 166 + \frac{5}{6}$. In other words, we find that $A_k/A_{k-1} \geq 1$ for all $k \leq 166$. Furthermore, equality in $A_k/A_{k-1} \geq 1$ only holds for $k = 166\frac{5}{6}$, and we have $A_k/A_{k-1} > 1$ for all $k \leq 166$. Conversely, we can use our work above to deduce that $A_k/A_{k-1} < 1$ for all $k > 166$. Therefore, we have $A_k > A_{k-1}$ for $k \leq 166$ and $A_k < A_{k-1}$ for $k > 166$. This means the sequence of A_k increases as k goes from 0 to 166, then decreases thereafter. Hence, A_k is maximized for $k = \boxed{166}$.

18.41 Since $1 \le r \le s \le t \le 4$, each of the quantities $r - 1$, $\frac{s}{r} - 1$, $\frac{t}{s} - 1$, and $\frac{4}{t} - 1$ is nonnegative. By QM-AM, we have

$$\sqrt{\frac{(r-1)^2 + (s/r - 1)^2 + (t/s - 1)^2 + (4/t - 1)^2}{4}} \ge \frac{r - 1 + s/r - 1 + t/s - 1 + 4/t - 1}{4} = \frac{r + s/r + t/s + 4/t - 4}{4}$$

$$= \frac{r + s/r + t/s + 4/t}{4} - 1.$$

Then by AM-GM, we have $\dfrac{r + s/r + t/s + 4/t}{4} \ge \sqrt[4]{r \cdot \dfrac{s}{r} \cdot \dfrac{t}{s} \cdot \dfrac{4}{t}} = \sqrt[4]{4} = \sqrt{2}$. Therefore,

$$\sqrt{\frac{(r-1)^2 + (s/r - 1)^2 + (t/s - 1)^2 + (4/t - 1)^2}{4}} \ge \sqrt{2} - 1.$$

Squaring both sides, we get $\dfrac{(r-1)^2 + (s/r - 1)^2 + (t/s - 1)^2 + (4/t - 1)^2}{4} \ge (\sqrt{2} - 1)^2 = 3 - 2\sqrt{2}$, so

$$(r-1)^2 + \left(\frac{s}{r} - 1\right)^2 + \left(\frac{t}{s} - 1\right)^2 + \left(\frac{4}{t} - 1\right)^2 \ge 12 - 8\sqrt{2}.$$

Equality occurs if and only if $r = s/r = t/s = 4/t = \sqrt{2}$. Solving for r, s, and t, we find $r = \sqrt{2}$, $s = 2$, and $t = 2\sqrt{2}$. Hence, the minimum value is $\boxed{12 - 8\sqrt{2}}$.

18.42 Since a and b are positive, and $x^2 > 0$, we can multiply both sides by $4abx^2$, to obtain the equivalent inequality $4ab(x + a)(x - b) \le (a + b)^2 x^2$. Subtracting $4ab(x + a)(x - b)$ from both sides gives us the equivalent inequality $(a + b)^2 x^2 - 4ab(x + a)(x - b) \ge 0$. Expanding and rearranging the left side gives us

$$(a + b)^2 x^2 - 4ab(x + a)(x - b) = (a^2 + 2ab + b^2)x^2 - 4ab[x^2 + (a - b)x - ab]$$
$$= (a^2 + 2ab + b^2)x^2 - 4abx^2 - 4ab(a - b)x + 4a^2 b^2$$
$$= (a^2 + b^2)x^2 + 2abx^2 - 4abx^2 - 4ab(a - b)x + 4a^2 b^2$$
$$= (a^2 - 2ab + b^2)x^2 - 4ab(a - b)x + 4a^2 b^2$$
$$= (a - b)^2 x^2 - 4ab(a - b)x + 4a^2 b^2$$
$$= [(a - b)x - 2ab]^2,$$

and we have $[(a - b)x - 2ab]^2 \ge 0$ from the Trivial Inequality. All our steps are reversible, so the desired inequality holds.

18.43 By the AM-HM inequality on the three numbers $a + b$, $a + c$, and $b + c$, we have

$$\frac{2a + 2b + 2c}{3} = \frac{(a + b) + (a + c) + (b + c)}{3} \ge \frac{3}{1/(a + b) + 1/(a + c) + 1/(b + c)}.$$

Therefore, $(a + b + c)\left(\dfrac{1}{a + b} + \dfrac{1}{a + c} + \dfrac{1}{b + c}\right) \ge \dfrac{9}{2}$, so $\dfrac{a + b + c}{a + b} + \dfrac{a + b + c}{a + c} + \dfrac{a + b + c}{b + c} \ge \dfrac{9}{2}$.

We can rewrite this inequality as $\dfrac{c}{a + b} + 1 + \dfrac{b}{a + c} + 1 + \dfrac{a}{b + c} + 1 \ge \dfrac{9}{2}$, so $\dfrac{c}{a + b} + \dfrac{b}{a + c} + \dfrac{a}{b + c} \ge \dfrac{9}{2} - 3 = \dfrac{3}{2}$.

18.44 The problem is symmetric in a, b, and c, so without loss of generality, let $a \le b \le c$. Let m be the smaller of $b - a$ and $c - b$. Then $m \le b - a$ and $m \le c - b$, and the smallest of $(a - b)^2$, $(a - c)^2$, and $(b - c)^2$ is m^2 (because $c - a \ge c - b$).

Now, we claim that $\dfrac{(c - a)^2}{4} \le \dfrac{a^2 + c^2}{2}$. To prove this, we must show that $\dfrac{a^2 + c^2}{2} - \dfrac{(c - a)^2}{4} \ge 0$. Indeed, we have

$$\frac{a^2 + c^2}{2} - \frac{(c - a)^2}{4} = \frac{2a^2 + 2c^2}{4} - \frac{a^2 - 2ac + c^2}{4} = \frac{a^2 + 2ac + c^2}{4} = \frac{(a + c)^2}{4} \ge 0,$$

which follows from the Trivial Inequality.

Since $m \leq b - a$ and $m \leq c - b$, we have $2m \leq c - a$, so $m \leq (c - a)/2$. Hence, we have

$$\frac{a^2 + b^2 + c^2}{2} \geq \frac{a^2 + c^2}{2} \geq \frac{(c - a)^2}{4} \geq m^2,$$

as desired.

18.45 For all real numbers x, we have $\lfloor x \rfloor > x - 1$. Therefore,

$$\left\lfloor \frac{a+b}{c} \right\rfloor + \left\lfloor \frac{a+c}{b} \right\rfloor + \left\lfloor \frac{b+c}{a} \right\rfloor > \frac{a+b}{c} - 1 + \frac{a+c}{b} - 1 + \frac{b+c}{a} - 1 = \frac{a}{b} + \frac{b}{a} + \frac{a}{c} + \frac{c}{a} + \frac{b}{c} + \frac{c}{b} - 3.$$

By AM-GM, we have $\frac{a}{b} + \frac{b}{a} \geq 2$, $\frac{a}{c} + \frac{c}{a} \geq 2$, and $\frac{b}{c} + \frac{c}{b} \geq 2$, so

$$\left\lfloor \frac{a+b}{c} \right\rfloor + \left\lfloor \frac{a+c}{b} \right\rfloor + \left\lfloor \frac{b+c}{a} \right\rfloor > 2 + 2 + 2 - 3 = 3.$$

Since the given expression is always an integer, it must be at least 4.

Furthermore, for $a = 3$, $b = 4$, and $c = 4$, we have

$$\left\lfloor \frac{a+b}{c} \right\rfloor + \left\lfloor \frac{a+c}{b} \right\rfloor + \left\lfloor \frac{b+c}{a} \right\rfloor = \left\lfloor \frac{7}{4} \right\rfloor + \left\lfloor \frac{7}{4} \right\rfloor + \left\lfloor \frac{8}{3} \right\rfloor = 1 + 1 + 2 = 4.$$

Therefore, the minimum value is $\boxed{4}$.

18.46 Let $P(x) = a_n x^n + a_{n-1} x^{n-1} + \cdots + a_1 x + a_0$. The inequality $P(1/x) \geq 1/P(x)$ is equivalent to

$$P(x) P\left(\frac{1}{x}\right) = (a_n x^n + a_{n-1} x^{n-1} + \cdots + a_1 x + a_0) \left(\frac{a_n}{x^n} + \frac{a_{n-1}}{x^{n-1}} + \cdots + \frac{a_1}{x} + a_0\right) \geq 1.$$

We are given that this inequality holds for $x = 1$, so

$$(a_n + a_{n-1} + \cdots + a_1 + a_0)(a_n + a_{n-1} + \cdots + a_1 + a_0) = (a_n + a_{n-1} + \cdots + a_1 + a_0)^2 \geq 1.$$

Then by Cauchy-Schwarz, for any $x > 0$, we have

$$P(x) P\left(\frac{1}{x}\right) = (a_n x^n + a_{n-1} x^{n-1} + \cdots + a_1 x + a_0) \left(\frac{a_n}{x^n} + \frac{a_{n-1}}{x^{n-1}} + \cdots + \frac{a_1}{x} + a_0\right)$$

$$\geq \left(\sqrt{a_n x^n \cdot \frac{a_n}{x^n}} + \sqrt{a_{n-1} x^{n-1} \cdot \frac{a_{n-1}}{x^{n-1}}} + \cdots + \sqrt{a_1 x \cdot \frac{a_1}{x}} + \sqrt{a_0 \cdot a_0}\right)^2$$

$$= (a_n + a_{n-1} + \cdots + a_1 + a_0)^2$$

$$\geq 1.$$

18.47 *Solution 1.* To deal with the square root in the denominator of the first fraction, we can rationalize it, to get

$$\frac{a}{a + \sqrt{(a+b)(a+c)}} = \frac{a}{\sqrt{(a+b)(a+c)} + a} \cdot \frac{\sqrt{(a+b)(a+c)} - a}{\sqrt{(a+b)(a+c)} - a} = \frac{a\sqrt{(a+b)(a+c)} - a^2}{(a+b)(a+c) - a^2} = \frac{a\sqrt{(a+b)(a+c)} - a^2}{ab + ac + bc}.$$

By AM-GM, we have $\sqrt{(a+b)(a+c)} \leq \dfrac{(a+b) + (a+c)}{2} = \dfrac{2a+b+c}{2}$, so

$$\frac{a\sqrt{(a+b)(a+c)} - a^2}{ab + ac + bc} \leq \frac{a \cdot (2a+b+c)/2 - a^2}{ab + ac + bc} = \frac{ab + ac}{2(ab + ac + bc)}.$$

Similarly, we have

$$\frac{b}{b + \sqrt{(a + b)(b + c)}} \le \frac{ab + bc}{2(ab + ac + bc)},$$

$$\frac{c}{c + \sqrt{(a + c)(b + c)}} \le \frac{ac + bc}{2(ab + ac + bc)}.$$

Adding all three inequalities, we get

$$\frac{a}{a + \sqrt{(a + b)(a + c)}} + \frac{b}{b + \sqrt{(a + b)(b + c)}} + \frac{c}{c + \sqrt{(a + c)(b + c)}} \le \frac{2ab + 2ac + 2bc}{2(ab + ac + bc)} = 1.$$

Solution 2. First, we claim that $\sqrt{(a + b)(a + c)} \ge \sqrt{ab} + \sqrt{ac}$. Squaring both sides, we get

$$a^2 + ab + ac + bc \ge ab + 2\sqrt{a^2bc} + ac,$$

which simplifies to $a^2 + bc \ge 2\sqrt{a^2bc}$. But this inequality follows from the AM-GM inequality on the numbers a^2 and bc.

Therefore,

$$\frac{a}{a + \sqrt{(a + b)(a + c)}} \le \frac{a}{a + \sqrt{ab} + \sqrt{ac}} = \frac{a}{\sqrt{a \cdot a} + \sqrt{ab} + \sqrt{ac}} = \frac{\sqrt{a}}{\sqrt{a} + \sqrt{b} + \sqrt{c}}.$$

Similarly, we have

$$\frac{b}{b + \sqrt{(a + b)(b + c)}} \le \frac{\sqrt{b}}{\sqrt{a} + \sqrt{b} + \sqrt{c}},$$

$$\frac{c}{c + \sqrt{(a + c)(b + c)}} \le \frac{\sqrt{c}}{\sqrt{a} + \sqrt{b} + \sqrt{c}}.$$

Adding all three inequalities, we get

$$\frac{a}{a + \sqrt{(a + b)(a + c)}} + \frac{b}{b + \sqrt{(a + b)(b + c)}} + \frac{c}{c + \sqrt{(a + c)(b + c)}} \le \frac{\sqrt{a} + \sqrt{b} + \sqrt{c}}{\sqrt{a} + \sqrt{b} + \sqrt{c}} = 1.$$

18.48 From the first equation, we have $y = 5 - x - z$. Substituting into the second equation, we get

$$x(5 - x - z) + xz + (5 - x - z)z = 3,$$

which simplifies as $x^2 + xz + z^2 - 5x - 5z + 3 = 0$. We can rewrite this equation as a quadratic in x:

$$x^2 + (z - 5)x + z^2 - 5z + 3 = 0.$$

The discriminant of this quadratic is

$$(z - 5)^2 - 4(z^2 - 5z + 3) = -3z^2 + 10z + 13 = -(z + 1)(3z - 13),$$

which must be nonnegative. Therefore, $z \le 13/3$.

Furthermore, for $z = 13/3$, we have the solution $(x, y, z) = (1/3, 1/3, 13/3)$. Hence, the maximum value of z is
$\boxed{13/3}$.

18.49 We might start by rationalizing denominators, which gives us

$$x = 1 + \frac{\sqrt{2}}{2} + \frac{\sqrt{3}}{3} + \frac{\sqrt{4}}{4} + \frac{\sqrt{5}}{5} + \cdots + \frac{\sqrt{1,000,000}}{1,000,000}.$$

That's not so helpful. So, we think about other strategies that have helped with nasty expressions. Perhaps we can compare this to other series that we can evaluate. However, picking a good series is a bit tricky. To help guide us, we think about all our strategies for tackling unusual series. At first, it might not seem that telescoping has a chance of helping, but our failed "rationalize the denominator" attempt above gives us an idea. If, instead of rationalizing the denominator of $1/\sqrt{n}$, we rationalize the denominator of $1/(\sqrt{n} + \sqrt{n-1})$, we get

$$\frac{1}{\sqrt{n} + \sqrt{n-1}} = \frac{1}{\sqrt{n} + \sqrt{n-1}} \cdot \frac{\sqrt{n} - \sqrt{n-1}}{\sqrt{n} - \sqrt{n-1}} = \sqrt{n} - \sqrt{n-1}.$$

This is a good candidate for telescoping! Using this, we have:

$$\sum_{n=1}^{1,000,000} \frac{1}{\sqrt{n} + \sqrt{n-1}} = \frac{1}{\sqrt{1} + \sqrt{0}} + \frac{1}{\sqrt{2} + \sqrt{1}} + \frac{1}{\sqrt{3} + \sqrt{2}} + \cdots + \frac{1}{\sqrt{1,000,000} + \sqrt{999,999}}$$

$$= (\sqrt{1} - \sqrt{0}) + (\sqrt{2} - \sqrt{1}) + (\sqrt{3} - \sqrt{2}) + \cdots + (\sqrt{1,000,000} - \sqrt{999,999})$$

$$= \sqrt{1,000,000} - \sqrt{0} = 1000.$$

We now have to relate this series to the one in the problem. Since we must find $\lfloor x \rfloor$, we must sandwich our series between two integers. So, we look for a way to sandwich $1/\sqrt{n}$ between two expressions that we can sum from 1 to 1,000,000. Our successful summation above is a clear guide. For every integer $n \geq 1$, we have

$$\frac{\sqrt{n} + \sqrt{n-1}}{2} < \sqrt{n} < \frac{\sqrt{n+1} + \sqrt{n}}{2}.$$

Taking the reciprocal of this inequality chain (and therefore reversing the directions of the inequality signs) gives us

$$\frac{2}{\sqrt{n} + \sqrt{n-1}} > \frac{1}{\sqrt{n}} > \frac{2}{\sqrt{n+1} + \sqrt{n}}.$$

Summing this inequality from $n = 1$ to $n = 1,000,000$ gives us

$$\sum_{n=1}^{1,000,000} \frac{2}{\sqrt{n} + \sqrt{n-1}} > \sum_{n=1}^{1,000,000} \frac{1}{\sqrt{n}} > \sum_{n=1}^{1,000,000} \frac{2}{\sqrt{n+1} + \sqrt{n}}.$$

The summation on the left is 2 times the sum we just telescoped. Similarly, the sum on the right telescopes to give us

$$2000 > \sum_{n=1}^{1,000,000} \frac{1}{\sqrt{n}} > 2\left(\sqrt{1,000,001} - \sqrt{1}\right).$$

We have $\sqrt{1,000,001} > \sqrt{1,000,000} = 1000$, so we have

$$2000 > \sum_{n=1}^{1,000,000} \frac{1}{\sqrt{n}} > 2\left(\sqrt{1,000,001} - \sqrt{1}\right) > 2(1000 - 1) = 1998.$$

Uh-oh. We placed our sum between two integers that are 2 apart, not 1. We'll have to tighten up one of our bounds. We look at the first and last terms of all three series.

The last terms of the three series are very, very close. But the first terms of the three series are, in order, 2, 1, and $2/(\sqrt{2} + 1) = 2(\sqrt{2} - 1)$. While 1 and $2(\sqrt{2} - 1) \approx 0.83$ are close, the fact that the first terms of the first two series are 1 apart lets us reduce our upper bound by 1, as needed. Specifically, since the first term of the desired sum is 1, we can replace the first term of our upper bound series with 1, and we have:

$$1 + \sum_{n=2}^{1,000,000} \frac{2}{\sqrt{n} + \sqrt{n-1}} > \sum_{n=1}^{1,000,000} \frac{1}{\sqrt{n}}.$$

Since

$$\sum_{n=2}^{1,000,000} \frac{2}{\sqrt{n} + \sqrt{n-1}} = \sum_{n=1}^{1,000,000} \frac{2}{\sqrt{n} + \sqrt{n-1}} - \frac{2}{\sqrt{1} + \sqrt{0}} = 1998,$$

we now have $1999 > \sum_{n=1}^{1,000,000} \frac{1}{\sqrt{n}} > 1998$, so $\lfloor x \rfloor = \boxed{1998}$.

18.50 First, we simplify the factor $a^5 - a^2 + 3$ by using a method similar to Simon's Favorite Factoring Trick. We want an expression that includes the terms a^5, a^2, and a constant, and that factors into two binomials. For example,

$$a^5 - a^3 - a^2 + 1 = a^3(a^2 - 1) - (a^2 - 1) = (a^3 - 1)(a^2 - 1).$$

This expression further factors as

$$(a^3 - 1)(a^2 - 1) = (a - 1)(a^2 + a + 1)(a - 1)(a + 1) = (a - 1)^2(a^2 + a + 1)(a + 1).$$

By the Trivial Inequality, we have $(a-1)^2 \geq 0$, and since a is nonnegative, we have $a^2 + a + 1 \geq 1 > 0$ and $a + 1 \geq 1 > 0$, so $a^5 - a^3 - a^2 + 1 = (a^3 - 1)(a^2 - 1) \geq 0$. (We could also have shown that $(a^3 - 1)(a^2 - 1) \geq 0$ by noting that for all a, either both $a^3 - 1$ and $a^2 - 1$ both equal 0, or they have the same sign.) It follows that

$$a^5 - a^2 + 3 \geq a^3 + 2.$$

Similarly, $b^5 - b^2 + 3 \geq b^3 + 2$ and $c^5 - c^2 + 3 \geq c^3 + 2$. Furthermore, since $a^3 + 2 \geq 2 > 0$, $b^3 + 2 \geq 2 > 0$, and $c^3 + 2 \geq 2 > 0$, we can take the product of all three inequalities, to get

$$(a^5 - a^2 + 3)(b^5 - b^2 + 3)(c^5 - c^2 + 3) \geq (a^3 + 2)(b^3 + 2)(c^3 + 2).$$

Now, it suffices to show that $(a^3 + 2)(b^3 + 2)(c^3 + 2) \geq (a + b + c)^3$.

Expanding both sides, this inequality becomes

$$a^3b^3c^3 + 2(a^3b^3 + a^3c^3 + b^3c^3) + 4(a^3 + b^3 + c^3) + 8 \geq a^3 + b^3 + c^3 + 3(a^2b + ab^2 + a^2c + ac^2 + b^2c + bc^2) + 6abc.$$

We may subtract $a^3 + b^3 + c^3$ from both sides, to get

$$a^3b^3c^3 + 2(a^3b^3 + a^3c^3 + b^3c^3) + 3(a^3 + b^3 + c^3) + 8 \geq 3(a^2b + ab^2 + a^2c + ac^2 + b^2c + bc^2) + 6abc.$$

We can prove this inequality by using AM-GM on appropriate groups of terms.

By AM-GM, we have

$$a^3b^3 + a^3 + 1 \geq 3\sqrt[3]{a^6b^3} = 3a^2b,$$
$$a^3b^3 + b^3 + 1 \geq 3\sqrt[3]{a^3b^6} = 3ab^2,$$
$$a^3c^3 + a^3 + 1 \geq 3\sqrt[3]{a^6c^3} = 3a^2c,$$
$$a^3c^3 + c^3 + 1 \geq 3\sqrt[3]{a^3c^6} = 3ac^2,$$
$$b^3c^3 + b^3 + 1 \geq 3\sqrt[3]{b^6c^3} = 3b^2c,$$
$$b^3c^3 + c^3 + 1 \geq 3\sqrt[3]{b^3c^6} = 3bc^2,$$
$$a^3b^3c^3 + a^3 + b^3 + c^3 + 1 + 1 \geq 6\sqrt[6]{a^6b^6c^6} = 6abc.$$

Adding all these inequalities, we get

$$a^3b^3c^3 + 2(a^3b^3 + a^3c^3 + b^3c^3) + 3(a^3 + b^3 + c^3) + 8 \geq 3(a^2b + ab^2 + a^2c + ac^2 + b^2c + bc^2) + 6abc,$$

as desired.

CHAPTER 19

Functional Equations

Exercises for Section 19.1

19.1.1 Since we are given $f(0) = 0$, we can start by substituting $x = 0$, to get $f(1) = 2f(0) + 3 = \boxed{3}$. In turn, we can substitute $x = 1$ to get $f(3) = 2f(1) + 3 = \boxed{9}$. Finally, substituting $x = 3$, we get $f(7) = 2f(3) + 3 = \boxed{21}$.

19.1.2 Substituting $n = 3$, we get $F(4) = \frac{F(3)F(2)+1}{F(1)} = \frac{1\cdot1+1}{1} = 2$. Continuing, we can substitute $n = 4$ and $n = 5$ to get

$$F(5) = \frac{F(4)F(3) + 1}{F(2)} = \frac{2 \cdot 1 + 1}{1} = 3,$$
$$F(6) = \frac{F(5)F(4) + 1}{F(3)} = \frac{3 \cdot 2 + 1}{1} = \boxed{7}.$$

19.1.3 Only rule (c) gives us direct information about $f(5,5)$. Taking $x = 4$ and $y = 5$, we get

$$f(5,5) = 5[f(4,5) + f(4,4)].$$

From rule (b), we have $f(4,5) = 0$, so $f(5,5) = 5f(4,4)$.

We can use rule (c) again. Taking $x = 3$ and $y = 4$, we get $f(4,4) = 4[f(3,4) + f(3,3)]$. Again from rule (b), we have $f(3,4) = 0$, so $f(4,4) = 4f(3,3)$. Similarly, we get $f(3,3) = 3f(2,2)$, and $f(2,2) = 2f(1,1)$ and $f(1,1) = 1$. Therefore,
$$f(5,5) = 5f(4,4) = 5 \cdot 4f(3,3) = 5 \cdot 4 \cdot 3f(2,2) = 5 \cdot 4 \cdot 3 \cdot 2f(1,1) = 5 \cdot 4 \cdot 3 \cdot 2 = \boxed{120}.$$

19.1.4 Setting $y = x$, we get $f(0) = (f(x))^2$. Setting $x = 0$, we get $f(0) = (f(0))^2$, so $f(0)[f(0) - 1] = 0$. Hence, $f(0) = 0$ or $f(0) = 1$. But $f(x)$ is never equal to 0, so $f(0) = 1$. Then $(f(x))^2 = 1$ for all x, so $f(x) = -1$ or $f(x) = 1$ for all x.

Setting $x = 2y$ in the given equation, we get $f(y) = f(2y)f(y)$. Since $f(y)$ is nonzero, we may divide both sides by $f(y)$ to get $f(2y) = 1$. Since this holds for all y, in particular, we have $f(1977) = \boxed{1}$. As a check, we note that the function f where $f(x) = 1$ for all x satisfies the given equation.

19.1.5 Setting $m = 0$, we get $f(n) = f(0) + f(n) - 3$, so $f(0) = 3$. Setting $m = 1$, we get

$$f(n + 1) = f(1) + f(n) + 3(4n - 1) = f(n) + 12n - 3.$$

Hence, we have $f(2) = f(1) + 12 \cdot 1 - 3$, $f(3) = f(2) + 12 \cdot 2 - 3$, $f(4) = f(3) + 12 \cdot 3 - 3$, and so on. In general, we see that for any positive integer n,

$$f(n) = 12 \cdot (n-1) - 3 + f(n-1) = 12 \cdot (n-1) - 3 + 12 \cdot (n-2) - 3 + f(n-2)$$
$$= \cdots$$
$$= 12 \cdot (n-1) - 3 + 12 \cdot (n-2) - 3 + \cdots + 12 \cdot 1 - 3 + f(1)$$
$$= 12[(n-1) + (n-2) + \cdots + 1] - 3(n-1) = 12 \cdot \frac{(n-1)n}{2} - 3(n-1)$$
$$= 6(n-1)n - 3(n-1) = 3(n-1)(2n-1).$$

In particular, $f(19) = 3 \cdot 18 \cdot 37 = \boxed{1998}$. Note that the function $f(n) = 3(n-1)(2n-1)$ does indeed satisfy the given equation.

Exercises for Section 19.2

19.2.1 Let $y = 2x$. Then $x = y/2$, so $f(y) = f(2x) = \frac{5}{2+x} = \frac{5}{2+y/2} = \frac{10}{4+y}$. We can express this using the variable x as $f(x) = \frac{10}{4+x}$. Then, we have $2f(x) = \boxed{\dfrac{20}{4+x}}$.

19.2.2 Taking $y = 0$, we get $2f(x) = 2x^2$, so $f(x) = \boxed{x^2}$. We then check this solution. If $f(x) = x^2$, then

$$f(x+2y) + f(x-2y) = (x+2y)^2 + (x-2y)^2 = x^2 + 4xy + 4y^2 + x^2 - 4xy + 4y^2 = 2x^2 + 8y^2,$$

so the function $f(x) = x^2$ works.

19.2.3 Taking $y = 1$, we get $f(x+1) = f(x)f(1) - f(x) + 1 = 2f(x) - f(x) + 1 = f(x) + 1$. Hence, $f(2) = f(1) + 1 = 3$, $f(3) = f(2) + 1 = 4$, and so on. By a straightforward induction argument, we have $f(x) = \boxed{x+1}$ for all positive integers x.

We then check this solution. If $f(x) = x + 1$, then $f(x+y) = x + y + 1$, and

$$f(x)f(y) - f(xy) + 1 = (x+1)(y+1) - (xy+1) + 1 = xy + x + y + 1 - xy - 1 + 1 = x + y + 1.$$

Thus, the function $f(x) = x + 1$ works.

19.2.4 Taking $y = 0$, we get $1 + f(x) = 2f(x)f(0)$. Solving for $f(x)$, we get $f(x) = \frac{1}{2f(0)-1}$. Therefore, $f(x)$ is a constant, say c. Substituting into the given equation, we get $1 + c = 2c^2$. This simplifies as $(c-1)(2c+1) = 0$, so $c = 1$ or $c = -1/2$. Therefore, the solutions are $f(x) = \boxed{1}$ and $f(x) = \boxed{-1/2}$.

19.2.5 Taking $b = 1$, we get

$$f(a+1) = f(a) + f(a)f(1) + f(1) = f(a) + (k-1)f(a) + k - 1 = kf(a) + k - 1.$$

Hence,

$$f(2) = kf(1) + k - 1 = k(k-1) + k - 1 = k^2 - 1,$$
$$f(3) = kf(2) + k - 1 = k(k^2 - 1) + k - 1 = k^3 - 1,$$
$$f(4) = kf(3) + k - 1 = k(k^3 - 1) + k - 1 = k^4 - 1,$$

and so on. By a straightforward induction argument, we have $f(n) = k^n - 1$ for all positive integers n. However, $f(n)$ is defined for all integers n, so we also need to find $f(n)$ for $n \leq 0$.

Taking $a = 0$ in the equation $f(a + 1) = kf(a) + k - 1$, we get $f(1) = kf(0) + k - 1$. Since $f(1) = k - 1$, we have $kf(0) = 0$, so $f(0) = 0$.

Now, in the original equation, take $a = n$ and $b = -n$, where n is a positive integer. Then

$$f(0) = f(n) + f(n)f(-n) + f(-n).$$

Since $f(0) = 0$, we have $f(n) + f(n)f(-n) + f(-n) = 0$. Solving for $f(-n)$ gives

$$f(-n) = \frac{-f(n)}{f(n) + 1} = \frac{-(k^n - 1)}{(k^n - 1) + 1} = \frac{1 - k^n}{k^n} = k^{-n} - 1.$$

Hence, $f(n) = \boxed{k^n - 1}$ for all integers n.

We then check this solution. If $f(n) = k^n - 1$, then $f(a + b) = k^{a+b} - 1$, and

$$f(a) + f(a)f(b) + f(b) = k^a - 1 + (k^a - 1)(k^b - 1) + k^b - 1 = k^a - 1 + k^{a+b} - k^a - k^b + 1 + k^b - 1 = k^{a+b} - 1.$$

Thus, the function $f(n) = k^n - 1$ works.

Exercises for Section 19.3

19.3.1 If a and b are nonzero, we can divide by a^2b^2 to find $\frac{f(a)}{a^2} = \frac{f(b)}{b^2}$. Letting $g(x) = \frac{f(x)}{x^2}$, we have $g(a) = g(b)$ for all nonzero a and b. Therefore, we must have $g(x) = c$ for some constant c. This gives us $f(x) = cx^2$ for all nonzero x. Since $f(2) \neq 0$, we have $c \neq 0$, so we have

$$\frac{f(5) - f(1)}{f(2)} = \frac{25c - c}{4c} = \boxed{6}.$$

19.3.2 Taking the logarithm with base 2 of both sides, we get $\log_2(y2^{f(x)}) = \log_2(x2^{f(y)})$. Applying the logarithm identity $\log_a bc = \log_a b + \log_a c$ makes our equation $\log_2 y + \log_2 2^{f(x)} = \log_2 x + \log_2 2^{f(y)}$. Since $\log_2 2^a = a$, we have $\log_2 y + f(x) = \log_2 x + f(y)$, so $f(x) - \log_2 x = f(y) - \log_2 y$. Therefore, $f(x) - \log_2 x$ is a constant, say c. Then $f(x) = \boxed{\log_2 x + c}$.

We then check this solution. If $f(x) = \log_2 x + c$, then

$$y2^{f(x)} = y2^{\log_2 x + c} = xy2^c, \qquad \text{and} \qquad x2^{f(y)} = x2^{\log_2 y + c} = xy2^c.$$

Thus, the function $f(x) = \log_2 x + c$ works.

19.3.3 The function $f(x) = 0$ is a solution. Otherwise, there is some z such that $f(z) \neq 0$, so we can divide both sides by $f(z)$ to get $\frac{f(x)}{y+1} = \frac{f(y)}{x+1}$. Multiplying both sides of this equation by $(x+1)(y+1)$, we get $(x+1)f(x) = (y+1)f(y)$. Thus, $(x + 1)f(x)$ is a constant, say c. Then $(x + 1)f(x) = c$, so $f(x) = \frac{c}{x+1}$.

If $f(x) = c/(x + 1)$, then in the original equation, the left side is

$$\frac{f(x)f(z)}{y + 1} = \frac{c^2}{(x + 1)(y + 1)(z + 1)},$$

and the right side is $\dfrac{f(y)f(z)}{x + 1} = \dfrac{c^2}{(x + 1)(y + 1)(z + 1)}$. Thus, all functions of the form $f(x) = \boxed{\dfrac{c}{x + 1}}$ work.

79.4

Exercises for Section 19.4

19.4.1 We have

$$f(f(t)) = \frac{\frac{t+\sqrt{3}}{1-t\sqrt{3}} + \sqrt{3}}{1 - \frac{t+\sqrt{3}}{1-t\sqrt{3}} \cdot \sqrt{3}} = \frac{t + \sqrt{3} + \sqrt{3} - 3t}{1 - t\sqrt{3} - t\sqrt{3} - 3} = \frac{-2t + 2\sqrt{3}}{-2 - 2t\sqrt{3}} = \frac{t - \sqrt{3}}{1 + t\sqrt{3}},$$

and

$$f(f(f(t))) = \frac{\frac{t-\sqrt{3}}{1+t\sqrt{3}} + \sqrt{3}}{1 - \frac{t-\sqrt{3}}{1+t\sqrt{3}} \cdot \sqrt{3}} = \frac{t - \sqrt{3} + \sqrt{3} + 3t}{1 + t\sqrt{3} - t\sqrt{3} + 3} = \frac{4t}{4} = t.$$

Hence, f is cyclic, with order $\boxed{3}$.

19.4.2 Substituting $\frac{1}{x}$ for x, we get $f\left(\frac{1}{x}\right) + f(x) = \frac{1}{x}$. But we are given that $f(\frac{1}{x}) + f(x) = x$, so we must have $x = \frac{1}{x}$ for every x in the domain of f. Then $x^2 = 1$, so $x = 1$ or $x = -1$. Therefore, the largest set of real numbers that can be the domain of f is $\boxed{\{1, -1\}}$. To show that this domain is attainable, take $f(1) = 1/2$ and $f(-1) = -1/2$. Then it is easy to check that the original equation is satisfied.

19.4.3 Substituting $1 - x$ for x, we get $2f(1 - x) + 3f(x) = 5(1 - x)^2 - 3(1 - x) - 11 = 5x^2 - 7x - 9$.

Thus, we have the following system of equations in $f(x)$ and $f(1 - x)$:

$$2f(x) + 3f(1 - x) = 5x^2 - 3x - 11,$$
$$3f(x) + 2f(1 - x) = 5x^2 - 7x - 9.$$

Subtracting twice the first equation from 3 times the second equation gives $5f(x) = 5x^2 - 15x - 5$, so $f(x) = \boxed{x^2 - 3x - 1}$.

We then check this solution. If $f(x) = x^2 - 3x - 1$, then

$$2f(x) + 3f(1 - x) = 2(x^2 - 3x - 1) + 3[(1 - x)^2 - 3(1 - x) - 1] = 5x^2 - 3x - 11.$$

Thus, the function $f(x) = x^2 - 3x - 1$ works.

19.4.4 First, we substitute $x/(x - 1)$ for x in the given equation. Then the expression $x/(x - 1)$ becomes

$$\frac{\frac{x}{x-1}}{\frac{x}{x-1} - 1} = \frac{x}{x - (x - 1)} = x,$$

so the given equation becomes

$$\frac{x}{x - 1} f\left(\frac{x}{x - 1}\right) + f(x) = \frac{2x}{x - 1}. \tag{19.1}$$

We also have the original $xf(x) + f\left(\frac{x}{x-1}\right) = 2x$. Multiplying this equation by $\frac{x}{x-1}$ gives $\frac{x^2}{x-1} f(x) + \frac{x}{x-1} f\left(\frac{x}{x-1}\right) = \frac{2x^2}{x-1}$. Subtracting Equation (19.1) from this equation eliminates $f\left(\frac{x}{x-1}\right)$ and leaves $\left(\frac{x^2}{x-1} - 1\right) f(x) = \frac{2x^2}{x-1} - \frac{2x}{x-1}$. Simplifying both sides gives $\left(\frac{x^2-x+1}{x-1}\right) f(x) = \frac{2x^2-2x}{x-1}$, so $f(x) = \boxed{\frac{2x^2-2x}{x^2-x+1}}$. This function does indeed satisfy the given functional equation.

19.4.5 Substituting $1/x$ for x in the given equation, we get $f\left(\frac{1}{x}\right) = (f(x))^2$. Squaring both sides, we get $\left(f\left(\frac{1}{x}\right)\right)^2 = (f(x))^4$. But $\left(f\left(\frac{1}{x}\right)\right)^2 = f(x)$. Hence, $(f(x))^4 = f(x)$, which implies $f(x)[(f(x))^3 - 1] = 0$, so $f(x) = 0$ or $f(x) = 1$ for all x. In either case, $(f(1977))^2 - f(1977) = \boxed{0}$.

95

Review Problems

19.18 Taking $x_1 = x_2 = x_3 = x_4 = x_5 = 0$, we get $f(0) = 5f(0) - 8$, so $f(0) = \boxed{2}$.

19.19 Taking $x = 500$ and $y = 600/500 = 6/5$, we get $f(xy) = f(600) = \dfrac{f(500)}{6/5} = \dfrac{3}{6/5} = \boxed{\dfrac{5}{2}}$.

19.20 Setting $y = 0$, we get $f(x) + f(x) + f(x) = 6x$, so $f(x) = \boxed{2x}$. We then check this solution. If $f(x) = 2x$, then $f(x) + f(x + y) + f(x + 2y) = 2x + 2(x + y) + 2(x + 2y) = 6x + 6y$, so the function $f(x) = 2x$ works.

19.21 Substituting $1/x$ for x in the given equation, we get $f(\frac{1}{x}) + 2f(x) = \frac{3}{x}$.

Thus, we have the following system of equations in $f(\frac{1}{x})$ and $f(x)$:

$$f(x) + 2f\left(\frac{1}{x}\right) = 3x,$$
$$2f(x) + f\left(\frac{1}{x}\right) = \frac{3}{x}.$$

Solving for $f(x)$, we get $f(x) = \dfrac{2 - x^2}{x}$.

We then check this solution. If $f(x) = (2 - x^2)/x$, then

$$f(x) + 2f\left(\frac{1}{x}\right) = \frac{2 - x^2}{x} + 2 \cdot \frac{2 - \frac{1}{x^2}}{\frac{1}{x}} = \frac{2 - x^2}{x} + 2 \cdot \frac{2x^2 - 1}{x} = \frac{2 - x^2 + 4x^2 - 2}{x} = 3x.$$

Thus, the function $f(x) = (2 - x^2)/x$ works.

Then the equation $f(x) = f(-x)$ becomes $\frac{2-x^2}{x} = \frac{2-(-x)^2}{-x} = -\frac{2-x^2}{x}$. Multiplying both sides by x, we get $2 - x^2 = x^2 - 2$, so $x^2 - 2 = 0$. Therefore, the solutions of $f(x) = f(-x)$ are $x = \sqrt{2}$ and $x = -\sqrt{2}$, for a total of $\boxed{2}$ solutions.

19.22 Taking $m = 1$, we get $a_{n+1} = 4^n a_1 a_n = 2 \cdot 4^n a_n = 2^{2n+1} a_n$.

Hence,

$$a_2 = 2^3 a_1 = 2^3 \cdot 2 = 2^4,$$
$$a_3 = 2^5 a_2 = 2^5 \cdot 2^4 = 2^9,$$
$$a_4 = 2^7 a_3 = 2^7 \cdot 2^9 = 2^{16},$$

and so on. We appear to have $a_n = 2^{n^2}$ for all n. We'll prove this with induction. We already have $a_1 = 2 = 2^1$. We assume that $a_k = 2^{k^2}$ for some positive integer k. We then have

$$a_{k+1} = 4^k a_1 a_k = 2^{2k}(2)(2^{k^2}) = 2^{k^2 + 2k + 1} = 2^{(k+1)^2},$$

so $a_n = 2^{n^2}$ for $n = k + 1$, and our induction is complete. We therefore have $a_n = \boxed{2^{n^2}}$ for all n.

We then check this solution. If $a_n = 2^{n^2}$, then the left side is $a_{m+n} = 2^{(m+n)^2} = 2^{m^2 + 2mn + n^2}$, and the right side is $4^{mn} a_m a_n = 2^{2mn} \cdot 2^{m^2} \cdot 2^{n^2} = 2^{m^2 + 2mn + n^2}$. Thus, $a_n = 2^{n^2}$ works.

19.23 We have

$$u(u(x)) = \frac{1}{1 - u(x)} = \frac{1}{1 - \frac{1}{1-x}} = \frac{1-x}{1-x-1} = -\frac{1-x}{x}$$

and

$$u(u(u(x))) = -\frac{1 - u(x)}{u(x)} = -\frac{1 - \frac{1}{1-x}}{\frac{1}{1-x}} = -\frac{1-x-1}{1} = x.$$

Hence, $u(x)$ is cyclic with order $\boxed{3}$.

19.24   Taking $y = 0$, we get $f(0) = f(x)$, so $f(x)$ is a constant, say c. Then the equation $f(xy) = (y^2 - y + 1)f(x)$ becomes $c = (y^2 - y + 1)c$ for all y. In such a case, the constant c must be 0, so the only solution is $f(x) = \boxed{0}$.

19.25   Taking $b = 1$, we get $f(a + 1) = f(a) + f(1) - 2f(a) = 1 - f(a)$ for all integers a. Hence,

$$f(2) = 1 - f(1) = 0,$$
$$f(3) = 1 - f(2) = 1,$$
$$f(4) = 1 - f(3) = 0,$$
$$f(5) = 1 - f(4) = 1,$$

and so on. By a straightforward induction argument, $f(n)$ is 0 when n is even, and 1 when n is odd.

We then check this solution. If a and b have opposite parity (meaning that one is even and the other is odd), then $a + b$ is odd, so $f(a + b) = 1$. Also, $f(a) + f(b) = 1$, since $f(a)$ and $f(b)$ are equal to 0 and 1, in some order, and $f(ab) = 0$, since ab is even. Therefore, $f(a) + f(b) - 2f(ab) = 1$, so the given equation is satisfied in this case.

Now assume that a and b have the same parity. If both a and b are even, then $f(a + b) = f(a) + f(b) - 2f(ab) = 0$. If both a and b are odd, then $a + b$ is even, so $f(a + b) = 0$, and $f(a) + f(b) - 2f(ab) = 1 + 1 - 2 = 0$. Hence, the given equation is also satisfied in this case, so our function works.

In particular, $f(2007) = \boxed{1}$.

19.26   Substituting $1 - z$ for z in the given equation, we get $f(1 - z) + (1 - z)f(z) = 1 + 1 - z = 2 - z$. Thus, we have the following system of equations in $f(z)$ and $f(1 - z)$:

$$f(z) + zf(1 - z) = 1 + z,$$
$$(1 - z)f(z) + f(1 - z) = 2 - z.$$

Multiplying the second equation by z, we get $(z - z^2)f(z) + zf(1 - z) = 2z - z^2$. Subtracting this equation from the equation $f(z) + zf(1 - z) = 1 + z$, we get $(1 - z + z^2)f(z) = 1 - z + z^2$.

The quadratic equation $1 - z + z^2 = 0$ has no real roots (which can be seen from completing the square, or from its negative discriminant), so we may divide both sides by $1 - z + z^2$ to find that $f(z) = \boxed{1}$ for all z.

We then check this solution. If $f(z) = 1$ for all z, then $f(z) + zf(1 - z) = 1 + z$. Thus, our solution works.

19.27   We have

$$x = f(f(x)) = \frac{\frac{x+1}{x+c} + 1}{\frac{x+1}{x+c} + c} = \frac{2x + c + 1}{(c + 1)x + c^2 + 1}.$$

Multiplying both sides by the denominator, we get $(c + 1)x^2 + (c^2 + 1)x = 2x + c + 1$. For this equation to hold for all x when both sides of the original equation are defined, the corresponding coefficients on both sides must match, so $c + 1 = 0$, $2 = c^2 + 1$, and $c + 1 = 0$. Therefore, the solution is $c = \boxed{-1}$.

19.28   Taking $y = 1$, we get $f(x) + f(1) = f(x + 1) - x - 1$, so $f(x + 1) = f(x) + x + 2$. Hence, we have $f(2) = f(1) + 3$, $f(3) = f(2) + 4$, $f(4) = f(3) + 5$, and so on. Therefore, for all positive integers n, we have

$$f(n) = (n + 1) + f(n - 1) = (n + 1) + n + f(n - 2)$$

$$\vdots$$

$$= (n + 1) + n + \cdots + 3 + f(1) = (n + 1) + n + \cdots + 3 + 1 = (n + 1) + n + \cdots + 3 + 2 + 1 - 2$$

$$= \frac{(n + 1)(n + 2)}{2} - 2 = \frac{n^2 + 3n + 2}{2} - 2 = \frac{n^2 + 3n - 2}{2}.$$

We have identified $f(n)$ for all positive integers n, so we must now solve for $f(n)$ where $n \leq 0$. Taking $x = y = 0$ in the given equation, we get $f(0) + f(0) = f(0) - 0 - 1$, so $f(0) = -1$, which agrees with the formula above. Now, let n be a positive integer, and take $x = -n$ and $y = n$ in the given equation. Then $f(-n) + f(n) = f(0) + n^2 - 1 = n^2 - 2$, so

$$f(-n) = n^2 - 2 - f(n) = n^2 - 2 - \frac{n^2 + 3n - 2}{2} = \frac{2n^2 - 4 - n^2 - 3n + 2}{2} = \frac{n^2 - 3n - 2}{2} = \frac{(-n)^2 + 3(-n) - 2}{2}.$$

Hence, the formula $f(n) = \frac{n^2+3n-2}{2}$ holds for all integers n.

We then check this solution. If $f(n) = \frac{n^2+3n-2}{2}$, then

$$f(x) + f(y) = \frac{x^2 + 3x - 2}{2} + \frac{y^2 + 3y - 2}{2} = \frac{x^2 + y^2 + 3x + 3y - 4}{2},$$

and

$$f(x+y) - xy - 1 = \frac{(x+y)^2 + 3(x+y) - 2}{2} - xy - 1 = \frac{x^2 + 2xy + y^2 + 3x + 3y - 2 - 2xy - 2}{2} = \frac{x^2 + y^2 + 3x + 3y - 4}{2}.$$

Thus, the function $f(n) = \frac{n^2+3n-2}{2}$ works, where n is an integer.

Now, we must solve the equation $f(n) = n$. This becomes $\frac{n^2+3n-2}{2} = n$, which simplifies as $(n-1)(n+2) = 0$. Therefore, the only solution other than $n = 1$ is $n = -2$, so there is only $\boxed{1}$ such solution.

19.29 Taking $x = 0$ and $y = 0$ in (iii), we get $(f(0))^2 = 2f(0)$, which simplifies as $f(0)[f(0) - 2] = 0$, so $f(0) = 0$ or $f(0) = 2$. But we are given that $f(0) \neq 0$, so $f(0) = 2$.

Taking $y = 1$ in rule (iii), we get $f(x)f(1) = f(x+1) + f(x-1)$, so $f(x+1) = 3f(x) - f(x-1)$ for all integers x.

Hence,

$$f(2) = 3f(1) - f(0) = 3 \cdot 3 - 2 = 7,$$
$$f(3) = 3f(2) - f(1) = 3 \cdot 7 - 3 = 18,$$
$$f(4) = 3f(3) - f(2) = 3 \cdot 18 - 7 = 47,$$
$$f(5) = 3f(4) - f(3) = 3 \cdot 47 - 18 = 123,$$
$$f(6) = 3f(5) - f(4) = 3 \cdot 123 - 47 = 322,$$
$$f(7) = 3f(6) - f(5) = 3 \cdot 322 - 123 = \boxed{843}.$$

19.30 Taking $n = 2$ and $n = 0$ in the given equation, we obtain the system of equations

$$a_2 - 3a_0 = 25,$$
$$a_0 - a_2 = 9.$$

Solving for a_0, we get $a_0 = \boxed{-17}$.

19.31 Substituting $-x$ for x, we get $-xf(-x) - 2xf(x) = -1$. Thus, we have the following system of equations in $f(x)$ and $f(-x)$:

$$xf(x) + 2xf(-x) = -1,$$
$$-2xf(x) - xf(-x) = -1.$$

Multiplying the second equation by 2, we get $-4xf(x) - 2xf(-x) = -2$. Adding this to the first equation, we get $-3xf(x) = -3$. Hence,

$$f(x) = \frac{-3}{-3x} = \boxed{\frac{1}{x}}.$$

We then check this solution. If $f(x) = 1/x$, then $xf(x) + 2xf(-x) = x \cdot \frac{1}{x} + 2x \cdot \frac{1}{-x} = -1$. Thus, the function $f(x) = 1/x$ works.

19.32 Taking $y = 0$, we get $-xf(0) = 0$. Since this is true for all x, we have $f(0) = 0$.

Now, let x and y be nonzero real numbers, and let $u = 2x$ and $v = 2y$, so $x = u/2$ and $y = v/2$, and u and v are nonzero. Then the given equation becomes

$$\frac{v}{2} \cdot f(u) - \frac{u}{2} \cdot f(v) = 8 \cdot \frac{u}{2} \cdot \frac{v}{2}\left(\frac{u^2}{4} - \frac{v^2}{4}\right),$$

which simplifies as $vf(u) - uf(v) = uv(u^2 - v^2)$. Dividing both sides by uv, we get

$$\frac{f(u)}{u} - \frac{f(v)}{v} = u^2 - v^2,$$

so $\frac{f(u)}{u} - u^2 = \frac{f(v)}{v} - v$. Therefore, $\frac{f(u)}{u} - u^2$ is a constant, say c. Then $f(u) = u^3 + cu$, or $f(x) = \boxed{x^3 + cx}$. (We showed above that $f(0) = 0$, so this formula also holds for $x = 0$.)

We then check this solution. If $f(x) = x^3 + cx$, then

$$yf(2x) - xf(2y) = y(8x^3 + 2cx) - x(8y^3 + 2cy) = 8x^3y - 8xy^3 = 8xy(x^2 - y^2).$$

Thus, the function $f(x) = x^3 + cx$ works.

Challenge Problems

19.33 If $x_i = x_3$ for all $i > 3$, then in particular, $x_4 = x_3$. But $x_4 = f(x_3) = 4x_3 - x_3^2$, so $4x_3 - x_3^2 = x_3$. This simplifies as $x_3^2 - 3x_3 = 0$, so $x_3 = 0$ or $x_3 = 3$. Note that the converse is true; if x_3 is one these values, then $f(x_3) = x_3$, so $x_i = x_3$ for all $i > 3$.

Now, $x_3 = f(x_2) = 4x_2 - x_2^2$. If $x_3 = 0$, then $x_2(4 - x_2) = 0$, so $x_2 = 0$ or $x_2 = 4$. Since x_2 and x_3 are distinct, we have $x_2 = 4$. Then $x_2 = f(x_1) = 4x_1 - x_1^2$, so $4x_1 - x_1^2 = 4$, so $(x_1 - 2)^2 = 0$. Therefore, we have $x_1 = 2$.

Otherwise, $x_3 = 3$, so $4x_2 - x_2^2 = 3$. This simplifies as $(x_2 - 1)(x_2 - 3) = 0$, so $x_2 = 1$ or $x_2 = 3$. Since x_2 and x_3 are distinct, we have $x_2 = 1$. Then $x_2 = f(x_1) = 4x_1 - x_1^2$, so $4x_1 - x_1^2 = 1$, which gives us $x_1^2 - 4x_1 + 1 = 0$. By the quadratic formula, we have $x_1 = 2 \pm \sqrt{3}$.

Therefore, there are $\boxed{3}$ values of x_1 that satisfy the given conditions, namely 2, $2 + \sqrt{3}$, and $2 - \sqrt{3}$.

19.34 Taking $x = 94$, we get $f(94) + f(93) = 94^2$, so $f(94) = 94^2 - f(93)$. Taking $x = 93$, we get $f(93) + f(92) = 93^2$, so

$$f(94) = 94^2 - f(93) = 94^2 - (93^2 - f(92)) = 94^2 - 93^2 + f(92).$$

We can continue these calculations as follows:

$$f(94) = 94^2 - 93^2 + f(92) = 94^2 - 93^2 + 92^2 - f(91) = 94^2 - 93^2 + 92^2 - 91^2 + f(90)$$

$$= \cdots$$

$$= 94^2 - 93^2 + 92^2 - 91^2 + \cdots + 22^2 - 21^2 + 20^2 - f(19) = 94^2 - 93^2 + 92^2 - 91^2 + \cdots + 22^2 - 21^2 + 20^2 - 94.$$

To compute this sum, we can pair the terms:

$$f(94) = 94^2 - 93^2 + 92^2 - 91^2 + \cdots + 22^2 - 21^2 + 20^2 - 94$$
$$= (94^2 - 93^2) + (92^2 - 91^2) + \cdots + (22^2 - 21^2) + 20^2 - 94$$
$$= (94 + 93)(94 - 93) + (92 + 91)(92 - 91) + \cdots + (22 + 21)(22 - 21) + 20^2 - 94$$
$$= 94 + 93 + 92 + 91 + \cdots + 22 + 21 + 20^2 - 94 = (94 + 93 + \cdots + 1) - (20 + 19 + \cdots + 1) + 20^2 - 94$$
$$= \frac{94 \cdot 95}{2} - \frac{20 \cdot 21}{2} + 20^2 - 94 = \boxed{4561}.$$

19.35 To get a feel for the problem, we compute $g^n(x)$ for the first few values of n:

$$g^1(x) = g(x) = \frac{x}{x + 1},$$
$$g^2(x) = g(g(x)) = g\left(\frac{x}{x + 1}\right) = \frac{x/(x + 1)}{x/(x + 1) + 1} = \frac{x}{2x + 1},$$
$$g^3(x) = g(g^2(x)) = g\left(\frac{x}{2x + 1}\right) = \frac{x/(2x + 1)}{x/(2x + 1) + 1} = \frac{x}{3x + 1},$$
$$g^4(x) = g(g^3(x)) = g\left(\frac{x}{3x + 1}\right) = \frac{x/(3x + 1)}{x/(3x + 1) + 1} = \frac{x}{4x + 1},$$

and so on. It appears that $g^n(x) = x/(nx + 1)$ for all positive integers n. We prove this using induction.

The formula is clearly true for the base case $n = 1$, so assume that it is true for some positive integer $n = k$, so $g^k(x) = x/(kx + 1)$. Then

$$g^{k+1}(x) = g(g^k(x)) = g\left(\frac{x}{kx + 1}\right) = \frac{x/(kx + 1)}{x/(kx + 1) + 1} = \frac{x}{x + kx + 1} = \frac{x}{(k + 1)x + 1}.$$

Hence, the formula is true for $n = k + 1$, and by induction, it is true for all positive integers n.

In particular, $g^{2000}(2) = \dfrac{2}{2000 \cdot 2 + 1} = \boxed{\dfrac{2}{4001}}.$

19.36 We have

$$F(F(z)) = \frac{F(z) + i}{F(z) - i} = \frac{\frac{z+i}{z-i} + i}{\frac{z+i}{z-i} - i} = \frac{z + i + i(z - i)}{z + i - i(z - i)} = \frac{(1 + i)z + 1 + i}{(1 - i)z - 1 + i} = \frac{(1 + i)z + (1 + i)}{(1 - i)z - (1 - i)}$$
$$= \frac{1 + i}{1 - i} \cdot \frac{z + 1}{z - 1} = \frac{(1 + i)(1 + i)}{(1 - i)(1 + i)} \cdot \frac{z + 1}{z - 1} = \frac{1 + 2i - 1}{2} \cdot \frac{z + 1}{z - 1} = i \cdot \frac{z + 1}{z - 1},$$

and

$$F(F(F(z))) = i \cdot \frac{F(z) + 1}{F(z) - 1} = i \cdot \frac{\frac{z+i}{z-i} + 1}{\frac{z+i}{z-i} - 1} = i \cdot \frac{z + i + z - i}{z + i - (z - i)} = i \cdot \frac{2z}{2i} = z.$$

Thus, F is cyclic with order 3, which means that the sequence $z_0, z_1, z_2, \ldots$ also has period 3. Then $z_{2002} = z_{1999} = z_{1996} = \cdots = z_1$, and

$$z_1 = F(z_0) = \frac{z_0 + i}{z_0 - i} = \frac{\frac{1}{137} + 2i}{\frac{1}{137}} = \boxed{1 + 274i}.$$

19.37 Setting $y = 0$ in the given functional equation, we get $f(x/2) = f(x)/2$. Since this holds for all real numbers x, we have $f(\frac{x+y}{2}) = \frac{f(x+y)}{2}$. Substituting into the given functional equation, we get $\frac{f(x+y)}{2} = \frac{f(x)+f(y)}{2}$, so $f(x + y) = f(x) + f(y)$ for all real numbers x, y.

We claim that $f(x_1 + x_2 + \cdots + x_n) = f(x_1) + f(x_2) + \cdots + f(x_n)$ for all real numbers $x_1, x_2, \ldots, x_n$. To see this, we apply $f(x + y) = f(x) + f(y)$ repeatedly:

$$f(x_1 + x_2 + \cdots + x_n) = f(x_1) + f(x_2 + \cdots + x_n)$$
$$= f(x_1) + f(x_2) + f(x_3 + \cdots + x_n) = \cdots$$
$$= f(x_1) + f(x_2) + \cdots + f(x_n).$$

Taking $x_1 = x_2 = \cdots = x_n = 1$ gives $f(n) = nf(1) = n$ for all positive integers n. Then $f(-n) + f(n) = f(0) = 0$, so $f(-n) = -f(n) = -n$, which shows that $f(n) = n$ for all integers n.

Finally, consider an arbitrary rational number a/b, where a and b are integers, and b is positive. Then $f(a/b + a/b + \cdots + a/b) = f(a)$, where the number a/b appears b times. But $f(a/b + a/b + \cdots + a/b) = f(a/b) + f(a/b) + \cdots + f(a/b) = bf(a/b)$, so $f(a/b) = f(a)/b = a/b$. Hence, $f(x) = x$ for all rational numbers x.

The next natural question to ask is, is $f(x) = x$ for all real numbers x? Perhaps surprisingly, the answer is no. There exist functions where $f(x)$ is not always equal to x, but describing such functions requires more advanced mathematics.

19.38 To find $f^{12}(x)$, we compute the first few iterates of f:

$$f(x) = \frac{1}{1+x},$$

$$f^2(x) = f\left(\frac{1}{1+x}\right) = \frac{1}{1 + \frac{1}{1+x}} = \frac{1+x}{1+x+1} = \frac{1+x}{2+x},$$

$$f^3(x) = f^2\left(\frac{1}{1+x}\right) = \frac{1 + \frac{1}{1+x}}{2 + \frac{1}{1+x}} = \frac{1+x+1}{2+2x+1} = \frac{2+x}{3+2x},$$

$$f^4(x) = f^3\left(\frac{1}{1+x}\right) = \frac{2 + \frac{1}{1+x}}{3 + 2 \cdot \frac{1}{1+x}} = \frac{2+2x+1}{3+3x+2} = \frac{3+2x}{5+3x},$$

$$f^5(x) = f^4\left(\frac{1}{1+x}\right) = \frac{3 + 2 \cdot \frac{1}{1+x}}{5 + 3 \cdot \frac{1}{1+x}} = \frac{3+3x+2}{5+5x+3} = \frac{5+3x}{8+5x}.$$

We see the Fibonacci numbers, defined by $F_0 = 0$, $F_1 = 1$ and $F_n = F_{n-1} + F_{n-2}$ for all $n \geq 2$, so $F_2 = 1$, $F_3 = 2$, $F_4 = 3$, $F_5 = 5$, $F_6 = 8$, and so on. It appears that

$$f^n(x) = \frac{F_n + F_{n-1}x}{F_{n+1} + F_n x}$$

for all positive integers n. We prove this using induction.

First, we prove the base case. When $n = 1$, the left side of our formula is $f(x) = \frac{1}{1+x}$, and the right side is

$$\frac{F_1 + F_0 x}{F_2 + F_1 x} = \frac{1}{1+x},$$

so the formula holds for $n = 1$. Now, assume that the formula holds for $n = k$ for some positive integer k, so

$$f^k(x) = \frac{F_k + F_{k-1}x}{F_{k+1} + F_k x}.$$

Then

$$f^{k+1}(x) = f^k(f(x)) = f^k\left(\frac{1}{1+x}\right) = \frac{F_k + F_{k-1} \cdot \frac{1}{1+x}}{F_{k+1} + F_k \cdot \frac{1}{1+x}} = \frac{F_k + F_k x + F_{k-1}}{F_{k+1} + F_{k+1}x + F_k} = \frac{F_{k+1} + F_k x}{F_{k+2} + F_{k+1}x},$$

where we use the relationship $F_{n-1} + F_n = F_{n+1}$ to simplify the fraction in the last step. Hence, the formula holds for $n = k + 1$, and by induction, it holds for all positive integers n.

In particular, $f^{12}(x) = \dfrac{F_{12} + F_{11}x}{F_{13} + F_{12}x} = \boxed{\dfrac{144 + 89x}{233 + 144x}}$.

19.39 The smallest power of 2 greater than 2002 is $2048 = 2^{11}$. Hence, we can take $a = 2002$, $b = 2048 - 2002 = 46$, and $n = 11$ to get $f(2002) + f(46) = 121$, so $f(2002) = 121 - f(46)$.

The smallest power of 2 greater than 46 is $64 = 2^6$. Hence, we can take $a = 46$, $b = 64 - 46 = 18$, and $n = 6$ to get $f(46) + f(18) = 36$, so $f(46) = 36 - f(18)$.

The smallest power of 2 greater than 18 is $32 = 2^5$. Hence, we can take $a = 18$, $b = 32 - 18 = 14$, and $n = 5$ to get $f(18) + f(14) = 25$, so $f(18) = 25 - f(14)$.

The smallest power of 2 greater than 14 is $16 = 2^4$. Hence, we can take $a = 14$, $b = 16 - 14 = 2$, and $n = 4$ to get $f(14) + f(2) = 16$, so $f(14) = 16 - f(2)$.

Finally, taking $a = b = 2$ and $n = 2$, we have $a + b = 2^n$, so $f(2) + f(2) = 4$, or $f(2) = 2$.

Hence,

$$f(2002) = 121 - f(46) = 121 - [36 - f(18)] = 85 + f(18) = 85 + [25 - f(14)] = 110 - f(14) = 110 - [16 - f(2)] = 94 + f(2) = \boxed{96}.$$

19.40 Taking $t = x$, we get $f(2x) - f(0) = 24x^2$, so $f(2x) = 24x^2 + f(0)$. Let $y = 2x$, so $x = y/2$. Then

$$f(y) = 24 \cdot \frac{y^2}{4} + f(0) = 6y^2 + f(0).$$

Let $c = f(0)$, a constant. Then $f(y) = 6y^2 + c$, or $f(x) = \boxed{6x^2 + c}$.

We then check this solution. If $f(x) = 6x^2 + c$, then

$$f(x + t) - f(x - t) = 6(x + t)^2 + c - 6(x - t)^2 - c = 6x^2 + 12xt + 6t^2 + c - 6x^2 + 12xt - 6t^2 - c = 24xt.$$

Thus, the function $f(x) = 6x^2 + c$ works.

19.41 Taking $x = 0$, we get $3p(0) = 0$, so $p(0) = 0$. Then by the Factor Theorem, we have $p(x) = xq(x)$ for some polynomial $q(x)$. Then the given equation becomes

$$(x + 3)xq(x) = x(x + 1)q(x + 1).$$

Dividing both sides by x, we get $(x + 3)q(x) = (x + 1)q(x + 1)$.

Taking $x = -1$, we get $2q(-1) = 0$. Then by the Factor Theorem, we have $q(x) = (x + 1)r(x)$ for some polynomial $r(x)$. Then the given equation becomes

$$(x + 3)(x + 1)r(x) = (x + 1)(x + 2)r(x + 1).$$

Dividing both sides by $x + 1$, we get $(x + 3)r(x) = (x + 2)r(x + 1)$.

Taking $x = -2$, we get $r(-2) = 0$. Then by the Factor Theorem, we have $r(x) = (x + 2)s(x)$ for some polynomial $s(x)$. Then the given equation becomes

$$(x + 3)(x + 2)s(x) = (x + 2)(x + 3)s(x + 1).$$

Dividing both sides by $(x + 2)(x + 3)$, we get $s(x) = s(x + 1)$. Then the polynomial $s(x)$ must be a constant, say c.

Then $r(x) = c(x + 2)$, $q(x) = (x + 1)r(x) = c(x + 1)(x + 2)$, and $p(x) = xq(x) = \boxed{cx(x + 1)(x + 2)}$.

We then check this solution. If $p(x) = cx(x + 1)(x + 2)$, then $(x + 3)p(x) = cx(x + 1)(x + 2)(x + 3)$, and $xp(x + 1) = cx(x + 1)(x + 2)(x + 3)$. Thus, the solution $p(x) = cx(x + 1)(x + 2)$ works.

19.42 Taking $b = 1$, we get $f(a + 1) - f(1) = -\frac{a}{a+1}$. Let $x = a + 1$, so $a = x - 1$. Then

$$f(x) - f(1) = -\frac{x - 1}{x} = \frac{1 - x}{x} = \frac{1}{x} - 1,$$

so $f(x) = \frac{1}{x} + f(1) - 1$. Let $c = f(1) - 1$, so $f(x) = \boxed{\frac{1}{x} + c}$.

We then check this solution. If $f(x) = \frac{1}{x} + c$, then

$$f(a + b) - f(b) = \frac{1}{a + b} + c - \frac{1}{b} - c = \frac{1}{a + b} - \frac{1}{b} = \frac{b}{b(a + b)} - \frac{a + b}{b(a + b)} = -\frac{a}{b(a + b)}.$$

Thus, the function $f(x) = \frac{1}{x} + c$ works.

19.43 Taking $n = 0$ in (i), we get $f(f(0)) = 0$. But from (iii), $f(0) = 1$, so $f(1) = 0$.

From (ii), we have $f(f(n+2)+2) = n$, so $f(f(f(n+2)+2)) = f(n)$. From (i), we have $f(f(f(n+2)+2)) = f(n+2)+2$, so

$$f(n + 2) + 2 = f(n),$$

so $f(n + 2) = f(n) - 2$ for all integers n.

Hence,

$$f(2) = f(0) - 2 = 1 - 2 = -1,$$
$$f(4) = f(2) - 2 = -1 - 2 = -3,$$
$$f(6) = f(4) - 2 = -3 - 2 = -5,$$

and so on. We can also go in the other direction:

$$f(-2) = f(0) + 2 = 1 + 2 = 3,$$
$$f(-4) = f(-2) + 2 = 3 + 2 = 5,$$
$$f(-6) = f(-4) + 2 = 5 + 2 = 7,$$

and so on. By a straightforward induction argument, we have $f(n) = 1 - n$ for all even integers n.

We can also do the same for odd integers:

$$f(3) = f(1) - 2 = 0 - 2 = -2,$$
$$f(5) = f(3) - 2 = -2 - 2 = -4,$$
$$f(7) = f(5) - 2 = -4 - 2 = -6,$$

and

$$f(-1) = f(1) + 2 = 0 + 2 = 2,$$
$$f(-3) = f(-1) + 2 = 2 + 2 = 4,$$
$$f(-5) = f(-3) + 2 = 4 + 2 = 6,$$

and so on. Hence, $f(n) = 1 - n$ for all integers n.

We then check this solution. If $f(n) = 1 - n$, then

$$f(f(n)) = 1 - f(n) = 1 - (1 - n) = n,$$

so (i) holds. Also, $f(f(n+2)+2) = f(1-(n+2)+2) = f(1-n) = 1-(1-n) = n$, and $f(0) = 1-0 = 1$, so (ii) and (iii) also hold. Hence, the solution $f(n) = 1-n$ works.

19.44 To find $f(4, 1981)$, we compute $f(x, y)$ incrementally on x, going from 0 to 4.

We are given that $f(0, y) = y + 1$ for all integers y, so we can fill in these values in the following table (the $x\backslash y$ in the upper left indicates that the numbers in the top row are y values and the numbers in the left column are x values):

$$
\begin{array}{c|ccccc}
x\backslash y & 0 & 1 & 2 & 3 & 4 & \cdots \\
\hline
0 & 1 & 2 & 3 & 4 & 5 & \cdots
\end{array}
$$

From (2), we have $f(1, 0) = f(0, 1) = 2$. Also, from (3), we have $f(1, y+1) = f(0, f(1, y)) = f(1, y) + 1$. Hence, $f(1, 1) = f(1, 0) + 1 = 3$, $f(1, 2) = f(1, 1) + 1 = 4$, and so on. Then, by a straightforward induction argument, we have $f(1, y) = y + 2$ for all y:

$$
\begin{array}{c|ccccc}
x\backslash y & 0 & 1 & 2 & 3 & 4 & \cdots \\
\hline
0 & 1 & 2 & 3 & 4 & 5 & \cdots \\
1 & 2 & 3 & 4 & 5 & 6 & \cdots
\end{array}
$$

From (2), we have $f(2, 0) = f(1, 1) = 3$. Also, from (3), we have $f(2, y+1) = f(1, f(2, y)) = f(2, y) + 2$. Hence, $f(2, 1) = f(2, 0) + 2 = 5$, $f(2, 2) = f(2, 1) + 2 = 7$, and so on, which means we have $f(2, y) = 2y + 3$ for all y:

$$
\begin{array}{c|ccccc}
x\backslash y & 0 & 1 & 2 & 3 & 4 & \cdots \\
\hline
0 & 1 & 2 & 3 & 4 & 5 & \cdots \\
1 & 2 & 3 & 4 & 5 & 6 & \cdots \\
2 & 3 & 5 & 7 & 9 & 11 & \cdots
\end{array}
$$

From (2), we have $f(3, 0) = f(2, 1) = 5$. Also, from (3), we have $f(3, y+1) = f(2, f(3, y)) = 2f(3, y) + 3$. Adding 3 to both sides, we get $f(3, y+1) + 3 = 2f(3, y) + 6 = 2[f(3, y) + 3]$. Hence, $f(3, y) + 3$ doubles for every increment of 1 in y. Since $f(3, 0) + 3 = 8 = 2^3$, we get

$$f(3, y) + 3 = 2^y \cdot 2^3 = 2^{y+3},$$

so $f(3, y) = 2^{y+3} - 3$ for all integers y:

$$
\begin{array}{c|cccccc}
x\backslash y & 0 & 1 & 2 & 3 & 4 & \cdots \\
\hline
0 & 1 & 2 & 3 & 4 & 5 & \cdots \\
1 & 2 & 3 & 4 & 5 & 6 & \cdots \\
2 & 3 & 5 & 7 & 9 & 11 & \cdots \\
3 & 2^3-3 & 2^4-3 & 2^5-3 & 2^6-3 & 2^7-3 & \cdots
\end{array}
$$

Finally, from (2), we have $f(4, 0) = f(3, 1) = 2^4 - 3$. Also, from (3), we have $f(4, y+1) = f(3, f(4, y)) = 2^{f(4,y)+3} - 3$. Adding 3 to both sides, we get $f(4, y+1) + 3 = 2^{f(4,y)+3}$.

Hence,

$$f(4, 1) + 3 = 2^{f(4,0)+3} = 2^{2^4} = 2^{2^{2^2}},$$

$$f(4, 2) + 3 = 2^{f(4,1)+3} = 2^{2^{2^{2^2}}},$$

$$f(4, 3) + 3 = 2^{f(4,2)+3} = 2^{2^{2^{2^{2^2}}}},$$

and so on. In general,

$$f(4, y) = 2^{2^{\cdot^{\cdot^{\cdot^2}}}} - 3,$$

where there are $y + 3$ 2s. In particular,

$$f(4, 1981) = \boxed{2^{2^{\cdot^{\cdot^{2}}}} - 3},$$

where there are 1984 2s.

The function f is known as *Ackermann's function*, and has applications in computer science.

19.45 Taking $z = 0$ results in both sides of the equation being equal to $p(x) + p(y) + p(x+y) + p(0)$, so this substitution does not give us any information. So, we substitute another constant, say $z = 1$, to get

$$p(x) + p(y) + p(1) + p(x + y + 1) = p(x + y) + p(x + 1) + p(y + 1) + p(0).$$

We can rearrange this equation as

$$[p(x + y + 1) - p(x + y)] + [p(1) - p(0)] = [p(x + 1) - p(x)] + [p(y + 1) - p(y)].$$

We can simplify this expression as follows. Let $p_1(x) = p(x + 1) - p(x)$, which is a polynomial. Then the above equation becomes $p_1(x + y) + p_1(0) = p_1(x) + p_1(y)$. Furthermore, let n be the degree of $p(x)$, so

$$p(x) = a_n x^n + a_{n-1} x^{n-1} + \cdots + a_1 x + a_0.$$

Then

$$
\begin{aligned}
p_1(x) &= p(x + 1) - p(x) \\
&= a_n(x + 1)^n + a_{n-1}(x + 1)^{n-1} + \cdots + a_1(x + 1) + a_0 - (a_n x^n + a_{n-1} x^{n-1} + \cdots + a_1 x + a_0) \\
&= a_n(x^n + n x^{n-1} + \cdots) + \cdots - a_n x^n - \cdots \\
&= n a_n x^{n-1} + \cdots.
\end{aligned}
$$

Thus, the degree of $p_1(x) = p(x + 1) - p(x)$ is $n - 1$, or one less than the degree of $p(x)$, unless $p(x)$ is constant, in which case $p_1(x)$ is zero.

Now, we go back to the equation $p_1(x + y) + p_1(0) = p_1(x) + p_1(y)$. Again, if we take $y = 0$, then both sides become $p_1(x) + p_1(0)$, so we substitute $y = 1$, to get

$$p_1(x + 1) + p_1(0) = p_1(x) + p_1(1).$$

We can rearrange this equation as $p_1(x + 1) - p_1(x) = p_1(1) - p_1(0)$.

Let $p_2(x) = p_1(x + 1) - p_1(x)$. Then the above equation becomes $p_2(x) = p_2(0)$, which means that the polynomial $p_2(x)$ is constant. Furthermore, as before, the degree of $p_2(x)$ is one less than the degree of $p_1(x)$, unless $p_1(x)$ is constant, in which case $p_2(x)$ is 0.

Therefore, the degree of $p_1(x)$ is at most 1, so the degree of $p(x)$ is at most 2. Hence, $p(x)$ is of the form $p(x) = ax^2 + bx + c$. Conversely, let $p(x) = ax^2 + bx + c$. Then

$$
\begin{aligned}
p(x) &+ p(y) + p(z) + p(x + y + z) \\
&= ax^2 + bx + c + ay^2 + by + c + az^2 + bz + c + a(x + y + z)^2 + b(x + y + z) + c \\
&= ax^2 + ay^2 + az^2 + bx + by + bz + 3c + a(x^2 + y^2 + z^2 + 2xy + 2xz + 2yz) + bx + by + bz + c \\
&= 2ax^2 + 2ay^2 + 2az^2 + 2axy + 2axz + 2ayz + 2bx + 2by + 2bz + 4c,
\end{aligned}
$$

and

$$
\begin{aligned}
p(x + y) &+ p(x + z) + p(y + z) + p(0) \\
&= a(x + y)^2 + b(x + y) + c + a(x + z)^2 + b(x + z) + c + a(y + z)^2 + b(y + z) + c + c \\
&= ax^2 + 2axy + ay^2 + bx + by + c + ax^2 + 2axz + az^2 + bx + bz + c + ay^2 + 2ayz + az^2 + by + bz + c + c \\
&= 2ax^2 + 2ay^2 + 2az^2 + 2axy + 2axz + 2ayz + 2bx + 2by + 2bz + 4c.
\end{aligned}
$$

Hence, the solution is all polynomials of the form $\boxed{ax^2 + bx + c}$.

19.46 We have an unusual argument to a function, so we check if it is cyclic. We let $g(x) = 1 - x$, and we have $g(g(x)) = 1 - (1 - x) = x$, so g is cyclic. Now, we know what to do. We let $x = t$, then let $x = 1 - t$, and we have the system of equations

$$f(1 - t) = f(t) + 1 - 2t,$$
$$f(t) = f(1 - t) + 2t - 1.$$

Uh-oh. If we eliminate $f(1 - t)$ from this system of equations, we get $0 = 0$.

But $g(x) = 1 - x$ is cyclic; how did our tactic fail us? Can we find some other way to use the cyclic nature of $1 - x$ to solve this functional equation?

We try and try to do so, but about the only thing we can produce is that same useless system of equations. This really "feels" like a cyclic function problem, though.

So, back to the drawing board with $f(1 - x) = f(x) + 1 - 2x$. We try rearranging the equation to get a fresh look. Bringing the function expressions to the same side gives $f(1 - x) - f(x) = 1 - 2x$. We then make the right look more like the left, writing $1 - 2x$ as $1 - x - x$. This gives $f(1 - x) - f(x) = (1 - x) - x$. This is interesting: it looks like our separation problems. We separate $1 - x$ and x to get $f(1 - x) - (1 - x) = f(x) - x$. But what does this tell us?

If we let $h(t) = f(t) - t$, we now know that $h(t) = h(1 - t)$. That's nice, but what does it mean? We can try a few values of t to get a feel for $h(t)$. We find $h(2) = h(-1)$, $h(3) = h(-2)$, $h(4) = h(-3)$. It looks like $h(x)$ is symmetric about $x = \frac{1}{2}$. Now that we know what to look for, we can prove it easily. We have $h(x + \frac{1}{2}) = h(1 - (x + \frac{1}{2})) = h(\frac{1}{2} - x)$. So, indeed, $h(x)$ is symmetric about $\frac{1}{2}$. But how does this help?

All we know about $h(x)$ is that it is symmetric about $x = \frac{1}{2}$. So, we think about functions $p(x)$ that are symmetric about some value of x. We have studied a special class of such functions: even functions are symmetric about $x = 0$. Since $h(x)$ is symmetric about $x = \frac{1}{2}$, we see that the graph of $h(x)$ is a horizontal shift by $\frac{1}{2}$ of the graph of some even function. In other words, there is some even function $p(x)$ such that $h(x) = p(x - \frac{1}{2})$. (Checking, we see that $h(4) = p(3\frac{1}{2}) = p(-3\frac{1}{2}) = p(-3 - \frac{1}{2}) = h(-3)$, as expected.)

So, for any even function $p(x)$, we can let $h(x) = p(x - \frac{1}{2})$. From $h(x) = f(x) - x$, we have $f(x) = h(x) + x = p(x - \frac{1}{2}) + x$, where p is an even function. But what do we know about p besides the fact that it is even?

Nothing! Every even function gives us a solution to the equation.

We test our solution to make sure it works. Letting $f(x) = p(x - \frac{1}{2}) + x$ in our equation gives $f(1 - x) = p(\frac{1}{2} - x) + 1 - x$. Because $p(x)$ is even, we have $p(\frac{1}{2} - x) = p(x - \frac{1}{2})$, and we also have $p(x - \frac{1}{2}) = f(x) - x$. Putting these together with our expression for $f(1 - x)$ gives

$$f(1 - x) = p\left(\frac{1}{2} - x\right) + 1 - x = p\left(x - \frac{1}{2}\right) + 1 - x = f(x) - x + 1 - x = f(x) + 1 - 2x.$$

Therefore, the solution to the functional equation is $\boxed{f(x) = p(x - \frac{1}{2}) + x, \text{ where } p(x) \text{ is any even function}}$.

CHAPTER 20

Some Advanced Strategies

Exercises for Section 20.1

20.1.1 Let the four integers be $a, b, c,$ and d so that we have

$$a + \frac{b+c+d}{3} = 29,$$

$$b + \frac{a+c+d}{3} = 23,$$

$$c + \frac{a+b+d}{3} = 21,$$

$$d + \frac{a+b+c}{3} = 17.$$

Multiplying each of these equations by 3, we get

$$3a + b + c + d = 87,$$
$$a + 3b + c + d = 69,$$
$$a + b + 3c + d = 63,$$
$$a + b + c + 3d = 51.$$

Adding these equations, we get $6a + 6b + 6c + 6d = 270$, so $a + b + c + d = 45$. Subtracting this equation from each of the four equations above, we get $2a = 42$, $2b = 24$, $2c = 18$, and $2d = 6$. Therefore, the four numbers are $\boxed{21, 12, 9, \text{ and } 3}$.

20.1.2 Cubing the equation $r + \frac{1}{r} = 3$, we get $r^3 + 3r + \frac{3}{r} + \frac{1}{r^3} = 27$. Therefore, we have $r^3 + \frac{1}{r^3} = 27 - 3\left(r + \frac{1}{r}\right) = 27 - 3 \cdot 3 = \boxed{18}$.

20.1.3 Finding x in terms of y, and then substituting for x, gets messy quickly. So, we try to write the given equation in terms of $x + \frac{1}{x}$, which will allow us to use a direct substitution to write it in terms of y.

Since y has both x and the reciprocal of x, we need to rewrite our equation so that for each term with x^n in the equation, a term with x^{-n} also appears. Clearly $x \neq 0$, so we can divide the given equation by x^2 to get such an equation:

$$x^2 + x - 4 + \frac{1}{x} + \frac{1}{x^2} = 0. \tag{20.1}$$

Squaring y will give us both an x^2 and a $\frac{1}{x^2}$ term. Since $y^2 = x^2 + 2 + \frac{1}{x^2}$, we have $x^2 + \frac{1}{x^2} = y^2 - 2$. Now, since $x + \frac{1}{x} = y$ and $x^2 + \frac{1}{x^2} = y^2 - 2$, we have

$$x^2 + x - 4 + \frac{1}{x} + \frac{1}{x^2} = \left(x^2 + \frac{1}{x^2}\right) + \left(x + \frac{1}{x}\right) - 4 = (y^2 - 2) + y - 4 = y^2 + y - 6,$$

so we can write Equation (20.1) as $y^2 + y - 6 = 0$.

20.1.4 Multiplying all five equations, we get $a^2b^2c^2d^2e^2 = 144$, so $abcde = \pm\sqrt{144} = \pm 12$.

If $abcde = 12$, then $a = \dfrac{12}{bcde} = \dfrac{12}{bc \cdot de} = \dfrac{12}{2 \cdot 4} = \dfrac{3}{2}$, so $b = \dfrac{1}{a} = \dfrac{2}{3}$, $c = \dfrac{2}{b} = 3$, $d = \dfrac{3}{c} = 1$, and $e = \dfrac{4}{d} = 4$.

If $abcde = -12$, then $a = -\dfrac{12}{bcde} = -\dfrac{12}{bc \cdot de} = -\dfrac{12}{2 \cdot 4} = -\dfrac{3}{2}$, so $b = \dfrac{1}{a} = -\dfrac{2}{3}$, $c = \dfrac{2}{b} = -3$, $d = \dfrac{3}{c} = -1$, and $e = \dfrac{4}{d} = -4$.

Therefore, the solutions are $(a, b, c, d, e) = \boxed{(3/2, 2/3, 3, 1, 4) \text{ and } (-3/2, -2/3, -3, -1, -4)}$.

20.1.5

(a) Adding all 100 equations and dividing by 3, we get $x_1 + x_2 + x_3 + \cdots + x_{100} = 0$. We can group all the terms in this sum, except x_1, in groups of three as follows:

$$x_1 + (x_2 + x_3 + x_4) + (x_5 + x_6 + x_7) + \cdots + (x_{98} + x_{99} + x_{100}) = 0.$$

Each group of three terms sums to 0, so this equation becomes $x_1 = 0$. We can isolate each of the variables in the same way. Hence, the only solution is $x_1 = x_2 = x_3 = \cdots = x_{100} = 0$.

(b) With 99 variables, the system becomes

$$x_1 + x_2 + x_3 = 0,$$
$$x_2 + x_3 + x_4 = 0,$$
$$\vdots$$
$$x_{97} + x_{98} + x_{99} = 0,$$
$$x_{98} + x_{99} + x_1 = 0,$$
$$x_{99} + x_1 + x_2 = 0.$$

This system has a nonzero solution. Let x_n be 0 if n is divisible by 3, 1 if n is 1 more than a multiple of 3, and -1 if n is 1 less than a multiple of 3. Then for each i, where $1 \le i \le 97$, the terms x_i, x_{i+1}, and x_{i+2} will be 0, 1, and -1 in some order, so $x_i + x_{i+1} + x_{i+2} = 0$. We also have $x_{98} + x_{99} + x_1 = x_{99} + x_1 + x_2 = 0$. So, it is possible to satisfy the system without setting all the variables equal to 0.

Exercises for Section 20.2

20.2.1 Let $x = 7^a$. Then the given equation becomes $x^2 = 45x + 196$. Rearranging this gives $x^2 - 45x - 196 = 0$, so $(x - 49)(x + 4) = 0$, which means $x = 49$ or $x = -4$. If $x = 49 = 7^a$, then $a = \log_7 49 = 2$. There are no solutions to $7^a = -4$, so the only solution is $a = \boxed{2}$.

20.2.2 Let $x = y^4$. Then the given equation becomes $x^2 + 108 = 21x$, which becomes $x^2 - 21x + 108 = 0$. Factoring gives $(x - 9)(x - 12) = 0$, so $x = 9$ or $x = 12$. Therefore, the real solutions are $y = \boxed{\pm\sqrt[4]{9} \text{ and } \pm\sqrt[4]{12}}$.

20.2.3

(a) Let $z = r^2 + 5r$. Then the given equation becomes $z(z + 3) = 4$, which simplifies as $(z - 1)(z + 4) = 0$, so $z = 1$ or $z = -4$.

If $z = 1$, then $r^2 + 5r = 1$, which simplifies as $r^2 + 5r - 1 = 0$. By the quadratic formula, we have $r = \frac{-5 \pm \sqrt{29}}{2}$.

If $z = -4$, then $r^2 + 5r = -4$, which simplifies as $(r + 1)(r + 4) = 0$. Hence, $r = -1$ or $r = -4$. Therefore, the solutions are

$$r = \boxed{-1, \quad -4, \quad \frac{-5 + \sqrt{29}}{2}, \quad \text{and} \quad \frac{-5 - \sqrt{29}}{2}}.$$

(b) First, we group the factors to create a common expression. Note that $x(x + 3) = x^2 + 3x$ and $(x + 1)(x + 2) = x^2 + 3x + 2$. Let $y = x^2 + 3x$. Then the given equation becomes $y(y + 2) + 1 = 379^2$, which simplifies as $(y + 1)^2 = 379^2$. Since x is a positive integer, y is a positive integer, so $y + 1 = 379$. But $y = x^2 + 3x$, so $x^2 + 3x - 378 = 0$, which gives $(x - 18)(x + 21) = 0$. Therefore, $x = \boxed{18}$.

20.2.4 Let $y = \sqrt{4 - x}$. Then $0 \le y \le \sqrt{4} = 2$, and $y^2 = 4 - x$, so $x = 4 - y^2$. Hence, the given equation becomes $(6y - 17)(4 - y^2) = 36y - 68$, which simplifies as $6y^3 - 17y^2 + 12y = 0$. Factoring gives $y(2y - 3)(3y - 4) = 0$, so $y = 0$, $y = 3/2$, or $y = 4/3$.

All these values satisfy $0 \le y \le 2$. We find that the corresponding solutions for x are $x = \boxed{4, 7/4, \text{ and } 20/9}$.

20.2.5 Let $y = 2x^2 - 3x - 19$. Then $4x^2 - 6x - 41 = 2(2x^2 - 3x - 19) - 3 = 2y - 3$, so the given equation becomes $2y - 3 + \frac{1}{y} = 0$. Multiplying by y, we get $2y^2 - 3y + 1 = 0$, so $(y - 1)(2y - 1) = 0$, which means $y = 1$ or $2y = 1$.

If $y = 1$, then $2x^2 - 3x - 19 = 1$, so $2x^2 - 3x - 20 = 0$. Factoring gives $(x - 4)(2x + 5) = 0$, so $x = 4$ or $x = -5/2$.

If $2y = 1$, then $2(2x^2 - 3x - 19) = 1$, so $4x^2 - 6x - 39 = 0$. By the quadratic formula, we have $x = \frac{3 \pm \sqrt{165}}{4}$.

Therefore, the solutions are $x = \boxed{4, \quad -\frac{5}{2}, \quad \frac{3 + \sqrt{165}}{4}, \quad \text{and} \quad \frac{3 - \sqrt{165}}{4}}$.

20.2.6 Let $a = \sqrt{x^2 + 17x + 59}$ and $b = \sqrt{x^2 + 17x - 85}$. Then the given equation becomes $a - 2b = 3$. Furthermore,

$$a^2 - b^2 = (x^2 + 17x + 59) - (x^2 + 17x - 85) = 144.$$

Substituting $a = 2b + 3$ into this equation, we get $(2b + 3)^2 - b^2 = 144$, which simplifies as $3b^2 + 12b - 135 = 0$. Factoring gives $3(b - 5)(b + 9) = 0$, so $b = 5$ or $b = -9$. Since $b = \sqrt{x^2 + 17x - 85}$, we must have $b \ge 0$, so b must be equal to 5. Therefore, $x^2 + 17x - 85 = 25$, so $x^2 + 17x - 110 = 0$. Factoring gives $(x - 5)(x + 22) = 0$, so the solutions are $x = \boxed{5 \text{ and } -22}$. (Checking reveals that both solutions work.)

20.2.7 We could solve the second equation for x or y and substitute, but that doesn't look like much fun. The first equation is homogeneous, so we let $x = yr$, which gives us $2y^4r^4 + 5y^4r^3 + 45y^4r = 34y^4r^2 + 18y^4$. If $y = 0$, then $x = 0$ in the original equation, but $(x, y) = (0, 0)$ doesn't satisfy $2x - 3y = 7$. Therefore, $y \ne 0$, so we can divide our equation by y^4 to give $2r^4 + 5r^3 + 45r = 34r^2 + 18$. Rearranging this equation, then factoring, gives $(r - 1)^2(r + 6)(2r - 3) = 0$, so r is 1, -6, or $3/2$.

If $r = 1$, we have $x = y$, so $2x - 3y = 7$ gives us $x = y = -7$.

If $r = -6$, then $x = -6y$, so $2x - 3y = 7$ gives us $-15y = 7$, which means $y = -7/15$ and $x = -6y = 14/5$.

If $r = 3/2$, then $x = 3y/2$, so $2x - 3y = 7$ gives us $0 = 7$, which has no solutions.

Therefore, the two solutions are $\boxed{(x, y) = (-7, -7) \text{ and } (x, y) = (14/5, -7/15)}$.

Exercises for Section 20.3

20.3.1 Let $ax^3 + bx^2 + 1 = (x^2 - x - 1)q(x)$, where $q(x)$ is a polynomial. Since $ax^3 + bx^2 + 1$ has degree 3 and $x^2 - x - 1$ has degree 2, the degree of $q(x)$ must be 1. Let $q(x) = cx + d$, so $ax^3 + bx^2 + 1 = (x^2 - x - 1)(cx + d)$. The x^3 term in the expansion of the right side is cx^3, so we have $c = a$. Similarly, the constant term in the expansion of the right side is $-d$, so $d = -1$, and we now have $ax^3 + bx^2 + 1 = (x^2 - x - 1)(ax - 1)$. The x term in the expansion of the right is $-ax + x$, so $a = 1$, and the quadratic term of the expansion is $-ax^2 - x^2 = -2x^2$, so $b = \boxed{-2}$.

20.3.2   Let
$$3x^2 + kxy - 2y^2 - 7x + 7y - 6 = (ax + by + c)(dx + ey + f),$$
for some integers a, b, c, d, e, and f. If we take $y = 0$, then the term kxy disappears, and the equation becomes $3x^2 - 7x - 6 = (ax + c)(dx + f)$. The quadratic $3x^2 - 7x - 6$ factors as $(x - 3)(3x + 2)$, so we set $a = 1$, $c = -3$, $d = 3$, and $f = 2$. This means
$$3x^2 + kxy - 2y^2 - 7x + 7y - 6 = (x + by - 3)(3x + ey + 2)$$
$$= 3x^2 + (3b + e)xy + bey^2 - 7x + (2b - 3e)y - 6.$$

for some constants b and e.

Since the corresponding coefficients of xy, y^2, and y on both sides must equal each other, we have that $k = 3b + e$, $be = -2$, and $2b - 3e = 7$. The last equation gives us $b = (3e + 7)/2$. Substituting this into $be = -2$, we have $(3e + 7)/2 \cdot e = -2$, which simplifies as $(e + 1)(3e + 4) = 0$. Therefore, $e = -1$ or $e = -4/3$. Since e is an integer, $e = -1$. Then $b = (3e + 7)/2 = 2$, so $k = 3b + e = \boxed{5}$.

20.3.3   Let $p(x) = 2x^4 + ax^3 + bx^2 + cx + d$. Then from the given values of $p(x)$, we obtain the system of equations
$$32 - 8a + 4b - 2c + d = 34,$$
$$2 - a + b - c + d = 10,$$
$$2 + a + b + c + d = 10,$$
$$32 + 8a + 4b + 2c + d = 34.$$

Moving all the constants to one side, we get the equations
$$-8a + 4b - 2c + d = 2,$$
$$-a + b - c + d = 8,$$
$$a + b + c + d = 8,$$
$$8a + 4b + 2c + d = 2.$$

Since $p(0) = d$, we just need to find d. Adding the second and third equations, and dividing by 2, we get $b + d = 8$. Adding the first and fourth equations, and dividing by 2, we get $4b + d = 2$. Subtracting $b + d = 8$ from this gives $3b = -6$, so $b = -2$. Therefore $d = 8 - b = \boxed{10}$.

20.3.4   Suppose that $x^3 + 3x^2 + 8x + 21$ has a pair of real roots whose product is 1. By Vieta's Formulas, the product of the roots is -21, so the third root must be -21. However, when we substitute -21 for x, we get $(-21)^3 + 3 \cdot (-21)^2 + 8 \cdot (-21) + 21 = -8085$, which is not zero. Therefore, the polynomial $x^3 + 3x^2 + 8x + 21$ does not have real roots whose product is 1.

20.3.5   First, we check if the quartic $z^4 + 4z^2 - z + 6 = 0$ has any integer roots. By the Rational Root Theorem, if there is an integer root, then it must be a factor of 6. Checking these values, we find that none of them is a root of the quartic, so there are no integer roots.

Next, we check if there are any rational roots. The given quartic is monic, so a rational root must also be an integer. However, there are no integer roots, so there are no rational roots either.

In the absence of rational roots, we try to factor the quartic using the Method of Undetermined Coefficients. Let
$$z^4 + 4z^2 - z + 6 = (az^2 + bz + c)(dz^2 + ez + f),$$

where $a, b, c, d, e,$ and f are integers. Then the coefficient of z^4 on the right side, when expanded, is ad, so $ad = 1$. Hence, we may assume that $a = d = 1$. Then the equation above becomes

$$z^4 + 4z^2 - z + 6 = (z^2 + bz + c)(z^2 + ez + f)$$
$$= z^4 + (b + e)z^3 + (be + c + f)z^2 + (bf + ce)z + cf.$$

Equating corresponding coefficients, we obtain the system of equations

$$b + e = 0,$$
$$be + c + f = 4,$$
$$bf + ce = -1,$$
$$cf = 6.$$

From the first equation, $e = -b$. Substituting this into the third equation, we get $bf - bc = -1$, so $b(f - c) = -1$. Hence, $b = 1$ or $b = -1$.

We take the case where $b = 1$. Then $e = -1$, and $f - c = -1$. Then from the second equation, we have $-1 + c + f = 4$, so $c + f = 5$. Solving for c and f, we find $f = 2$ and $c = 3$. Note that these values satisfy the fourth equation. Thus, we have the factorization $z^4 + 4z^2 - z + 6 = (z^2 + z + 3)(z^2 - z + 2)$. Using the quadratic formula, we find the solutions are

$$z = \boxed{\frac{-1 + i\sqrt{11}}{2}, \quad \frac{-1 - i\sqrt{11}}{2}, \quad \frac{1 + i\sqrt{7}}{2}, \quad \text{and} \quad \frac{1 - i\sqrt{7}}{2}}.$$

Exercises for Section 20.4

20.4.1 Let $y = x - 1$, $a = r - 1$, $b = s - 1$ and $c = t - 1$. Then $x = y + 1$, and $a, b,$ and c are the roots of

$$(y + 1)^3 - 2(y + 1)^2 + 3(y + 1) - 4 = y^3 + y^2 + 2y - 2.$$

Then by Vieta's Formulas, we have $ab + ac + bc = 2$ and $abc = 2$, so

$$\frac{1}{r-1} + \frac{1}{s-1} + \frac{1}{t-1} = \frac{1}{a} + \frac{1}{b} + \frac{1}{c} = \frac{ab + ac + bc}{abc} = \frac{2}{2} = \boxed{1}.$$

20.4.2 From the given equations, $p, q,$ and r are the roots of the cubic $x^3 + sx + t - 1 = 0$. Hence, from Vieta's Formulas, we have $p + q + r = 0$ and $pq + pr + qr = s$. Squaring the equation $p + q + r = 0$, we get

$$p^2 + q^2 + r^2 + 2pq + 2pr + 2qr = 0,$$

so $p^2 + q^2 + r^2 + 2s = 0$. Therefore, we have $p^2 + q^2 + r^2 = \boxed{-2s}$.

20.4.3 Let r be a root of $f(x) = 0$, so $a_n r^n + a_{n-1} r^{n-1} + \cdots + a_1 r + a_0 = 0$. Note that $r \neq 0$ since $a_0 \neq 0$. Then

$$g\left(\frac{1}{r}\right) = a_0 \left(\frac{1}{r}\right)^n + a_1 \left(\frac{1}{r}\right)^{n-1} + \cdots + a_{n-1}\left(\frac{1}{r}\right) + a_n = \frac{a_0 + a_1 r + \cdots + a_{n-1} r^{n-1} + a_n r^n}{r^n} = 0.$$

Hence, each root of $f(x)$ is the reciprocal of a root of $g(x)$.

20.4.4 Since $a, b,$ and c are the roots of $f(x)$, and $f(x)$ is monic, we have $f(x) = (x - a)(x - b)(x - c)$. Therefore, we have $f(-1) = (-1 - a)(-1 - b)(-1 - c) = (-1)^3(1 + a)(1 + b)(1 + c)$, which means $-f(-1) = (1 + a)(1 + b)(1 + c)$.

20.4.5 *Solution 1: The hard way.* Let $p(x) = ax^2 + bx + c$. Then from the given values of $p(x)$, we obtain the system of equations

$$a + b + c = 1,$$
$$4a + 2b + c = 3,$$
$$9a + 3b + c = 2.$$

Subtracting the first equation from the second equation, we get $3a + b = 2$. Subtracting the second equation from the third equation, we get $5a + b = -1$. Solving for a and b, we find $a = -3/2$ and $b = 13/2$. Then from the first equation, $c = -4$, so

$$p(x) = -\frac{3}{2}x^2 + \frac{13}{2}x - 4 = \frac{-3x^2 + 13x - 8}{2}.$$

Then

$$p(p(x)) = p\left(\frac{-3x^2 + 13x - 8}{2}\right) = \frac{-3(\frac{-3x^2+13x-8}{2})^2 + 13(\frac{-3x^2+13x-8}{2}) - 8}{2} = -\frac{27}{8}x^4 + \frac{117}{4}x^3 - \frac{729}{8}x^2 + \frac{481}{4}x - 54.$$

Thus, the equation $p(p(x)) = x$ simplifies as $27x^4 - 234x^3 + 729x^2 - 954x + 432 = 0$. This quartic looks difficult to work with, until we realize that $p(p(1)) = p(1) = 1$, $p(p(2)) = p(3) = 2$, and $p(p(3)) = p(2) = 3$, so $x = 1, 2$, and 3 must be roots of this quartic. By Vieta's Formulas, the sum of the roots of this quartic is $234/27 = 26/3$, so the fourth root is $26/3 - 1 - 2 - 3 = \boxed{8/3}$.

Solution 2: The easy way. Since $p(1) = 1$, $x = 1$ is a root of the polynomial $p(x) - x$. Furthermore, $p(x) - x$ is a quadratic. Therefore, there exist constants a and k such that $p(x) - x = a(x - 1)(x - k)$. Then $p(k) - k = 0$, so $p(k) = k$.

Now, observe that $p(p(1)) = p(1) = 1$, $p(p(k)) = p(k) = k$, $p(p(2)) = p(3) = 2$, and $p(p(3)) = p(2) = 3$. Therefore, $x = 1$, $x = k$, $x = 2$, and $x = 3$ are the four roots of the quartic polynomial $p(p(x)) - x = 0$. Hence, k is the fourth non-integer root we seek.

In the equation $p(x) - x = a(x - 1)(x - k)$, taking $x = 2$, we get $3 - 2 = a(2 - k)$, so $2a - ak = 1$. Taking $x = 3$, we get $2 - 3 = 2a(3 - k)$, so $6a - 2ak = -1$. Multiplying $2a - ak = 1$ by 2, we get $4a - 2ak = 2$, so $2ak = 4a - 2$. Therefore, from $6a - 2ak = -1$, we have $-1 = 6a - 2ak = 6a - (4a - 2) = 2a + 2$, so $a = -3/2$. Substituting this into $2a - ak = 1$ gives $k = \boxed{8/3}$.

20.4.6 Let $Q(x) = x(x + 1)P(x) - 1$. Then $Q(1) = Q(2) = \cdots = Q(10) = 0$, so by the Factor Theorem, we have

$$Q(x) = (x - 1)(x - 2) \cdots (x - 10)R(x)$$

for some polynomial $R(x)$. In addition, we have $Q(-1) = Q(0) = -1$, so taking $x = -1$ in the equation above, we get

$$-1 = (-2)(-3) \cdots (-11)R(-1) \quad \Rightarrow \quad R(-1) = -\frac{1}{11!},$$

and taking $x = 0$, we get

$$-1 = (-1)(-2) \cdots (-10)R(0) \quad \Rightarrow \quad R(0) = -\frac{1}{10!}.$$

The polynomial $R(x)$ of smallest degree that satisfies both conditions is a linear polynomial, so let $R(x) = ax + b$. Then substituting $x = -1$ and $x = 0$, we obtain the system of equations

$$-a + b = R(-1) = -\frac{1}{11!},$$
$$b = R(0) = -\frac{1}{10!}.$$

Taking the difference of the equations, we get $a = \dfrac{1}{11!} - \dfrac{1}{10!} = \dfrac{1}{11!} - \dfrac{11}{11!} = -\dfrac{10}{11!}$.

Now, $Q(11) = 11 \cdot 12P(11) - 1 = 132P(11) - 1$. But

$$Q(11) = 10 \cdot 9 \cdots 1 \cdot R(11) = 10!(11a + b) = 10!\left(-11 \cdot \frac{10}{11!} - \frac{1}{10!}\right) = 10!\left(-\frac{10}{10!} - \frac{1}{10!}\right) = 10!\left(-\frac{11}{10!}\right) = -11,$$

so $132P(11) - 1 = -11$, which means $P(11) = (1 - 11)/132 = -10/132 = \boxed{-5/66}$.

Exercises for Section 20.5

20.5.1   Let r be the common root, so $r^2 + ar + 1 = 0$ and $r^2 + r + a = 0$. Taking the difference of these two equations, we get $ar + 1 - r - a = 0$. Factoring the left side gives $(a - 1)(r - 1) = 0$, so either $a = 1$ or $r = 1$.

If $a = 1$, then both polynomials become $x^2 + x + 1$, so they have the same roots. If $r = 1$, then $1 + a + 1 = 0$, so $a = -2$. Therefore, the solutions are $a = \boxed{1 \text{ and } -2}$.

20.5.2   Let r and s be the two common roots of $x^3 + ax^2 + 11x + 6$ and $x^3 + bx^2 + 14x + 8$. Then r and s are also roots of their difference, $(a - b)x^2 - 3x - 2$.

Taking the difference of the two cubics eliminated the x^3 term. We can also eliminate the constant term by multiplying the first cubic by 4 and the second cubic by 3, and then taking the difference. This gives us the cubic

$$4(x^3 + ax^2 + 11x + 6) - 3(x^3 + bx^2 + 14x + 8) = x^3 + (4a - 3b)x^2 + 2x = x[x^2 + (4a - 3b)x + 2].$$

Neither r nor s can be 0 (otherwise, the constant terms of the two original cubics in the problem would also be 0), so r and s are the roots of $x^2 + (4a - 3b)x + 2$.

Now we look at the two quadratics that we have derived, $(a - b)x^2 - 3x - 2$ and $x^2 + (4a - 3b)x + 2$. Since r and s are the roots of both quadratics, one must be a constant multiple of the other. Furthermore, the constant term is 2 in one and -2 in the other, so the constant multiple must be -1. Comparing the coefficients of x^2 and x, we obtain the system of equations

$$a - b = -1,$$
$$4a - 3b = 3.$$

Solving, we get $\boxed{a = 6 \text{ and } b = 7}$.

Checking, we find that

$$x^3 + 6x^2 + 11x + 6 = (x + 1)(x + 2)(x + 3),$$
$$x^3 + 7x^2 + 14x + 8 = (x + 1)(x + 2)(x + 4),$$

so the common roots are -1 and -2.

Exercises for Section 20.6

20.6.1   Multiplying both sides of the equation $\frac{1}{x} + \frac{1}{y} + \frac{1}{z} = 0$ by xyz, we get $xy + xz + yz = 0$. Then squaring the equation $x + y + z = 1$, we get $x^2 + y^2 + z^2 + 2xy + 2xz + 2yz = 1$, so $x^2 + y^2 + z^2 = \boxed{1}$.

20.6.2   Squaring the equation $x^2 + y^2 = 21$, we get $x^4 + 2x^2y^2 + y^4 = 441$. Therefore,

$$x^4 + y^4 = 441 - 2x^2y^2 = 441 - 2 \cdot 5^2 = \boxed{391}.$$

20.6.3   Let $s = x + y$ and $p = xy$. Then $s^2 = x^2 + 2xy + y^2 = 34 + 2p$, so $p = (s^2 - 34)/2$, and

$$x^3 + y^3 = (x + y)(x^2 - xy + y^2) = s(34 - p),$$

so $s(34 - p) = 98$. Substituting $p = (s^2 - 34)/2$ into this equation, and simplifying, we get $s^3 - 102s + 196 = 0$.

First, we search for integer roots. By the Rational Root Theorem, if there is an integer root, it must be a factor of 196. Checking these values, we find that the cubic is 0 for $s = 2$, so by the Factor Theorem, it is divisible by $s - 2$. Factoring out $s - 2$, we get $(s - 2)(s^2 + 2s - 98) = 0$. By the quadratic formula, the roots of the quadratic are $-1 \pm 3\sqrt{11}$.

If $s = 2$, then $p = (s^2 - 34)/2 = -15$. Hence, by Vieta's Formulas, x and y are the roots of the quadratic $t^2 - 2t - 15$, which factors as $(t - 5)(t + 3)$. This gives us the solutions $(x, y) = (5, -3)$ and $(-3, 5)$.

If $s = -1 + 3\sqrt{11}$, then $p = (s^2 - 34)/2 = 33 - 3\sqrt{11}$. Hence, x and y are the roots of the quadratic

$$t^2 - (-1 + 3\sqrt{11})t + 33 - 3\sqrt{11}.$$

The discriminant of this quadratic is $(-1 + 3\sqrt{11})^2 - 4(33 - 3\sqrt{11}) = -32 + 6\sqrt{11} < 0$, so it has no real roots.

Finally, if $s = -1 - 3\sqrt{11}$, then $p = (s^2 - 34)/2 = 33 + 3\sqrt{11}$. Hence, x and y are the roots of the quadratic

$$t^2 - (-1 - 3\sqrt{11})t + 33 + 3\sqrt{11}.$$

The discriminant of this quadratic is $(-1 - 3\sqrt{11})^2 - 4(33 + 3\sqrt{11}) = -32 - 6\sqrt{11} < 0$, so it has no real roots.

Therefore, the only real solutions are $(x, y) = \boxed{(5, -3) \text{ and } (-3, 5)}$.

20.6.4   We already proved that two of a, b, c, and d must sum to 0. We are given that $a + b + c + d = 0$. Therefore, if two of the numbers sum to 0, then the other two numbers also sum to 0.

20.6.5   Squaring the equation $x + y + z = 2$, we get

$$x^2 + y^2 + z^2 + 2xy + 2xz + 2yz = 4,$$

so $14 + 2(xy + xz + yz) = 4$, which gives $xy + xz + yz = (4 - 14)/2 = -5$. Hence, by Vieta's Formulas, x, y, and z are the roots of the cubic $t^3 - 2t^2 - 5t + 6$, which factors as $(t - 1)(t - 3)(t + 2)$. Therefore, the solutions are $(x, y, z) = \boxed{(1, 3, -2) \text{ and all of its permutations}}$.

20.6.6   As in previous similar problems, we would like to find a cubic whose roots are x, y, and z. To do this, we seek the values of $x + y + z$ (which is given), $xy + xz + yz$, and xyz.

Squaring the equation $x + y + z = 6$, we get

$$36 = x^2 + y^2 + z^2 + 2xy + 2xz + 2yz = 26 + 2(xy + xz + yz),$$

so $xy + xz + yz = (36 - 26)/2 = 5$.

To get xyz, a term with degree 3, we cube the equation $x + y + z = 6$, to get

$$216 = (x + y + z)^3 = x^3 + y^3 + z^3 + 3x^2y + 3xy^2 + 3x^2z + 3xz^2 + 3y^2z + 3yz^2 + 6xyz. \tag{20.2}$$

We are given that $x^3 + y^3 + z^3 = 90$, so we must determine the value of $x^2y + xy^2 + x^2z + xz^2 + y^2z + yz^2$.

If we take the product of $x + y + z = 6$ and $xy + xz + yz = 5$, then we get

$$x^2y + xy^2 + x^2z + xz^2 + y^2z + yz^2 + 3xyz = 30.$$

Therefore, $x^2y + xy^2 + x^2z + xz^2 + y^2z + yz^2 = 30 - 3xyz$. Substituting into Equation (20.2) above, we get

$$216 = 90 + 3(30 - 3xyz) + 6xyz = -3xyz + 180,$$

which means $xyz = -12$.

Hence, by Vieta's Formulas, x, y, and z are the roots of $t^3 - 6t^2 + 5t + 12$, which factors as $(t-3)(t-4)(t+1)$. Therefore, the solutions are $(x, y, z) = \boxed{(3, 4, -1) \text{ and all of its permutations}}$.

Review Problems

20.34 Adding all three equations, we get $(2x + 2y + 2z)(x + y + z) = 18$, so $2(x + y + z)^2 = 18$, which means $(x + y + z)^2 = 9$, or $x + y + z = \pm 3$.

If $x + y + z = 3$, then the first equation becomes $(3 - z)(3) = 15$, so $z = -2$. Similarly, the second and third equations tell us that $x = 1$ and $y = 4$.

If $x + y + z = -3$, then the first equation becomes $(-3 - z)(-3) = 15$, so $z = 2$. Similarly, the second and third equations tell us that $x = -1$ and $y = -4$.

Therefore, the solutions are $(x, y, z) = \boxed{(1, 4, -2) \text{ and } (-1, -4, 2)}$.

20.35 When the right side is expanded, the coefficient of x^4 is p^2, so $p^2 = 1$. Hence, we can take $p = 1$. (If $p = -1$, we can multiply $px^2 + qx + r$ by -1.) Then

$$x^4 + 4x^3 - 2x^2 - 12x + 9 = (x^2 + qx + r)^2 = x^4 + 2qx^3 + (q^2 + 2r)x^2 + 2qrx + r^2.$$

Equating the coefficients of x^3 on either side, we get $2q = 4$, so $q = 2$. Then equating the coefficients of x on either side, we get $2qr = 4r = -12$, so $r = -3$. To check that $(p, q, r) = \boxed{(1, 2, -3)}$ works, we expand $(x^2 + 2x - 3)^2$, to get $x^4 + 4x^3 - 2x^2 - 12x + 9$, so these constants work. Note that $(p, q, r) = (-1, -2, 3)$ is also a valid solution.

20.36 Let $x^2 + 3xy + x + my - m = (ax + by + c)(dx + ey + f)$, where a, b, c, d, e, and f are integers. Note that the coefficient of x^2 in the right side, when expanded, is ad, so $ad = 1$. Hence, we take $a = d = 1$. We also note that there is no y^2 term, so either $b = 0$ or $e = 0$. Without loss of generality, we let $b = 0$, and we have

$$x^2 + 3xy + x + my - m = (x + c)(x + ey + f) = x^2 + exy + (c + f)x + cey + cf.$$

We then see that $e = 3$, so from the y term, we have $m = 3c$. From the constant term (that is, the term without x or y), we have $cf = -m$, so $cf = -3c$. Hence, either $c = 0$ or $f = -3$.

If $c = 0$, then $m = 0$, and $x^2 + 3xy + x = x(x + 3y + 1)$.

If $f = -3$, then $c = 1 - f = 4$, so $m = ce = 12$, and $x^2 + 3xy + x + 12y - 12 = (x + 4)(x + 3y - 3)$.

Hence, the possible values of m are $m = \boxed{0 \text{ and } 12}$.

20.37 Since $P(x)$ has the four roots $1/4$, $1/2$, 2, and 4, we have

$$P(x) = k\left(x - \frac{1}{4}\right)\left(x - \frac{1}{2}\right)(x - 2)(x - 4)$$

for some constant k. Taking $x = 1$, we get $P(1) = k \cdot \frac{3}{4} \cdot \frac{1}{2} \cdot (-1) \cdot (-3) = \frac{9k}{8}$. Since $P(1) = 1$, we have $k = \frac{8}{9}$. Therefore, we have $P(0) = \frac{8}{9} \cdot \left(-\frac{1}{4}\right) \cdot \left(-\frac{1}{2}\right) \cdot (-2) \cdot (-4) = \boxed{\frac{8}{9}}$.

20.38 By Vieta's Formulas, a, b, and c are the roots of the cubic $t^3 - 4t^2 - 4t + 21$. Using the usual factoring techniques, we find that this cubic factors as $(t - 3)(t^2 - t - 7)$. Hence, the solutions are

$$(a, b, c) = \boxed{\left(3, \ \frac{1 + \sqrt{29}}{2}, \ \frac{1 - \sqrt{29}}{2}\right) \text{ and all of its permutations}}.$$

20.39 Let $u = \sqrt{4x-3}$. Then the given equation becomes $u + \frac{10}{u} = 7$. Multiplying both sides by u and simplifying, we get $u^2 - 7u + 10 = 0$. Factoring gives $(u-2)(u-5) = 0$, so $u = 2$ or $u = 5$.

If $u = 2$, then $\sqrt{4x-3} = 2$. Squaring both sides, we get $4x - 3 = 4$, so $x = 7/4$. If $u = 5$, then $\sqrt{4x-3} = 5$. Squaring both sides, we get $4x - 3 = 25$, so $x = 7$. Therefore, the solutions are $x = \boxed{7 \text{ and } 7/4}$.

20.40 Adding all 17 equations and dividing by 3, we get

$$a_1 + a_2 + a_3 + \cdots + a_{17} = \frac{1}{3} \cdot \frac{17 \cdot 18}{2} = 51.$$

To isolate a_{17}, we need to find the right combination of the given equations. Note that each of the given equations contains three terms, and 17 is one less than a multiple of 3. So by adding another term of a_{17}, we obtain the sum

$$a_1 + a_2 + a_3 + \cdots + a_{16} + a_{17} + a_{17},$$

which contains 18 terms, and 18 is a multiple of 3. Thus, we can try to group these 18 terms accordingly.

We find

$$a_1 + a_2 + a_3 + \cdots + a_{16} + a_{17} + a_{17} = (a_{17} + a_1 + a_2) + (a_3 + a_4 + a_5) + (a_6 + a_7 + a_8)$$
$$+ (a_9 + a_{10} + a_{11}) + (a_{12} + a_{13} + a_{14}) + (a_{15} + a_{16} + a_{17})$$
$$= 17 + 3 + 6 + 9 + 12 + 15 = 62.$$

We also have $a_1 + a_2 + a_3 + \cdots + a_{16} + a_{17} + a_{17} = 51 + a_{17}$. Therefore, $a_{17} = 62 - 51 = \boxed{11}$.

20.41 We want to find all real numbers x, y, and z such that $x = yz$, $y = xz$, and $z = xy$. Multiplying all three equations, we get $xyz = x^2y^2z^2$, so $x^2y^2z^2 - xyz = 0$, which means $xyz(xyz - 1) = 0$.

If any of x, y, or z is equal to 0, then all three variables must be 0, so we have the solution $(x, y, z) = (0, 0, 0)$. Otherwise, $xyz = 1$. Then $1 = xyz = x \cdot x = x^2$, so $x = \pm 1$. By symmetry, $y = \pm 1$ and $z = \pm 1$. (Here, all the $\pm$ signs are independent.)

For any such choice of x and y (among the values 1 and -1), we must have $z = xy$, which gives the solutions $(x, y, z) = \boxed{(0,0,0), (1,1,1), (1,-1,-1), (-1,1,-1), \text{ and } (-1,-1,1)}$. It is easy to check that all these solutions work.

20.42 Let $y = x^2 + 18x + 45$. Then $x^2 + 18x + 30 = y - 15$, so the given equation becomes $y - 15 = 2\sqrt{y}$. Since $\sqrt{y} \geq 0$, the equation $y - 15 = 2\sqrt{y}$ means that $y \geq 15$. Squaring both sides, we get $y^2 - 30y + 225 = 4y$, which becomes $(y-9)(y-25) = 0$, so $y = 9$ or $y = 25$. Only $y = 25$ satisfies the condition $y \geq 15$.

Then $x^2 + 18x + 45 = 25$, which becomes $x^2 + 18x + 20 = 0$. The discriminant of this quadratic is $18^2 - 4 \cdot 20 = 244 > 0$, so its roots are real. By Vieta's Formulas, the product of the roots is $\boxed{20}$.

20.43 Let the roots be r, r, s, and s. Then

$$x^4 - 16x^3 + 94x^2 + px + q = (x-r)^2(x-s)^2 = (x^2 - 2rx + r^2)(x^2 - 2sx + s^2)$$
$$= x^4 - (2r + 2s)x^3 + (r^2 + 4rs + s^2)x^2 - (2r^2s + 2rs^2)x + r^2s^2.$$

Equating coefficients of like terms, we obtain the system of equations

$$-2r - 2s = -16,$$
$$r^2 + 4rs + s^2 = 94,$$
$$-2r^2s - 2rs^2 = p,$$
$$r^2s^2 = q.$$

From the first equation, we have $r + s = 8$. Squaring this equation, we get $r^2 + 2rs + s^2 = 64$. Subtracting this from the second equation, we get $2rs = 30$, so $rs = 15$. Therefore,

$$p + q = -2r^2s - 2rs^2 + r^2s^2 = -2(r+s)rs + (rs)^2 = -2 \cdot 8 \cdot 15 + 15^2 = \boxed{-15}.$$

20.44

(a) Let $y = x - 2$, and let $a = r - 2$, $b = s - 2$, and $c = t - 2$. Then $x = y + 2$, and a, b, and c are the roots of

$$2(y+2)^3 - 3(y+2)^2 + 2(y+2) - 1 = 2y^3 + 9y^2 + 14y + 7.$$

By Vieta's Formulas, we have $ab + ac + bc = 7$ and $abc = -7/2$. Therefore, we have

$$\frac{1}{r-2} + \frac{1}{s-2} + \frac{1}{t-2} = \frac{1}{a} + \frac{1}{b} + \frac{1}{c} = \frac{ab+ac+bc}{abc} = \frac{7}{-7/2} = \boxed{-2}.$$

(b) Let $y = 2x + 1$, and let $a = 2r + 1$, $b = 2s + 1$, and $c = 2s + 1$. Then $x = (y-1)/2$, and a, b, and c are the roots of

$$2\left(\frac{y-1}{2}\right)^3 - 3\left(\frac{y-1}{2}\right)^2 + 2\left(\frac{y-1}{2}\right) - 1 = \frac{y^3 - 6y^2 + 13y - 12}{4}.$$

By Vieta's Formulas, we have $ab + ac + bc = 13$ and $abc = 12$. Therefore, we have

$$\frac{1}{2r+1} + \frac{1}{2s+1} + \frac{1}{2t+1} = \frac{1}{a} + \frac{1}{b} + \frac{1}{c} = \frac{ab+ac+bc}{abc} = \boxed{\frac{13}{12}}.$$

20.45 Let the roots of $P(x) = 0$ be r, s, and t. Then by Vieta's Formulas, we have $rst = 1$, so $rs = 1/t$. Then

$$Q(rs) = Q\left(\frac{1}{t}\right) = \frac{1}{t^3} - \frac{1}{t} - 1 = \frac{1 - t^2 - t^3}{t^3} = -\frac{t^3 + t^2 - 1}{t^3} = -\frac{P(t)}{t^3} = 0.$$

Thus, rs is a root of $Q(x) = 0$.

20.46 Multiplying the two given equations, we get

$$26 \cdot 28 = (a + b + c)\left(\frac{1}{a} + \frac{1}{b} + \frac{1}{c}\right) = \frac{a}{b} + \frac{b}{a} + \frac{a}{c} + \frac{c}{a} + \frac{b}{c} + \frac{c}{b} + 3.$$

Therefore, $\dfrac{a}{b} + \dfrac{b}{a} + \dfrac{a}{c} + \dfrac{c}{a} + \dfrac{b}{c} + \dfrac{c}{b} = \boxed{725}$.

20.47 To deal with the square roots, let $a = \sqrt{x}$, $b = \sqrt{y}$, and $c = \sqrt{z}$. Then the given system of equations becomes

$$a + b + c = 10,$$
$$a^2 + b^2 + c^2 = 38,$$
$$ab + ac + bc = 30.$$

Squaring the first equation, we get $a^2 + b^2 + c^2 + 2ab + 2ac + 2bc = 100$. But we also must have $a^2 + b^2 + c^2 + 2ab + 2ac + 2bc = 38 + 2 \cdot 30 = 98$, so the given system has $\boxed{\text{no solutions}}$.

20.48 To introduce symmetry into the given equation, we can rewrite it as $[(x-4) + 1]^4 + [(x-4) - 1]^4 = -8$. Hence, let $y = x - 4$, so the given equation becomes

$$-8 = (y+1)^4 + (y-1)^4 = (y^4 + 4y^3 + 6y^2 + 4y + 1) + (y^4 - 4y^3 + 6y^2 - 4y + 1) = 2y^4 + 12y^2 + 2,$$

which simplifies as $(y^2 + 1)(y^2 + 5) = 0$. The solutions are $y = \pm i$ and $y = \pm i\sqrt{5}$. Hence, the solutions in x are $x = \boxed{4 + i, 4 - i, 4 + i\sqrt{5}, \text{ and } 4 - i\sqrt{5}}$.

20.49 Let $a = \sqrt{2x^2 - 3xy + 64}$ and $b = \sqrt{2x^2 - 3xy}$. Then from the given equation, we have $a + b = 16$. Also,

$$(a + b)(a - b) = a^2 - b^2 = (2x^2 - 3xy + 64) - (2x^2 - 3xy) = 64,$$

so $a - b = 64/(a + b) = 4$. Subtracting this equation from the equation $a + b = 16$, we get $2b = 16 - 4 = 12$. Therefore, $\sqrt{2x^2 - 3xy} = b = \boxed{6}$.

20.50 Let $a = x + y$ and $b = xy$. Then the equations can be rewritten as $a + b = 71$ and $ab = 880$. Then by Vieta's Formulas, a and b are the roots of the quadratic $t^2 - 71t + 880$, which factors as $(t - 16)(t - 55)$. Therefore, a and b are equal to 16 and 55 in some order.

If $x + y = a = 55$ and $xy = b = 16$, then x and y are the roots of the quadratic $t^2 - 55t + 16$. This quadratic does not have roots that are integers.

If $x + y = a = 16$ and $xy = b = 55$, then x and y are the roots of the quadratic $t^2 - 16t + 55$, which factors as $(t - 5)(t - 11)$. Therefore, x and y are equal to 5 and 11 in some order, so $x^2 + y^2 = \boxed{146}$.

See if you can find another solution by applying Simon's Favorite Factoring Trick to $xy + x + y = 71$.

20.51 By Vieta's Formulas, we have $a + b + c + d = 0$. Therefore, $\frac{a+b+c}{d^2} = \frac{-d}{d^2} = -\frac{1}{d}$. Similarly, the other roots are $-\frac{1}{c}, -\frac{1}{b}$, and $-\frac{1}{a}$.

Let $y = -\frac{1}{x}$. Then $x = -\frac{1}{y}$, and we have

$$x^4 - bx - 3 = \left(-\frac{1}{y}\right)^4 + \frac{b}{y} - 3 = \frac{1}{y^4} + \frac{b}{y} - 3 = \frac{1 + by^3 - 3y^4}{y^4}.$$

Thus, $-\frac{1}{a}, -\frac{1}{b}, -\frac{1}{c}, -\frac{1}{d}$ are roots of the polynomial $\boxed{3y^4 - by^3 - 1}$.

20.52 If the two equations $x^3 - 7x^2 + px + q = 0$ and $x^3 - 9x^2 + px + r = 0$ have two roots in common, then they are also the roots of the equation

$$(x^3 - 7x^2 + px + q) - (x^3 - 9x^2 + px + r) = 0.$$

Simplifying the left side gives $2x^2 + q - r = 0$. Hence, by Vieta's Formulas, the sum of the two common roots is 0.

But the sum of the three roots of $x^3 - 7x^2 + px + q = 0$ is 7, so the third root is 7. Similarly, the third root of $x^3 - 9x^2 + px + r = 0$ is 9. Therefore, $(x_1, x_2) = \boxed{(7, 9)}$.

20.53 Let $y = \sqrt[4]{x}$, so $x = y^4$. Then the given equation becomes $y = \frac{12}{7-y}$, which simplifies as $(y - 3)(y - 4) = 0$, so $y = 3$ or $y = 4$. Then $x = 3^4 = 81$ or $x = 4^4 = 256$, so the sum of the real solutions is $81 + 256 = \boxed{337}$.

20.54 Expanding the left side of the second equation makes the equation $x^2 + 2xy + \frac{y^2}{2} = 7$. So, the left sides of both equations are homogeneous. We make a single homogeneous equation by subtracting 12 times the second equation from 7 times the first equation, which eliminates the constant terms and leaves $2x^2 - 3xy + y^2 = 0$. Factoring this equation gives $(2x - y)(x - y) = 0$, so $2x = y$ or $x = y$.

Letting $y = 2x$ in the first equation gives $12x^2 = 12$, from which we find $x = \pm 1$, which gives us the solutions $(x, y) = (1, 2)$ and $(x, y) = (-1, -2)$.

Letting $y = x$ in the first equation gives $6x^2 = 12$, from which we find $x = \pm\sqrt{2}$, which gives us the solutions $(x, y) = (\sqrt{2}, \sqrt{2})$ and $(x, y) = (-\sqrt{2}, -\sqrt{2})$.

We therefore have the four solutions $(x, y) = \boxed{(1, 2); (-1, -2); (\sqrt{2}, \sqrt{2}); \text{ or } (-\sqrt{2}, -\sqrt{2})}$.

20.55 Let $g(x) = f(x) - x$, so

$$g(n) = \frac{n^2 + 1}{n} - n = \frac{1}{n}$$

for $1 \le n \le 10$. Then, let $h(x) = xg(x) - 1$, so $h(n) = ng(n) - 1 = 0$ for $1 \le n \le 10$. By the Factor Theorem, $h(x)$ is divisible by $(x - 1)(x - 2) \cdots (x - 10)$, so

$$h(x) = (x - 1)(x - 2) \cdots (x - 10)q(x)$$

for some polynomial $q(x)$. Since the degree of f is 9, the degree of g is also 9, and the degree of h is 10. Therefore, $q(x)$ is some constant, say c:

$$h(x) = c(x - 1)(x - 2) \cdots (x - 10).$$

Taking $x = 0$, we get $h(0) = c(-1) \cdot (-2) \cdots (-10) = 10!c$. But $h(0) = 0 \cdot g(0) - 1 = -1$, so $c = -1/10!$.

Then taking $x = 11$, we get $h(11) = -\frac{1}{10!} \cdot 10 \cdot 9 \cdots 1 = -1$. But $h(11) = 11g(11) - 1$, so $g(11) = 0$, and finally $g(11) = f(11) - 11$, so $f(11) = \boxed{11}$.

Challenge Problems

20.56 Since a polynomial in x is a linear combination of the powers of x, we list the first few powers of $\sqrt{2} + \sqrt{3}$ to see if we can isolate $\sqrt{2}$:

$$(\sqrt{2} + \sqrt{3})^1 = \sqrt{2} + \sqrt{3},$$
$$(\sqrt{2} + \sqrt{3})^2 = 2 + 2\sqrt{6} + 3 = 5 + 2\sqrt{6},$$
$$(\sqrt{2} + \sqrt{3})^3 = (\sqrt{2} + \sqrt{3})(5 + 2\sqrt{6}) = 5\sqrt{2} + 4\sqrt{3} + 5\sqrt{3} + 6\sqrt{2} = 11\sqrt{2} + 9\sqrt{3}.$$

We note that $(\sqrt{2} + \sqrt{3})^3 - 9(\sqrt{2} + \sqrt{3}) = 11\sqrt{2} + 9\sqrt{3} - 9(\sqrt{2} + \sqrt{3}) = 2\sqrt{2}$. Hence, we may take $f(x) = \frac{x^3 - 9x}{2} = \boxed{\frac{1}{2}x^3 - \frac{9}{2}x}$.

20.57 Let $p = xyz$. Then $\frac{1}{x} + \frac{1}{y} + \frac{1}{z} = \frac{xy + xz + yz}{xyz} = \frac{xy + xz + yz}{p}$, so from $\frac{1}{x} + \frac{1}{y} + \frac{1}{z} = 1$, we have $xy + xz + yz = p$.

Therefore, by Vieta's Formulas, x, y, and z are the roots of the cubic $t^3 - t^2 + pt - p = 0$, which factors as $(t^2 + p)(t - 1) = 0$. Hence, one of x, y, and z must be equal to 1.

20.58 Note that $x + y + z = 2007a - 2007b + 2007b - 2007c + 2007c - 2007a = 0$. Squaring $x + y + z = 0$, we get $x^2 + y^2 + z^2 + 2xy + 2xz + 2yz = 0$. Therefore,

$$\frac{x^2 + y^2 + z^2}{xy + xz + yz} = \frac{-2xy - 2xz - 2yz}{xy + xz + yz} = \boxed{-2}.$$

20.59 Let $a = 2^x - 4$ and $b = 4^x - 2$. Then we can rewrite the given equation as

$$a^3 + b^3 = (a + b)^3 = a^3 + 3a^2b + 3ab^2 + b^3,$$

so $3a^2b + 3ab^2 = 0$, which means $3ab(a + b) = 0$. Hence, either $a = 0$, $b = 0$, or $a + b = 0$.

If $a = 2^x - 4 = 0$, then $x = \log_2 4 = 2$. If $b = 4^x - 2 = 0$, then $x = \log_4 2 = 1/2$.

Finally, we consider the case $a + b = 2^x + 4^x - 6 = 0$. Let $y = 2^x$. Then $4^x = (2^2)^x = 2^{2x} = y^2$, so the equation $2^x + 4^x - 6 = 0$ becomes $y^2 + y - 6 = 0$. Factoring gives $(y - 2)(y + 3) = 0$, so $y = 2$ or $y = -3$. Since $y = 2^x > 0$ for all x, we have $y = 2^x = 2$, so $x = 1$. (We can also argue that the function $f(x) = 2^x + 4^x - 6$ is increasing, and since $f(1) = 0$, $x = 1$ must be the only solution of $f(x) = 0$.)

Therefore, the solutions are $x = \boxed{1/2, 1, \text{ and } 2}$.

20.60 Since $abcd = 1$, we have

$$a + b + c + d = abcd \left(\frac{1}{a} + \frac{1}{b} + \frac{1}{c} + \frac{1}{d} \right) = abc + abd + acd + bcd.$$

Thus, we have information about three of the coefficients of the following polynomial:

$$f(x) = (x - a)(x - b)(x - c)(x - d)$$
$$= x^4 - (a + b + c + d)x^3 + (ab + ac + ad + bc + bd + cd)x^2 - (abc + abd + acd + bcd)x + abcd.$$

We can write $f(x)$ in the form
$$f(x) = x^4 - px^3 + qx^2 - px + 1,$$
where $p = a + b + c + d = abc + abd + acd + bcd$ and $q = ab + ac + ad + bc + bd + cd$.

Now, let r be a root of $f(x) = 0$ (so r is one of a, b, c, or d), so $r^4 - pr^3 + qr^2 - pr + 1 = 0$. We wish to show that $ad = bc = 1$, so we need to show that if a is a root, then so is $1/a$. Therefore, we consider $f(1/r)$.

$$f\left(\frac{1}{r}\right) = \frac{1}{r^4} - \frac{p}{r^3} + \frac{q}{r^2} - \frac{p}{r} + 1 = \frac{1 - pr + qr^2 - pr^3 + r^4}{r^4} = 0.$$

Hence, if r is a root of $f(x) = 0$, then so is $1/r$. Therefore, the numbers a, b, c, and d are equal to $\frac{1}{a}$, $\frac{1}{b}$, $\frac{1}{c}$, and $\frac{1}{d}$ in some order.

But $a > b > c > d > 0$, and so $\frac{1}{d} > \frac{1}{c} > \frac{1}{b} > \frac{1}{a}$. Therefore, $a = 1/d$ and $b = 1/c$, which means $ad = bc = 1$.

20.61 Expanding the equation, we obtain the sixth-degree polynomial $x^6 + x^5 + x^4 - 28x^3 + x^2 + x + 1 = 0$. Observe that the coefficients are symmetric, in the sense that the coefficient of x^6 is equal to the constant coefficient, the coefficient of x^5 is equal to the coefficient of x, and so on. Therefore, as described in the text, if r is a root of the polynomial, then so is $\frac{1}{r}$. As we saw in the text, we can let $y = x + \frac{1}{x}$ and write such a polynomial as a polynomial in y

First, we divide both sides by x^3, to get

$$x^3 + \frac{1}{x^3} + x^2 + \frac{1}{x^2} + x + \frac{1}{x} - 28 = 0.$$

Let $y = x + \frac{1}{x}$. Our first goal is to express $x^2 + \frac{1}{x^2}$ and $x^3 + \frac{1}{x^3}$ in terms of $y = x + \frac{1}{x}$.

Squaring the equation $y = x + \frac{1}{x}$, we get

$$y^2 = x^2 + 2 + \frac{1}{x^2},$$

so $x^2 + \frac{1}{x^2} = y^2 - 2$. Cubing the equation $y = x + \frac{1}{x}$, we get

$$y^3 = x^3 + 3x^2 \cdot \frac{1}{x} + 3x \cdot \frac{1}{x^2} + \frac{1}{x^3} = x^3 + \frac{1}{x^3} + 3x + \frac{3}{x} = x^3 + \frac{1}{x^3} + 3y,$$

so $x^3 + \frac{1}{x^3} = y^3 - 3y$. Hence, we have

$$x^3 + \frac{1}{x^3} + x^2 + \frac{1}{x^2} + x + \frac{1}{x} - 28 = (y^3 - 3y) + (y^2 - 2) + y - 28 = y^3 + y^2 - 2y - 30,$$

so we must have $y^3 + y^2 - 2y - 30 = 0$. Factoring gives $(y-3)(y^2 + 4y + 10) = 0$. The discriminant of the quadratic is $4^2 - 4 \cdot 10 = -24 < 0$, so it has no real roots. Therefore, $y = x + \frac{1}{x} = 3$, which simplifies as $x^2 - 3x + 1 = 0$. By the quadratic formula, the solutions are

$$x = \boxed{\frac{3 \pm \sqrt{5}}{2}}.$$

20.62 Let $a = (x + \frac{1}{x})^3$, so $a^2 = (x + \frac{1}{x})^6$. Let $b = x^3 + \frac{1}{x^3}$, so $b^2 = x^6 + \frac{1}{x^6} + 2$. Hence, $f(x)$ may be rewritten as

$$f(x) = \frac{a^2 - b^2}{a + b} = a - b = \left(x + \frac{1}{x}\right)^3 - x^3 - \frac{1}{x^3} = x^3 + 3x^2 \cdot \frac{1}{x} + 3x \cdot \frac{1}{x^2} + \frac{1}{x^3} - x^3 - \frac{1}{x^3} = 3\left(x + \frac{1}{x}\right).$$

By the AM-GM inequality, we have $x + \frac{1}{x} \geq 2$, with equality if and only if $x = 1$. Therefore, the minimum value of $f(x)$ is $\boxed{6}$.

20.63 From the second equation, we have $\frac{1}{z} = 5 - x$, so $z = \frac{1}{5-x}$. From the third equation, we have $y = 29 - \frac{1}{x} = \frac{29x-1}{x}$. Hence,

$$1 = xyz = x \cdot \frac{29x-1}{x} \cdot \frac{1}{5-x} = \frac{29x-1}{5-x}.$$

Then $29x - 1 = 5 - x$, so $30x = 6$, or $x = \frac{1}{5}$. We then find $y = 24$ and $z = \frac{5}{24}$. Therefore, $z + \frac{1}{y} = \frac{5}{24} + \frac{1}{24} = \boxed{\frac{1}{4}}$.

Here is an alternative solution that takes advantage of the symmetry among the equations. Let $a = z + \frac{1}{y}$. Then multiplying the equations $x + \frac{1}{z} = 5$, $y + \frac{1}{x} = 29$, and $z + \frac{1}{y} = a$, and expanding, we get

$$145a = \left(x + \frac{1}{z}\right)\left(y + \frac{1}{x}\right)\left(z + \frac{1}{y}\right) = xyz + x + y + z + \frac{1}{x} + \frac{1}{y} + \frac{1}{z} + \frac{1}{xyz}.$$

We know that $xyz = 1$, and that

$$x + y + z + \frac{1}{x} + \frac{1}{y} + \frac{1}{z} = \left(x + \frac{1}{z}\right) + \left(y + \frac{1}{x}\right) + \left(z + \frac{1}{y}\right) = 5 + 29 + a = a + 34.$$

Therefore, $a + 36 = 145a$, which implies that $a = 36/144 = 1/4$.

20.64 We can rewrite the first equation as $x^2 + y^2 = 3xy - 5$. Squaring this equation, we get $x^4 + 2x^2y^2 + y^4 = 9x^2y^2 - 30xy + 25$. But rearranging the given $7x^2y^2 - x^4 - y^4 = 155$ gives $x^4 + y^4 = 7x^2y^2 - 155$, and substituting this above gives $7x^2y^2 - 155 + 2x^2y^2 = 9x^2y^2 - 30xy + 25$, which simplifies as $30xy = 180$, so $xy = 6$. Hence, $x^2 + y^2 = 3 \cdot 6 - 5 = 13$.

Let $s = x + y$. Then $s^2 = x^2 + 2xy + y^2 = 13 + 2 \cdot 6 = 25$, so $s = \pm 5$. If $s = 5$, then by Vieta's Formulas, x and y are the roots of $t^2 - 5t + 6$, which factors as $(t-2)(t-3)$. This gives the solutions $(x, y) = (2, 3)$ and $(3, 2)$. Similarly, if $s = -5$, then x and y are the roots of $t^2 + 5t + 6$ which factors as $(t+2)(t+3)$. This gives the solutions $(x, y) = (-2, -3)$ and $(-3, -2)$. We check that each of these four solutions work, so the solutions are $(x, y) = \boxed{(2, 3), (3, 2), (-2, -3), \text{ and } (-3, -2)}$.

20.65 Let $x = a + 1$, $y = b + 1$, and $z = c + 1$, so $a = x - 1$, $b = y - 1$ and $c = z - 1$.

Then $-1 = a + b + c = (x-1) + (y-1) + (z-1) = x + y + z - 3$, so $x + y + z = 2$, and

$$-1 = ab + ac + bc = (x-1)(y-1) + (x-1)(z-1) + (y-1)(z-1) = xy + xz + yz - 2(x+y+z) + 3,$$

so $xy + xz + yz = -4 + 2(x + y + z) = 0$.

Therefore,

$$\frac{a}{a+1} + \frac{b}{b+1} + \frac{c}{c+1} = \frac{x-1}{x} + \frac{y-1}{y} + \frac{z-1}{z} = 1 - \frac{1}{x} + 1 - \frac{1}{y} + 1 - \frac{1}{z}$$

$$= 3 - \frac{1}{x} - \frac{1}{y} - \frac{1}{z} = 3 - \frac{xy + xz + yz}{xyz}$$

$$= \boxed{3}.$$

20.66 By Vieta's Formulas applied to $x^3 + rx^2 + sx + t$, we have $t = -(a+b)(a+c)(b+c)$. By Vieta's Formulas applied to $x^3 + 3x^2 + 4x - 11$, we have $a + b + c = -3$, so $a + b = -3 - c$, $a + c = -3 - b$, and $b + c = -3 - a$. Therefore,

$$t = -(a+b)(b+c)(c+a) = -(-3-c)(-3-a)(-3-b).$$

Since a, b, and c are the roots of $x^3 + 3x^2 + 4x - 11 = 0$, we have

$$(x-a)(x-b)(x-c) = x^3 + 3x^2 + 4x - 11.$$

Taking $x = -3$, we get $(-3-a)(-3-b)(-3-c) = (-3)^3 + 3(-3)^2 + 4(-3) - 11 = -23$. Therefore, $t = \boxed{23}$.

20.67

(a) Let $f(t) = (t-x)(t-y)(t-z) = t^3 - (x+y+z)t^2 + (xy + xz + yz)t - xyz$. Note that the coefficients of $f(t)$ alternate in sign, so $f(t)$ does not have any negative roots. But all the roots of $f(t)$, namely x, y, and z are real, so they must be all positive.

(b) Yes. For example, let $x = 1$, $y = 1+i$, and $z = 1-i$. Then $x+y+z = 3$, $xy+xz+yz = 1+i+1-i+(1+i)(1-i) = 4$, and $xyz = (1+i)(1-i) = 2$.

20.68 We have $P(x,y) = (x-y)Q(x,y)$, so switching the roles of x and y, we get $P(y,x) = (y-x)Q(y,x)$. But P is symmetric, so $(x-y)Q(x,y) = (y-x)Q(y,x)$, or

$$(x-y)[Q(x,y) + Q(y,x)] = 0$$

for all x, y. Hence, for all $x \neq y$, we have $Q(x,y) = -Q(y,x)$. Since Q is a polynomial, we have $Q(x,y) = -Q(y,x)$ for all x, y (even if $x = y$), so $Q(x,x) = 0$. Therefore, by the Factor Theorem, $x - y$ is a factor of $Q(x,y)$, which implies that $(x-y)^2$ is a factor of $P(x,y)$.

20.69 First, we generate a polynomial whose roots are $\frac{1}{p}$, $\frac{1}{q}$, and $\frac{1}{r}$. Let $y = \frac{1}{x}$. Then $x = \frac{1}{y}$, so

$$x^3 - 2x + 5 = \frac{1}{y^3} - \frac{2}{y} + 5 = \frac{1 - 2y^2 + 5y^3}{y^3}.$$

Hence, $\frac{1}{p}$, $\frac{1}{q}$, and $\frac{1}{r}$ are the roots of $5y^3 - 2y^2 + 1 = 0$.

By Vieta's Formulas, $\frac{1}{p} + \frac{1}{q} + \frac{1}{r} = \frac{2}{5}$, so $\frac{1}{p} + \frac{1}{q} = \frac{2}{5} - \frac{1}{r}$, $\frac{1}{p} + \frac{1}{r} = \frac{2}{5} - \frac{1}{q}$, and $\frac{1}{q} + \frac{1}{r} = \frac{2}{5} - \frac{1}{p}$. So, let $z = \frac{2}{5} - y$. Then $y = \frac{2}{5} - z$, and

$$5y^3 - 2y^2 + 1 = 5\left(\frac{2}{5} - z\right)^3 - 2\left(\frac{2}{5} - z\right)^2 + 1 = -5z^3 + 4z^2 - \frac{4}{5}z + 1 = \frac{-25z^3 + 20z^2 - 4z + 5}{5}.$$

Thus, $\frac{1}{p} + \frac{1}{q}$, $\frac{1}{p} + \frac{1}{r}$, and $\frac{1}{q} + \frac{1}{r}$ are roots of the polynomial $\boxed{25z^3 - 20z^2 + 4z - 5 = 0}$.

20.70 We want to use all our tools for polynomials, so we note that for $x \neq 0$, we can multiply our equation by x to get $xP(x) = 1$ for $x = 1, 2, 4, \ldots, 2^n$.

Let $Q(x) = xP(x) - 1$, so the degree of $Q(x)$ is $n + 1$. Then $Q(k) = 0$ for $k = 1, 2, 4, \ldots, 2^n$, so by the Factor Theorem, we have

$$Q(x) = (x - 1)(x - 2)(x - 4) \cdots (x - 2^n)R(x)$$

for some polynomial $R(x)$. Taking the degree of both sides, we get $\deg Q = n + 1 + \deg R$, so $\deg R = 0$, which means that $R(x)$ is a constant. Letting $R(x) = c$, we have $Q(x) = c(x - 1)(x - 2)(x - 4) \cdots (x - 2^n)$. We have $Q(0) = 0P(0) - 1 = -1$, and from the formula above, we have

$$Q(0) = c(-1)(-2)(-4) \cdots (-2^n) = (-1)^{n+1}2^0 \cdot 2^1 \cdot 2^2 \cdots 2^n c = (-1)^{n+1}2^{n(n+1)/2}c.$$

Since $Q(0) = -1$, we have $c = \dfrac{(-1)^n}{2^{n(n+1)/2}}$, so $Q(x) = \dfrac{(-1)^n}{2^{n(n+1)/2}}(x - 1)(x - 2)(x - 4) \cdots (x - 2^n)$.

We wish to find $P(0)$, but substituting $x = 0$ into $Q(x) = xP(x) - 1$ only gives us $Q(0) = -1$. Note that $P(0)$ is the constant term of $P(x)$, so it is also the coefficient of x in $xP(x) - 1 = Q(x)$. From Vieta's Formulas, the coefficient of x in $(x - 1)(x - 2)(x - 4) \cdots (x - 2^n)$ is

$$(-2)(-4)(-8) \cdots (-2^n) + (-1)(-4)(-8) \cdots (-2^n) + (-1)(-2)(-8) \cdots (-2^n) + \cdots + (-1)(-2)(-4) \cdots (-2^{n-1})$$

$$= (-1)(-2)(-4)(-8) \cdots (-2^n)\left(\frac{1}{-1} + \frac{1}{-2} + \cdots + \frac{1}{-2^n}\right)$$

$$= (-1)^n 2^0 \cdot 2^1 \cdot 2^2 \cdots 2^n \left(\frac{1}{1} + \frac{1}{2} + \cdots + \frac{1}{2^n}\right)$$

$$= (-1)^n 2^{n(n+1)/2}\left(2 - \frac{1}{2^n}\right),$$

so the coefficient of x in $Q(x) = c(x - 1)(x - 2)(x - 4) \cdots (x - 2^n)$ is $c(-1)^n 2^{n(n+1)/2}\left(2 - \dfrac{1}{2^n}\right) = \boxed{2 - \dfrac{1}{2^n}}$.

20.71 From the condition given in the problem, we have

$$x_k = x_1 + x_2 + \cdots + x_{k-1} + x_{k+1} + \cdots + x_{100} - k$$

for all $1 \le k \le 100$. Let $s = x_1 + x_2 + \cdots + x_{100}$. Then we may rewrite this condition as $x_k = s - x_k - k$, so $2x_k = s - k$. Summing over $1 \le k \le 100$, we get

$$2x_1 + 2x_2 + \cdots + 2x_{100} = 100s - (1 + 2 + \cdots + 100) = 100s - \frac{100 \cdot 101}{2} = 100s - 5050,$$

so we have $2s = 100s - 5050$. Therefore, we have $s = 5050/98 = 2525/49$, so, $x_{50} = \dfrac{s - 50}{2} = \dfrac{2525/49 - 50}{2} = \boxed{\dfrac{75}{98}}$.

20.72 Let $a = (x + y)^{1/3}$ and $b = (x - y)^{1/3}$. Then $(x^2 - y^2)^{1/3} = ab$, so the first equation may be rewritten as $a^2 + 2b^2 = 3ab$, which becomes $a^2 - 3ab + 2b^2 = 0$, so $(a - b)(a - 2b) = 0$.

If $a = b$, then $(x + y)^{1/3} = (x - y)^{1/3}$. Cubing this equation, we get $x + y = x - y$, so $y = 0$. Then from the second equation, we have $x = 13/3$.

If $a = 2b$, then $(x + y)^{1/3} = 2(x - y)^{1/3}$. Cubing this equation, we get $x + y = 8x - 8y$, so $7x = 9y$. Combining this with the equation $3x - 2y = 13$, we get $x = 9$ and $y = 7$.

Therefore, the solutions are $(x, y) = \boxed{(13/3, 0) \text{ and } (9, 7)}$.

20.73 The polynomial $p(x, y) = x - y$ is a counterexample. The polynomial $(p(x, y))^2$ is symmetric, because

$$(p(x, y))^2 = (x - y)^2 = (p(y, x))^2,$$

but $p(y, x) = y - x \ne p(x, y) = x - y$, so $p(x, y)$ is not symmetric.

20.74   In a symmetric polynomial in the variables x and y, every term must be of the form cx^ny^m. If $n = m$, then $cx^ny^m = cx^ny^n = c(xy)^n = cp^n$. Otherwise, $n \neq m$. Since the polynomial is symmetric, there must also be a term of the form cx^my^n. If $n > m$, then

$$cx^ny^m + cx^my^n = cx^my^m(x^{n-m} + y^{n-m}) = cp^m(x^{n-m} + y^{n-m}),$$

and if $n < m$, then

$$cx^ny^m + cx^my^n = cx^ny^n(y^{m-n} + x^{m-n}) = cp^n(x^{m-n} + y^{m-n}).$$

Hence, to prove we can express the whole polynomial in terms of s and p, it suffices to prove we can express $x^n + y^n$ in terms of s and p, for all $n \geq 1$. We prove this using induction.

Note that $x + y = s$ and $x^2 + y^2 = (x + y)^2 - 2xy = s^2 - 2p$. Hence, we assume that both $x^n + y^n$ and $x^{n+1} + y^{n+1}$ can be expressed in terms of s and p, where n is a positive integer, and will prove that this implies that $x^{n+2} + y^{n+2}$ can be expressed in terms of s and p. To obtain $x^{n+2} + y^{n+2}$, we take the product of $s = x + y$ and $x^{n+1} + y^{n+1}$:

$$s(x^{n+1} + y^{n+1}) = (x + y)(x^{n+1} + y^{n+1}) = x^{n+2} + y^{n+2} + x^{n+1}y + xy^{n+1} = x^{n+2} + y^{n+2} + xy(x^n + y^n) = x^{n+2} + y^{n+2} + p(x^n + y^n).$$

Therefore,

$$x^{n+2} + y^{n+2} = s(x^{n+1} + y^{n+1}) - p(x^n + y^n).$$

Thus, if $x^n + y^n$ and $x^{n+1} + y^{n+1}$ can be expressed in terms of s and p, then so can $x^{n+2} + y^{n+2}$. Hence, by induction, $x^n + y^n$ can be expressed in terms of s and p for all $n \geq 1$.

20.75   As in previous similar problems, we would like to establish a cubic whose roots are a, b, and c, and to do this, we require the values of $a + b + c$ (which is given), $ab + ac + bc$, and abc.

Squaring the equation $1 = a+b+c$, we get $1 = a^2+b^2+c^2+2ab+2ac+2bc = 2+2(ab+ac+bc)$, so $ab+ac+bc = -1/2$.

Squaring the equation $2 = a^2 + b^2 + c^2$, we get

$$4 = a^4 + b^4 + c^4 + 2a^2b^2 + 2a^2c^2 + 2b^2c^2 = 3 + 2(a^2b^2 + a^2c^2 + b^2c^2),$$

so $a^2b^2 + a^2c^2 + b^2c^2 = 1/2$.

Squaring the equation $-1/2 = ab + ac + bc$, we get

$$\frac{1}{4} = a^2b^2 + a^2c^2 + b^2c^2 + 2a^2bc + 2ab^2c + 2abc^2 = \frac{1}{2} + 2(a^2bc + ab^2c + abc^2),$$

so $a^2bc + ab^2c + abc^2 = -1/8$. But $a^2bc + ab^2c + abc^2 = abc(a + b + c) = abc$, so $abc = -1/8$.

Thus, by Vieta's Formulas, a, b, and c are the roots of the equation

$$x^3 - x^2 - \frac{1}{2}x + \frac{1}{8} = \frac{8x^3 - 8x^2 - 4x + 1}{8}.$$

By the Rational Root Theorem, if this cubic has a rational root, then it must be of the form $1/b$, where b divides 8. Checking these values, we find that $x = -1/2$ is a root, so by the Factor Theorem, $2x + 1$ is a factor. Dividing by $2x + 1$, we find

$$8x^3 - 8x^2 - 4x + 1 = (2x + 1)(4x^2 - 6x + 1).$$

We can then use the quadratic formula to find the roots of the quadratic factor.

Therefore, the solutions are $\boxed{(a,b,c) = \left(-\dfrac{1}{2}, \dfrac{3 + \sqrt{5}}{4}, \dfrac{3 - \sqrt{5}}{4}\right) \text{ and all of its permutations}}$.

20.76   Algebraically, the condition that two of x, y, and z be equal is equivalent to $(x - y)(y - z)(z - x) = 0$. Expanding the left side gives

$$(x - y)(yz - xy - z^2 + xz) = xyz - x^2y - xz^2 + x^2z - y^2z + xy^2 + yz^2 - xyz$$
$$= xy^2 + yz^2 + zx^2 - x^2y - y^2z - z^2x,$$

so we expect this expression to appear as a consequence of the given condition.

First, we multiply the given equation by $(y + z)(z + x)(x + y)$ to get rid of the fractions and leave

$$x(y - z)(z + x)(x + y) + y(z - x)(x + y)(y + z) + z(x - y)(y + z)(z + x) = 0.$$

We'd like to show that this equation implies that $(x - y)(y - z)(z - x) = 0$ when $x + y + z \neq 0$. So, we try factoring the left side. We already have pretty strong clues what three of the factors are. The left side is 0 for $x = y$, $y = z$, and $z = x$, so we have

$$x(y - z)(z + x)(x + y) + y(z - x)(x + y)(y + z) + z(x - y)(y + z)(z + x) = (x - y)(y - z)(z - x)f(x, y, z)$$

for some polynomial $f(x, y, z)$. The degree of the left side is 4, so $f(x, y, z)$ must be linear. We have another clue what $f(x, y, z)$ is in the condition that $x + y + z \neq 0$. So, we check if $x + y + z$ is a factor of the left side. When $x = -y - z$, the left side is

$$(-y - z)(y - z)(-y)(-z) + y(2z + y)(-z)(y + z) + z(-2y - z)(y + z)(-y)$$
$$= -yz(y + z)(y - z) - yz(y + z)(2z + y) - yz(y + z)(-2y - z)$$
$$= -yz(y + z)(y - z + 2z + y - 2y - z)$$
$$= 0.$$

This tells us that $x + y + z$ is indeed a factor, so we have

$$x(y - z)(z + x)(x + y) + y(z - x)(x + y)(y + z) + z(x - y)(y + z)(z + x) = c(x - y)(y - z)(z - x)(x + y + z)$$

for some constant c. Therefore, we have $c(x - y)(y - z)(z - x)(x + y + z) = 0$. If $x + y + z \neq 0$, then some two of x, y, and z must be equal.

20.77 By Vieta's Formulas, the product of the roots is -1984. Therefore, if the product of two of the roots is -32, then the product of the other two roots is $-1984/(-32) = 62$. Hence, the given quartic factors as

$$x^4 - 18x^3 + kx^2 + 200x - 1984 = (x^2 + ax - 32)(x^2 + bx + 62) = x^4 + (a + b)x^3 + (ab + 30)x^2 + (62a - 32b)x - 1984$$

for some constants a and b.

Equating corresponding coefficients of the x^3 and x terms, we obtain the system of equations $a + b = -18$ and $62a - 32b = 200$. Solving for a and b, we find $a = -4$ and $b = -14$. Therefore, $k = ab + 30 = \boxed{86}$.

20.78 We have

$$p(2) - p(0) = 6 \cdot 0^2 + 12 \cdot 0 + 8,$$
$$p(4) - p(2) = 6 \cdot 2^2 + 12 \cdot 2 + 8,$$
$$\vdots$$
$$p(2n) - p(2n - 2) = 6 \cdot (2n - 2)^2 + 12 \cdot (2n - 2) + 8.$$

Adding all these equations, we get

$$p(2n) - p(0) = 6[0^2 + 2^2 + \cdots + (2n - 2)^2] + 12[0 + 2 + \cdots + (2n - 2)] + 8n$$
$$= 24[1^2 + 2^2 + \cdots + (n - 1)^2] + 24[1 + 2 + \cdots + (n - 1)] + 8n$$
$$= 24 \cdot \frac{(n - 1)n(2n - 1)}{6} + 24 \cdot \frac{(n - 1)n}{2} + 8n = 4(n - 1)n(2n - 1) + 12(n - 1)n + 8n$$
$$= 8n^3 - 12n^2 + 4n + 12n^2 - 12n + 8n = 8n^3.$$

Substituting $x = 2n$, we get $p(x) - p(0) = x^3$. Hence, $p(x)$ must be of the form $\boxed{x^3 + c}$, where c is a constant. Checking, we find $p(x + 2) - p(x) = (x + 2)^3 + c - x^3 - c = 6x^2 + 12x + 8$, so all solutions of this form work.

20.79   Since $a_{2n-i} = a_i$ for all $0 \le i \le 2n$, we can rewrite $f(x)$ as

$$f(x) = a_{2n}x^{2n} + a_{2n-1}x^{2n-1} + \cdots + a_1x + a_0 = a_0x^{2n} + a_1x^{2n-1} + \cdots + a_nx^n + \cdots + a_1x + a_0.$$

Hence, dividing by x^n, the equation $f(x) = 0$ becomes

$$a_0x^n + a_1x^{n-1} + \cdots + a_n + \cdots + \frac{a_1}{x^{n-1}} + \frac{a_0}{x^n} = 0.$$

(Note that 0 cannot be a root of $f(x) = 0$, since $a_0 = a_{2n} \ne 0$, so dividing by x does not affect the roots of the equation.) We can rewrite this equation as

$$a_0\left(x^n + \frac{1}{x^n}\right) + a_1\left(x^{n-1} + \frac{1}{x^{n-1}}\right) + \cdots + a_n = 0.$$

We want to show that this equation can be expressed in the form $g(x + \frac{1}{x})$, that is, as a polynomial in $x + \frac{1}{x}$. To do so, it suffices to show that $x^n + \frac{1}{x^n}$ can be expressed as a polynomial in $x + \frac{1}{x}$ for all integers $n \ge 1$. We'll do so with induction.

Let $p_1(t) = t$, so $p_1(x + \frac{1}{x}) = x + \frac{1}{x}$. To obtain $x^2 + \frac{1}{x^2}$, we square $x + \frac{1}{x}$ to get $\left(x + \frac{1}{x}\right)^2 = x^2 + 2 + \frac{1}{x^2}$, so $x^2 + \frac{1}{x^2} = \left(x + \frac{1}{x}\right)^2 - 2$. Thus, let $p_2(t) = t^2 - 2$, so $x^2 + \frac{1}{x^2} = p_2\left(x + \frac{1}{x}\right)$.

We now have a pair of base cases, so we move on to the inductive step. We assume that for some positive integer n, we have

$$x^n + \frac{1}{x^n} = p_n\left(x + \frac{1}{x}\right) \quad \text{and} \quad x^{n+1} + \frac{1}{x^{n+1}} = p_{n+1}\left(x + \frac{1}{x}\right)$$

for some polynomials $p_n(t)$ and $p_{n+1}(t)$. We wish to show that we can express $x^{n+2} + \frac{1}{x^{n+2}}$ as a polynomial $p_{n+2}(t)$. To obtain $x^{n+2} + \frac{1}{x^{n+2}}$, we can multiply $x^{n+1} + \frac{1}{x^{n+1}}$ by $x + \frac{1}{x}$, to get

$$\left(x + \frac{1}{x}\right)\left(x^{n+1} + \frac{1}{x^{n+1}}\right) = x^{n+2} + \frac{1}{x^{n+2}} + x^n + \frac{1}{x^n},$$

so

$$x^{n+2} + \frac{1}{x^{n+2}} = \left(x + \frac{1}{x}\right)\left(x^{n+1} + \frac{1}{x^{n+1}}\right) - \left(x^n + \frac{1}{x^n}\right) = tp_{n+1}(t) - p_n(t).$$

We then let $p_{n+2}(t) = tp_{n+1}(t) - p_n(t)$, so

$$x^{n+2} + \frac{1}{x^{n+2}} = p_{n+2}\left(x + \frac{1}{x}\right).$$

Thus, $x^{n+2} + \frac{1}{x^{n+2}}$ can be expressed as a polynomial in $x + \frac{1}{x}$, so our induction is complete. This means that $x^n + \frac{1}{x^n}$ can be expressed as a polynomial in $x + \frac{1}{x}$ for all integers $n \ge 1$.

The first few polynomials $p_n(t)$ are listed below:

n	$p_n(t)$
1	t
2	$t^2 - 2$
3	$t^3 - 3t$
4	$t^4 - 4t^2 + 2$
5	$t^5 - 5t^3 + 5t$
6	$t^6 - 6t^4 + 9t^2 - 2$

These polynomials are related to a famous family of polynomials called *Chebyshev polynomials of the first kind*.

20.80   We can produce $ax^5 + by^5$ by multiplying $ax^4 + by^4 = 42$ by $x + y$. This gives us $ax^5 + bx^5 + ax^4y + bxy^4 = 42(x + y)$. Therefore, we have $ax^5 + by^5 + xy(ax^3 + by^3) = 42(x + y)$. We are given $ax^3 + by^3 = 16$, so we have $ax^5 + by^5 + 16xy = 42(x + y)$. So, if we can find xy and $x + y$, we can determine $ax^5 + by^5$.

Our success multiplying $ax^4 + by^4 = 42$ by $x + y$ suggests multiplying our other equations by $x + y$. Multiplying $ax^3 + by^3 = 16$ by $x + y$ and rearranging gives $ax^4 + by^4 + (ax^2 + bx^2)xy = 16(x + y)$. Since $ax^2 + by^2 = 7$ and $ax^4 + by^4 = 42$, we have $42 + 7xy = 16(x + y)$. Similarly, multiplying $ax^2 + by^2 = 7$ by $x + y$, we get $ax^3 + by^3 + (ax + by)xy = 7(x + y)$, from which we have $16 + 3xy = 7(x + y)$. Therefore, letting $s = x + y$ and $p = xy$, we have the system of equations $42 + 7p = 16s$, $16 + 3p = 7s$. Solving this system gives $s = -14$ and $p = -38$.

Finally, we have $ax^5 + by^5 = 42(x + y) - 16xy = 42s - 16p = \boxed{20}$.

20.81 To simplify the expression for y_i, we can use difference of cubes:

$$y_i = x_i^2 + x_i + 1 = \frac{x_i^3 - 1}{x_i - 1}.$$

Since x_i is a root of $x^3 - 2x^2 - x + 1 = 0$, we have $x_i^3 = 2x_i^2 + x_i - 1$, so

$$y_i = \frac{x_i^3 - 1}{x_i - 1} = \frac{2x_i^2 + x_i - 2}{x_i - 1}.$$

But $y_i = x_i^2 + x_i + 1$, so $x_i^2 = y_i - x_i - 1$. Hence,

$$y_i = \frac{2x_i^2 + x_i - 2}{x_i - 1} = \frac{2(y_i - x_i - 1) + x_i - 2}{x_i - 1} = \frac{2y_i - x_i - 4}{x_i - 1}.$$

Multiplying both sides by $x_i - 1$, we get $2y_i - x_i - 4 = y_i(x_i - 1) = x_i y_i - y_i$. Solving for x_i, we get

$$x_i = \frac{3y_i - 4}{y_i + 1}.$$

Hence, y_1, y_2, and y_3 are the roots of

$$x^3 - 2x^2 - x + 1 = \left(\frac{3y - 4}{y + 1}\right)^3 - 2\left(\frac{3y - 4}{y + 1}\right)^2 - \frac{3y - 4}{y + 1} + 1$$

$$= \frac{(3y - 4)^3 - 2(3y - 4)^2(y + 1) - (3y - 4)(y + 1)^2 + (y + 1)^3}{(y + 1)^3}$$

$$= \frac{7y^3 - 77y^2 + 168y - 91}{(y + 1)^3}$$

$$= \frac{7(y^3 - 11y^2 + 24y - 13)}{(y + 1)^3}.$$

Thus, the monic cubic we seek is $\boxed{y^3 - 11y^2 + 24y - 13}$.

20.82 The equation we're trying to solve has the quadratic expression $z^2 - 3z + 1$ repeated, so we set that expression equal to another variable to get a simpler look at the equation. Letting $y = z^2 - 3z + 1$, we have $y^2 - 3y + 1 = z$. Interestingly, our new equation looks exactly like our substitution equation, but with the variables reversed. We therefore have a symmetric-looking system of equations:

$$y = z^2 - 3z + 1,$$
$$z = y^2 - 3y + 1.$$

Nothing stands out immediately, so we rewrite the system with the y's on the left and the z's on the right:

$$y = z^2 - 3z + 1,$$
$$y^2 - 3y + 1 = z.$$

CHAPTER 20. SOME ADVANCED STRATEGIES

Here, the idea that jumps out is adding the equations, since the result will be a perfect square on both sides of the equation. Following this impulse gives us $y^2 - 2y + 1 = z^2 - 2z + 1$, which leads to $(y-1)^2 = (z-1)^2$. This equation means that either $y - 1 = z - 1$, which gives $y = z$, or $y - 1 = -(z - 1)$, which gives $y = 2 - z$.

When $y = z$, the equation $y = z^2 - 3z + 1$ gives us $z = z^2 - 3z + 1$, which implies $z^2 - 4z + 1 = 0$. This yields the solutions $2 \pm \sqrt{3}$. When $y = 2 - z$, we have $2 - z = z^2 - 3z + 1$, which implies $z^2 - 2z - 1 = 0$. This gives us the solutions $1 \pm \sqrt{2}$. Therefore, our four solutions are $\boxed{z = 2 + \sqrt{3}, 2 - \sqrt{3}, 1 + \sqrt{2}, 1 - \sqrt{2}}$.

20.83 We seek the product

$$q(r_1)q(r_2)q(r_3)q(r_4)q(r_5) = (r_1^2 - 2)(r_2^2 - 2)(r_3^2 - 2)(r_4^2 - 2)(r_5^2 - 2).$$

In theory, we could expand this product and use Vieta's Formulas to compute it numerically, but this would be excruciating. An easier approach would be to find a polynomial whose roots are $r_1^2, r_2^2, \ldots, r_5^2$.

Let $y = x^2$. Our goal is to transform $p(x)$ into a polynomial in y that will have roots $r_1^2, r_2^2, \ldots, r_5^2$. With $y = x^2$, the equation $p(x) = x^5 + x^2 + 1 = 0$ can be rewritten as $xy^2 + y + 1 = 0$. To express everything in terms of y, we can move the term xy^2 to the other side, to get $y + 1 = -xy^2$, and then square both sides. This gives us $y^2 + 2y + 1 = (xy^2)^2 = x^2 y^4$. Since $x^2 = y$, we then have $y^2 + 2y + 1 = y^5$.

Therefore, $r_1^2, r_2^2, \ldots, r_5^2$ are the roots of the equation $y^5 - y^2 - 2y - 1 = 0$. In other words,

$$y^5 - y^2 - 2y - 1 = (y - r_1^2)(y - r_2^2)(y - r_3^2)(y - r_4^2)(y - r_5^2).$$

Taking $y = 2$, we get

$$(2 - r_1^2)(2 - r_2^2)(2 - r_3^2)(2 - r_4^2)(2 - r_5^2) = 2^5 - 2^2 - 2 \cdot 2 - 1 = 23.$$

Therefore, $q(r_1)q(r_2)q(r_3)q(r_4)q(r_5) = (r_1^2 - 2)(r_2^2 - 2)(r_3^2 - 2)(r_4^2 - 2)(r_5^2 - 2) = \boxed{-23}$.

20.84 The presence of the expression $x^2 + y^2$ suggests we might try using complex numbers.

Let $z = x + yi$, where x and y are real numbers. Then $z^2 = (x + yi)^2 = x^2 - y^2 + 2xyi$. Thus, we see the real part $x^2 - y^2$ in the definition of A, and the imaginary part $2xy$ in the definition of B. Furthermore,

$$\frac{1}{z} = \frac{1}{x + yi} = \frac{x - yi}{(x + yi)(x - yi)} = \frac{x - yi}{x^2 + y^2} = \frac{x}{x^2 + y^2} - \frac{y}{x^2 + y^2}i.$$

We can tie all of this together as follows. If (x, y) is on both A and B, then

$$x^2 - y^2 + 2xyi + \frac{yi}{x^2 + y^2} = \frac{x}{x^2 + y^2} + 3i.$$

The real parts of this equation give us A and the imaginary parts give us B. Rearranging this equation gives

$$x^2 - y^2 + 2xyi = \frac{x - yi}{x^2 + y^2} + 3i.$$

From our expressions above for z and $\frac{1}{z}$, we can rewrite this equation as

$$z^2 = \frac{1}{z} + 3i.$$

Thus, the points where A and B intersect correspond to the solutions of the equation $z^2 = \frac{1}{z} + 3i$.

Now we look at curves C and D. Because several of the terms have degree 3, we compute z^3:

$$z^3 = (x + yi)^3 = x^3 + 3x^2 yi + 3x(yi)^2 + (yi)^3 = x^3 + 3x^2 yi - 3xy^2 - y^3 i = (x^3 - 3xy^2) + (3x^2 y - y^3)i.$$

We recognize the real part $x^3 - 3xy^2$ in the definition of C, and the imaginary part $3x^2y - y^3$ in the definition of D. Similar to our representation of A and B, we can capture both C and D with the equation

$$x^3 - 3xy^2 + 3y + (3x^2y - 3x - y^3)i = 1.$$

The real parts give us C and the imaginary parts give us D. Rearranging this equation gives

$$x^3 - 3xy^2 + (3x^2y - y^3)i + 3(y - xi) = 1.$$

Since $y - xi = -i(x + yi)$, we can rewrite this equation as $z^3 - 3iz = 1$. Thus, points at which C and D intersect correspond to the solutions of the equation $z^3 - 3iz = 1$. Clearly, $z = 0$ is not a solution to this equation, so we can divide by z to get $z^2 - 3i = \frac{1}{z}$, which is the same equation that corresponds to the intersection of A and B. Therefore, the points at which A and B intersect are the same as the points at which C and D intersect.

20.85 Squaring both sides of the first equation, we get $a^2 = 4 + \sqrt{5 + a}$, so $a^2 - 4 = \sqrt{5 + a}$. Squaring both sides of this equation, we get $a^4 - 8a^2 + 16 = a + 5$, so $a^4 - 8a^2 - a + 11 = 0$. Thus, we have established a as a root of this polynomial.

If we perform the same algebraic manipulations on the other three equations, we get

$$b^4 - 8b^2 - b + 11 = 0,$$
$$c^4 - 8c^2 + c + 11 = 0,$$
$$d^4 - 8d^2 + d + 11 = 0.$$

Let $f(x) = x^4 - 8x^2 - x + 11$, and $g(x) = x^4 - 8x^2 + x + 11$. Then a and b are two roots of $f(x) = 0$, and c and d are two roots of $g(x) = 0$. But $f(-x) = x^4 - 8x^2 + x + 11 = g(x)$. Hence, the roots of $f(x) = 0$ are a, b, $-c$, and $-d$. Then by Vieta's Formulas, we have $ab(-c)(-d) = abcd = \boxed{11}$.

20.86 All four equations may be expressed in a single statement:

$$\frac{x^2}{t - 1^2} + \frac{y^2}{t - 3^2} + \frac{z^2}{t - 5^2} + \frac{w^2}{t - 7^2} = 1$$

for $t = 2^2, 4^2, 6^2$, and 8^2.

Multiplying both sides by $(t - 1^2)(t - 3^2)(t - 5^2)(t - 7^2)$, we get

$$x^2(t - 3^2)(t - 5^2)(t - 7^2) + y^2(t - 1^2)(t - 5^2)(t - 7^2) + z^2(t - 1^2)(t - 3^2)(t - 7^2) + w^2(t - 1^2)(t - 3^2)(t - 5^2)$$
$$= (t - 1^2)(t - 3^2)(t - 5^2)(t - 7^2).$$

Before expanding both sides, there are two things we should note. First, the equation, as a whole, represents a quartic in t, whose roots are $t = 2^2, 4^2, 6^2$, and 8^2. Second, the coefficient of t^3 on the left side is $x^2 + y^2 + z^2 + w^2$. With these observations in mind, we expand both sides, but only keep track of the terms that have degree 3 or 4 (in t):

$$(x^2 + y^2 + z^2 + w^2)t^3 + \cdots = t^4 - (1^2 + 3^2 + 5^2 + 7^2)t^3 + \cdots,$$

which becomes

$$t^4 - (x^2 + y^2 + z^2 + w^2 + 1^2 + 3^2 + 5^2 + 7^2)t^3 + \cdots = 0.$$

By Vieta's Formulas, the sum of the roots is $x^2 + y^2 + z^2 + w^2 + 1^2 + 3^2 + 5^2 + 7^2$. Hence,

$$x^2 + y^2 + z^2 + w^2 + 1^2 + 3^2 + 5^2 + 7^2 = 2^2 + 4^2 + 6^2 + 8^2,$$

so $x^2 + y^2 + z^2 + w^2 = (2^2 + 4^2 + 6^2 + 8^2) - (1^2 + 3^2 + 5^2 + 7^2) = 120 - 84 = \boxed{36}$.

www.artofproblemsolving.com

The Art of Problem Solving (AoPS) is:

- # Books

  For over 24 years, the classic *Art of Problem Solving* books have been used by students as a resource for the American Mathematics Competitions and other national and local math events.

  > *Every school should have this in their math library.*
  > – Paul Zeitz, past coach of the U.S. International Mathematical Olympiad team

  The Art of Problem Solving Introduction and Intermediate texts, together with our *Prealgebra*, *Precalculus*, and *Calculus* texts, form a complete curriculum for outstanding math students in grades 5-12.

  > *The new book [Introduction to Counting & Probability] is great. I have started to use it in my classes on a regular basis. I can see the improvement in my kids over just a short period.*
  > – Jeff Boyd, 4-time MATHCOUNTS National Competition winning coach

- # Classes

  The Art of Problem Solving offers online classes on topics such as number theory, counting, geometry, algebra, and more at beginning, intermediate, and Olympiad levels.

  > *All the children were very engaged. It's the best use of technology I have ever seen.*
  > – Mary Fay-Zenk, coach of National Champion California MATHCOUNTS teams

- # Online Community

  As of November 2017, the Art of Problem Solving Forum has over 150,000 members who have posted over 6,300,000 messages on our discussion board. Members can also participate in any of our free "Math Jams."

  > *I'd just like to thank the coordinators of this site for taking the time to set it up... I think this is a great site, and I bet just about anyone else here would say the same...*
  > – AoPS Community Member

- # Resources

  We have links to summer programs, book resources, problem sources, national and local competitions, scholarship listings, a math wiki, and a LaTeX tutorial.

  > *I'd like to commend you on your wonderful site. It's informative, welcoming, and supportive of the math community. I wish it had been around when I was growing up.*
  > – AoPS Community Member

- # ...and more!

Membership is **FREE**! Come join the Art of Problem Solving community today!